Multimodal Remote Sensing Fusion and Classification

Earth Observation

Multimodal Remote Sensing Fusion and Classification

Algorithms and Applications

Edited by

Man-On Pun

School of Science and Engineering
The Chinese University of Hong Kong, Shenzhen
Shenzhen, China

Xiaokang Zhang

School of Artificial Intelligence
Wuhan University
Wuhan, China

Series Editor

George P. Petropoulos

ELSEVIER

Elsevier
Radarweg 29, PO Box 211, 1000 AE Amsterdam, Netherlands
125 London Wall, London EC2Y 5AS, United Kingdom
50 Hampshire Street, 5th Floor, Cambridge, MA 02139, United States

ISBN: 978-0-443-29152-4

For information on all Elsevier publications
visit our website at https://www.elsevier.com/books-and-journals

Publisher: Peter Llewellyn
Acquisitions Editor: Menka Mehta
Editorial Project Manager: Sneha Kashyap
Production Project Manager: Aishwarya R
Cover Designer: Raman Kumar

Typeset by VTeX

To my wife, Ying, and our children, Charlotte and Vincent

— Man-On Pun

To all contributors who made this volume possible

— Xiaokang Zhang

Contents

Contributors

Hongruixuan Chen Graduate School of Frontier Sciences, The University of Tokyo, Kashiwa, Japan
RIKEN Center for Advanced Intelligence Project (AIP), RIKEN, Tokyo, Japan

Yang Chen Beidou Research Institute, School of Geography, South China Normal University, Nanhai District, Foshan, China

Teng Fei School of Resources and Environmental Sciences, Wuhan University, Wuhan, China

Bo Huang Department of Geography, The University of Hong Kong, Hong Kong, China

Xiaolu Jiang Department of Geography, The University of Hong Kong, Hong Kong, China

Zihan Kan Department of Geography and Resource Management, The Chinese University of Hong Kong, Shatin, Hong Kong Special Administrative Region of China
Institute of Space and Earth Information Science, The Chinese University of Hong Kong, Shatin, Hong Kong Special Administrative Region of China

Dong Liu School of Humanities and Social Science, The Chinese University of Hong Kong, Shenzhen, China

Yuxuan Liu State Key Laboratory of Spatial Datum, Chinese Academy of Surveying and Mapping (CASM), Beijing, China

Xianping Ma Faculty of Geosciences and Engineering, Southwest Jiaotong University, Chengdu, China

Man-On Pun School of Science and Engineering, The Chinese University of Hong Kong, Shenzhen, China

Wenzhong Shi The Department of Land Surveying and Geo-Informatics, The Hong Kong Polytechnic University, Hong Kong SAR, China

Jian Song RIKEN Center for Advanced Intelligence Project (AIP), RIKEN, Tokyo, Japan

Haoyu Wang School of Information and Control Engineering, China University of Mining and Technology, Xuzhou, China

Lukang Wang College of Geodesy and Geomatics, Shandong University of Science and Technology, Qingdao, China

Ziyao Wang School of Science and Engineering, The Chinese University of Hong Kong, Shenzhen, China

Ziyi Wang School of Science and Engineering, The Chinese University of Hong Kong, Shenzhen, China

Qianqian Wu School of Science and Engineering, The Chinese University of Hong Kong, Shenzhen, China

Junshi Xia RIKEN Center for Advanced Intelligence Project (AIP), RIKEN, Tokyo, Japan

Naoto Yokoya Graduate School of Frontier Sciences, The University of Tokyo, Kashiwa, Japan
RIKEN Center for Advanced Intelligence Project (AIP), RIKEN, Tokyo, Japan

Min Zhang School of Geosciences and Info-Physics, Central South University, Changsha, China

Xiaokang Zhang School of Artificial Intelligence, Wuhan University, Wuhan, China

Yan Zhang Institute of Space and Earth Information Science, The Chinese University of Hong Kong, Shatin, Hong Kong Special Administrative Region of China

Qiming Zheng Department of Geography and Resource Management, The Chinese University of Hong Kong, Shatin, Hong Kong Special Administrative Region of China

Ningning Zhu College of Geography and Planning, Chengdu University of Technology, Chengdu, China

About the editors

Man-On Pun

Prof. Man-On Pun received his BEng degree in Electronic Engineering from The Chinese University of Hong Kong (CUHK) in 1996, the MEng degree in Computer Science from University of Tsukuba, Japan in 1999 and the PhD degree in Electrical Engineering from University of Southern California (USC) in Los Angeles, USA in 2006, respectively. He was a postdoctoral research associate at Princeton University, Princeton, NJ, USA from 2006 to 2008.

Currently, he is an associate professor with the School of Science and Engineering, The Chinese University of Hong Kong, Shenzhen (CUHKSZ), China. Prior to joining CUHKSZ in 2015, he held research positions at Huawei (USA) in New Jersey, USA, Mitsubishi Electric Research Labs (MERL) in Boston, USA and Sony in Tokyo, Japan. His research interests include machine learning, satellite remote sensing, and the Internet of Things.

Xiaokang Zhang

Prof. Xiaokang Zhang received the PhD degree in photogrammetry and remote sensing from Wuhan University, Wuhan, China, in 2018. From 2019 to 2022, he was a Postdoctoral Research Associate with The Hong Kong Polytechnic University, Hong Kong, and The Chinese University of Hong Kong, Shenzhen, Shenzhen, China.

He is currently an associate professor with the School of Artificial Intelligence, Wuhan University, Wuhan, China. He has authored or coauthored more than 50 scientific publications in international journals and conferences. His research interests include remote sensing image analysis, computer vision and machine learning.

Preface

The idea of this edited book, *Multimodal Remote Sensing Data Fusion for Classification: Algorithms and Applications*, originated from a simple but growing recognition: no single remote sensing modality can tell the full story of the Earth's surface. Combining optical, SAR, LiDAR, hyperspectral, and thermal infrared data offers a far richer, more reliable understanding. Yet, bringing these modalities together both algorithmically and practically remains a significant challenge for researchers, engineers, and graduate students alike.

When we first invited colleagues to contribute to this volume, the response was enthusiastic. Leading researchers from The University of Hong Kong, The Chinese University of Hong Kong (Shenzhen), Wuhan University, The Hong Kong Polytechnic University, The University of Tokyo, and the Chinese Academy of Surveying and Mapping, among others, generously agreed to share their expertise. Their willingness to participate, despite their already demanding schedules, humbled us.

However, as is often the case with multiauthor, multiinstitution projects, the path from proposal to completion proved longer than any of us anticipated. Deadlines were extended, manuscripts were revised multiple times, and unexpected personal and professional commitments delayed progress. There were moments when the completion of this book felt uncertain. But the contributors never gave up. Their patience, persistence, and dedication to scholarly excellence kept the project alive.

Today, with genuine relief and deep gratitude, we are able to present this finished work.

We owe an enormous debt of thanks to every chapter author. In particular, we thank Professor Bo Huang (Chapter 1), Dr. Yuxuan Liu (Chapter 4), Professor Haoyu Wang (Chapter 5), Professor Min Zhang (Chapter 6), Professor Ningning Zhu (Chapter 7), Dr. Hongruixuan Chen (Chapter 8), Professor Dong Liu (Chapter 9), Professor Xianping Ma (Chapters 2 and 3), and Dr. Qianqian Wu (Chapter 10). Each of you shaped this book with your insight, rigor, and willingness to endure multiple rounds of editing. This volume is as much yours as it is ours.

We also wish to thank the editorial and production team at Elsevier, in particular Ms. Sneha Kashyap and Ms. Aleksandra Packowska. Your professionalism, patience, and gentle reminders kept us on track even when we fell behind. We are especially grateful for your understanding when the manuscript was delayed more than once.

Finally, we thank the readers, the graduate students, researchers, and practitioners, who will use this book to advance remote sensing science. We hope the algorithms, case studies, and future perspectives presented here will help you solve real world problems, from carbon neutrality monitoring to disaster response and smart city planning.

This book took longer to finish than we ever imagined. But now that it is done, we believe, and we hope you will agree, that it was worth the wait.

Man-On Pun

Xiaokang Zhang

CHAPTER 1

Introduction – understanding multimodal remote sensing

Bo Huang and Xiaolu Jiang

Department of Geography, The University of Hong Kong, Hong Kong, China

1.1 Introduction

Leveraging satellite and airborne sensors, remote sensing technology provides consistent and extensive observations of the Earth's surface and atmosphere over broad temporal and spatial scales [1]. This technology facilitates thorough analysis and offers valuable insights into both long-term trends and short-term fluctuations by tracking environmental changes over time [2]. Remotely sensed images of the Earth are available for wide applications such as, but not limited to, land use/cover mapping [3], urban planning [4], characterizing crop growth [5], tracking natural disasters [6] [7], and monitoring water environment [8].

The successful launch of the Earth observation satellite, Landsat-1, marked the beginning of a new era in remote sensing, which was commenced by NASA in 1967 [9]. This landmark event marked a shift from primarily ground-based observations to a space-borne perspective, vastly expanding our ability to monitor and understand the Earth's surface and atmosphere. Over the past 50 years, remote sensing has experienced unparalleled rapid progress, driven by rapid advancements in both sensor technology and platform capabilities [10]. Early sensors were often limited in their resolutions, but technological innovations led to the development of increasingly sophisticated sensors capable of capturing more detailed and nuanced information, often with a higher frequency of data collection [11]. One key advancement has been the transition from single-band panchromatic data to multispectral sensors, and further to hyperspectral sensors [12]. This progression has significantly expanded the spectral range of data collection, enabling finer distinctions between different land cover types and species. Additionally, satellites and airborne sensors with increasingly finer spatial resolutions have become available, allowing us to map Earth's features with greater details and accuracy [13] [14]. The development of satellite constellations with frequent revisit times has enabled continuous monitoring of dynamic environmental processes [15]. The introduction of active sensors, such as synthetic aperture radar (SAR) and light detection and ranging (LiDAR), revolutionized remote sensing by allowing us to penetrate cloud cover and darkness, map

Multimodal Remote Sensing Fusion and Classification
https://doi.org/10.1016/B978-0-44-329152-4.00008-8

terrain elevation in detail, identify surface changes, and monitor natural disasters [16]. However, while single-sensor techniques have advanced significantly and have proven valuable, they often fall short when addressing complex and interconnected environmental issues [17] [18]. This is where the power of multimodal remote sensing truly comes into play.

The growing need for more comprehensive and accurate Earth observations has given rise to the increasing importance of multimodal remote sensing. By fusing data from multiple modalities, we can gain a far more comprehensive and accurate understanding of Earth's complex systems, enabling deeper insights, more reliable analyses, and ultimately, more effective decision-making for a wide range of environmental, societal, and economic challenges [19]. Data acquired from multiple sensor types and data sources provide critical information about the structure (using optical or SAR), high-resolution elevation (via LiDAR), and material content (from multispectral and hyperspectral sensors) of objects captured in images [20]. When considered collectively, this complementary information is invaluable for characterizing specific land uses in their environmental contexts. Specifically, fusing data from multiple sensors with different resolutions can significantly enhance the spatial detail of objects and improve the accuracy of analysis. This approach is particularly beneficial when dealing with complex or heterogeneous environments [21]. By combining data from different modalities, we can overcome limitations inherent in individual sensor types, leading to more comprehensive and reliable results. For example, integrating optical imagery with SAR data can significantly improve cloud cover mitigation. While optical imagery is often compromised by cloud cover, SAR's ability to penetrate clouds allows us to acquire valuable information about the underlying land cover, even in challenging weather conditions [22]. The extension of multimodal access leads to the ever-growing capability of remote sensing big data applications, particularly for tracking complex climate changes [23], advancing urban sensing capabilities in the development of smart cities [17], disaster monitoring and mitigation [24], and carbon management [25].

Despite of remarkable advancements, multimodal remote sensing, with its promise of greater insight and accuracy, faces unique challenges. Multimodal remote sensing data are usually acquired in a variety of formats with different sensors and imaging mechanisms for the same ground objects [26] [27]. These differences, along with the need to ensure sensor compatibility and calibration across various platforms and manufacturers, pose significant challenges in data processing, requiring complex preprocessing steps to align and standardize datasets for effective integration and analysis [10]. In addition, managing the vast and diverse data volumes generated by multiple sensor types requires efficient storage and high-performance computing resources [28]. Variations in spatial resolution, spectral characteristics, and temporal frequency necessitate the development of sophisticated algorithms and techniques tailored to harmonize these disparate datasets, ensuring that the

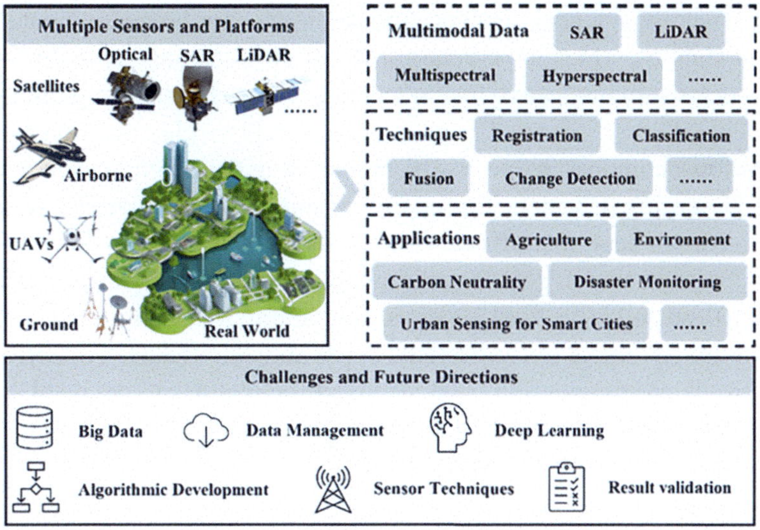

Figure 1.1:
Illustration of Multimodal Remote Sensing Systems.

full potential of multimodal remote sensing can be harnessed for accurate and meaningful insights [20] [29]. Developing robust algorithms for fusing and classifying multimodal datasets remains a significant hurdle, as these algorithms must effectively handle the complexities of diverse data sources and preserve the unique characteristics of each modality [30]. Moreover, interpreting and validating the results of multimodal analyses often demand expertise across multiple scientific disciplines and rely on the availability of reliable ground truth data, posing additional challenges [31]. As we move towards a more sustainable and informed future, addressing these specific challenges is paramount to unlocking the full potential of multimodal remote sensing [32]. With continuous technological advancements, research, and innovation, we are poised to create smarter and more integrated Earth observation systems, generate real-time insights and predictions, and enable precise monitoring and responses [33] [34].

This chapter offers a comprehensive overview of multimodal remote sensing, aiming to deepen the reader's understanding of its latest techniques, applications, and emerging trends in this rapidly evolving field. The organization encompasses data processing, fusion, classification, and change detection techniques, as well as applications, challenges, and future directions within multimodal remote sensing systems, as illustrated in Fig. 1.1.

Chapter 1 serves as an introduction, guiding readers through the foundational concepts and significance of multimodal remote sensing systems. After this chapter, the book progresses through a series of topics that build on these foundational principles. Specifically, Chapter 2 builds on this by covering essential concepts and techniques for processing data from multiple remote sensing modalities, which are crucial for integrating and analyzing diverse data types. Chapters 3 and 4 delve into innovative fusion techniques for effectively combining data from various sources and different sensors. Chapters 5 and 6 explore classification algorithms and change detection techniques, respectively, highlighting their significance in the field of multimodal remote sensing. Chapters 7 through 9 examine the contributions of multimodal remote sensing to carbon neutrality efforts, disaster monitoring and responsiveness, and smart city applications, emphasizing its role in environmental management and urban sustainability. Finally, Chapter 10 concludes the book by discussing future trends and technologies poised to transform multimodal remote sensing and its application landscape. Collectively, these chapters offer a structured progression from technical foundations to practical applications, underscoring the crucial role of multimodal remote sensing in addressing contemporary challenges and advancing sustainable development.

1.2 Multimodal remote sensing systems

Multimodality in remote sensing emphasizes the use of data collected from multiple sensor types, including optical systems (such as panchromatic, multispectral and hyperspectral cameras), SAR, and LiDAR sensors [35]. These sensors are deployed on multiple platforms, including satellites, airborne systems, unmanned aerial vehicles (UAVs), and ground-based systems [36]. The diversity of these imaging technologies facilitates collaborative observation, allowing for comprehensive data collection and integration across different wavelengths and conditions. Recent advancements have prioritized the acquisition of high spatial, temporal, and spectral resolution data, which are essential for in-depth analyses of remote sensing imagery and big data analytics [37].

1.2.1 Sensor types and platforms

Multimodal remote sensing utilizes a variety of sensor types, each employing distinct imaging technologies, which capture remote sensing data with a range of advantages and disadvantages [26]. Optical sensors capture images using visible and near-infrared wavelengths, offering high spatial resolution and rich spectral information [38]. These sensors rely on solar illumination, operating through a passive imaging approach to acquire information about planets and objects. While powerful, optical sensors are often limited by cloud cover, spectral similarities among land covers, and atmospheric conditions that can affect reflectance measurements [18]. Active remote sensing involves capturing signals at microwave wavelengths with greater intensity than passive microwave sensors by utilizing an internal energy to illuminate the objects [39]. SAR sensors use microwaves to penetrate clouds and darkness, providing valuable information about surface features, even in challenging weather conditions [40]. However, SAR data can suffer from issues like speckle noise, geometric distortions due to topography, limited availability of dense time-series data, and often lower spatial resolution compared to optical sensors [41]. LiDAR sensors emit laser pulses to measure the distance to objects, creating detailed 3D models of the terrain [42]. LiDAR data is invaluable for understanding topography, mapping urban infrastructure, and assessing the impacts of natural disasters [43]. However, it can be limited by cost and may require specialized processing techniques [44].

Both passive and active sensing devices are carried on a variety of platforms to conduct remote measurement operations [45]. Satellites provide broad coverage and long-term monitoring capabilities, making them well-suited for large-scale studies and tracking long-term changes, though their spatial resolution is relatively coarse [46]. Airborne systems, such as manned aircraft, offer higher spatial resolution and flexibility than satellites, allowing for targeted data acquisition over specific areas [47]. However, they have limitations, including higher operational costs and more complex logistics compared to some other platforms, which can restrict the frequency and scale of data collection missions. UAVs provide extremely high spatial resolution data, ideal for detailed mapping of small areas [48]. However, their limited payload and short flight endurance present challenges for large-scale implementation. Ground-based systems are used for localized data collection and for calibrating and validating data from other platforms [49]. Recent advancements have improved the operational time, reliability, simplicity, and resolution of these platforms [50], allowing them to be tailored for specific studies or precise measurements.

1.2.2 The power of multimodality

The diverse data acquired by these platforms provide complementary information that, when exploited jointly, enhance our understanding and capabilities for monitoring Earth and its environment [10]. The advent of advanced sensors and platforms, like hyperspectral sensors and satellite constellations with frequent revisit times, has greatly expanded our data acquisition capabilities [51]. Hyperspectral sensors capture data across a wide spectrum of wavelengths, enabling detailed spectral analysis that can distinguish between different materials and surface types with high precision [52] [53]. This allows for more accurate environmental monitoring, agricultural assessment, and mineral exploration [54] [55]. Meanwhile, satellite constellations offer frequent revisit times, ensuring continuous and up-to-date monitoring of dynamic processes and changes on Earth's surface, which is crucial for timely applications such as disaster response, climate change tracking, and urban development monitoring [56] [57].

Integrating data from multiple sensors allows us to overcome the inherent limitations of individual sensor types, resulting in datasets that are more continuous and seamless in both temporal and spatial dimensions [58]. This integration significantly enhances the accuracy and spatial details of analyses, especially in complex or heterogeneous environments [59]. Insights that would be difficult to obtain from a single data source can be uncovered. For instance, optical sensors, which rely on visible and near-infrared light, are often impeded by cloud cover or darkness, which can obscure the data collected. By integrating optical imagery with SAR data, which is capable of penetrating clouds and capturing images regardless of lighting conditions, it is possible to mitigate these limitations [60] [61]. Multispectral and hyperspectral data, when combined with topographical data from LiDAR or thermal imagery, enable researchers to gain a nuanced understanding of the landscape [62] [63]. In urban settings, combining high-resolution optical data with LiDAR can enhance the mapping of building heights, vegetation, and surface temperatures, leading to more robust urban planning and management strategies [64] [65]. Similarly, in ecological studies, integrating data from different sensors can reveal patterns of vegetation health and biodiversity that are critical for conservation and resource management efforts [66]. The integration of these advanced technologies facilitates the collection of high-resolution, multitemporal, and multispectral data, greatly enhancing the scope and reliability of remote sensing applications. This advancement underscores the critical role of innovative sensor technologies in expanding our ability to understand and manage the changing environment.

1.3 Data processing, fusion, and analysis

In the current era of remote sensing big data, effectively managing the vast amount of information and transforming it into actionable insights holds significant scientific importance for the diverse applications of multimodal remote sensing [13]. The rapid growth in multimodal remote sensing data presents significant challenges due to its diverse sources, structures, formats, scales, and resolutions [37].

1.3.1 Data processing and integration

Processing multimodal remote sensing data involves several essential steps. Image registration is fundamental, particularly due to the nonlinear radiation intensity differences and geometric distortions inherent in images captured across different spectral regions or by various sensors [67] [68]. Achieving high-precision alignment of these images is challenging because of the difficulty in identifying common features across different modalities [69]. Consequently, developing robust matching techniques that can effectively accommodate these complexities inherent in multimodal data is essential for successful image registration [70]. Following registration, preprocessing steps like noise reduction, radiometric correction, and geometric correction enhance data quality and facilitate feature extraction, laying the groundwork for effective data fusion, classification, and change detection [27].

1.3.2 Fusion techniques

Multimodal data fusion in remote sensing aims to integrate data from various sensor modalities to generate more comprehensive and accurate information than could be obtained from any individual data source alone [20][71]. This process has significantly evolved over the years, from simple image enhancements to sophisticated decision-level integration techniques [72].

Initially, fusion efforts focused heavily on improving image resolution, especially through pansharpening, which combines high-resolution panchromatic images with lower-resolution multispectral images [73]. These early approaches laid the foundation for enhancing spatial detail while maintaining spectral fidelity [74]. As remote sensing technology advanced, the fusion process began accommodating data with varying spatial, temporal, and spectral resolutions [75] [76] [77]. Traditional approaches, such as weighted averaging [78], spatial unmixing [79], and wavelet transforms [80], played crucial roles in efforts to fuse images at the pixel level [81] [82]. However, they often fell short in handling complex, nonlinear data relationships and capturing high-level semantic features, which limited their efficacy in heterogeneous environments [83] [84].

The rise of deep learning has significantly transformed fusion techniques, introducing automated and efficient methods for feature-level and decision-level fusion [85] [86]. This has allowed for sophisticated handling of complex environmental changes and broadened the possibilities for integrating multimodal imagery at varying resolutions [87]. One of the foundational approaches in deep learning for fusion is sparse representation [88]. In multimodal data fusion, sparse representation enhances and recovers images from diverse sources by concentrating on the most relevant features and eliminating redundant information, which is especially beneficial for handling spatially and spectrally varied data [89]. Convolutional neural networks (CNNs) have played a crucial role in automatically extracting image features and enabling sophisticated processing during fusion [74] [90]. CNNs can be designed to process and integrate data from different sources, such as optical imagery and SAR data [91]. Additionally, the introduction of Transformers, which utilize self-attention mechanisms, has further revolutionized deep learning and multimodal fusion [92]. Recent advancements driven by modern algorithms and deep learning have turned multimodal fusion into a highly sophisticated field.

1.3.3 Classification and change detection

The integration of various modalities, including multispectral and hyperspectral data, SAR, and LiDAR, significantly enhances object identification and analysis capabilities in remote sensing, particularly for land use and land cover classification as well as change detection [93] [94] [95].

Land use and land cover classification involves identifying and mapping various land cover types such as forests, agricultural areas, vegetation species, and urban environments. By fusing data from different sensor modalities, the analysis benefits from improved classification accuracy [96]. The integration of optical and SAR imagery, for example, has improved classification accuracy by overcoming cloud cover limitations in optical data [97], while LiDAR data has been deployed to enhance object recognition and estimate biophysical parameters by adding a three-dimensional perspective to traditional 2D imagery [98]. Additionally, combining hyperspectral imagery with LiDAR has significantly improved classification tasks by leveraging detailed spectral information alongside structural insights, thus enabling finer differentiation of materials and land cover types [99]. By merging spectral information from optical sensors with structural data from LiDAR and SAR, classification accuracy and details are improved, reducing the risk of misclassification in complex environments [100]. Enhanced classification capabilities support applications in land use management, biodiversity conservation, and agricultural monitoring [101].

Change detection is a critical area where multimodal remote sensing enables precise monitoring of temporal changes [102]. By integrating diverse datasets, subtle and rapid landscape transformations like deforestation, urban expansion, and postdisaster impacts are effectively captured [103] [104]. This integration enhances the ability to distinguish actual changes from sensor-related discrepancies, providing essential insights for environmental monitoring and strategic planning in dynamic Earth processes [105]. Recent developments have also expanded fusion applications to involve the integration of space-borne stereo imagery and digital surface models (DSMs) for precise change detection, offering capabilities to monitor landscape alterations effectively over time [106]. In summary, the analysis techniques that follow multimodal image fusion significantly enhance the capabilities for land cover classification and change detection, leading to more informed environmental management and planning strategies.

1.4 Applications and challenges

Multimodal remote sensing leverages the integration of diverse datasets such as multispectral, multitemporal, multiscale, SAR, LiDAR, and hyperspectral data to provide multidimensional measurements and understanding of the Earth and its various features [107]. This capability enhances the precision and efficiency of analyses across various domains, supporting applications in urban analysis [108], classification of agricultural crops [109], assessing forest attributes [110], disaster management [111], and environmental studies [112].

1.4.1 Key applications

In agriculture, the fusion of different data modalities improves crop monitoring, health assessment, and yield prediction, thus advancing precision farming practices [113]. Additionally, in disaster management, the combination of SAR and optical data enhances our ability to assess and respond to natural catastrophes, providing critical information even under challenging conditions like cloud cover or nighttime scenarios [114]. Urban planning also greatly benefits from multimodal approaches, as these data-integrated insights aid in monitoring urban growth, evaluating infrastructure health, and supporting the sustainable development of smart cities [115]. These applications collectively demonstrate the versatility and impact of multimodal remote sensing, facilitating informed decision-making and strategic planning across numerous sectors, from forestry and carbon monitoring to security and public health.

This book delves into the applications of multimodal remote sensing with a focus on carbon neutrality, disaster monitoring, and urban sensing for smart cities. In the pursuit of carbon neutrality, multimodal remote sensing plays a pivotal role by enabling the accurate estimation of carbon stocks and emissions through the integration of LiDAR and optical data [25]. This enhances our ability to monitor forest biomass and health, which are crucial for carbon offset initiatives and climate policy formulation. In disaster monitoring, the fusion of multispectral optical images and SAR data facilitates rapid assessments of affected areas, regardless of weather conditions, thus supporting effective emergency response and recovery planning [24]. Meanwhile, in the context of urban sensing for smart cities, multimodal remote sensing provides detailed insights into urban dynamics and infrastructure integrity, aiding urban planners in resource management, sustainability enhancement, and improving the quality of life for residents [101] [116]. The forthcoming chapters will explore these applications in detail, highlighting methodologies and case studies that demonstrate their transformative potential and practical implementation.

1.4.2 Challenges and future directions

The integration of multiple sensor modalities in remote sensing provides considerable advantages that extend the applications of this technology. However, it also introduces distinct challenges that must be addressed to fully realize its potential [117]. These challenges encompass data management, algorithm development, and the validation of results.

One of the primary challenges in multimodal remote sensing is managing the substantial data storage and processing requirements needed to handle the large volumes of data generated by various sensors. Efficient storage solutions and high-performance computing resources are essential for managing this data volume and complexity, facilitating real-time or near-real-time analysis to support timely decision-making [118]. To achieve shorter revisit cycles and greater coverage to compensate for the limitations of individual sensors, there are over 200 on-orbit satellite sensors capturing multimodal remote sensing data [119]. These sensors deliver data with varying resolutions, time frames, and types, making it essential to harmonize these differences for effective data integration. Accurate calibration techniques are necessary to align this data, often involving complex preprocessing steps to correct for sensor biases and environmental conditions. Achieving seamless interoperability between diverse sensor systems requires rigorous standardization protocols and methodologies. NASA's Earth Observing System Data and Information System (EOSDIS) has made significant strides in managing a growing archive of remote sensing data [120]. However, this is still in the early stages, and challenges remain that need to be addressed in future research.

Developing algorithms for data fusion and classification that can handle diverse datasets is a significant challenge. These algorithms must be able to extract meaningful information from each modality without losing the unique characteristics of individual data sources [121]. Most multimodal research operates under the assumption that all modalities are present, aligned, and noiseless during both training and testing phases. However, in real-world scenarios, it's common to encounter situations where one or more modalities are missing, noisy, or misaligned during either training or testing, or both [122]. Furthermore, the acquisition of large-scale, noise-free annotations is costly, resulting in challenges such as insufficient annotated datasets and the need to manage inconsistent labels. These issues represent significant hurdles in effectively harnessing multimodal data, necessitating the development of robust methodologies that can handle incomplete, noisy, and misaligned data while still maintaining performance and accuracy [107]. Additionally, interpreting and validating the results obtained from multimodal remote sensing analyses is inherently challenging due to the complexity and uncertainty of data involved [123]. Validation requires reliable ground truth data, which can be difficult and expensive to obtain at a large scale. Furthermore, interpreting the integrated results to extract actionable insights demands expertise across multiple scientific domains [31]. Ensuring the reliability and validity of these interpretations is essential for their practical application in real-world scenarios.

Looking to the future, addressing these challenges will involve the development of more sophisticated data management systems, improved sensor technologies, and advanced algorithms tailored to the needs of multimodal data. Fully leveraging big spatiotemporal data involves integrating conventional sources, such as remote sensing images, with new data sources, including cellular signaling, social media, smart cards, review forums, video cameras, smart equipment, and online taxi services, and developing enhanced methods for their analysis [13]. Furthermore, as artificial intelligence and machine learning technologies continue to advance, they will likely play a pivotal role in overcoming many of the current challenges associated with multimodal remote sensing, enabling more robust and scalable applications across diverse fields [124]. Through overcoming these challenges, multimodal remote sensing has the potential to significantly enhance our understanding of Earth systems, providing critical insights for environmental monitoring, urban development, disaster response, and numerous other applications. As such, continued innovation and research in this area are imperative for unlocking its full potential.

1.5 Conclusion

Multimodal remote sensing has emerged as a transformative force in Earth observation, providing an unparalleled capacity for data integration and analysis. By utilizing the strengths of diverse datasets such as multispectral, SAR, LiDAR, and hyperspectral data we achieve a nuanced, multidimensional understanding of complex Earth systems. The integration of these multimodal data sources leads to more comprehensive datasets and increased accuracy in various analyses, including land use and land cover classification and change detection.

The power of multimodal remote sensing extends to complex practical applications. From precision agriculture and forest monitoring to urban planning and carbon neutrality initiatives, the insights derived from this approach support informed decision-making and strategic planning across a multitude of sectors. However, to fully realize this potential, we must address ongoing challenges such as managing vast data volumes, ensuring sensor compatibility, and developing robust algorithms. Advancements in data management systems, sensor technologies, and artificial intelligence will be crucial for overcoming these hurdles.

This book delves into these techniques, applications and challenges in detail, highlighting the transformative impact of multimodal remote sensing on our ability to monitor, understand, and manage our dynamic planet. As research and innovation in this field continue, multimodal remote sensing promises even greater advancements, leading to a more sustainable and resilient future.

References

[1] C. Elachi, J.J. van Zyl, Introduction to the Physics and Techniques of Remote Sensing, John Wiley and Sons, 2021.

[2] B. Zhang, et al., Remotely sensed big data: evolution in model development for information extraction [point of view], Proceedings of the IEEE 107 (12) (Dec. 2019) 2294–2301, https://doi.org/10.1109/JPROC.2019.2948454.

[3] B. Huang, B. Zhao, Y. Song, Urban land-use mapping using a deep convolutional neural network with high spatial resolution multispectral remote sensing imagery, Remote Sensing of Environment 214 (Sep. 2018) 73–86, https://doi.org/10.1016/j.rse.2018.04.050.
[4] H.M. Pham, Y. Yamaguchi, T.Q. Bui, A case study on the relation between city planning and urban growth using remote sensing and spatial metrics, Landscape and Urban Planning 100 (3) (Apr. 2011) 223–230, https://doi.org/10.1016/j.landurbplan.2010.12.009.
[5] M. Weiss, F. Jacob, G. Duveiller, Remote sensing for agricultural applications: a meta-review, Remote Sensing of Environment 236 (Jan. 2020) 111402.
[6] C.J. Van Westen, Remote sensing for natural disaster management, International Archives of Photogrammetry and Remote Sensing 33 (B7/4; PART 7) (2000) 1609–1617.
[7] N.Y.D. Twumasi, Z. Shao, A. Orhan, Remote sensing and GIS methods in urban disaster monitoring and management – an overview, International Journal of Trend in Scientific Research and Development 3 (2019) 918–926.
[8] J. Chen, et al., Remote sensing big data for water environment monitoring: current status, challenges, and future prospects, Earths Future 10 (2) (2022) e2021EF002289, https://doi.org/10.1029/2021EF002289.
[9] M.A. Wulder, et al., Current status of Landsat program, science, and applications, Remote Sensing of Environment 225 (May 2019) 127–147, https://doi.org/10.1016/j.rse.2019.02.015.
[10] D. Hong, J. Hu, J. Yao, J. Chanussot, X.X. Zhu, Multimodal remote sensing benchmark datasets for land cover classification with a shared and specific feature learning model, ISPRS Journal of Photogrammetry and Remote Sensing 178 (Aug. 2021) 68–80, https://doi.org/10.1016/j.isprsjprs.2021.05.011.
[11] W. Fu, J. Ma, P. Chen, F. Chen, Remote sensing satellites for digital earth, in: Manual of Digital Earth, 2020, pp. 55–123.
[12] R.J. Zomer, A. Trabucco, S.L. Ustin, Building spectral libraries for wetlands land cover classification and hyperspectral remote sensing, Journal of Environmental Management 90 (7) (May 2009) 2170–2177, https://doi.org/10.1016/j.jenvman.2007.06.028.
[13] B. Huang, J. Wang, Big spatial data for urban and environmental sustainability, Geo-Spatial Information Science 23 (2) (Apr. 2020) 125–140, https://doi.org/10.1080/10095020.2020.1754138.
[14] D.A. Holland, D.S. Boyd, P. Marshall, Updating topographic mapping in Great Britain using imagery from high-resolution satellite sensors, ISPRS Journal of Photogrammetry and Remote Sensing 60 (3) (May 2006) 212–223, https://doi.org/10.1016/j.isprsjprs.2006.02.002.
[15] S.S. Lee, J.P. Kim, E. You, J.-H. Youn, H.-H. Shin, Satellite constellation method to achieve desired revisit performance for multiple targets, Journal of Applied Remote Sensing 18 (2) (2024) 024509.
[16] P. Hyde, R. Dubayah, W. Walker, J.B. Blair, M. Hofton, C. Hunsaker, Mapping forest structure for wildlife habitat analysis using multi-sensor (LiDAR, SAR/InSAR, ETM+, Quickbird) synergy, Remote Sensing of Environment 102 (1) (May 2006) 63–73, https://doi.org/10.1016/j.rse.2006.01.021.
[17] A.R. Javed, et al., Future smart cities: requirements, emerging technologies, applications, challenges, and future aspects, Cities 129 (Oct. 2022) 103794, https://doi.org/10.1016/j.cities.2022.103794.
[18] B. Huang, Y. Li, X. Han, Y. Cui, W. Li, R. Li, Cloud removal from optical satellite imagery with SAR imagery using sparse representation, IEEE Geoscience and Remote Sensing Letters 12 (5) (May 2015) 1046–1050, https://doi.org/10.1109/LGRS.2014.2377476.
[19] X. Sun, et al., From single- to multi-modal remote sensing imagery interpretation: a survey and taxonomy, Science China. Information Sciences 66 (4) (Apr. 2023) 140301, https://doi.org/10.1007/s11432-022-3588-0.
[20] M. Dalla Mura, S. Prasad, F. Pacifici, P. Gamba, J. Chanussot, J.A. Benediktsson, Challenges and opportunities of multimodality and data fusion in remote sensing, Proceedings of the IEEE 103 (9) (Sep. 2015) 1585–1601, https://doi.org/10.1109/JPROC.2015.2462751.
[21] X. Jiang, B. Huang, Unmixing-based spatiotemporal image fusion accounting for complex land cover changes, IEEE Transactions on Geoscience and Remote Sensing 60 (2022) 1–10, https://doi.org/10.1109/TGRS.2022.3173172.

[22] H. Liu, B. Huang, J. Cai, Thick cloud removal under land cover changes using multisource satellite imagery and a spatiotemporal attention network, IEEE Transactions on Geoscience and Remote Sensing 61 (2023) 1–8, https://doi.org/10.1109/TGRS.2023.3236106.

[23] H.-D. Guo, L. Zhang, L.-W. Zhu, Earth observation big data for climate change research, Advances in Climate Change Research 6 (2) (Jun. 2015) 108–117, https://doi.org/10.1016/j.accre.2015.09.007.

[24] Z. Miao, et al., Integrating data modality and statistical learning methods for earthquake-induced landslide susceptibility mapping, Applied Sciences 12 (3) (Jan. 2022) 1760, https://doi.org/10.3390/app12031760.

[25] N.N. Le, et al., Learning from multimodal and multisensor earth observation dataset for improving estimates of mangrove soil organic carbon in Vietnam, International Journal of Remote Sensing 42 (18) (Sep. 2021) 6866–6890, https://doi.org/10.1080/01431161.2021.1945158.

[26] B. Chen, B. Huang, B. Xu, Multi-source remotely sensed data fusion for improving land cover classification, ISPRS Journal of Photogrammetry and Remote Sensing 124 (Feb. 2017) 27–39, https://doi.org/10.1016/j.isprsjprs.2016.12.008.

[27] B. Zhu, L. Zhou, S. Pu, J. Fan, Y. Ye, Advances and challenges in multimodal remote sensing image registration, IEEE Journal on Miniaturization for Air and Space Systems 4 (2) (Jun. 2023) 165–174, https://doi.org/10.1109/JMASS.2023.3244848.

[28] L. Wang, Y. Ma, J. Yan, V. Chang, A.Y. Zomaya, pipsCloud: high performance cloud computing for remote sensing big data management and processing, Future Generations Computer Systems 78 (Jan. 2018) 353–368, https://doi.org/10.1016/j.future.2016.06.009.

[29] Y. Zhao, B. Huang, A hybrid image fusion model for generating high spatial-temporal-spectral resolution data using OLI-MODIS-Hyperion satellite imagery, International Journal of Geological and Environmental Engineering 11 (Sep. 2017) 854–859, https://doi.org/10.5281/zenodo.1132545.

[30] S.K. Roy, A. Deria, D. Hong, B. Rasti, A. Plaza, J. Chanussot, Multimodal fusion transformer for remote sensing image classification, IEEE Transactions on Geoscience and Remote Sensing 61 (2023) 1–20, https://doi.org/10.1109/TGRS.2023.3286826.

[31] X. Zhu, et al., A novel framework to assess all-round performances of spatiotemporal fusion models, Remote Sensing of Environment 274 (Jun. 2022) 113002, https://doi.org/10.1016/j.rse.2022.113002.

[32] P. Ghamisi, et al., Multisource and multitemporal data fusion in remote sensing: a comprehensive review of the state of the art, IEEE Geoscience and Remote Sensing Magazine 7 (1) (Mar. 2019) 6–39, https://doi.org/10.1109/MGRS.2018.2890023.

[33] C. Persello, et al., Deep learning and Earth observation to support the sustainable development goals: current approaches, open challenges, and future opportunities, IEEE Geoscience and Remote Sensing Magazine 10 (2) (2022) 172–200.

[34] H. Guo, Big Earth data: a new frontier in Earth and information sciences, Big Earth Data 1 (1–2) (Dec. 2017) 4–20, https://doi.org/10.1080/20964471.2017.1403062.

[35] R. Booysen, R. Gloaguen, S. Lorenz, R. Zimmermann, P.A.M. Nex, Geological remote sensing, in: D. Alderton, S.A. Elias (Eds.), Encyclopedia of Geology, second edition, Academic Press, Oxford, 2021, pp. 301–314, https://doi.org/10.1016/B978-0-12-409548-9.12127-X.

[36] S.S. Panda, M.N. Rao, Remote sensing systems – platforms and sensors: aerial, satellite, UAV, optical, radar, and LiDAR, in: Remote Sensing Handbook, Volume I, CRC Press, 2024, pp. 3–86. (Accessed 5 September 2024).

[37] Y. Zhang, Y. Liu, H. Zhang, G. Ma, Multimodal remote sensing image matching combining learning features and Delaunay triangulation, IEEE Transactions on Geoscience and Remote Sensing 60 (2022) 1–7, https://doi.org/10.1109/TGRS.2022.3229366.

[38] S. Aggarwal, Earth resource satellites, in: Satellite Remote Sensing and GIS Applications in Agricultural Meteorology, 2003, pp. 39–65. (Accessed 4 September 2024).

[39] S. Agrawal, G.B. Khairnar, A comparative assessment of remote sensing imaging techniques: optical, SAR and LiDAR, The International Archives of the Photogrammetry, Remote Sensing and Spatial Information Sciences 42 (2019) 1–6.

[40] D.L. Evans, et al., Seasat—a 25-year legacy of success, Remote Sensing of Environment 94 (3) (Feb. 2005) 384–404, https://doi.org/10.1016/j.rse.2004.09.011.
[41] N. Joshi, et al., A review of the application of optical and radar remote sensing data fusion to land use mapping and monitoring, Remote Sensing 8 (1) (Jan. 2016) 70, https://doi.org/10.3390/rs8010070.
[42] U. Wandinger, Introduction to Lidar, in: C. Weitkamp (Ed.), Lidar, vol. 102, in: Springer Series in Optical Sciences, vol. 102, Springer-Verlag, New York, 2005, pp. 1–8, https://doi.org/10.1007/0-387-25101-4-1.
[43] W.Y. Yan, A. Shaker, N. El-Ashmawy, Urban land cover classification using airborne LiDAR data: a review, Remote Sensing of Environment 158 (2015) 295–310.
[44] J.U. Eitel, et al., Beyond 3-D: the new spectrum of LiDAR applications for Earth and ecological sciences, Remote Sensing of Environment 186 (2016) 372–392.
[45] J.D. Rudd, G.T. Roberson, J.J. Classen, Application of satellite, unmanned aircraft system, and ground-based sensor data for precision agriculture: a review, in: 2017 ASABE Annual International Meeting, American Society of Agricultural and Biological Engineers, 2017, p. 1, https://elibrary.asabe.org/abstract.asp?aid=47844. (Accessed 23 September 2024).
[46] G. Balsamo, et al., Satellite and in situ observations for advancing global Earth surface modelling: a review, Remote Sensing 10 (12) (2018) 2038.
[47] M. Pepe, L. Fregonese, M. Scaioni, Planning airborne photogrammetry and remote-sensing missions with modern platforms and sensors, European Journal of Remote Sensing 51 (1) (Jan. 2018) 412–436, https://doi.org/10.1080/22797254.2018.1444945.
[48] K.K. Singh, A.E. Frazier, A meta-analysis and review of unmanned aircraft system (UAS) imagery for terrestrial applications, International Journal of Remote Sensing 39 (15–16) (Aug. 2018) 5078–5098, https://doi.org/10.1080/01431161.2017.1420941.
[49] T.J. Wagner, P.M. Klein, D.D. Turner, A new generation of ground-based mobile platforms for active and passive profiling of the boundary layer, Bulletin of the American Meteorological Society 100 (1) (2019) 137–153.
[50] H. Jafarbiglu, A. Pourreza, A comprehensive review of remote sensing platforms, sensors, and applications in nut crops, Computers and Electronics in Agriculture 197 (Jun. 2022) 106844, https://doi.org/10.1016/j.compag.2022.106844.
[51] M.S. Wong, X. Zhu, S. Abbas, C.Y.T. Kwok, M. Wang, Optical remote sensing, in: W. Shi, M.F. Goodchild, M. Batty, M.-P. Kwan, A. Zhang (Eds.), Urban Informatics, Springer, Singapore, 2021, pp. 315–344, https://doi.org/10.1007/978-981-15-8983-6-20.
[52] B. Zou, X. Jiang, H. Feng, Y. Tu, C. Tao, Multisource spectral-integrated estimation of cadmium concentrations in soil using a direct standardization and Spiking algorithm, Science of the Total Environment 701 (2020) 134890.
[53] A.F.H. Goetz, Three decades of hyperspectral remote sensing of the Earth: a personal view, Remote Sensing of Environment 113 (Sep. 2009) S5–S16, https://doi.org/10.1016/j.rse.2007.12.014.
[54] S. Peyghambari, Y. Zhang, Hyperspectral remote sensing in lithological mapping, mineral exploration, and environmental geology: an updated review, Journal of Applied Remote Sensing 15 (3) (2021) 031501.
[55] T. Adao, et al., Hyperspectral imaging: a review on UAV-based sensors, data processing and applications for agriculture and forestry, Remote Sensing 9 (11) (2017) 1110.
[56] C.S. Ruf, et al., A new paradigm in Earth environmental monitoring with the CYGNSS small satellite constellation, Scientific Reports 8 (1) (2018) 8782.
[57] Y. Xue, Y. Li, J. Guang, X. Zhang, J. Guo, Small satellite remote sensing and applications – history, current and future, International Journal of Remote Sensing 29 (15) (Aug. 2008) 4339–4372, https://doi.org/10.1080/01431160801914945.
[58] H. Liu, et al., Reconstruction of seamless harmonized Landsat Sentinel-2 (HLS) time series via self-supervised learning, Remote Sensing of Environment 308 (2024) 114191.
[59] H. Xue, et al., DeepFusion: a deep learning framework for the fusion of heterogeneous sensory data, in: Proceedings of the Twentieth ACM International Symposium on Mobile Ad Hoc Networking and Computing, ACM, Catania, Italy, Jul. 2019, pp. 151–160, https://doi.org/10.1145/3323679.3326513.

[60] L.H. Hughes, D. Marcos, S. Lobry, D. Tuia, M. Schmitt, A deep learning framework for matching of SAR and optical imagery, ISPRS Journal of Photogrammetry and Remote Sensing 169 (2020) 166–179.

[61] H. McNairn, C. Champagne, J. Shang, D. Holmstrom, G. Reichert, Integration of optical and Synthetic Aperture Radar (SAR) imagery for delivering operational annual crop inventories, ISPRS Journal of Photogrammetry and Remote Sensing 64 (5) (2009) 434–449.

[62] Q. Li, F.K.K. Wong, T. Fung, Mapping multi-layered mangroves from multispectral, hyperspectral, and LiDAR data, Remote Sensing of Environment 258 (2021) 112403.

[63] R.M. Turner, M.M. MacLaughlin, S.R. Iverson, Identifying and mapping potentially adverse discontinuities in underground excavations using thermal and multispectral UAV imagery, Engineering Geology 266 (2020) 105470.

[64] K.K. Singh, J.B. Vogler, D.A. Shoemaker, R.K. Meentemeyer, LiDAR-Landsat data fusion for large-area assessment of urban land cover: balancing spatial resolution, data volume and mapping accuracy, ISPRS Journal of Photogrammetry and Remote Sensing 74 (2012) 110–121.

[65] G. Sohn, I. Dowman, Data fusion of high-resolution satellite imagery and LiDAR data for automatic building extraction, ISPRS Journal of Photogrammetry and Remote Sensing 62 (1) (2007) 43–63.

[66] J. Cavender-Bares, et al., Integrating remote sensing with ecology and evolution to advance biodiversity conservation, Nature Ecology & Evolution 6 (5) (2022) 506–519.

[67] Y. Zhang, C. Lan, H. Zhang, G. Ma, H. Li, Multimodal remote sensing image matching via learning features and attention mechanism, IEEE Transactions on Geoscience and Remote Sensing 62 (2024) 1–20, https://doi.org/10.1109/TGRS.2023.3348980.

[68] Y. Ye, J. Shan, L. Bruzzone, L. Shen, Robust registration of multimodal remote sensing images based on structural similarity, IEEE Transactions on Geoscience and Remote Sensing 55 (5) (May 2017) 2941–2958, https://doi.org/10.1109/TGRS.2017.2656380.

[69] H. Zhang, et al., Registration of multimodal remote sensing image based on deep fully convolutional neural network, IEEE Journal of Selected Topics in Applied Earth Observations and Remote Sensing 12 (8) (2019) 3028–3042.

[70] Y. Ye, L. Bruzzone, J. Shan, F. Bovolo, Q. Zhu, Fast and robust matching for multimodal remote sensing image registration, IEEE Transactions on Geoscience and Remote Sensing 57 (11) (Nov. 2019) 9059–9070, https://doi.org/10.1109/TGRS.2019.2924684.

[71] C. Song, B. Huang, L. Ke, Modeling and analysis of lake water storage changes on the Tibetan Plateau using multi-mission satellite data, Remote Sensing of Environment 135 (Aug. 2013) 25–35, https://doi.org/10.1016/j.rse.2013.03.013.

[72] S. Kalamkar, G.M. A, Multimodal image fusion: a systematic review, Decision Analytics Journal 9 (Dec. 2023) 100327, https://doi.org/10.1016/j.dajour.2023.100327.

[73] C. Pohl, J.L. Van Genderen, Review article Multisensor image fusion in remote sensing: concepts, methods and applications, International Journal of Remote Sensing 19 (5) (Jan. 1998) 823–854, https://doi.org/10.1080/014311698215748.

[74] J. Cai, B. Huang, Super-resolution-guided progressive pansharpening based on a deep convolutional neural network, IEEE Transactions on Geoscience and Remote Sensing 59 (6) (2020) 5206–5220.

[75] Y. Zhao, D. Liu, A robust and adaptive spatial-spectral fusion model for PlanetScope and Sentinel-2 imagery, GIScience and Remote Sensing 59 (1) (Dec. 2022) 520–546, https://doi.org/10.1080/15481603.2022.2036054.

[76] F. Gao, et al., Toward mapping crop progress at field scales through fusion of Landsat and MODIS imagery, Remote Sensing of Environment 188 (2017) 9–25.

[77] B. Huang, H. Zhang, H. Song, J. Wang, C. Song, Unified fusion of remote-sensing imagery: generating simultaneously high-resolution synthetic spatial-temporal-spectral Earth observations, Remote Sensing Letters 4 (6) (Jun. 2013) 561–569, https://doi.org/10.1080/2150704X.2013.769283.

[78] F. Gao, J. Masek, M. Schwaller, F. Hall, On the blending of the Landsat and MODIS surface reflectance: predicting daily Landsat surface reflectance, IEEE Transactions on Geoscience and Remote Sensing 44 (8) (2006) 2207–2218.

[79] B. Huang, H. Zhang, Spatio-temporal reflectance fusion via unmixing: accounting for both phenological and land-cover changes, International Journal of Remote Sensing 35 (16) (Aug. 2014) 6213–6233, https://doi.org/10.1080/01431161.2014.951097.
[80] K. Amolins, Y. Zhang, P. Dare, Wavelet based image fusion techniques – an introduction, review and comparison, ISPRS Journal of Photogrammetry and Remote Sensing 62 (4) (Sep. 2007) 249–263, https://doi.org/10.1016/j.isprsjprs.2007.05.009.
[81] X. Jiang, B. Huang, Y. Zhao, Spatiotemporal image fusion with spectrally preserved pre-prediction: tackling complex land cover changes, IEEE Transactions on Geoscience and Remote Sensing 62 (2024) 4–14. (Accessed 10 September 2024).
[82] Z.H. Nezhad, A. Karami, R. Heylen, P. Scheunders, Fusion of hyperspectral and multispectral images using spectral unmixing and sparse coding, IEEE Journal of Selected Topics in Applied Earth Observations and Remote Sensing 9 (6) (2016) 2377–2389.
[83] Z. Shao, J. Cai, Remote sensing image fusion with deep convolutional neural network, IEEE Journal of Selected Topics in Applied Earth Observations and Remote Sensing 11 (5) (2018) 1656–1669.
[84] X. Zhu, E.H. Helmer, F. Gao, D. Liu, J. Chen, M.A. Lefsky, A flexible spatiotemporal method for fusing satellite images with different resolutions, Remote Sensing of Environment 172 (2016) 165–177.
[85] X. Ma, X. Zhang, M.-O. Pun, M. Liu, A multilevel multimodal fusion transformer for remote sensing semantic segmentation, IEEE Transactions on Geoscience and Remote Sensing 62 (2024) 1–15. (Accessed 24 September 2024).
[86] Q. Tang, J. Liang, F. Zhu, A comparative review on multi-modal sensors fusion based on deep learning, Signal Processing 213 (Dec. 2023) 109165, https://doi.org/10.1016/j.sigpro.2023.109165.
[87] J. Li, et al., Deep learning in multimodal remote sensing data fusion: a comprehensive review, International Journal of Applied Earth Observation and Geoinformation 112 (Aug. 2022) 102926, https://doi.org/10.1016/j.jag.2022.102926.
[88] B. Huang, H. Song, Spatiotemporal reflectance fusion via sparse representation, IEEE Transactions on Geoscience and Remote Sensing 50 (10) (2012) 3707–3716.
[89] B. Huang, H. Song, H. Cui, J. Peng, Z. Xu, Spatial and spectral image fusion using sparse matrix factorization, IEEE Transactions on Geoscience and Remote Sensing 52 (3) (2013) 1693–1704.
[90] H. Song, Q. Liu, G. Wang, R. Hang, B. Huang, Spatiotemporal satellite image fusion using deep convolutional neural networks, IEEE Journal of Selected Topics in Applied Earth Observations and Remote Sensing 11 (3) (2018) 821–829.
[91] G. Scarpa, M. Gargiulo, A. Mazza, R. Gaetano, A CNN-based fusion method for feature extraction from sentinel data, Remote Sensing 10 (2) (2018) 236.
[92] J. Li, et al., Fusion of optical and SAR images based on deep learning to reconstruct vegetation NDVI time series in cloud-prone regions, International Journal of Applied Earth Observation and Geoinformation 112 (2022) 102818.
[93] X. Wu, D. Hong, J. Chanussot, Convolutional neural networks for multimodal remote sensing data classification, IEEE Transactions on Geoscience and Remote Sensing 60 (2021) 1–10.
[94] X. Wang, M. Xu, X. Xiong, C. Ning, Remote sensing scene classification using heterogeneous feature extraction and multi-level fusion, IEEE Access 8 (2020) 217628–217641.
[95] S. Mahyoub, A. Fadil, E.M. Mansour, H. Rhinane, F. Al-Nahmi, Fusing of optical and synthetic aperture radar (SAR) remote sensing data: a systematic literature review (SLR), The International Archives of the Photogrammetry, Remote Sensing and Spatial Information Sciences 42 (2019) 127–138.
[96] T. Lorieul, E. Cole, B. Deneu, M. Servajean, P. Bonnet, A. Joly, Overview of GeoLifeCLEF 2022: predicting species presence from multi-modal remote sensing, bioclimatic and pedologic data, in: CLEF (Working Notes), 2022, pp. 1940–1956, https://ceur-ws.org/Vol-3180/paper-155.pdf. (Accessed 11 September 2024).
[97] A. Shakya, M. Biswas, M. Pal, CNN-based fusion and classification of SAR and Optical data, International Journal of Remote Sensing 41 (22) (Nov. 2020) 8839–8861, https://doi.org/10.1080/01431161.2020.1783713.

[98] G. Chen, G.J. Hay, A support vector regression approach to estimate forest biophysical parameters at the object level using airborne Lidar transects and QuickBird data, Photogrammetric Engineering and Remote Sensing 77 (7) (Jul. 2011) 733–741, https://doi.org/10.14358/PERS.77.7.733.
[99] M. Schmitt, X.X. Zhu, Data fusion and remote sensing: an ever-growing relationship, IEEE Geoscience and Remote Sensing Magazine 4 (4) (Dec. 2016) 6–23, https://doi.org/10.1109/MGRS.2016.2561021.
[100] P. Iervolino, R. Guida, D. Riccio, R. Rea, A novel multispectral, panchromatic and SAR data fusion for land classification, IEEE Journal of Selected Topics in Applied Earth Observations and Remote Sensing 12 (10) (2019) 3966–3979.
[101] D. Hong, et al., Cross-city matters: a multimodal remote sensing benchmark dataset for cross-city semantic segmentation using high-resolution domain adaptation networks, Remote Sensing of Environment 299 (2023) 113856.
[102] X. Zhang, W. Yu, M.-O. Pun, Multilevel deformable attention-aggregated networks for change detection in bitemporal remote sensing imagery, IEEE Transactions on Geoscience and Remote Sensing 60 (2022) 1–8.
[103] D. Lee, Y. Choi, A learning strategy for Amazon deforestation estimations using multi-modal satellite imagery, Remote Sensing 15 (21) (Jan. 2023) 5167, https://doi.org/10.3390/rs15215167.
[104] R. Fan, F. Li, W. Han, J. Yan, J. Li, L. Wang, Fine-scale urban informal settlements mapping by fusing remote sensing images and building data via a transformer-based multimodal fusion network, IEEE Transactions on Geoscience and Remote Sensing 60 (2022) 1–6.
[105] X. Zhang, W. Shi, Z. Lv, F. Peng, Land cover change detection from high-resolution remote sensing imagery using multitemporal deep feature collaborative learning and a semi-supervised Chan-Vese model, Remote Sensing 11 (23) (2019) 2787.
[106] R.V. Patil, D.J. Pete, Image change detection using stereo imagery and digital surface mode, in: 2015 International Conference on Information Processing (ICIP), IEEE, 2015, pp. 192–197. (Accessed 11 September 2024).
[107] N. Kieu, K. Nguyen, A. Nazib, T. Fernando, C. Fookes, S. Sridharan, Multimodal co-learning meets remote sensing: taxonomy, state of the art, and future works, IEEE Journal of Selected Topics in Applied Earth Observations and Remote Sensing (2024), https://ieeexplore.ieee.org/abstract/document/10474099/. (Accessed 11 September 2024).
[108] B. Shrestha, H. Stephen, S. Ahmad, Impervious surfaces mapping at city scale by fusion of radar and optical data through a random forest classifier, Remote Sensing 13 (15) (2021) 3040.
[109] A. Orynbaikyzy, U. Gessner, C. Conrad, Crop type classification using a combination of optical and radar remote sensing data: a review, International Journal of Remote Sensing 40 (17) (Sep. 2019) 6553–6595, https://doi.org/10.1080/01431161.2019.1569791.
[110] R.N. Treuhaft, B.E. Law, G.P. Asner, Forest attributes from radar interferometric structure and its fusion with optical remote sensing, Bioscience 54 (6) (Jun. 2004) 561–571, https://doi.org/10.1641/0006-3568(2004)054[0561:FAFRIS]2.0.CO;2.
[111] B. Gokaraju, R.A. Nobrega, D.A. Doss, A.C. Turlapaty, R.C. Tesiero, Data fusion of multi-source satellite data sets for cost-effective disaster management studies, in: SoutheastCon 2017, IEEE, 2017, pp. 1–5. (Accessed 9 September 2024).
[112] F. Balik Sanli, Y. Kurucu, M.T. Esetlili, Determining land use changes by radar-optic fused images and monitoring its environmental impacts in Edremit region of western Turkey, Environmental Monitoring and Assessment 151 (2009) 45–58.
[113] P. Karmakar, S.W. Teng, M. Murshed, S. Pang, Y. Li, H. Lin, Crop monitoring by multimodal remote sensing: a review, Remote Sensing Applications: Society and Environment (2023) 101093.
[114] S. Plank, A. Twele, S. Martinis, Landslide mapping in vegetated areas using change detection based on optical and polarimetric SAR data, Remote Sensing 8 (4) (2016) 307.
[115] X. Wu, J. Ou, Y. Wen, X. Liu, J. He, J. Zhang, Developing a data-fusing method for mapping fine-scale urban three-dimensional building structure, Sustainable Cities and Society 80 (May 2022) 103716, https://doi.org/10.1016/j.scs.2022.103716.

[116] Y. Zhang, X. Dong, L. Shang, D. Zhang, D. Wang, A multi-modal graph neural network approach to traffic risk forecasting in smart urban sensing, in: 2020 17th Annual IEEE International Conference on Sensing, Communication, and Networking (SECON), IEEE, 2020, pp. 1–9. (Accessed 11 September 2024).

[117] D. Lahat, T. Adali, C. Jutten, Multimodal data fusion: an overview of methods, challenges, and prospects, Proceedings of the IEEE 103 (9) (Sep. 2015) 1449–1477, https://doi.org/10.1109/JPROC.2015.2460697.

[118] M.K. Pekturk, M. Unal, Performance-aware high-performance computing for remote sensing big data analytics, in: Data Mining, BoD-Books on Demand, 2018. (Accessed 15 September 2024).

[119] Y. Ma, et al., Remote sensing big data computing: challenges and opportunities, Future Generations Computer Systems 51 (Oct. 2015) 47–60, https://doi.org/10.1016/j.future.2014.10.029.

[120] J. Behnke, NASA's Earth Observing System Data and Information System (EOSDIS), 2017. (Accessed 15 September 2024).

[121] X. Du, A. Zare, Multiresolution multimodal sensor fusion for remote sensing data with label uncertainty, IEEE Transactions on Geoscience and Remote Sensing 58 (4) (Apr. 2020) 2755–2769, https://doi.org/10.1109/TGRS.2019.2955320.

[122] Y. Zhao, B. Huang, H. Song, A robust adaptive spatial and temporal image fusion model for complex land surface changes, Remote Sensing of Environment 208 (Apr. 2018) 42–62, https://doi.org/10.1016/j.rse.2018.02.009.

[123] J. Wang, B. Huang, A rigorously-weighted spatiotemporal fusion model with uncertainty analysis, Remote Sensing 9 (10) (Oct. 2017) 990, https://doi.org/10.3390/rs9100990.

[124] L. Zhang, L. Zhang, Artificial intelligence for remote sensing data analysis: a review of challenges and opportunities, IEEE Geoscience and Remote Sensing Magazine 10 (2) (Jun. 2022) 270–294, https://doi.org/10.1109/MGRS.2022.3145854.

CHAPTER 2

Multimodal data preprocessing

Man-On Pun[a] and Xianping Ma[b]

[a]School of Science and Engineering, The Chinese University of Hong Kong, Shenzhen, China [b]Faculty of Geosciences and Engineering, Southwest Jiaotong University, Chengdu, China

2.1 Introduction

2.1.1 General process of data preprocessing

Remote sensing multimodal data preprocessing involves various specific technologies. In this chapter, they are broadly grouped into four key categories: calibration, enhancement, registration, and alignment. These steps are interconnected and often need to be performed in a structured sequence to ensure the data is optimized for analysis.

The first step, calibration, focuses on standardizing data from different sources. This process corrects sensor-specific biases, removes radiometric and atmospheric distortions, and ensures that measurements reflect consistent and accurate physical values [8,58]. By harmonizing datasets from various modalities or platforms, calibration lays the foundation for seamless integration and comparison. Following calibration, enhancement improves the data's quality and clarity for further processing [27]. This step may include reducing noise, enhancing contrast, or applying transformations to highlight specific features of interest, such as vegetation or urban structures. Enhancement ensures that the data is not only visually clearer but also analytically richer, making it more effective for subsequent processing steps.

Once the data is calibrated and enhanced, registration is performed to achieve spatial alignment across datasets. This process corrects for geometric distortions and ensures that corresponding features in different datasets are accurately matched [43,70]. Registration is particularly critical when working with multimodal or multitemporal data, as it ensures that the same geographic areas align correctly despite differences in sensor characteristics, acquisition angles, or temporal intervals. Finally, alignment integrates the datasets at a fine scale, ensuring consistency across spatial and temporal resolutions. Alignment often requires subpixel-level precision to harmonize data from diverse sources or acquisition times. This step addresses variations in resolution and ensures the datasets are ready for advanced analysis [76,82].

Multimodal Remote Sensing Fusion and Classification
https://doi.org/10.1016/B978-0-44-329152-4.00009-X

Table 2.1 provides a concise overview to offer readers an initial understanding of key preprocessing techniques. Together, these preprocessing steps prepare multimodal remote sensing data for sophisticated analytical tasks, such as data fusion, classification, or change detection. By ensuring the data is consistent, high-quality, and spatially accurate, these processes enable researchers to extract reliable insights from complex datasets. Calibration, enhancement, registration, and alignment form the backbone of preprocessing, transforming raw multimodal data into a cohesive and actionable dataset for advanced applications.

Table 2.1: Comparison of Calibration, Enhancement, Registration, and Alignment.

Aspect	Calibration	Enhancement	Registration	Alignment
Goal	Correct sensor/system errors	Improve data quality for analysis	Match datasets in a common space	Refine matching of features within datasets
Focus	Sensor/system accuracy	Visual and analytical clarity	Dataset consistency	Local feature correspondence
Domain	Sensor operation and raw data	Image enhancement and noise reduction	Multiple datasets (often spatial)	Spatial/geometric feature matching
Stage	Early (sensor correction)	Early to mid (postcalibration)	Mid	Substep of registration
Examples	Radiometric calibration, geometric calibration	Noise removal, contrast adjustment	Image-to-image registration	Feature/keypoint alignment

2.1.2 Importance of data preprocessing

Multimodal data preprocessing plays a vital role in remote sensing applications by enabling the integration of various sensor data sources, which often provide complementary information that enhances the accuracy and depth of analysis [17,46]. Remote sensing data from different modalities, such as optical imagery, LiDAR, radar, and hyperspectral sensors, differ in terms of spatial resolution, spectral range, and acquisition time. These differences make it challenging to directly combine these datasets, and thus, effective preprocessing is essential to ensure their compatibility and maximize their potential for meaningful insights [22].

Calibration, as a critical aspect of preprocessing, corrects various sensor-related distortions such as radiometric, geometric, and spectral inaccuracies. Radiometric calibration ensures that pixel values represent true surface reflectance, correcting for sensor imperfections and atmospheric conditions that may distort the raw data [59,73]. Geometric calibration, on the other hand, addresses spatial misalignments by adjusting pixel locations to ensure that datasets from different sensors are accurately aligned in space [67,68]. This process is fundamental for data fusion, as misalignments can lead to errors in the final analysis and interpretation. Without proper calibration, data fusion could be severely compromised, leading to unreliable results. Data enhancement, another crucial component, improves the quality and interpretability of remote sensing data. Enhancement techniques, such as contrast adjustment, noise reduction, and feature amplification, are used to emphasize critical information and suppress unwanted noise or distortions. For example, image enhancement algorithms can enhance the visibility of land surface features in optical images, remove noise from radar or LiDAR data, and improve the clarity of hyperspectral data. These techniques facilitate better human interpretation and enhance the performance of machine learning algorithms by improving the quality of input data.

In multimodal remote sensing, registration and alignment are essential for achieving spatial consistency between datasets from different sensors, which is critical for accurate data fusion. Data registration ensures that corresponding features across multiple modalities, such as optical, radar, or LiDAR data, are spatially aligned by transforming them into a common coordinate system [26,78]. This transformation typically involves geometric operations like translation, rotation, scaling, and sometimes more complex nonlinear deformations. Subpixel-level accuracy is often required to achieve precise alignment, especially when working with high-resolution data or when registering images captured at different times or from different viewing angles. Misalignments can occur due to factors such as sensor discrepancies, terrain variability, or atmospheric conditions, making the alignment process crucial for eliminating distortions and preserving spatial integrity. Furthermore, alignment is not just about aligning pixels, but ensuring that the spatial relationships between various features, such as roads, buildings, and vegetation, are maintained across datasets. Accurate registration and alignment are crucial for effective data fusion, allowing for improved applications such as land cover classification, change detection, environmental monitoring, and disaster management. The registration and alignment process support the integration of complementary information from multiple sources, ultimately enabling more robust and informed decision-making based on remote sensing data.

In recent years, deep learning has significantly enhanced traditional preprocessing techniques, enabling automatic correction and registration of multimodal data. For example, convolutional neural networks (CNNs) have been used to remove sensor-specific noise from data and improve calibration accuracy [36]. Moreover, deep learning-based models for registration use feature extraction methods to align data more accurately than traditional methods [28,47]. These advancements help to automate preprocessing tasks that were once time-consuming and labor-intensive, enabling the processing of large-scale multimodal datasets. By combining traditional methods with modern deep learning techniques, preprocessing frameworks are now better equipped to handle the challenges posed by heterogeneous datasets. This enables more accurate, reliable, and efficient analyses for a wide range of remote sensing applications.

2.2 Data calibration

Remote Sensing Data Calibration refers to the process of correcting raw data acquired by remote sensing instruments to ensure its accuracy, reliability, and compatibility with scientific analyses. Calibration adjusts the data to reflect true physical measurements by addressing systematic errors introduced by the sensors, environmental conditions, or data acquisition processes. It is a foundational step in the preprocessing workflow, enabling consistent comparisons across datasets, time periods, and instruments.

Generally, the key types of calibration include radiometric calibration, geometric calibration, spectral calibration, temporal calibration, and, in some cases, polarimetric calibration and thermal calibration, depending on the sensor and data type [8,58].

Radiometric calibration adjusts the sensor's raw digital output, often measured in digital numbers, to represent physical quantities such as radiance or reflectance. This involves correcting for sensor nonlinearity, atmospheric interference, and variations in sun angle, ensuring consistency across datasets. Radiometric calibration is critical for accurate spectral analysis and applications like vegetation health monitoring and water quality assessment.

Geometric calibration, on the other hand, ensures spatial accuracy by correcting distortions caused by sensor optics, platform motion, and Earth's curvature. It aligns imagery to a geographic coordinate system, enabling accurate mapping and integration with other geospatial datasets.

Spectral calibration focuses on ensuring that each sensor's spectral bands accurately measure the intended wavelengths [62]. This is especially important for hyperspectral and multispectral sensors used in applications like mineral mapping and vegetation classification. Temporal calibration addresses sensor degradation over time, ensuring consistency in data collected across different time periods. This is vital for long-term environmental monitoring, such as tracking deforestation or glacier retreat. Polarimetric calibration, specific to synthetic aperture radar (SAR) systems, corrects distortions in polarization measurements, allowing for precise analysis in urban mapping or agricultural studies. Thermal calibration is essential for thermal sensors, converting raw data to accurate surface temperature measurements, which is important for urban heat analysis and water body monitoring.

Radiometric calibration ensures accurate data values before spatial adjustments are made. Geometric calibration involves resampling the image (e.g., interpolation), which can alter pixel values. Applying radiometric calibration afterward could inadvertently adjust these interpolated values, leading to inaccuracies. The order reflects the logical progression of ensuring data fidelity (radiometric) before positional accuracy (geometric). This sequence is critical for ensuring that remote sensing data is both radiometrically accurate and geometrically precise, making it suitable for scientific analysis and decision-making.

Table 2.2: Comparison of Calibration Technologies.

Aspect	Radiometric Calibration	Geometric Calibration
Objective	Convert raw data to physical quantities like radiance or reflectance	Correct spatial distortions for accurate positioning
Key Corrections	Sensor nonlinearity, atmospheric effects, sun angle variations	Platform motion, sensor optics, Earth's curvature
Applications	Vegetation health, water quality, atmospheric studies	Mapping, integration with GIS, object detection
Data Type	Multispectral, hyperspectral, thermal sensors	All spatial imagery types
Dependency	Independent of spatial geometry	Requires radiometric calibration as a precursor

Table 2.2 provides a clear and concise comparison of radiometric and geometric calibration technologies, highlighting their unique objectives, corrections, applications, and dependencies. Finally, calibration is often conducted in conjunction with validation, which involves verifying the calibrated data against independent ground truth measurements. These steps form the backbone of accurate and meaningful remote sensing analyses.

2.2.1 Radiometric calibration

Radiometric calibration is a critical step in processing remote sensing data, ensuring that the measured values accurately represent the physical properties of the observed surface. The process begins with the conversion of raw digital numbers (DN) recorded by the sensor into at-sensor radiance (L_λ), which quantifies the amount of electromagnetic energy received by the sensor in a specific wavelength. This conversion corrects for sensor-specific biases and enables comparisons across datasets.

The conversion from DN to radiance is performed using a linear scaling equation:

$$L_\lambda = \text{Gain} \cdot \text{DN} + \text{Offset}, \tag{2.1}$$

where Gain and Offset are calibration coefficients provided in the sensor metadata. Alternatively, many sensors define minimum and maximum radiance values ($L_{\text{min}}, L_{\text{max}}$) and the corresponding DN range ($\text{DN}_{\text{min}}, \text{DN}_{\text{max}}$). In such cases, radiance can be computed using:

$$L_\lambda = L_{\text{min}} + \frac{(\text{DN} - \text{DN}_{\text{min}})}{\text{DN}_{\text{max}} - \text{DN}_{\text{min}}} \cdot (L_{\text{max}} - L_{\text{min}}). \tag{2.2}$$

This step ensures that the raw sensor data is transformed into a meaningful physical quantity, enabling further analysis and application.

The next step in radiometric calibration is atmospheric calibration, which accounts for the effects of the atmosphere on the signal received by the sensor. The atmosphere introduces distortions through scattering, absorption, and emission, all of which need to be mitigated to retrieve the surface reflectance (ρ_λ)—the fraction of incoming solar radiation reflected by the Earth's surface.

To address atmospheric scattering, the path radiance (L_p) is subtracted from the at-sensor radiance:

$$L_{\lambda,\mathrm{corr}} = L_\lambda - L_\mathrm{p}, \tag{2.3}$$

where $L_{\lambda,\mathrm{corr}}$ represents the corrected radiance, while L_p is typically estimated using atmospheric models or methods like dark-object subtraction, where regions assumed to have near-zero reflectance are used to infer scattering effects.

Following this, surface reflectance (ρ_λ) is calculated to normalize for solar illumination and atmospheric conditions. The equation is:

$$\rho_\lambda = \frac{\pi \cdot (L_\lambda - L_\mathrm{p}) \cdot d^2}{E_\lambda \cdot \cos(\theta)}, \tag{2.4}$$

where d is the Earth-Sun distance in astronomical units, E_λ is the solar irradiance at the top of the atmosphere for the specific wavelength, and θ is the solar zenith angle.

These two steps, namely conversion to radiance and atmospheric calibration, are fundamental for ensuring that remote sensing data accurately reflects the surface properties, setting the stage for reliable analysis and interpretation in applications such as land cover mapping, environmental monitoring, and resource management. In addition, there are several calibration options depending on the scene, including vicarious calibration, cross-calibration, and temporal calibration.

Vicarious calibration uses in-situ measurements of well-characterized Earth surfaces to refine and validate sensor calibration. These surfaces, such as deserts or specific calibration sites, are chosen for their homogeneity and known reflectance properties. The observed radiance is compared against the predicted radiance computed using surface reflectance and atmospheric conditions:

$$L_\mathrm{measured} = L_\mathrm{predicted} \pm \Delta. \tag{2.5}$$

Here, Δ represents the calibration error, which is minimized by adjusting the sensor's calibration parameters.

Cross-calibration aligns data from different sensors to ensure consistency across platforms. This is especially crucial for long-term studies requiring data from multiple sensors.

A common approach involves using overlapping observations of the same target from two sensors. By comparing their radiometric responses, a correction function can be derived:

$$L_{\text{sensor1}} = k \cdot L_{\text{sensor2}} + c, \tag{2.6}$$

where k is a scaling factor and c is an offset, both determined empirically.

Temporal calibration monitors the stability of a sensor's radiometric response over its operational lifespan. This step accounts for potential degradation caused by environmental factors, such as exposure to cosmic radiation or temperature fluctuations. Trends in radiance measurements over time are analyzed, and corrections are applied to maintain consistency:

$$L_{\lambda,\text{corrected}} = L_{\lambda} \cdot f(t), \tag{2.7}$$

where $f(t)$ is a time-dependent correction factor derived from sensor performance monitoring.

2.2.2 Geometric calibration

Geometric calibration is another critical step in remote sensing data preprocessing, designed to correct spatial distortions and ensure that the spatial representation of the acquired data accurately reflects the real-world geometry. These corrections are necessary to align the data with a defined map projection or coordinate system. Geometric calibration typically includes sensor-specific calibrations, systematic distortion removal, and ground control point (GCP)-based calibrations [57,74].

Sensor-specific calibrations address distortions inherent to the imaging system, such as those caused by the sensor's geometry, optics, or platform motion. For instance, optical sensors may introduce radial or tangential lens distortions, which can be modeled and corrected using formulas such as:

$$\begin{aligned} x_{\text{corrected}} &= x \cdot (1 + k_1 r^2 + k_2 r^4 + k_3 r^6), \\ y_{\text{corrected}} &= y \cdot (1 + k_1 r^2 + k_2 r^4 + k_3 r^6), \end{aligned} \tag{2.8}$$

where x and y are the original image coordinates, r is the radial distance from the optical center, and k_1, k_2, k_3 are distortion coefficients determined through calibration.

For satellite sensors, platform motion such as pitch, roll, and yaw can lead to distortions that are corrected using information from onboard attitude control systems.

Systematic distortion removal corrects predictable errors due to Earth's rotation, curvature, or sensor movement during data acquisition. For example, the Earth rotation effect during image capture can cause eastward displacements, particularly in line-scanning sensors. These displacements can be modeled and corrected using:

$$\Delta x = \omega \cdot R \cdot \cos(\phi) \cdot \Delta t, \tag{2.9}$$

where ω is the Earth's rotational velocity, R is the Earth's radius, ϕ is the latitude, and Δt is the integration time.

Other systematic errors, such as panoramic distortions in wide-field sensors, are corrected using geometric transformation models.

GCP-based calibrations involve aligning image features with known geographic locations. A set of GCPs with precise coordinates in the image and the real world is used to calculate transformation parameters. Common models include affine transformations for small corrections and polynomial or rational functions for more complex distortions, such as:

$$\begin{aligned} x' &= a_0 + a_1 x + a_2 y + a_3 x^2 + a_4 xy + a_5 y^2, \\ y' &= b_0 + b_1 x + b_2 y + b_3 x^2 + b_4 xy + b_5 y^2, \end{aligned} \tag{2.10}$$

where (x, y) are the image coordinates, (x', y') are the corrected coordinates, and a_i, b_i are coefficients determined through least-squares fitting.

Once the geometric transformation model is established, resampling methods such as nearest neighbor, bilinear interpolation, or cubic convolution are applied to map the corrected coordinates to a regular grid. This step ensures the final data is geometrically aligned and ready for further analysis.

Geometric calibration is essential for achieving spatial accuracy in remote sensing data. Through sensor-specific calibrations, systematic distortion removal, GCP-based calibrations, and resampling, geometric calibration aligns imagery to a consistent spatial framework. This ensures compatibility with geographic information systems (GIS), improves feature extraction accuracy, and supports reliable multitemporal and multimodal data analysis.

2.2.3 Other calibrations

There are additional calibration methods used in remote sensing to ensure data quality and usability besides radiometric and geometric calibration. Some notable methods include temporal, spectral, thermal, and polarimetric calibrations [4,11,18].

Temporal calibration ensures consistency in remote sensing data acquired at different times by accounting for changes in sensor performance and external conditions. This type of calibration is particularly useful in monitoring vegetation dynamics (e.g., NDVI time-series), change detection for disaster response (e.g., floods or deforestation), and urban expansion studies. Some common remote sensing data formats that involve temporal calibration include MODIS time-series products (e.g., MOD09A1) and Landsat temporal imagery (e.g., Landsat Collection 2 Level-1).

Spectral calibration ensures that sensors accurately measure the wavelengths of reflected or emitted radiation. This is particularly critical for multispectral and hyperspectral sensors where precise band placement is essential for distinguishing materials or surface types. Spectral calibration aligns observed spectra to known reference spectra using correction coefficients. The observed wavelength, λ, is adjusted using a correction factor, $\Delta\lambda$:

$$\lambda' = \lambda + \Delta\lambda, \tag{2.11}$$

where λ' is the corrected wavelength. Spectral calibration is widely used in mineral mapping with hyperspectral data, water quality assessment using spectral indices, and crop classification in agriculture. Typical remote sensing data formats for spectral calibration include hyperspectral imagery (e.g., AVIRIS, Hyperion) and multispectral imagery (e.g., Sentinel-2, Landsat OLI).

Thermal calibration ensures the accuracy of temperature measurements derived from thermal infrared sensors. This process typically uses onboard blackbody radiators for reference to correct any thermal sensor bias. Thermal calibration corrects for sensor bias to derive accurate brightness temperatures. It is crucial for studies like urban heat island effects, sea surface temperature monitoring, and wildfire detection and assessment. Thermal calibration is typically applied to MODIS thermal products (e.g., MOD11A1), Landsat TIRS data, and NOAA AVHRR thermal imagery.

Polarimetric calibration corrects for distortions in the polarization properties of radar signals, ensuring accurate representation of scattering mechanisms. This is essential for fully polarimetric synthetic aperture radar (SAR) data. The scattering matrix **S** is calibrated by removing channel imbalances using a calibration matrix **C**:

$$\mathbf{S}_{cal} = \mathbf{C}\mathbf{S}_{raw}, \tag{2.12}$$

where $\mathbf{S}_{raw}$ is the raw scattering matrix and $\mathbf{S}_{cal}$ is the calibrated scattering matrix. Polarimetric calibration is vital for forest biomass estimation, sea ice classification, and soil moisture estimation. Remote sensing data formats for polarimetric calibration include fully polarimetric SAR data such as RADARSAT-2, ALOS PALSAR, and other SAR data sources.

In conclusion, temporal, spectral, thermal, and polarimetric calibrations also play important roles in preprocessing remote sensing data for diverse applications. Each method addresses specific challenges, ensuring data quality and enabling precise analysis across domains such as environmental monitoring, agriculture, and urban studies.

2.3 Data enhancement

Remote sensing data enhancement is essential for improving the quality and interpretability of remotely sensed images. Key technologies in this domain include noise reduction, radiometric enhancement, transformation-based enhancement, spatial enhancement and so on. Noise reduction methods, including median and Wiener filtering, help remove noise caused by sensor imperfections or atmospheric conditions, preserving important details [48]. Radiometric enhancement techniques, such as histogram equalization and contrast stretching, adjust pixel intensity values to improve the visibility of features, making them easier to analyze [24,63]. Transformation-based enhancement methods, such as Principal Component Analysis (PCA) and Independent Component Analysis (ICA), transform multispectral or hyperspectral data into more interpretable formats, aiding in feature extraction and dimensionality reduction [7,40]. Spatial enhancement methods, like edge enhancement or smoothing filters, improve image resolution and clarity, crucial for applications such as object detection and land-cover classification. In particular, superresolution increases the spatial resolution of images beyond their original limits, often by reconstructing high-resolution images from low-resolution inputs. It can be seen as a subset of spatial enhancement techniques that specifically focus on improving spatial resolution [1,39].

The necessity of data enhancement lies in its ability to improve the accuracy and efficiency of remote sensing data analysis. By enhancing image quality, these technologies make it easier to extract meaningful information for applications like environmental monitoring, disaster management, and urban planning.

2.3.1 Noise reduction

Noise reduction is an essential aspect of data enhancement. Remote sensing images often contain noise due to sensor imperfections, atmospheric interference, or environmental conditions. Noise reduction techniques, such as median filtering and Wiener filtering, help to remove unwanted noise while preserving the important details of the image. These methods are vital for improving the signal quality, especially when working with low-quality images or when trying to extract subtle features that might be overshadowed by noise.

Median Filtering is a nonlinear filter that replaces each pixel's value with the median of the neighboring pixels, effectively removing salt-and-pepper noise. The median filter is given by:

$$I_{\text{enhanced}}(x, y) = \text{median}(I(x', y')), \tag{2.13}$$

where (x', y') are the coordinates of the neighboring pixels within a specified window.

Wiener Filtering is a signal processing technique used to enhance data by reducing noise and improving the quality of a signal. It operates under the principle of minimizing the mean squared error (MSE) between the desired clean signal and the estimated signal. The Wiener filter uses statistical knowledge about the signal and noise, such as their power spectral densities or autocorrelations, to achieve this goal.

For a noisy observation $Y(f) = X(f) + N(f)$, where $X(f)$ is the original signal, $N(f)$ is the noise, and f represents frequency, the Wiener filter $H(f)$ is defined as:

$$H(f) = \frac{S_X(f)}{S_X(f) + S_N(f)}. \tag{2.14}$$

Here, $S_X(f)$ is the power spectral density (PSD) of the signal, and $S_N(f)$ is the power spectral density (PSD) of the noise.

The filtered output $X_{\text{filtered}}(f)$ is then given by:

$$X_{\text{filtered}}(f) = H(f) \cdot Y(f). \tag{2.15}$$

Wiener filtering effectively balances noise suppression and signal preservation, making it a robust choice for many enhancement tasks.

2.3.2 Radiometric enhancement

Radiometric enhancement focuses on improving the radiometric properties of the image, such as contrast, brightness, and dynamic range, to make features more distinguishable. Techniques like histogram equalization and linear contrast stretching are commonly used to adjust pixel intensity values, ensuring that relevant features are clearly visible and easier to analyze. These methods are particularly important when dealing with raw satellite or aerial imagery, where certain features might be obscured by poor contrast or underexposure.

Histogram Equalization adjusts the image's contrast by redistributing the intensity values of pixels. It aims to create a uniform distribution of pixel values, making the image more visually appealing and revealing hidden features. The transformation function for HE is defined as:

$$T(r) = \frac{(r - r_{\min})}{(r_{\max} - r_{\min})} \times (L - 1), \tag{2.16}$$

where $r_{\min}$ and $r_{\max}$ are the minimum and maximum pixel values in the original image, and L is the number of available intensity levels.

Linear Contrast Stretching stretches the pixel values of an image to enhance its contrast, making it easier to distinguish between different objects. The linear stretch formula is:

$$I_{\text{enhanced}} = \frac{I - I_{\min}}{I_{\max} - I_{\min}} \times (I_{\max} - I_{\min}), \tag{2.17}$$

where I_{enhanced} is the enhanced intensity, and $I_{\min}$ and $I_{\max}$ are the minimum and maximum intensities of the input image.

2.3.3 Transformation-based enhancement

Transformation-based enhancement methods, including Principal Component Analysis (PCA) and Independent Component Analysis (ICA), offer ways to transform multispectral or hyperspectral data into a more usable format. PCA reduces dimensionality while preserving the most significant features, making it easier to analyze the data. ICA, on the other hand, identifies independent components within the data, which can be beneficial for signal separation or feature extraction. Additionally, Fourier transforms are used for analyzing the frequency components of an image, which is especially useful for removing periodic noise or enhancing specific spatial frequencies.

PCA is used to reduce the dimensionality of multispectral data while preserving as much variance as possible. By transforming the original data into a new coordinate system, PCA emphasizes the most significant features. The PCA transformation is given by:

$$\mathbf{Z} = \mathbf{V}^T (\mathbf{X} - \boldsymbol{\mu}), \tag{2.18}$$

where $\mathbf{X}$ is the original data matrix, $\mathbf{V}$ is the matrix of eigenvectors, and $\boldsymbol{\mu}$ is the mean of the data.

ICA is similar to PCA but focuses on finding statistically independent components in the data, making it useful for feature extraction and mixing of signals. The ICA transformation is expressed as:

$$\mathbf{X} = \mathbf{W} \cdot \mathbf{S}, \tag{2.19}$$

where $\mathbf{X}$ is the observed mixed signal, $\mathbf{S}$ is the source signal, and $\mathbf{W}$ is the mixing matrix.

Fourier transform is used to analyze the frequency components of an image. It is often applied to remove periodic noise or enhance spatial frequencies [45,54]. The 2D Fourier transform is expressed as:

$$\mathcal{F}(f(x, y)) = \int_{-\infty}^{\infty} \int_{-\infty}^{\infty} f(x, y) e^{-2\pi i (ux+vy)} dxdy, \tag{2.20}$$

where $f(x, y)$ is the image function, and u, v are the frequency domain coordinates.

These classical enhancement techniques are widely used to improve remote sensing data for various applications such as land use classification, environmental monitoring, and disaster management.

2.3.4 Spatial enhancement

Spatial enhancement aims to improve the spatial resolution and clarity of features within an image. Methods such as edge enhancement and smoothing filters (e.g., Gaussian filters) are applied to highlight important features and reduce unnecessary details. This enhancement is critical in applications such as object detection, land-cover classification, and monitoring changes in environmental landscapes, where fine spatial resolution can make a significant difference in the accuracy of analysis.

Edge enhancement techniques, such as high-pass filtering, focus on amplifying the edges or boundaries in an image, which can help in object detection. This is typically achieved by subtracting a low-pass filtered version of the image from the original:

$$I_{\text{enhanced}} = I - \text{LPF}(I), \tag{2.21}$$

where $\text{LPF}(I)$ represents the low-pass filtered image.

Smoothing or low-pass filters are applied to reduce noise and blur the image. The most common filter used for this purpose is the Gaussian filter, which is defined as:

$$G(x, y) = \frac{1}{2\pi\sigma^2} e^{-\frac{x^2+y^2}{2\sigma^2}}, \tag{2.22}$$

where σ is the standard deviation of the filter, and x and y are the coordinates of the pixel.

2.3.4.1 Superresolution

Superresolution refers specifically to techniques designed to increase the spatial resolution. It typically involves the application of advanced algorithms or models, such as deep learning-based methods, that infer high-frequency details (such as fine textures and edges) in images, resulting in more detailed representations. Therefore, this chapter provides a focused technical introduction and discusses several advanced methods based on deep learning.

Broadly speaking, existing superresolution methods can be classified into four main approaches: image reconstruction [23,49], sparse representation [69,77], regression [2,60], and deep learning (DL) [9,20,25,29,72,79]. Over the past few years, significant advancements have been made with the rise of deep learning approaches. Among the most prominent deep learning models for superresolution are Convolutional Neural Networks (CNNs), Transformers, and Diffusion Models. Each of these approaches offers unique advantages in the context of remote sensing data enhancement.

CNN-based superresolution methods have been the most widely adopted techniques for image resolution enhancement. These methods leverage the hierarchical structure of CNNs to learn low-level image features and progressively map them to high-resolution outputs. CNN-based superresolution typically relies on deep networks with multiple layers, such as the popular VDSR model [25], which improves image quality by increasing the depth of the network. The key benefit of CNNs lies in their ability to learn spatial hierarchies effectively, allowing for a strong representation of local patterns. Recent developments, such as the use of residual learning [52] and generative adversarial networks [71], have further improved superresolution performance by enhancing texture details and reducing artifacts. However, one limitation of CNN-based approaches is their difficulty in capturing long-range dependencies and global contextual information, which can affect the quality of superresolution images in complex scenes.

Transformers, originally developed for natural language processing, have been increasingly adopted for image-based tasks due to their ability to capture long-range dependencies and global context through self-attention mechanisms [10,64]. In the context of superresolution, Transformer-based methods have shown significant improvements over CNNs, particularly when dealing with images that require the model to learn intricate global relations. One of the most influential transformer-based methods for superresolution is the Vision Transformer (ViT) [10], which divides images into patches and applies self-attention to capture long-range dependencies across the entire image. Transformer-based superresolution methods, such as SwinIR [32], have demonstrated state-of-the-art performance in preserving fine details and improving image quality, especially in larger images where contextual information plays a crucial role. Furthermore, it has also been suggested to integrate transformer-based self-attention modules into CNNs to improve global information recognition. For instance, NLSA [37] enhanced nonlocal attention by employing spherical locality-sensitive hashing, enabling the model to focus on the most relevant regions while disregarding unrelated areas. However, transformer models generally require more computational resources and larger datasets for training compared to CNNs. Additionally, transformers can be more complex to implement and optimize.

Diffusion models represent a newer class of generative models that have shown great potential in various image generation and enhancement tasks. They work by gradually adding noise to the image during the training process and then learning to reverse this process to generate high-quality images. In superresolution tasks, they are trained to model the distribution of high-resolution images and progressively denoise lower-resolution inputs to recover fine details. They have been shown to generate images with more natural textures and fewer artifacts compared to traditional CNN-based methods. However, diffusion models tend to be computationally expensive and slow to train, which limits their application in real-time superresolution tasks.

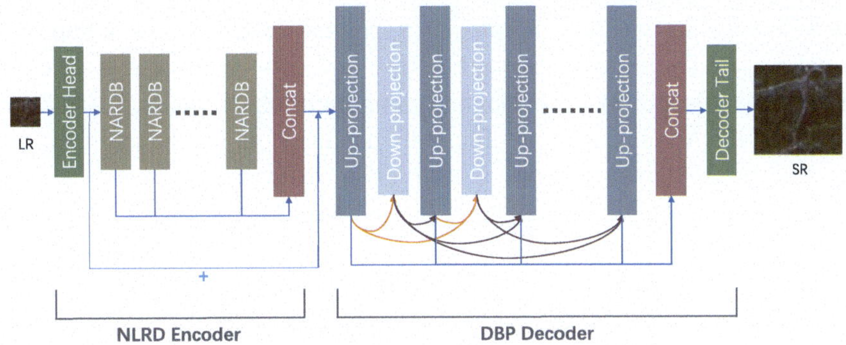

Figure 2.1:
The framework of the GCRND [55].

2.3.4.2 Global context-driven convolutional neural network

GCRND [55] introduces a global context-driven residual dense network for remote sensing superresolution. It incorporates nonlocal sparse self-attention (NLSA) into the residual dense encoder to strengthen global contextual information extraction. Additionally, backprojection blocks are utilized in the decoder to refine reconstruction, iteratively feeding reconstruction errors through up-sampling and down-sampling stages to recover detailed information from the encoder.

Framework: As illustrated in Fig. 2.1, the GCRND comprises an NLRD Encoder and a DBP Decoder. In the NLRD Encoder, shallow convolutional features produced by the encoder head are processed through a series of nonlocal attentive residual dense blocks (NARDBs). These NARDBs leverage a contiguous memory mechanism to maintain feed-forward characteristics while capturing both local and global textural features. The multilevel features extracted by the NLRD Encoder are then concatenated along the channel axis and refined through convolution operations. In the DBP Decoder, the encoded feature maps are used to reconstruct high-resolution images through iterative up- and down-projection processes. This mechanism maps encoded features to higher-resolution feature maps and repeatedly converts them back to lower-resolution, enhancing detail recovery. Finally, all high-resolution features generated during the up-projection steps are concatenated and processed by the decoder tail via convolution operations to produce the output image at the desired resolution.

NLRD Encoder: The NLRD Encoder is designed to capture enriched local-global features in remote sensing images. It consists of three main components: an encoder head, multiple NARDBs, and a feature fusion module. The encoder head, comprising two convolutional layers, extracts shallow features from the input images. These features are then passed through a series of NARDBs for multilevel representation learning. The feature fusion module, built with convolutional layers, integrates the multilevel features to enhance their representational power. The structure of the NARDB is shown in Fig. 2.2. As the core element of the NLRD Encoder, the NARDB is designed using hierarchical dense residual learning combined with the NLSA mechanism to efficiently capture both local and global contextual information.

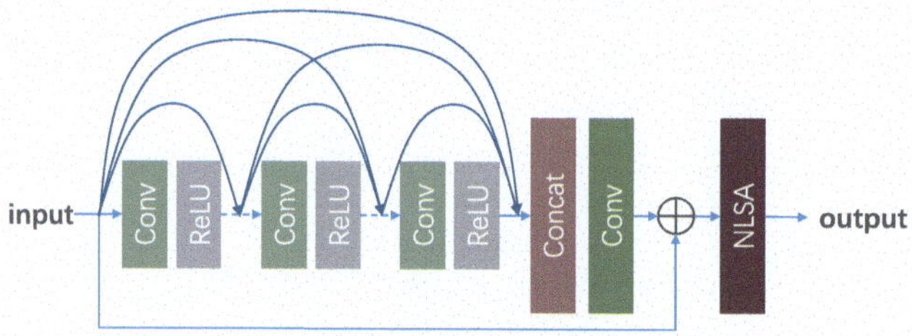

Figure 2.2:
The architecture of the NARDBs [55].

Dense Residual Learning: The dense residual learning in the NARDB can be expressed as follows:

$$F_i = Conv([R_J; R_{J-1}; \cdots; R_1; F_{i-1}]) + F_{i-1}, \tag{2.23}$$

where $[\cdot]$ denotes the concatenation operation, F_i represents the output of the i-th NARDB, and R_j refers to the feature maps produced by the j-th convolutional layer followed by a ReLU activation function, where $j \in 1, \cdots, J$. In dense residual learning, the outputs from all convolutional blocks in the previous stage are concatenated with the next-stage blocks. Subsequently, an NLSA module is applied to capture global contextual information based on the integrated local features.

NLSA: Inspired by the fact that remote sensing images often contain repeated patterns, such as vegetation, roads, wastelands, and mountains, NLSA is used to capture global contextual information for enhanced feature extraction. Specifically, NLSA is an advanced nonlocal operation that partitions the input features into hash buckets, thereby reducing the computational cost of attention.

As depicted in Fig. 2.3, the feature maps are processed using the spherical locality-sensitive hashing (LSH) technique to form the attention bucket. The LSH algorithm partitions the feature vector x into m distinct hash buckets by calculating the angular distances between elements, ensuring that similar features are grouped together in the same bucket. Specifically, LSH begins by randomly rotating a cross-polytope inscribed within a hypersphere and projecting the feature tensor onto this hypersphere. The rotation helps define a unique mapping for each input element. Subsequently, the algorithm assigns each feature to the hash bucket corresponding to the nearest vertex of the rotated polytope, ensuring that features

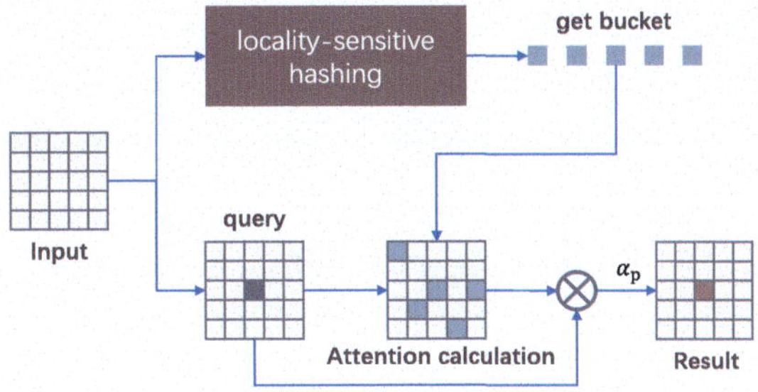

Figure 2.3:
The process of the NLSA for each feature map [55].

with similar angular relationships are placed in the same bucket. This process increases the efficiency and robustness of the attention mechanism by ignoring irrelevant or noisy elements that are less correlated with the features of interest. Mathematically, the tensor x is projected onto the hypersphere, and a random rotation is applied using a matrix $M \in R^{c \times m}$, where c is the feature dimension, and m is the number of hash buckets. The following equation represents this process:

$$x' = M\left(\frac{x}{\|x\|_2}\right), \tag{2.24}$$

and subsequently, the hash code is defined as

$$hash(x) = \arg\max_p(x'). \tag{2.25}$$

As a result, features that share the same hash code are assigned to the same hash bucket. Specifically, for a feature x_p at location p, the set of indices corresponding to the elements in the same bucket can be defined as $\eta_p = q \mid \text{hash}(x_q) = \text{hash}(x_p)$, where η_p represents the index set of all elements whose hash codes match that of x_p. This set effectively identifies the features that are contextually related to the feature at location p.

Denote by a_p the NLSA output of location p, compute a_p is computed as follows:

$$a_p = \sum_{q \in \eta_p} \frac{s(x_q, x_p)}{\sum_{\hat{q} \in \eta_p} s(x_p, x_{\hat{q}})} t(x_q), \tag{2.26}$$

where $t(\cdot)$ represents the feature transformation function, while $s(\cdot, \cdot)$ measures the similarity between features. Additionally, the set η_p refers to the indices of the nonzero elements in the vector $[s(x_p, x_1), \ldots, s(x_p, x_n)] \in \mathbb{R}^n$, which represents the similarity values between the feature at position p and all other features in the image. Since η_p identifies the pixel locations that are relevant for the query at p, a sparsity constraint can be applied within the NLSA framework by limiting the number of nonzero entries in this set. This constraint effectively controls the number of elements the attention mechanism focuses on, either by reducing the number of nonzero entries to a specified chunk size or to match the size of the attention bucket.

Finally, the outputs of all NARDBs are integrated before being fed into the decoder:

$$E_{output} = Conv([N_T; N_{T-1}; \cdots ; N_1]) + N_0, \tag{2.27}$$

where E_{output}, N_T, and N_0 are the output of the encoder, the T-th NARDB, and the encoder head, respectively.

DBP Decoder: Unlike most existing decoders in encoder-decoder architectures that directly reconstruct high-resolution images through a simple, feed-forward process involving progressive convolution and upsampling, GCRDN introduced the DBP Decoder to better preserve high resolution details during image reconstruction, following a method similar to that proposed in [20]. The DBP Decoder focuses on enhancing feature sampling at multiple depths, while distributing the task of evaluating reconstruction error across several stages. This approach enables the DBP Decoder to learn distinct up- and down-projection operators, which helps retain crucial high resolution components throughout the reconstruction process. In practice, the feature maps are iteratively upscaled and downscaled within the DBP Decoder. Specifically, the up-projection module takes the down-projected features from previous stages as input, while the down-projection module processes the feature maps generated by each up-projection stage. This iterative process ensures that the high resolution details are effectively captured and refined during the decoding process.

2.3.4.3 Adaptive semantic-enhanced denoising diffusion probabilistic model

Recently, the denoising diffusion probabilistic model (DDPM) [21,41] has emerged as a promising generative model, showing impressive performance in computer vision tasks, particularly in image synthesis, where it delivers high-quality reconstructions. The DDPM consists of two main components: a diffusion process and a reverse process. In the diffusion process, noise is progressively added to the images, while the reverse process involves generating superresolved (SR) images by projecting noise back into the original image space. DDPM is trained by optimizing a variant of the variational lower bound, offering advantages over GANs, such as reduced susceptibility to mode collapse and the ability to overcome the limitations of PSNR-based loss functions typically used in discriminative models [21].

However, one challenge with DDPM-based methods is that the high-frequency details generated during the reconstruction may not perfectly align with the target image, leading to performance degradation. This issue arises because the U-Net-based conditional noise predictor (CNP) tends to focus excessively on high-frequency image components that are local and geographically specific, while neglecting the broader, global context of the image [65]. For generative models, relying solely on high-frequency information is insufficient, as real-world images consist of both high- and low-frequency signals. Even though the loss function guides the CNP towards the true data distribution, the model's structure inherently limits its ability to fully capture global information. This overemphasis on local details results in significant noise and texture inconsistencies during the noise prediction stage, leading to undesirable artifacts in the generated images. This issue is especially problematic for complex remote sensing image generation tasks, where global context plays a more critical role. Remote sensing images typically exhibit global similarities, such as consistent textures, colors, and shapes. Therefore, further research focused on incorporating both high- and low-frequency components is essential to improve the quality of reconstructed HR images and better meet the requirements of downstream remote sensing applications.

ASDDPM [56] introduces an efficient solution, the adaptive semantic-enhanced DDPM (ASDDPM), to address previous challenges using a dual-decoder strategy within the CNP. Recognizing the semantic gap between the encoder and decoder, we focus on enhancing the decoder's ability to reconstruct fine-grained details from low-resolution images. The model incorporates an adaptive diffusion Transformer decoder (ADTD), which integrates global context and long-range dependencies to better generate detailed information. This approach ensures that the model produces more realistic noise distributions during training and testing. Additionally, multihead self-attention (MSA) modules are introduced to capture low-frequency information [42], while a residual feature fusion method enables effective feature exchange between the decoders across multiple levels.

Diffusion Process: Given a data distribution $x_0 \sim q(x_0)$, the diffusion process creates a sequence $x_1, x_2, \ldots, x_T$, conditioned on x_0 where T is the number of diffusion steps. The process can be expressed as:

$$q(x_1, \ldots, x_T|x_0) = \prod_{t=1}^{T} q(x_t|x_{t-1}), \tag{2.28}$$

where

$$q(x_t|x_{t-1}) = \mathcal{N}(x_t; \sqrt{1-\beta_t}x_{t-1}, \beta_t \boldsymbol{I}), \tag{2.29}$$

with $\boldsymbol{I}$ and $\mathcal{N}$ denote an identity matrix and the Gaussian distribution, respectively.

According to a variance schedule $\{\beta_1, \ldots, \beta_T\}$ with $\beta_t \in \{1, 0\}$ for $t = 1, 2, \cdots, T$, Gaussian noise $\epsilon \in \mathcal{N}(0, \boldsymbol{I})$ is inserted into the data in the sequence. In other words, each x_t is calculated from x_{t-1}. With $a_t = 1 - \beta_t$ and $\bar{a}_t = \prod_{s=0}^{t} a_s$, it can be shown that $q(x_t|x_0)$ takes the following form [21,41]:

$$q(x_t|x_0) = \mathcal{N}\left(x_t; \sqrt{\bar{a}_t}x_0, (1-\bar{a}_t)\boldsymbol{I}\right). \tag{2.30}$$

As a result, every x_t can be directly calculated through x_0 using the following formula:

$$x_t = \sqrt{\bar{a}_t}x_0 + \sqrt{1-\bar{a}_t}\epsilon. \tag{2.31}$$

The reverse process converts the data from $p_\theta(x_T)$ to $p_\theta(x_0)$ parameterized by θ. The prior distribution is chosen as $p(x_t) = \mathcal{N}(x_t; 0, \boldsymbol{I})$ since $q(x_t)$ approximately follows $\mathcal{N}(x_t; 0, \boldsymbol{I})$ in the diffusion process. As a result, the learnable transition kernel is given by:

$$p_\theta(x_0, \ldots, x_{T-1}|x_T) = \prod_{t=1}^{T} p_\theta(x_{t-1}|x_t), \tag{2.32}$$

where

$$p_\theta(x_{t-1}|x_t) = \mathcal{N}(x_{t-1}; \mu_\theta(x_t, t), \sigma_\theta(x_t, t)^2\boldsymbol{I}), \tag{2.33}$$

with θ representing the model parameters. Furthermore, μ_θ and σ_θ denote the mean and variance of the Gaussian distribution, respectively.

The model is trained to minimize the discrepancy between the reverse Markov chain and the actual time-reversed forward chain. This difference is typically measured using the Kullback-Leibler (KL) divergence, which quantifies the difference between the log-likelihoods $p_\theta(x_0, x_1, \ldots, x_T)$ and $q(x_0, x_1, \ldots, x_T)$ as:

$$\begin{aligned}\mathbb{E}\left[-\log p_\theta(x_0)\right] &\le KL\left(p_\theta(x_0, \ldots, x_T), q(x_0, \ldots, x_T)\right)\\ &= \mathbb{E}_q\Big[D_{KL}\left(q(x_T|x_0)||p(x_T)\right) +\\ &\sum_{t>1} D_{KL}\left(q(x_{t-1}|x_t, x_0)||p_\theta(x_{t-1}|x_t)\right) - \log p_\theta(x_0|x_1)\Big].\end{aligned}$$

For simplicity, the loss function, denoted as ℓ, can be expressed as follows by omitting terms that do not depend on θ:

$$\min_\theta \ell_{t-1} = \mathbb{E}_{x_0,\epsilon,t}\left[\|\epsilon - \epsilon_\theta(x_t, t)\|^2\right], \tag{2.34}$$

where ϵ_θ is the noise predictor.

Finally, assuming $z \sim \mathcal{N}(0, \boldsymbol{I})$, we can express x_{t-1} as:

$$x_{t-1} = \mu_\theta(x_t, t) + \sigma_\theta(x_t, t)^2 z. \tag{2.35}$$

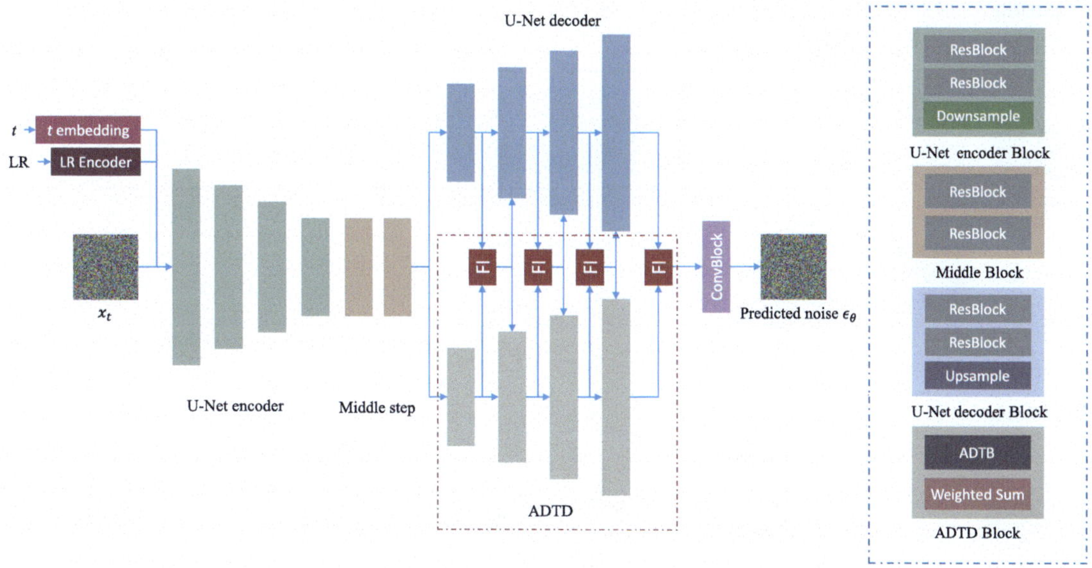

Figure 2.4:
The architecture of the CNP in the ASDDPM [56].

Conditional noise predictor: The Conditional Noise Predictor (CNP) is designed to estimate the noise ϵ_θ introduced at each diffusion step by utilizing the input x_t. As depicted in Fig. 2.4, the CNP consists of three main modules: an encoder, a middle processing step, and a dual decoder structure. Initially, the input x_t, which is the output from a pretrained low-resolution (LR) encoder, along with the diffusion time embedding which is encoded through positional encoding [64], is fed into a U-Net encoder. The low-resolution encoder, being a pretrained superresolution model, helps stabilize the training process and accelerate convergence by providing coarse-level feature representations. These features are then passed through a middle module, consisting of two residual blocks, where further refinement occurs.

The processed features from the middle module are subsequently input into the dual decoder, which combines a U-Net decoder and an Adaptive Diffusion Transformer Decoder (ADTD). The decoders are linked via feature integration (FI) modules [53], facilitating the fusion of residual features at each layer. This setup enables the model to prioritize important features while suppressing irrelevant ones adaptively through the FI modules. In essence, the CNP is capable of processing x_t and predicting the corresponding noise by leveraging both high- and low-frequency information. This fusion of U-Net and Transformer architectures significantly improves the reconstruction quality of superresolution images.

Training and Inference: The training begins by calculating the difference between the upsampled low-resolution (LR) and high-resolution (HR) images, which is then augmented

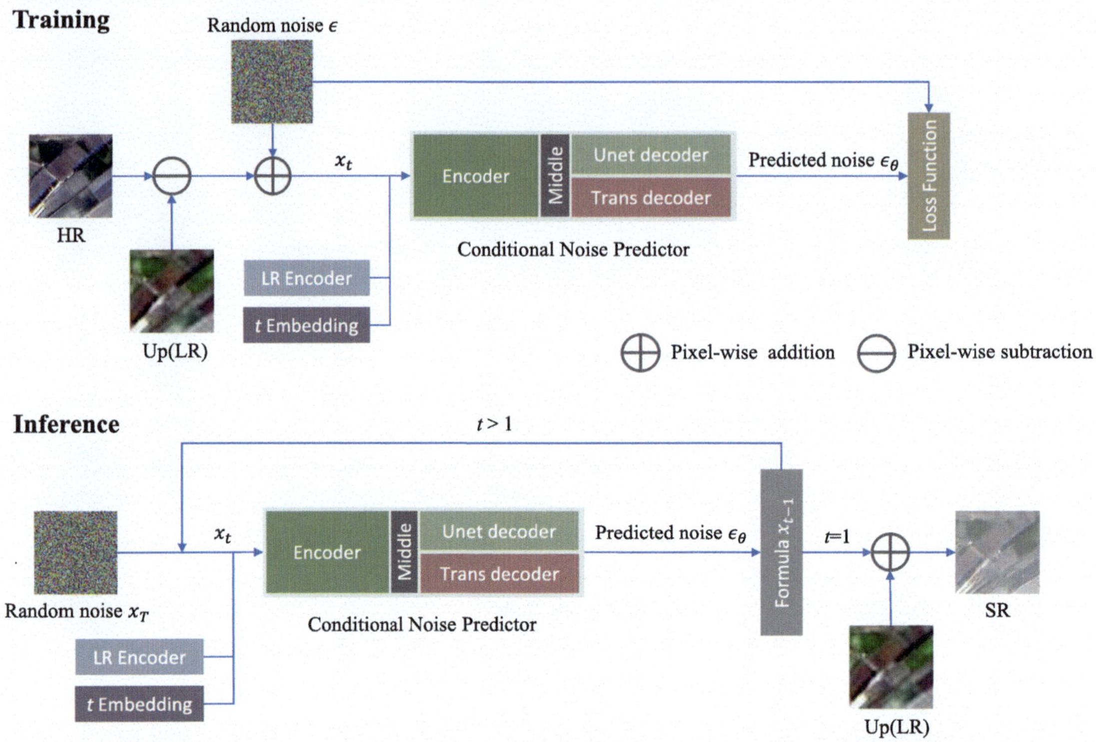

Figure 2.5:
The process of Inference and Training in ASDDPM [56].

with random noise ϵ. This noisy sum, denoted as x_t, is fed into the Conditional Noise Predictor (CNP) along with the time step t and the LR features extracted by the LR encoder. Specifically, ϵ is sampled from a standard Gaussian distribution, while t is drawn from the integer set $\{1, \ldots, T\}$, where T represents the total number of diffusion steps. The CNP is trained by minimizing the loss function as described in Eq. (2.34). During inference, Gaussian noise x_T is first generated and passed through the reverse process. At each step, the LR encoder extracts features from the LR image, which are combined with the corresponding x_t and input into the CNP. The CNP predicts the noise, which is used to generate the previous timestep x_{t-1} for $t > 1$ using Eq. (2.35). Finally, the predicted x_0 is added to the upsampled LR image to produce the superresolved (SR) image. The process of training and inference is shown in Fig. 2.5.

Validation on Semantic Segmentation Task: ASDDPM is one of the first to assess the impact of superresolution on semantic segmentation tasks. To test the effectiveness of superresolution images, two high-resolution datasets, Vaihingen and Potsdam, were used. The high-resolution images were downsampled by a factor of 4 through bicubic interpolation, followed by the application of various single-image superresolution models. Image quality was evaluated using PSNR, SSIM, LPIPS, and FID metrics. The superresolution images were then input into a pretrained U-Netformer [66] for segmentation performance evaluation, with metrics such as overall accuracy (OA), mean intersection over union (mIoU), and mean F1 score (mF1). The experimental results emphasize the importance of using high-quality superresolution methods as a valuable data enhancement technique.

Data enhancement plays an important role in remote sensing by improving the quality and usability of data, which is often affected by factors such as sensor limitations, environmental conditions, or low resolution. Through various techniques such as radiometric, spatial, and spectral enhancement, data enhancement aims to improve the accuracy, clarity, and detail of the captured information, enabling more precise analysis and decision-making. For instance, spatial enhancement techniques, such as superresolution, help increase the resolution of images, while spectral enhancement improves the representation of certain features in different bands. The integration of data enhancement with deep learning techniques has significantly advanced the capabilities of remote sensing systems. Looking forward, the future of data enhancement in remote sensing will likely see greater convergence with advanced deep learning techniques.

2.4 Data registration

Data registration is a basic step in remote sensing that ensures spatial alignment and consistency across datasets obtained from various sensors, platforms, or time frames. It involves the transformation of disparate datasets into a common coordinate framework, allowing for meaningful integration and analysis. The necessity of registration arises from the inherent differences in sensor characteristics, acquisition geometries, and environmental conditions, which lead to spatial discrepancies in the collected data. Without proper registration, these discrepancies can result in misinterpretation or inaccuracies in applications such as change detection, land use analysis, and environmental monitoring.

The main technologies for data registration can be broadly categorized into feature-based, area-based, and hybrid methods [81]. Feature-based methods rely on identifying and matching salient features, such as points, edges, or regions, between datasets. Techniques like scale-invariant feature transform (SIFT) [33] and speeded-up robust features (SURF) [3] are widely used for their robustness to scale and rotation variations. Area-based methods, on the other hand, involve the direct comparison of intensity or pixel values over predefined regions using metrics such as mutual information or normalized cross-correlation. Hybrid methods combine the strengths of both approaches to improve robustness and accuracy.

Advances in deep learning have significantly enhanced the performance of data registration. Convolutional neural networks and transformer-based architectures are now employed to automate feature extraction and matching, even for complex multimodal data. These methods not only improve registration accuracy but also reduce the dependency on manual intervention, enabling efficient handling of large-scale datasets.

2.4.1 Feature-based method

Feature-based registration is a widely used technique in data registration that relies on the detection, extraction, and matching of distinctive features within datasets. These features serve as identifiable landmarks that can guide the alignment process. The method is particularly effective when datasets have clear geometric or structural patterns.

Feature-based registration involves several steps. First, features such as points, edges, or corners are detected in the datasets using algorithms like SIFT, SURF, or the Harris Corner Detector [44]. Second, the detected features are described using robust descriptors to facilitate matching. For example, SIFT generates scale- and rotation-invariant descriptors. Third, feature matching is performed by establishing correspondences between features in different datasets using similarity metrics such as Euclidean distance. Next, transformation parameters (e.g., translation, rotation, scaling) are computed to align the datasets based on matched features. Random Sample Consensus (RANSAC) [16] is often used at this stage to reject outliers and ensure robust estimation. Finally, the computed transformation is applied to one dataset to align it with the reference dataset.

Feature-based registration encompasses several classic methods. SIFT detects key points and describes them in a manner invariant to scale and rotation. For a key point (x, y), the SIFT descriptor is computed from a gradient orientation histogram over a local image patch. SURF is similar to SIFT but optimized for speed, utilizing integral images and approximations to accelerate feature detection and description. Oriented FAST and Rotated BRIEF (ORB) [51] combines the FAST keypoint detector [50] and BRIEF descriptors [5] to provide a fast and efficient alternative to SIFT and SURF. Harris Corner Detection identifies corners by measuring the response of the second-moment matrix:

$$M = \begin{bmatrix} I_x^2 & I_x I_y \\ I_x I_y & I_y^2 \end{bmatrix}, \tag{2.36}$$

where I_x and I_y are image gradients. The corner response R is computed as:

$$R = \det(M) - k(\text{trace}(M))^2, \tag{2.37}$$

where k is a constant (typically 0.04–0.06).

Once feature correspondences are identified, the next step is to estimate the transformation that aligns the datasets. The transformation is generally represented as a matrix H that maps corresponding points from one dataset to the other. For rigid transformations (translation and rotation), the matrix can be expressed as:

$$x' = Hx, \tag{2.38}$$

where $x = [x, y, 1]^\top$ represents the homogeneous coordinates of a point in the reference dataset, x' is its corresponding point in the target dataset, and H is the transformation matrix. For a rigid transformation, the matrix is given by:

$$H = \begin{bmatrix} \cos\theta & -\sin\theta & t_x \\ \sin\theta & \cos\theta & t_y \\ 0 & 0 & 1 \end{bmatrix}, \tag{2.39}$$

where θ is the rotation angle and t_x, t_y are translations. For an affine transformation, the matrix is:

$$H = \begin{bmatrix} a & b & t_x \\ c & d & t_y \\ 0 & 0 & 1 \end{bmatrix}, \tag{2.40}$$

where a, b, c, d represent scaling, rotation, and shear. A more generalized transformation for perspective changes, called homography, is expressed as:

$$H = \begin{bmatrix} h_{11} & h_{12} & h_{13} \\ h_{21} & h_{22} & h_{23} \\ h_{31} & h_{32} & h_{33} \end{bmatrix}. \tag{2.41}$$

In remote sensing, the feature-based registration method is especially useful when the images being compared are of the same area but may differ due to factors like changes in lighting, atmospheric conditions, or sensor types. Therefore, it is widely used for tasks like change detection, where images of the same area taken at different times are compared, or for multisensor image fusion, where data from different sensors (e.g., optical and radar) need to be combined. It is also used in applications like creating digital elevation models (DEM) from stereo imagery or aligning high-resolution panchromatic images with multispectral ones. This method is effective because it focuses on robust, recognizable features that are less affected by environmental variations.

2.4.2 Area-based method

The area-based registration is a technique used to align two or more images by comparing the intensity values in local regions or patches of the datasets. Unlike feature-based methods, which rely on distinct features such as edges, points, or contours, area-based methods work by evaluating the similarity between regions in the images. These methods are particularly useful when the images do not have clearly identifiable features, or when the changes between images are more global rather than localized [6,14,61].

The fundamental idea behind area-based registration is to define a similarity measure that quantifies the similarity between two image patches and use this measure to align the images. This approach generally involves two main stages: (1) defining the similarity measure, and (2) using optimization techniques to estimate the transformation parameters that align the images based on the selected similarity metric.

In remote sensing, the area-based method is widely used for the registration of multispectral, hyperspectral, and multitemporal images. These types of images often contain similar features over large areas, but the intensity values might differ due to changes in sensor properties, environmental conditions, or viewing angles. The goal of the area-based method is to align such images so that corresponding areas match as closely as possible.

One common method used in area-based registration is the Normalized Cross-Correlation (NCC). The NCC measure computes the similarity between two image patches I and J as follows:

$$NCC(I,J) = \frac{\sum_{x,y}(I(x,y)-\mu_I)(J(x,y)-\mu_J)}{\sqrt{\sum_{x,y}(I(x,y)-\mu_I)^2 \sum_{x,y}(J(x,y)-\mu_J)^2}}, \tag{2.42}$$

where $I(x, y)$ and $J(x, y)$ represent the pixel intensities of the patches at position (x, y), and μ_I and μ_J are the mean pixel intensities of patches I and J, respectively. The value of NCC ranges from -1 (no correlation) to 1 (perfect correlation), and higher values indicate better alignment of the patches.

Another commonly used similarity measure is the Sum of Squared Differences (SSD), which computes the total squared difference in pixel intensities between two image patches:

$$SSD(I,J) = \sum_{x,y}[I(x,y) - J(x,y)]^2. \tag{2.43}$$

In SSD, a smaller value indicates higher similarity between the patches, and thus, better alignment.

For images with significantly different intensity distributions, such as multispectral or multimodal images, Mutual Information (MI) is often used as a similarity metric. MI quantifies the amount of information shared between the images and is particularly effective when images from different sensors or imaging modalities need to be registered. The Mutual

Information between two images I and J is given by:

$$MI(I, J) = H(I) + H(J) - H(I, J), \tag{2.44}$$

where $H(I)$ and $H(J)$ are the entropies of images I and J, and $H(I, J)$ is the joint entropy of both images. Maximizing the MI between the images ensures the best alignment, especially when the images exhibit different intensities or features due to varying sensor types or environmental factors.

Once the similarity measure is chosen, optimization methods such as gradient descent or the Lucas-Kanade method are used to estimate the transformation parameters. For rigid transformations, such as translation or rotation, the transformation can be expressed as:

$$T(x, y) = (x + t_x, y + t_y), \tag{2.45}$$

where t_x and t_y are the translation parameters in the x and y directions, respectively. For affine transformations, the transformation can be modeled by a matrix:

$$T(x, y) = \begin{bmatrix} a & b & t_x \\ c & d & t_y \\ 0 & 0 & 1 \end{bmatrix}, \tag{2.46}$$

where a, b, c, d are the affine parameters, and t_x, t_y are the translation components. For more complex, nonrigid transformations, such as elastic deformations, spline-based methods like B-splines or thin-plate splines (TPS) are used to model the local deformations and align the images accordingly.

The area-based method is effective in aligning images taken from different sensors or at different times, helping to create accurate composite images or detect subtle changes in the environment over time. Moreover, since area-based methods do not require the identification of specific features, they can be applied in situations where other methods, like feature-based registration, may not be effective.

2.4.3 Learning-based method

In remote sensing, the learning-based method refers to techniques that leverage machine learning and deep learning models to analyze and process remote sensing data. The advantages of deep learning over machine learning are more pronounced, as discussed in the Section of superresolution. Recently, deep learning-based registration methods are becoming increasingly popular due to their ability to automatically align and register images by learning complex patterns in the data. These methods use deep learning models to map relationships between different image datasets, improving accuracy and efficiency over traditional methods.

Unlike conventional methods that rely on manually designed features, deep learning-based approaches can automatically extract relevant features and perform registration directly from raw pixel data. This capability is especially useful in scenarios where the images may have different resolutions, viewpoints, or sensor types [15,30,81].

One common approach in deep learning-based registration is the use of Convolutional Neural Networks (CNNs), which have been adapted for image registration tasks [75]. In this method, a CNN is trained to directly learn the spatial transformation that best aligns the images. The transformation can be rigid (translation and rotation) or nonrigid (such as scaling and warping), depending on the complexity of the registration task. The loss function used to train the CNN often incorporates a similarity measure between the registered images, such as the Mean Squared Error (MSE) or normalized cross-correlation (NCC).

Another approach is Spatial Transformer Networks (STNs) [13,31]. They introduce a learnable spatial transformation module that applies spatial transformations (e.g., affine or projective) to the input image in an end-to-end learning framework. The spatial transformer network outputs the transformed image, and the loss function measures how well the transformation aligns the images. A common objective function used in STNs for registration is:

$$L = \|I_{fixed} - T(I_{moving})\|_2^2, \tag{2.47}$$

where I_{fixed} and I_{moving} are the fixed and moving images, and $T(I_{moving})$ represents the transformed moving image.

Generative Adversarial Networks (GANs) are also employed for image registration tasks. GANs consist of two networks: a generator and a discriminator [35,80]. The generator creates transformed images, while the discriminator assesses how well the transformation aligns the images. This adversarial training helps refine the registration process to produce more accurate alignments. The loss function for GAN-based registration often combines an adversarial loss with a pixel-wise similarity loss, such as:

$$L = L_{adv}(G, D) + \lambda L_{sim}(I_{fixed}, G(I_{moving})), \tag{2.48}$$

where L_{adv} is the adversarial loss, G is the generator, D is the discriminator, L_{sim} is the similarity loss, and λ is a regularization parameter.

Lastly, Siamese Networks have been used for deep learning-based registration [12,34]. A Siamese network consists of two identical neural networks that share the same weights. These networks are trained to extract feature representations from the fixed and moving images and then compare the features to learn the transformation required to align the images. The loss function is typically a combination of feature similarity and spatial alignment, expressed as:

$$L = \| f_{fixed}(x) - f_{moving}(x) \|_2^2, \quad (2.49)$$

where f_{fixed} and f_{moving} are the feature maps of the fixed and moving images, respectively.

Deep learning-based registration methods provide significant improvements in robustness and automation, particularly in handling complex and large datasets, making them highly effective for remote sensing applications such as multitemporal image alignment, multisensor fusion, and georeferencing.

The significance of data registration lies in its ability to unlock the full potential of multimodal and temporal remote sensing data. Properly registered data facilitate advanced analyses, such as multisensor data fusion, which combines complementary information from optical, radar, and hyperspectral sensors for enriched insights. By maintaining spatial coherence, data registration continues to be a foundational element of remote sensing, serving as a critical bridge between raw sensor outputs and meaningful, actionable insights. Temporal data registration enables precise tracking of environmental changes over time, crucial for disaster management, agricultural monitoring, and urban planning. Moreover, the integration of artificial intelligence with edge computing offers exciting potential for achieving real-time registration of data acquired from an ever-expanding array of diverse sensors.

2.4.4 Data alignment

Data alignment ensures that corresponding features from different datasets are precisely positioned in the same spatial framework. Unlike data registration, which focuses on the general spatial alignment of entire datasets, data alignment is concerned with achieving high accuracy in the spatial correspondence of specific features within the data. This is especially important when dealing with datasets that originate from different sensors, acquisition times, or modalities. Data alignment is a refined process that adjusts for minor discrepancies that may exist even after the data has been registered, often involving more advanced techniques for fine-tuning the spatial positioning of features such as buildings, roads, vegetation, and other critical objects.

The necessity of data alignment becomes evident in several remote sensing applications. In change detection, for instance, it is critical that the features in the datasets are precisely aligned to identify small but significant changes between different temporal images. Similarly, in multisensor fusion, such as combining data from optical, radar, or LiDAR sensors, ensuring that corresponding features are spatially aligned is essential for generating accurate, integrated products. Without proper alignment, discrepancies between datasets can lead to errors in feature extraction, classification, and other analytical tasks, undermining the quality and reliability of the results.

The significance of data alignment extends to various fields of remote sensing, including land cover classification, urban monitoring, environmental change tracking, and disaster response. In these contexts, even slight misalignments can result in inaccurate analyses and decision-making. For example, in the monitoring of urban growth, precise alignment of satellite images from different periods allows for accurate detection of new developments, while in agriculture, alignment of multitemporal data helps in tracking crop growth and predicting yields.

Main technologies for data alignment include advanced feature matching algorithms, which use keypoints, edges, or textures to align datasets, and optimization techniques, which refine the alignment process by minimizing discrepancies between corresponding features. Recent advances in machine learning and deep learning have further enhanced data alignment, offering the potential for more automated and accurate alignment processes, especially in the context of complex, large-scale datasets.

2.5 Conclusion

Multimodal data preprocessing is a basic component in the remote sensing field, enabling the integration of diverse data sources to achieve comprehensive and actionable insights. Given the variety of sensors, acquisition platforms, and data types involved, preprocessing techniques play an essential role in ensuring that multimodal datasets are aligned, calibrated, and enhanced for effective analysis. These preprocessing steps, including calibration, enhancement, registration, and alignment, form the foundation for the accurate and reliable integration of multimodal data. Each of these steps addresses specific challenges, such as sensor inaccuracies, data inconsistencies, and spatial misalignments, which are common in remote sensing applications.

Calibration techniques, such as radiometric and geometric calibration, are crucial for correcting sensor-specific errors and ensuring that the data is in a standardized format for comparison and fusion. Radiometric calibration corrects for variations in sensor response, ensuring that pixel values reflect accurate measurements of the Earth's surface, while geometric calibration addresses distortions in spatial information, ensuring proper alignment with real-world coordinates. These steps are essential for maintaining the integrity of the data, especially when integrating data from different sensors or acquisition platforms.

Enhancement techniques, including radiometric enhancement, spatial enhancement, and noise reduction, improve the quality of the data, making it more suitable for analysis. These techniques allow for the refinement of spatial resolution, the reduction of noise, and the adjustment of image contrast, all of which enhance the clarity and accuracy of the data, facilitating more precise interpretation and analysis.

Registration and alignment, as the final preprocessing steps, are essential for ensuring that multimodal datasets are spatially consistent and correspond to one another. While data registration involves matching datasets to a common coordinate system, data alignment focuses on achieving high precision in aligning specific features, such as roads, buildings, and vegetation. These processes are particularly important when working with multitemporal, multispectral, or multisensor data, as they ensure that corresponding features are properly aligned, allowing for accurate comparisons and integrated analyses.

The integration of deep learning techniques with traditional data preprocessing methods is a promising direction for the future of multimodal data analysis. Deep learning can automate many of the preprocessing tasks, improving efficiency and accuracy. For example, deep learning-based methods can be employed for feature extraction, alignment, and even data enhancement, enabling the processing of large-scale datasets with minimal manual intervention. Furthermore, advancements in unsupervised and self-supervised learning could reduce the reliance on labeled data, making preprocessing more accessible and scalable for a broader range of remote sensing applications [19,38].

In conclusion, multimodal data preprocessing is an indispensable step in remote sensing, laying the groundwork for accurate data analysis and decision-making. As remote sensing technologies continue to advance, the development of more sophisticated preprocessing methods, particularly those leveraging deep learning, will be crucial in enhancing the accuracy, efficiency, and scalability of multimodal data analysis in various applications, from environmental monitoring to urban planning and disaster response.

References

[1] Nour Aburaed, Mohammed Q. Alkhatib, Stephen Marshall, Jaime Zabalza, Hussain Al Ahmad, A review of spatial enhancement of hyperspectral remote sensing imaging techniques, IEEE Journal of Selected Topics in Applied Earth Observations and Remote Sensing 16 (2023) 2275–2300.

[2] Eirikur Agustsson, Radu Timofte, Luc Van Gool, Regressor basis learning for anchored super-resolution, in: 2016 23rd International Conference on Pattern Recognition (ICPR), IEEE, 2016, pp. 3850–3855.

[3] Herbert Bay, Tinne Tuytelaars, Luc Van Gool, SURF: speeded up robust features, in: Computer Vision–ECCV 2006: 9th European Conference on Computer Vision, Graz, Austria, May 7–13, 2006. Proceedings, Part I 9, Springer, 2006, pp. 404–417.

[4] M. Blázquez, A new approach to spatio-temporal calibration of multi-sensor systems, in: Proceedings of the International Archives of the Photogrammetry, Remote Sensing and Spatial Information Sciences, Beijing, China, 2008, pp. 3–13.

[5] Michael Calonder, Vincent Lepetit, Christoph Strecha, Pascal Fua, BRIEF: binary robust independent elementary features, in: Computer Vision–ECCV 2010: 11th European Conference on Computer Vision, Heraklion, Crete, Greece, September 5-11, 2010, Proceedings, Part IV 11, Springer, 2010, pp. 778–792.
[6] C.P. Dalmiya, V.S. Dharun, A survey of registration techniques in remote sensing images, Indian Journal of Science and Technology 8 (26) (2015) 1–7.
[7] Urška Demšar, Paul Harris, Chris Brunsdon, A. Stewart Fotheringham, Sean McLoone, Principal component analysis on spatial data: an overview, Annals of the Association of American Geographers 103 (1) (2013) 106–128.
[8] Magdeleine Dinguirard, Philip N. Slater, Calibration of space-multispectral imaging sensors: a review, Remote Sensing of Environment 68 (3) (1999) 194–205.
[9] Chao Dong, Chen Change Loy, Kaiming He, Xiaoou Tang, Image super-resolution using deep convolutional networks, IEEE Transactions on Pattern Analysis and Machine Intelligence 38 (2) (2015) 295–307.
[10] Alexey Dosovitskiy, An image is worth 16x16 words: transformers for image recognition at scale, arXiv preprint, arXiv:2010.11929, 2020.
[11] Oleg Dubovik, Zhengqiang Li, Michael I. Mishchenko, Didier Tanré, Yana Karol, Bojan Bojkov, Brian Cairns, David J. Diner, W. Reed Espinosa, Philippe Goloub, et al., Polarimetric remote sensing of atmospheric aerosols: instruments, methodologies, results, and perspectives, Journal of Quantitative Spectroscopy and Radiative Transfer 224 (2019) 474–511.
[12] Rongbo Fan, Bochuan Hou, Jinbao Liu, Jianhua Yang, Zenglin Hong, Registration of multiresolution remote sensing images based on L2-Siamese model, IEEE Journal of Selected Topics in Applied Earth Observations and Remote Sensing 14 (2020) 237–248.
[13] Yibo Fan, Feng Wang, Haipeng Wang, A transformer-based coarse-to-fine wide-swath SAR image registration method under weak texture conditions, Remote Sensing 14 (5) (2022) 1175.
[14] Ruitao Feng, Qingyun Du, Xinghua Li, Huanfeng Shen, Robust registration for remote sensing images by combining and localizing feature- and area-based methods, ISPRS Journal of Photogrammetry and Remote Sensing 151 (2019) 15–26.
[15] Ruitao Feng, Huanfeng Shen, Jianjun Bai, Xinghua Li, Advances and opportunities in remote sensing image geometric registration: a systematic review of state-of-the-art approaches and future research directions, IEEE Geoscience and Remote Sensing Magazine 9 (4) (2021) 120–142.
[16] Martin A. Fischler, Robert C. Bolles, Random sample consensus: a paradigm for model fitting with applications to image analysis and automated cartography, Communications of the ACM 24 (6) (1981) 381–395.
[17] Hassan Ghassemian, A review of remote sensing image fusion methods, Information Fusion 32 (2016) 75–89.
[18] Luis Guanter, Víctor Estellés, Jose Moreno, Spectral calibration and atmospheric correction of ultra-fine spectral and spatial resolution remote sensing data. Application to CASI-1500 data, Remote Sensing of Environment 109 (1) (2007) 54–65.
[19] Wei Han, Xiaohan Zhang, Yi Wang, Lizhe Wang, Xiaohui Huang, Jun Li, Sheng Wang, Weitao Chen, Xianju Li, Ruyi Feng, et al., A survey of machine learning and deep learning in remote sensing of geological environment: challenges, advances, and opportunities, ISPRS Journal of Photogrammetry and Remote Sensing 202 (2023) 87–113.
[20] Muhammad Haris, Gregory Shakhnarovich, Norimichi Ukita, Deep back-projection networks for super-resolution, in: Proceedings of the IEEE Conference on Computer Vision and Pattern Recognition, 2018, pp. 1664–1673.
[21] Jonathan Ho, Ajay Jain, Pieter Abbeel, Denoising diffusion probabilistic models, in: H. Larochelle, M. Ranzato, R. Hadsell, M.F. Balcan, H. Lin (Eds.), Advances in Neural Information Processing Systems, vol. 33, Curran Associates, Inc., 2020, pp. 6840–6851.
[22] Danfeng Hong, Lianru Gao, Naoto Yokoya, Jing Yao, Jocelyn Chanussot, Qian Du, Bing Zhang, More diverse means better: multimodal deep learning meets remote-sensing imagery classification, IEEE Transactions on Geoscience and Remote Sensing 59 (5) (2020) 4340–4354.

[23] Junjun Jiang, Xiang Ma, Chen Chen, Tao Lu, Zhongyuan Wang, Jiayi Ma, Single image super-resolution via locally regularized anchored neighborhood regression and nonlocal means, IEEE Transactions on Multimedia 19 (1) (2016) 15–26.
[24] Nur Huseyin Kaplan, Isin Erer, Kumlu Deniz, Image enhancement methods for remote sensing: a survey, in: Recent Remote Sensing Sensor Applications-Satellites and Unmanned Aerial Vehicles (UAVs), IntechOpen, 2021.
[25] Jiwon Kim, Jung Kwon Lee, Kyoung Mu Lee, Accurate image super-resolution using very deep convolutional networks, in: Proceedings of the IEEE Conference on Computer Vision and Pattern Recognition, 2016, pp. 1646–1654.
[26] Rakesh Kumar Mishra, Yun Zhang, A review of optical imagery and airborne LiDAR data registration methods, The Open Remote Sensing Journal 5 (1) (2012).
[27] V. Lalitha, B. Latha, A review on remote sensing imagery augmentation using deep learning, Materials Today: Proceedings 62 (2022) 4772–4778.
[28] Wooju Lee, Donggyu Sim, Seoung-Jun Oh, A CNN-based high-accuracy registration for remote sensing images, Remote Sensing 13 (8) (2021) 1482.
[29] Sen Lei, Zhenwei Shi, Wenjing Mo, Transformer-based multistage enhancement for remote sensing image super-resolution, IEEE Transactions on Geoscience and Remote Sensing 60 (2021) 1–11.
[30] Liangzhi Li, Ling Han, Mingtao Ding, Hongye Cao, Huijuan Hu, A deep learning semantic template matching framework for remote sensing image registration, ISPRS Journal of Photogrammetry and Remote Sensing 181 (2021) 205–217.
[31] Chenbin Liang, Yunyun Dong, Changjun Zhao, Zengguo Sun, A coarse-to-fine feature match network using transformers for remote sensing image registration, Remote Sensing 15 (13) (2023) 3243.
[32] Jingyun Liang, Jiezhang Cao, Guolei Sun, Kai Zhang, Luc Van Gool, Radu Timofte, SwinIR: image restoration using Swin Transformer, in: Proceedings of the IEEE/CVF International Conference on Computer Vision, 2021, pp. 1833–1844.
[33] Tony Lindeberg, Scale invariant feature transform, Scholarpedia 7 (5) (2012) 10491, https://doi.org/10.4249/scholarpedia.10491.
[34] Yuyan Liu, Xiaoying Gong, Jiaxuan Chen, Shuang Chen, Yang Yang, Rotation-invariant Siamese network for low-altitude remote-sensing image registration, IEEE Journal of Selected Topics in Applied Earth Observations and Remote Sensing 13 (2020) 5746–5758.
[35] Luca Maggiolo, David Solarna, Gabriele Moser, Sebastiano Bruno Serpico, Registration of multisensor images through a conditional generative adversarial network and a correlation-type similarity measure, Remote Sensing 14 (12) (2022) 2811.
[36] Shaohui Mei, Jingyu Ji, Junhui Hou, Xu Li, Qian Du, Learning sensor-specific spatial-spectral features of hyperspectral images via convolutional neural networks, IEEE Transactions on Geoscience and Remote Sensing 55 (8) (2017) 4520–4533.
[37] Yiqun Mei, Yuchen Fan, Yuqian Zhou, Image super-resolution with non-local sparse attention, in: Proceedings of the IEEE/CVF Conference on Computer Vision and Pattern Recognition, 2021, pp. 3517–3526.
[38] Mohammed Abdulmajeed Moharram, Divya Meena Sundaram, Land use and land cover classification with hyperspectral data: a comprehensive review of methods, challenges and future directions, Neurocomputing 536 (2023) 90–113.
[39] A. Mookambiga, V. Gomathi, Comprehensive review on fusion techniques for spatial information enhancement in hyperspectral imagery, Multidimensional Systems and Signal Processing 27 (2016) 863–889.
[40] Ganesh R. Naik, Dinesh K. Kumar, An overview of independent component analysis and its applications, Informatica 35 (1) (2011).
[41] Alexander Quinn Nichol, Prafulla Dhariwal, Improved denoising diffusion probabilistic models, in: International Conference on Machine Learning, PMLR, 2021, pp. 8162–8171.
[42] Namuk Park, Songkuk Kim, How do vision transformers work?, arXiv preprint, arXiv:2202.06709, 2022.

[43] Sourabh Paul, Umesh C. Pati, A comprehensive review on remote sensing image registration, International Journal of Remote Sensing 42 (14) (2021) 5396–5432.
[44] Yijian Pei, Hao Wu, Jiang Yu, Guanghui Cai, Effective image registration based on improved Harris corner detection, in: 2010 International Conference on Information, Networking and Automation (ICINA), vol. 1, IEEE, 2010, pp. 93–96.
[45] M.J. Persky, A review of spaceborne infrared Fourier transform spectrometers for remote sensing, Review of Scientific Instruments 66 (10) (1995) 4763–4797.
[46] Cle Pohl, John L. Van Genderen, Review article multisensor image fusion in remote sensing: concepts, methods and applications, International Journal of Remote Sensing 19 (5) (1998) 823–854.
[47] Dou Quan, Huiyuan Wei, Shuang Wang, Yu Gu, Biao Hou, Licheng Jiao, A novel coarse-to-fine deep learning registration framework for multi-modal remote sensing images, IEEE Transactions on Geoscience and Remote Sensing (2023).
[48] Behnood Rasti, Yi Chang, Emanuele Dalsasso, Loic Denis, Pedram Ghamisi, Image restoration for remote sensing: overview and toolbox, IEEE Geoscience and Remote Sensing Magazine 10 (2) (2021) 201–230.
[49] Chao Ren, Xiaohai He, Truong Q. Nguyen, Single image super-resolution via adaptive high-dimensional non-local total variation and adaptive geometric feature, IEEE Transactions on Image Processing 26 (1) (2016) 90–106.
[50] Edward Rosten, Tom Drummond, Fusing points and lines for high performance tracking, in: Tenth IEEE International Conference on Computer Vision (ICCV'05) Volume 1, vol. 2, IEEE, 2005, pp. 1508–1515.
[51] Ethan Rublee, Vincent Rabaud, Kurt Konolige, Gary Bradski, Orb: an efficient alternative to sift or surf, in: 2011 International Conference on Computer Vision, IEEE, 2011, pp. 2564–2571.
[52] Francesco Salvetti, Vittorio Mazzia, Aleem Khaliq, Marcello Chiaberge, Multi-image super resolution of remotely sensed images using residual attention deep neural networks, Remote Sensing 12 (14) (2020) 2207.
[53] Daniel Seichter, Mona Köhler, Benjamin Lewandowski, Tim Wengefeld, Horst-Michael Gross, Efficient RGB-D semantic segmentation for indoor scene analysis, in: 2021 IEEE International Conference on Robotics and Automation (ICRA), 2021, pp. 13525–13531.
[54] Xianyu Su, Wenjing Chen, Fourier transform profilometry:: a review, Optics and Lasers in Engineering 35 (5) (2001) 263–284.
[55] Jialu Sui, Xianping Ma, Xiaokang Zhang, Man-On Pun, GCRDN: global context-driven residual dense network for remote sensing image superresolution, IEEE Journal of Selected Topics in Applied Earth Observations and Remote Sensing 16 (2023) 4457–4468.
[56] Jialu Sui, Xianping Ma, Xiaokang Zhang, Man-On Pun, Hao Wu, Adaptive semantic-enhanced denoising diffusion probabilistic model for remote sensing image super-resolution, IEEE Journal of Selected Topics in Applied Earth Observations and Remote Sensing (2024).
[57] K.N. Tahar, An evaluation on different number of ground control points in unmanned aerial vehicle photogrammetric block, The International Archives of the Photogrammetry, Remote Sensing and Spatial Information Sciences 40 (2013) 93–98.
[58] P.M. Teillet, D.N.H. Horler, N.T. O'Neill, Calibration, validation, and quality assurance in remote sensing: a new paradigm, Canadian Journal of Remote Sensing 23 (4) (1997) 401–414.
[59] K. Thorne, B. Markharn, P. Slater Barker, S.J.P.E. Biggar, Radiometric calibration of Landsat, Photogrammetric Engineering and Remote Sensing 63 (7) (1997) 853–858.
[60] Radu Timofte, Vincent De Smet, Luc Van Gool, A+: adjusted anchored neighborhood regression for fast super-resolution, in: Asian Conference on Computer Vision, Springer, 2014, pp. 111–126.
[61] Ms. Priyanka S. Tondewad, Ms. Manisha P. Dale, Remote sensing image registration methodology: review and discussion, Procedia Computer Science 171 (2020) 2390–2399.
[62] Simon A. Trim, Kimberley Mason, Andreas Hueni, Spectroradiometer spectral calibration, ISRF shapes, and related uncertainties, Applied Optics 60 (18) (2021) 5405–5417.
[63] Thaweesak Trongtirakul, Werapon Chiracharit, Susan Imberman, Sos Agaian, Fractional contrast stretching for image enhancement of aerial and satellite images, The Journal of Imaging Science and Technology 63 (6) (2019).

[64] A. Vaswani, Attention is all you need, in: Advances in Neural Information Processing Systems, 2017.
[65] Haohan Wang, Xindi Wu, Zeyi Huang, Eric P. Xing, High-frequency component helps explain the generalization of convolutional neural networks, in: Proceedings of the IEEE/CVF Conference on Computer Vision and Pattern Recognition, 2020, pp. 8684–8694.
[66] Libo Wang, Rui Li, Ce Zhang, Shenghui Fang, Chenxi Duan, Xiaoliang Meng, Peter M. Atkinson, UNetFormer: a UNet-like transformer for efficient semantic segmentation of remote sensing urban scene imagery, ISPRS Journal of Photogrammetry and Remote Sensing 190 (2022) 196–214.
[67] Mi Wang, Bo Yang, Fen Hu, Xi Zang, On-orbit geometric calibration model and its applications for high-resolution optical satellite imagery, Remote Sensing 6 (5) (2014) 4391–4408.
[68] Mi Wang, Yuan Tian, Yufeng Cheng, Development of on-orbit geometric calibration for high resolution optical remote sensing satellite, Geomatics and Information Science of Wuhan University 42 (11) (2017) 1580–1588.
[69] Shenlong Wang, Lei Zhang, Yan Liang, Quan Pan, Semi-coupled dictionary learning with applications to image super-resolution and photo-sketch synthesis, in: 2012 IEEE Conference on Computer Vision and Pattern Recognition, IEEE, 2012, pp. 2216–2223.
[70] Shuang Wang, Dou Quan, Xuefeng Liang, Mengdan Ning, Yanhe Guo, Licheng Jiao, A deep learning framework for remote sensing image registration, ISPRS Journal of Photogrammetry and Remote Sensing 145 (2018) 148–164.
[71] Xuan Wang, Lijun Sun, Abdellah Chehri, Yongchao Song, A review of GAN-based super-resolution reconstruction for optical remote sensing images, Remote Sensing 15 (20) (2023) 5062.
[72] Yi Xiao, Qiangqiang Yuan, Kui Jiang, Jiang He, Xianyu Jin, Liangpei Zhang, EDiffSR: an efficient diffusion probabilistic model for remote sensing image super-resolution, IEEE Transactions on Geoscience and Remote Sensing (2023).
[73] Xiaoxiong Xiong, William Barnes, An overview of MODIS radiometric calibration and characterization, Advances in Atmospheric Sciences 23 (2006) 69–79.
[74] Bo Yang, Mi Wang, Wen Xu, Deren Li, Jianya Gong, Yingdong Pi, Large-scale block adjustment without use of ground control points based on the compensation of geometric calibration for ZY-3 images, ISPRS Journal of Photogrammetry and Remote Sensing 134 (2017) 1–14.
[75] Famao Ye, Yanfei Su, Hui Xiao, Xuqing Zhao, Weidong Min, Remote sensing image registration using convolutional neural network features, IEEE Geoscience and Remote Sensing Letters 15 (2) (2018) 232–236.
[76] Nicholas E. Young, Ryan S. Anderson, Stephen M. Chignell, Anthony G. Vorster, Rick Lawrence, Paul H. Evangelista, A survival guide to Landsat preprocessing, Ecology 98 (4) (2017) 920–932.
[77] Roman Zeyde, Michael Elad, Matan Protter, On single image scale-up using sparse-representations, in: International Conference on Curves and Surfaces, Springer, 2012, pp. 711–730.
[78] Han Zhang, Weiping Ni, Weidong Yan, Deliang Xiang, Junzheng Wu, Xiaoliang Yang, Hui Bian, Registration of multimodal remote sensing image based on deep fully convolutional neural network, IEEE Journal of Selected Topics in Applied Earth Observations and Remote Sensing 12 (8) (2019) 3028–3042.
[79] Wenjian Zhang, Zheng Tan, Qunbo Lv, Jiaao Li, Baoyu Zhu, Yangyang Liu, An efficient hybrid CNN-transformer approach for remote sensing super-resolution, Remote Sensing 16 (5) (2024) 880.
[80] Xinyue Zhang, Chengcai Leng, Yameng Hong, Zhao Pei, Irene Cheng, Anup Basu, Multimodal remote sensing image registration methods and advancements: a survey, Remote Sensing 13 (24) (2021) 5128.
[81] Bai Zhu, Liang Zhou, Simiao Pu, Jianwei Fan, Yuanxin Ye, Advances and challenges in multimodal remote sensing image registration, IEEE Journal on Miniaturization for Air and Space Systems 4 (2) (2023) 165–174.
[82] Zhe Zhu, Change detection using landsat time series: a review of frequencies, preprocessing, algorithms, and applications, ISPRS Journal of Photogrammetry and Remote Sensing 130 (2017) 370–384.

CHAPTER 3

Fusion techniques for multimodal remote sensing

Man-On Pun[a] **and Xianping Ma**[b]

[a]*School of Science and Engineering, The Chinese University of Hong Kong, Shenzhen, China* [b]*Faculty of Geosciences and Engineering, Southwest Jiaotong University, Chengdu, China*

3.1 Introduction

Multimodal remote sensing data fusion for classification is a cutting-edge approach in geospatial analysis that leverages the strengths of different sensing modalities to enhance the accuracy and reliability of classification tasks. In the remote sensing community, the term "classification" is commonly referred to the term "semantic segmentation," which is more frequently used in the community of computer vision. In this chapter, the term "classification" is used to refer to the process of assigning each pixel to a specific semantic category corresponding to a land cover type [14]. This distinction arises because image classification typically involves predicting a single label that describes the primary geographical scene of an image, necessitating the extraction of image features. From the model's perspective, this process primarily involves image encoding. In contrast, semantic segmentation requires both the extraction and restoration of image features, as it assigns a label to each pixel in the image. As a result, it involves both image encoding and decoding, which introduces additional complexities and necessitates more diverse strategies in the multimodal data fusion process. In other words, the framework of semantic segmentation tasks allows for a more comprehensive introduction of fusion techniques for multimodal remote sensing [26]. By integrating data from various sensors, such as optical, radar, and LiDAR, this technique allows for a more comprehensive representation of the observed environment [12,60]. Each sensing modality captures distinct aspects of the target area, and their fusion enables the extraction of richer, more informative features that are crucial for precise classification. This approach not only improves the differentiation of complex land cover types but also mitigates the limitations inherent in using a single sensor type. As a result, multimodal data fusion has become an essential tool in advancing the field of remote sensing, with applications ranging from environmental monitoring to urban planning and beyond [7].

Multimodal Remote Sensing Fusion and Classification
https://doi.org/10.1016/B978-0-44-329152-4.00010-6

Generally, the multimodal fusion techniques in remote sensing can be broadly categorized into four main levels: pixel-level fusion, feature-level fusion, decision-level fusion, and mixed fusion.

- Pixel-Level Fusion: In pixel-level fusion, data from multiple sensors are combined at the pixel level. This approach also known as early fusion merges raw data from different modalities, typically after they have been preprocessed to align spatial and spectral resolutions. Pixel-level fusion can enhance the spatial and spectral detail in the final output, making it particularly useful in applications with high-resolution imagery. However, it also requires careful calibration and alignment of the input data to ensure consistency across modalities.
- Feature-Level Fusion: Feature-level fusion involves extracting and combining features from the different data modalities before applying a classification model. This method aims to capitalize on the strengths of each sensor by integrating features that capture different aspects of the observed scene, such as texture, shape, or spectral signatures. Feature-level fusion is beneficial in providing a richer set of information for classification, but it can be computationally intensive and requires sophisticated feature extraction techniques.
- Decision-Level Fusion: Decision-level fusion, also known as late fusion, operates at a higher level of abstraction, where the outputs or decisions from separate classifiers are combined to make a final decision. The different classifiers are trained on data from different modalities. This method is robust to variations in the input data, which means specific classifiers can be designed based on the characteristics of the multimodal data. It can also be implemented using various techniques, such as majority voting, weighted averaging, or more complex decision rules. The Decision-level fusion is often used when the individual modalities have been processed independently and need to be integrated without reprocessing the raw data.
- Mixed Fusion: Mixed fusion, also known as hybrid fusion, is an approach in multimodal data fusion that combines different levels of fusion within a single framework. The goal of mixed fusion is to leverage the strengths of each fusion level to achieve more robust and accurate results in applications. This technique is particularly useful in complex remote sensing scenarios where data from multiple sensors or modalities vary significantly in terms of resolution, scale, or the type of information they provide, and can significantly enhance the accuracy and reliability of the analysis.

The above three fusion techniques are shown in Fig. 3.1 where we use different colors to represent different modal information, and different textures to represent different encoders. It illustrates in a succinct way the main stages of information fusion between these three technologies. Fig. 3.1 illustrates the first three fusion techniques, with different colors representing various modalities, and different textures indicating different encoders. The

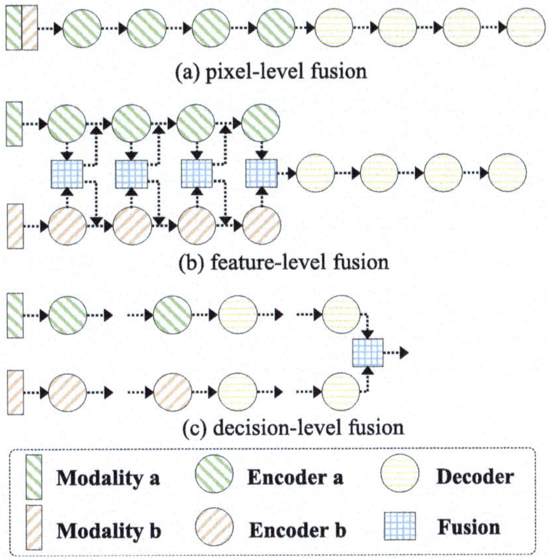

Figure 3.1:
(a) Pixel-level fusion: Directly stacking the data before encoding, also known as early fusion;
(b) Feature-level fusion: Fusion occurring at different scales within separate encoder layers;
(c) Decision-level fusion: Fusion applied after decoding or before the final classification, also referred to as late fusion [34].

figure succinctly highlights the key stages of information fusion across these three fusion techniques. Overall, each of these fusion methods offers distinct advantages and is suited to different types of remote sensing applications. The choice of fusion method depends on factors such as the specific characteristics of the data, the computational resources available, and the desired outcome of the classification task.

This chapter begins by exploring advancements in deep learning-based models that are versatile and not tied to specific fusion strategies. These models, including convolutional neural networks (CNNs) and Transformers, serve as foundational tools applicable across various fusion approaches. Following this, we provide a brief introduction to the fundamental pixel-level fusion technique. We then delve into the feature-level fusion strategy, highlighting it as a key focus of this chapter. The emphasis on feature-level fusion stems from its role in the Specific designation of model structure based on feature fusion, making it a core concept that can be readily extended to decision-level fusion and mixed fusion techniques. Finally, we cover decision-level fusion and hybrid fusion, which should be easier to grasp after the detailed discussion on feature-level fusion.

3.1.1 Deep learning-based techniques

Recent advancements in Earth observation technology have significantly increased access to high-resolution remote sensing data across various modalities, including optical, multispectral, and hyperspectral imagery, synthetic aperture radar (SAR), and light detection and ranging (LiDAR). The effective integration of these multimodal datasets enables more comprehensive characterizations of the land surface, benefiting numerous tasks in geoscience research. These tasks include change detection [31,51], land cover mapping [20,27], object extraction [38,61], and several other applications [41,48,57]. Remote sensing data classification, also refer to as semantic segmentation, a pixel-wise classification task aimed at categorizing each pixel into a specific land cover type, has garnered significant attention in recent years. Various methods have been proposed in the literature, utilizing techniques such as random forests [13], support vector machines [11,30], and conditional random fields [25]. However, these traditional approaches are limited by their inadequate ability to extract abstract and semantic features. Recently, deep learning (DL)-based methods have been developed for remote sensing semantic segmentation, particularly through the use of CNNs [1,10,15,19,21,34,35,54,58,64,65,69]. CNNs offer several strengths that make them particularly effective for remote sensing tasks. Firstly, CNNs excel at extracting local features from images. The convolutional layers apply filters across small regions of the input, capturing essential features such as edges, textures, and patterns and these localized features are crucial for understanding the structure of images. Secondly, CNNs automatically learn hierarchical representations of the input data, which also called multiscale features. Lower layers capture simple features like edges and corners, while deeper layers combine these into more complex features, such as shapes and objects, and even more abstract representations. This hierarchical feature extraction is critical for recognizing ground object patterns and making accurate predictions. In CNNs, the same set of filters is applied across the entire image, which means the network shares parameters across different locations. This reduces the number of parameters compared to fully connected networks, making CNNs more efficient and less prone to overfitting. Furthermore, CNNs are naturally translation-invariant due to their use of pooling layers, which downsample the feature maps. This means that the network's ability to recognize objects is robust to changes in their position within the image. As a result, CNNs can generalize well across different spatial locations of features. This capability is particularly important for remote sensing impacts, since in earth surface, similar ground objects tend to appear in different geographical locations. These strengths make CNNs powerful tools for a wide range of applications in computer vision and remote sensing, contributing to their widespread adoption in both academic research and industry. Meanwhile, although these CNN-based methods have achieved outstanding performance, they are also limited by the convolutional operation's tendency to aggregate information from a smaller receptive field, which results in capturing local details while neglecting long-range dependencies.

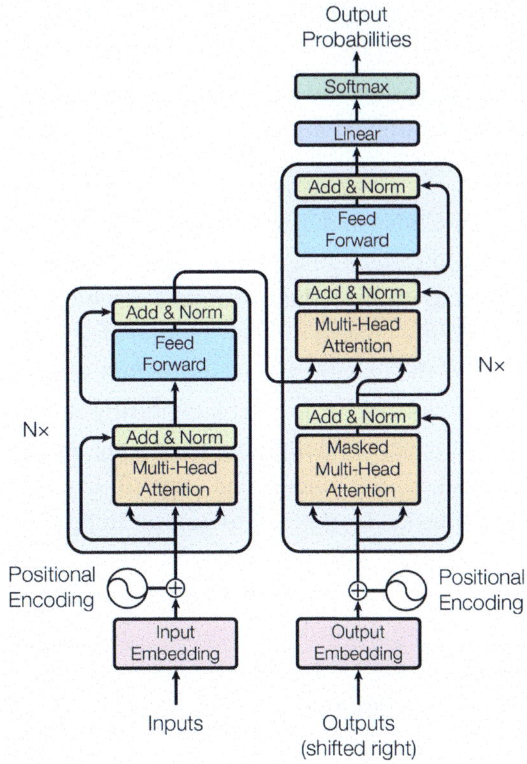

Figure 3.2:
The architecture of Transformer network [53].

Interestingly, the computer vision community has encountered challenges similar to those in remote sensing. To cope with this problem, vision Transformer (Vit) [9] and Swin Transformer [28] were developed for computer vision applications to compute the relationships between each pair of elements in the feature map by capitalizing on the self-attention mechanism [53], which enhances the modeling of global contextual information. The self-attention mechanism was proposed in Transformer network, which have emerged as a powerful and versatile architecture as shown in Fig. 3.2. One of the core strengths of Transformers is their use of the attention mechanism, particularly multihead self-attention, which allows the model to focus on relevant parts of the input sequence regardless of their position. This enables the Transformer to capture long-range dependencies and relationships within the whole image, making it highly effective for tasks where context across the entire input is important. Furthermore, Transformers can be scaled up by increasing the number of layers, attention heads, and model parameters. This scalability has led to the development of large language models and large vision models like GPT [44], and Segment Anything Model

(SAM) [24], which have set new benchmarks in many fields. These strengths have made Transformer-based networks the architecture of choice for many state-of-the-art models in remote sensing tasks, driving innovation and improving performance across a broad range of tasks. In particular, we place significant emphasis on the Transformers' global modeling capabilities.

With the powerful backbones of CNNs and Transformers, semantic segmentation for single modality [5,6,29,45] has achieved remarkable success in the computer vision field. However, limited attention has been given to multimodal fusion tasks, which emphasize integrating information from different modalities based on their unique characteristics [3,15,47]. Unlike single-modal data constrained by the capabilities of individual sensors, multimodal data provides a more comprehensive view of the target by capturing it from various perspectives. This allows for improved semantic segmentation by leveraging the complementary properties of different modalities. Nevertheless, the inherent incompatibility of multimodal data presents significant challenges for fusion tasks. Compared to natural images, high-resolution remote sensing images exhibit greater spectral heterogeneity and more complex spatial structures [20,26]. Additionally, ground objects in remote sensing data display significant variations in scale and shape, further complicating the task of accurately locating and recognizing these objects. Consequently, CNN and Transformer-based models adapted from computer vision still face challenges in effectively learning discriminative fused features for remote sensing. To address the multimodal fusion problem in this domain, different fusion techniques have been explored as we discussed above based on the advancing backbones.

3.1.2 Single-modal remote sensing data classification

The foundational work [29] introduced the first CNN-based end-to-end model for semantic segmentation, known as the Fully Convolutional Network (FCN). However, FCN faces challenges such as blurred edges and inaccurate segmentation, primarily due to its overly simplistic upsampling operation in the decoder. To address this issue, UNet [45] employs a classic encoder-decoder architecture with an expanding path in the decoder. Specifically, UNet's encoder extracts multiscale features through gradual downsampling, while its decoder progressively restores spatial resolution to capture more contextual semantic information. Despite their strong performance, CNN-based methods struggle to effectively capture global semantic information and long-range dependencies within input images, owing to the limited receptive fields inherent in CNNs [37,40]. To address this challenge, the Transformer architecture was introduced in [53] as an effective solution for capturing long-range relationships, initially in the field of natural language processing. Building on this foundation, a Transformer-based model called ViT was developed in [9] for computer vision applications. Leveraging its sequence-to-sequence modeling capabilities, ViT significantly improves the extraction of global contextual information compared to CNN-based methods, leading to a substantial performance boost. This breakthrough has inspired a wave of subsequent works [6,10,19,34,50,55,59,62] that build on the strengths of Transformers for various computer vision or remote sensing tasks. These techniques serve as the foundation for multimodal remote sensing image classification.

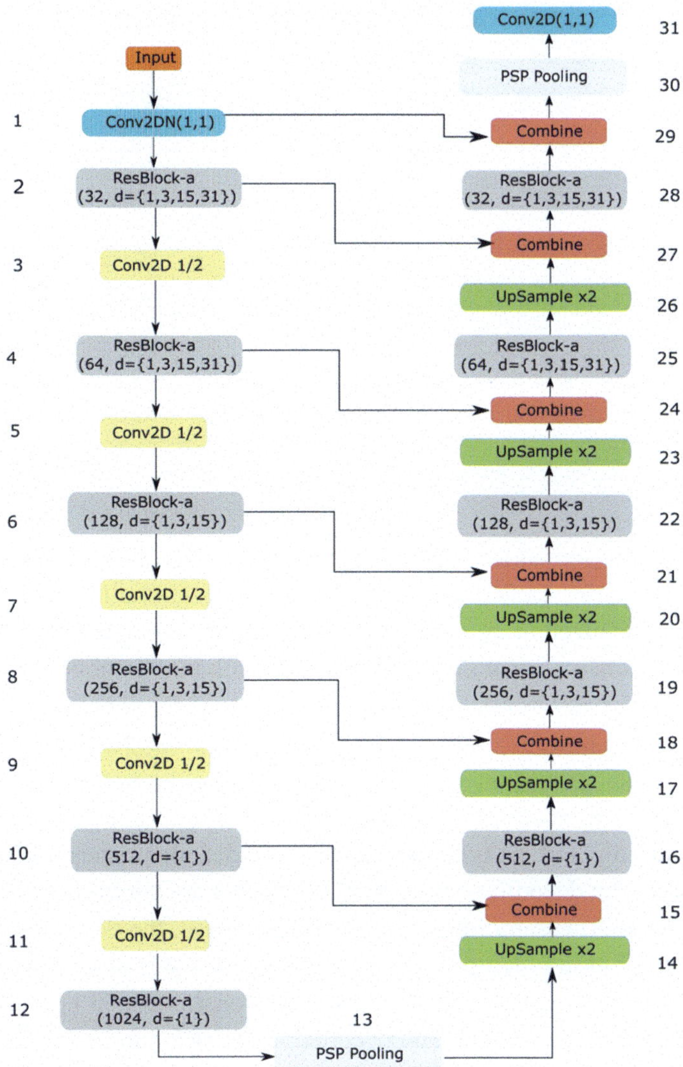

Figure 3.3:
Overview of the ResUNet-a network [8]. The left (downward) branch represents the encoder part of the architecture, while the right (upward) branch represents the decoder. The final convolutional layer contains the same number of channels as there are distinct classes.

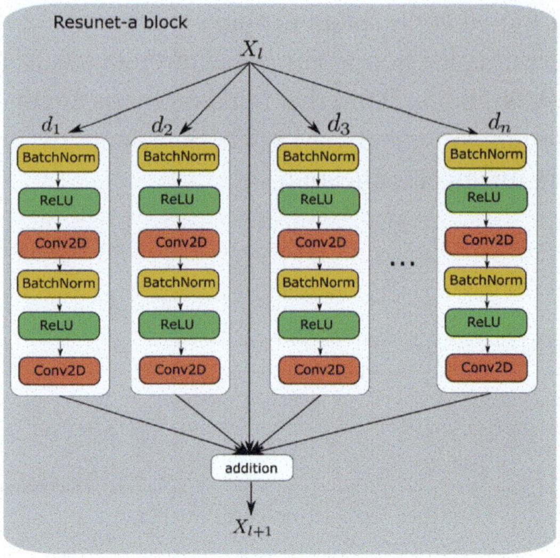

Figure 3.4:
Building block of the ResUNet-a network [8]. Each unit within the residual block utilizes the same number of filters as the other units, but features different dilation rates denoted by d_i.

Fig. 3.3 presents a classic semantic segmentation network, namely ResUNet-a [8] based on CNN. This network features stacked layers of modified residual building blocks (ResBlock-a) as shown in Fig. 3.4 organized in an encoder-decoder architecture similar to UNet. Initially, the input undergoes a convolutional layer to increase the channel numbers to the desired initial filter size. This step is essential for preparing the input for subsequent processing. Following this, the network employs many residual blocks to extract features. The residual block incorporates four parallel atrous convolutions to improve the limitation of the receptive field. The outputs from these convolutions are then added to the initial input, adhering to the principles of residual learning [16]. At both the encoder and decoder stages, a PSPooling operator [66] is applied. During the decoding phase, residual blocks are similarly stacked. Meanwhile, unlike the convolutional downsampling in the encoding phase, the decoding phase gradually restores the spatial resolution of features through upsampling while incorporating the residual features from the encoding phase. This network demonstrates that how a remote sensing image is encoded into abstract features and decoded into a structured classification map. Despite technological advancements, this framework remains broadly applicable, allowing for improved results by refining the individual modules within it. More importantly, multimodal fusion technology integrates multisource information at different stages within this encoder-decoder framework. By first mastering the fundamentals of the single-modal framework, the readers can more intuitively grasp the concepts of multimodal fusion.

3.1.3 Multimodal remote sensing data classification based on CNN

Advancements in earth observation technology have led to the increasing availability of multimodal remote sensing data, such as optical images, multispectral images, and digital surface model (DSM) data. DSM, which provides height information for ground objects, is particularly valuable for identifying features with consistent heights, like roads and buildings in small geographical areas. Additionally, DSM data offers clear boundary information, which is beneficial across various categories and enhances overall segmentation performance [22,34]. To leverage multimodal data in remote sensing, innovative fusion schemes have been developed within the deep learning framework to integrate information from multiple sensors. For example, ResUNet-a was introduced in [8], where Red-Green-Blue (RGB) images and DSM data were combined into synthesized four-channel inputs. This approach is referred to as pixel-level fusion, which relies on precise alignment of RGB images and DSM data. Clearly, such a basic fusion of multimodal data struggles to effectively manage the heterogeneous statistical properties and varying noise levels across different modalities [4,39]. To address this challenge, FuseNet [15] introduced a straightforward two-branch network architecture to fuse RGB and DSM data. In this approach, RGB and DSM data are processed in parallel by separate branches during the encoding phase, and the feature maps generated by each branch are combined using element-wise addition after each convolutional block. This approach is referred to as feature-level fusion which contains two encoder to extract multimodal features, respectively. Building on the concept of FuseNet, vFuseNet [1] employed multiscale fusion strategies. However, vFuseNet's reliance on simple element-wise addition for combining multimodal feature maps limits its effectiveness. To overcome these limitations, CMGFNet [22] introduced a gated fusion module designed to combine two modalities for building extraction, allowing the network to adaptively learn discriminative features and filter out irrelevant information. IIHN [17] developed a hypergraph network incorporating an interpretable intuition mechanism, while CIMFNet [68] and ABHNet [32] explored feature fusion at adjacent levels. CIMFNet utilized a cross-layer gate fusion mechanism, and ABHNet leveraged attention mechanisms and residual connections. These approaches investigate multimodal fusion based on CNN networks from various perspectives, including fusion stages, strategies, and other critical aspects.

3.1.4 Multimodal remote sensing data classification based on CNN and Transformer

More recently, the Transformer architecture has been considered for multimodal fusion in semantic segmentation for its outstanding capability in extracting contextual information [5,9,19,28] and fusing different modalities [18,33,34,43,46,52,63]. For instance, TransFuser was developed in [43] to combine feature maps using multiple Transformers after each convolutional block. Furthermore, SwinFusion [33] was proposed to first extract shallow-level features using convolutional layers before generating deep-level features using the SA-based Swin Transformer [28], which has shown that the Transformer can also serve as the backbone for multimodal fusion tasks. In remote sensing, CMFNet [34] improved the skip connections by exploiting a crossmodal multiscale fusion Transformer, which allows the CMFNet to learn robust representations about surface objects of significant variations in scale. EDFT [63] constructed a two-branch network based on Swin Transformer and fused multimodal features by a depth-aware self-attention module. Similarly, MFTransNet [18] constructed a two-branch network based on CNN while fusing features with an self-attention module followed by a channel attention module and a spatial attention module. FTransUNet [36] introduced an advanced fusion strategy that enhances existing techniques by integrating the strengths of CNNs and Transformers in a multilevel framework. These methods further refine the fusion strategy, delving deeply into the roles of different network architectures in the fusion of remote sensing image features.

3.2 Pixel-level fusion methods

3.2.1 Advantages and limitations

Pixel-level fusion, also known as pixel-level fusion, is a technique in the remote sensing field that involves the integration of raw data from multiple sensors at the pixel level before any feature extraction or classification is performed. This method directly combines the pixel values from different modalities, ensuring that the fused data contains a rich set of information from all input sources, leading to more detailed and informative outputs. Therefore, pixel-level fusion can significantly enhance the accuracy of classification tasks by combining complementary information from different sensors. Fig. 3.5 presents an early fusion example of optical image and DSM Data. As shown in this figure, 4-channel multimodal data is created by simply stacking the 3-channel RGB data and 1-channel DSM data in channel dimension. In ResUNet-a [8], this stacked multimodal data can be directly fed into the first layer of the model for dimensional preprocessing. The same network can be applied to 3-channel RGB data as well. This method also significantly simplifies network complexity. Specifically, the single-modal semantic segmentation method can accomplish this fusion task directly.

Despite it has great advantages, pixel-level fusion relies on the following key prerequisites. The first key problem for pixel-level fusion is data alignment. A crucial aspect of pixel-level fusion is the precise spatial alignment of data from different sensors. Since pixel-level fusion operates directly on raw data, any misalignment can lead to significant errors. Techniques like image registration [42,56] are often employed to ensure that the data from different modalities are accurately aligned. The second problem is resolution matching since in many

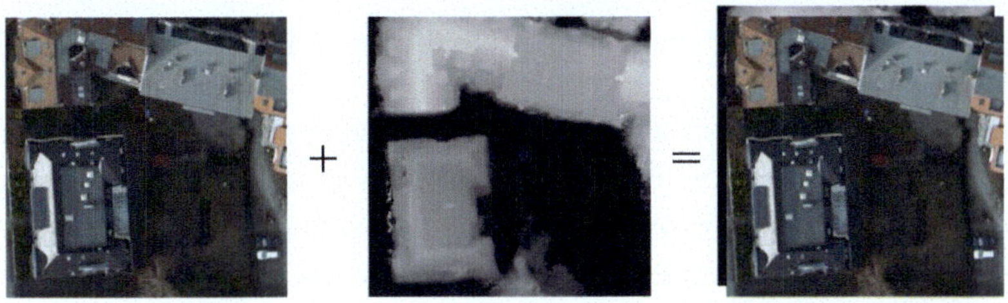

Figure 3.5:
Illustration of Pixel-level fusion. By stacking the three channels of RGB data and a single channel of DSM data, a four-channel RGB-DSM data is generated. This approach can be similarly applied to any multichannel data in multimodal fusion tasks.

cases, sensors may operate at different spatial resolutions. For example, optical sensors might provide high-resolution imagery, while SAR data might be available at a lower resolution. Before pixel-level fusion, resolution matching is necessary to ensure that all data sources contribute effectively to the fused image. The third problem is data normalization. Different sensors capture data in varying ranges and units (e.g., radiance, reflectance, or height). Therefore, normalization techniques should be applied to bring all data to a common scale, facilitating effective fusion.

Ensuring accurate alignment and normalization of multimodal data is crucial but challenging, especially when dealing with data from sensors with different resolutions and noise characteristics. Meanwhile, pixel-level fusion requires processing large amounts of data, which can be computationally intensive, especially for high-resolution or hyperspectral datasets. These advantages and limitations of pixel-level fusion influence the remote sensing scenarios where it is most suitable.

3.2.2 Real world applications

Pixel-level fusion has been widely applied in remote sensing, particularly in tasks that require a high level of detail and precision. Here are some examples of how pixel-level fusion is used with specific remote sensing data:

- Optical and LiDAR Data Fusion: LiDAR provides accurate elevation data, while optical sensors offer detailed visual information. Pixel-level fusion of these data types is often used in urban planning and 3D modeling. In urban area mapping, optical images can capture building facades, while LiDAR data provides the height and shape of buildings. Fusing these datasets at the pixel level enables the creation of detailed 3D models of urban environments, which are crucial for city planning and infrastructure development.

- Optical and SAR Data Fusion: Fusion of optical imagery with SAR data is commonly used in land cover classification. Optical sensors provide rich spectral information, while SAR data offers additional information on surface roughness and texture. Pixel-level fusion of these datasets can improve classification accuracy by leveraging the complementary nature of the data. In flood mapping, optical images can capture surface water bodies clearly, while SAR data can penetrate clouds and capture surface changes during and after flood events. By fusing these data at the pixel level, more accurate and reliable flood maps can be generated.
- Multispectral and Hyperspectral Data Fusion: Multispectral and hyperspectral sensors capture data at different spectral resolutions. While multispectral data covers broader spectral bands, hyperspectral data provides detailed spectral signatures. Pixel-level fusion of these datasets can enhance the spectral resolution, aiding in tasks like mineral exploration and vegetation analysis. In precision agriculture, fusing multispectral data, such as visible and near-infrared bands, with hyperspectral data allows for more accurate monitoring of crop health, soil properties, and nutrient levels.
- Panchromatic and Multispectral Data Fusion (Pansharpening): Pansharpening is a specific type of pixel-level fusion where high-resolution panchromatic images are fused with lower-resolution multispectral images to produce high-resolution color images. This technique is extensively used in satellite imaging. In land use and land cover mapping, pansharpening can be used to enhance the spatial resolution of multispectral images, allowing for more precise identification and classification of land cover types.

In conclusion, pixel-level fusion methods play a critical role in the remote sensing field, enabling the integration of diverse datasets to produce more accurate and detailed analyses. By effectively combining the strengths of various sensors, pixel-level fusion enhances the capability to monitor and understand the Earth's surface, supporting a wide range of applications from environmental monitoring to urban planning.

3.3 Feature-level fusion methods

Feature-level fusion involves integrating features extracted from multiple modalities after they have been processed by feature extraction algorithms. Unlike pixel-level fusion, which combines raw data at the pixel level, feature-level fusion operates on higher-level abstractions or representations derived from the data. In a deep learning-based framework, this fusion technique typically fuses multiscale features extracted by the encoders. This approach also aims to leverage complementary information from different sensors to enhance the overall analysis and interpretation.

Feature-level fusion begins with the extraction of features from individual modalities using various algorithms or models. This can include spatial, spectral, or temporal features, depending on the type of data and the application. Once features are extracted, they are represented in a common feature space. Conventionally, this may involve dimensionality reduction techniques, such as Principal Component Analysis (PCA) or t-Distributed Stochastic Neighbor Embedding (t-SNE), to align features from different modalities. However, in a deep learning framework, multimodal features are extracted in an end-to-end manner. That means all these steps can be accomplished using the encoder-decoder structure, as exemplified by ResUNet-a [8]. This kind of data-driven approach eliminates the need for manually designed feature extraction algorithms, and can benefit from the rapid development of deep learning technologies, such as CNNs, Transformers and large models. After representation, features from different sources are combined using various fusion mechanisms. These can include concatenation, summation, or more complex strategies like attention mechanisms [18,34,36] or gated fusion [22,67]. The goal is to generate unified fused features that contains the strengths of each modality. This part is the main difference between multimodal fusion methods and single-modal methods. Finally, the fused features are then used as input for subsequent decoding to give the final segmentation map. Feature-level fusion captures complex interactions between modalities, making more effective use of their complementary information compared to pixel-level fusion. However, it also needs to ensure that features from different modalities are aligned and comparable. This is a challenging problem, particularly when dealing with data from sensors with different resolutions or characteristics. Additionally, the feature extraction and fusion process involves complex processing and sophisticated algorithms, which can increase computational demands. Next, we give two examples to demonstrate different fusion strategies and how the fused features are utilized after fusion.

3.3.1 Crossmodal multiscale fusion

CMFNet introduces a crossmodal multiscale fusion network that leverages both CNN and transformer architectures. Specifically, the transformer is used for multimodal fusion, and its output is further refined by a decoder as residual connections. Experimental results demonstrate that this fusion approach significantly improves segmentation performance by effectively utilizing crossmodal multiscale data sources. The key contributions of CMFNet are summarized as follows [34]:

- A novel transformer-based crossmodal network with multiscale skip fusion is introduced to reduce decoding ambiguity by integrating complementary multimodal information. This crossmodal multiscale skip fusion not only performs multimodal fusion but also implements the residual function simultaneously;

- This work establishes a CrossModal Attention (CMA) layer and proposes a Crossmodal Multiscale Transformer (CMTrans) for semantic segmentation. CMTrans effectively models long-range dependencies across multiscale feature maps from different remote sensing modalities. The fused features are further refined by a Multiscale Context Augmented Transformer (MCATrans), which uses a multiscale self-attention mechanism to learn more discriminative and distinguishable integrated representations for semantic segmentation. By integrating CMTrans and MCATrans, CMFNet, a crossmodal and multiscale fusion network for multimodal remote sensing classification is introduced.

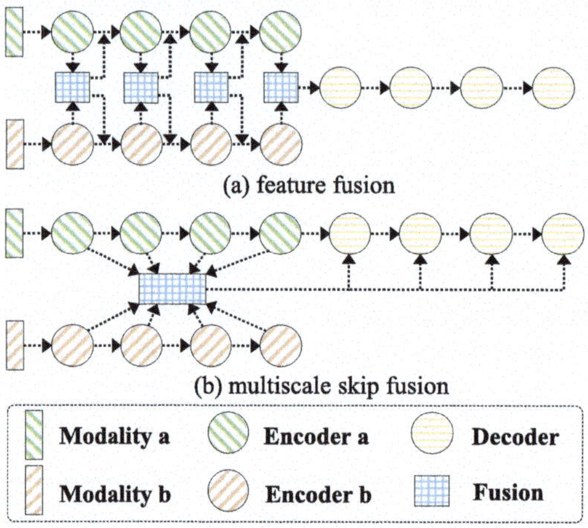

Figure 3.6:
This fusion approach first performs crossmodal multiscale information fusion, followed by combining the fused information with the decoder features as residual connections.

CMFNet is the first transformer-based architecture for crossmodal multiscale fusion in the field of remote sensing. In typical feature-level fusion networks, the fused features are usually fed into the next encoder layer. However, as shown in Fig. 3.6, multiscale skip fusion further uses the fused features as residual connections. It can more effectively leverages the fused information, enhancing overall network performance.

CMFNet begins by separately extracting RGB and DSM features using two branches of the VGG-16 [49]. In this setup, the first four layers of VGG-16, a five-layer encoder, are employed to extract multiscale features from each branch. These features are then fused by CMTrans, a crossmodal and multilevel fusion module. Following this, MCATrans is introduced to refine the multiscale fusion of multimodal features and filter the representations using a multiscale skip fusion strategy. Finally, the features with the richest semantic information, obtained from the fifth encoder, are fused through a channel reweighting module called Combined Squeeze-and-Excitation (CSE). This design enables CMFNet to achieve strong performance while minimizing computational complexity.

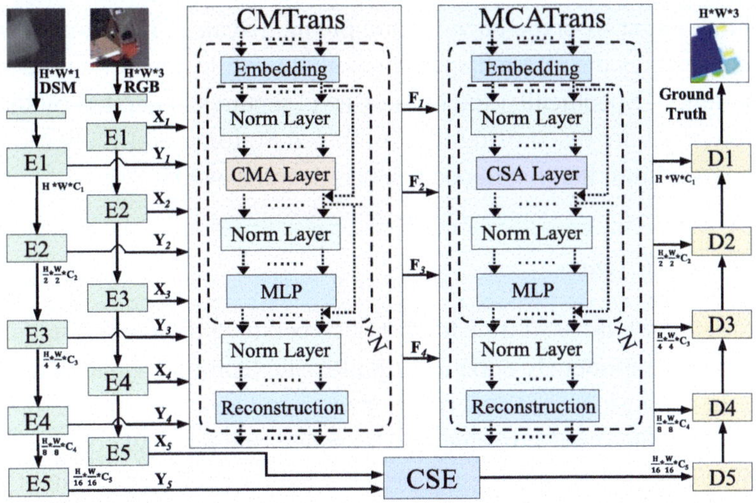

Figure 3.7:
Illustration of CMFNet [34]. The CMTrans fuses feature maps from different modalities by leveraging multiscale crossmodal information simultaneously. The resulting fused features are then input into the decoder at the corresponding scale, where pixel-level classification is performed.

3.3.1.1 Network frame

Fig. 3.7 illustrates the framework of the CMFNet. The RGB and DSM images are utilized as the two modalities of input data. Given RGB images $X \in \mathbb{R}^{H \times W \times 3}$ and the corresponding DSM depth data $Y \in \mathbb{R}^{H \times W \times 1}$, a dual-branch encoder first extracts multilevel features from each modality. Recognizing that semantic segmentation of remote sensing images requires dense pixel-wise classification, CMFNet employs an encoder-decoder architecture. SegNet, as reported in [2], is selected as the network backbone due to its ability to produce outputs with the same resolution as the inputs, thus avoiding performance degradation associated with direct upsampling. The encoder in SegNet utilizes the classical convolutional layers of VGG-16 [49]. A dual-branch encoder architecture is designed to process each input data source independently, using two VGG-16 branches to extract features from the RGB and DSM images, respectively. For an image of size $H \times W$, the downsampled feature maps generated by the i-th encoder layer are of size $\frac{H}{2^{i-1}} \times \frac{W}{2^{i-1}}$, where i represents the layer index, ranging from $i = 1, 2, \cdots, 5$.

Following the encoding process, the multilevel feature maps are fed into the CMTrans for crossmodal feature fusion. The output of CMTrans comprises a set of fused feature maps. These fused features are further enhanced by MCATrans, a self-attention-based transformer, and then passed to the decoder for upsampling and classification. The decoder layer includes convolutional blocks, normalization layers, ReLU blocks, and an upsampling block.

Leveraging the final output from the encoder and the multiscale skip fusion at corresponding scales, the decoder restores the original spatial resolution while converting the encoded features into final labels. Specifically, the highest-level features generated by the fifth encoder layer from both branches are fused using the Combined Squeeze-and-Excitation (CSE) module and then passed into the first decoder layer, denoted as D5.

3.3.1.2 CMTrans and MCATrans

CMTrans is a cross-attention-based transformer designed to explore crossmodal and multiscale features by continuously enhancing multiscale features extracted from the multimodal data. In contrast, MCATrans functions as a self-attention-based transformer, focusing on reinforcing the fused information and generating residual connections. Both CMTrans and MCATrans follow a four-step process: feature embedding, attention block, MultiLayer Perceptron (MLP), and reconstruction. Let $\mathbf{X}_i$ and $\mathbf{Y}_i$ denote the outputs of the first four encoder layers of the RGB and DSM branches, respectively, with dimensions $C_i \times \frac{256}{2^{i-1}} \times \frac{256}{2^{i-1}}$, where $C_i = 64 \times i$ is the channel dimension of the i-th encoder layer output, for $i = 1, 2, 3, 4$. Tokenization begins by flattening the extracted features into sequences of 2D patches using a convolutional layer with kernel and stride sizes of $m_i \times m_i$, where $m_i = \frac{32}{2^i}$. This process downscales the 2D feature patches to $16 \times 16 \times C_i$ through feature embedding, maintaining the channel dimension C_i to preserve as much information as possible throughout both CMTrans and MCATrans. Following tokenization, eight tokens are generated: four for RGB, denoted $\mathbf{R}_i \in \mathbb{R}^{P \times C_i}$, and four for DSM, denoted $\mathbf{D}_i \in \mathbb{R}^{P \times C_i}$, where $P = 256$ and $i = 1, 2, 3, 4$.

CMTrans: The eight tokens are subsequently input into the CMA layer, followed by processing through a MultiLayer Perceptron (MLP). A residual structure, based on independent simple copying, is used to encode channel relationships and dependencies. The CMA Layer operates with ten inputs: eight tokens, $\mathbf{R}i$ and $\mathbf{D}i$, serve as queries, while two concatenated tokens, $\mathbf{R}_\Sigma$ and $\mathbf{D}_\Sigma$, are used as keys and values, respectively. The tokens $\mathbf{R}_\Sigma$ and $\mathbf{D}_\Sigma$, with dimensions $P \times C_\Sigma$ where $C_\Sigma = \sum C_i$, are constructed by concatenating the four $\mathbf{R}i$ and $\mathbf{D}i$ tokens.

To fully utilize multihead attention, the tokens are divided into four nonoverlapping segments, each with $\frac{C_i}{4}$ channels. For example, $\mathbf{R}i$ is split into four segments, $\tilde{\mathbf{R}}^{(j)}i \in \mathbb{R}^{P \times \frac{C_i}{4}}$, used as queries for the i-th layer, where $j = 1, 2, 3, 4$. Similarly, the key and value are also split into four segments, each containing nonoverlapping information with the same number of channels at different scales, denoted as $\tilde{\mathbf{R}}^{(j)}{}_\Sigma \in \mathbb{R}^{P \times \frac{C_\Sigma}{4}}$. Since the operations across all four heads are identical, the superscript $^{(j)}$ will be omitted for simplicity in the following discussions. The multilevel query, concatenated key, and value for each head are denoted

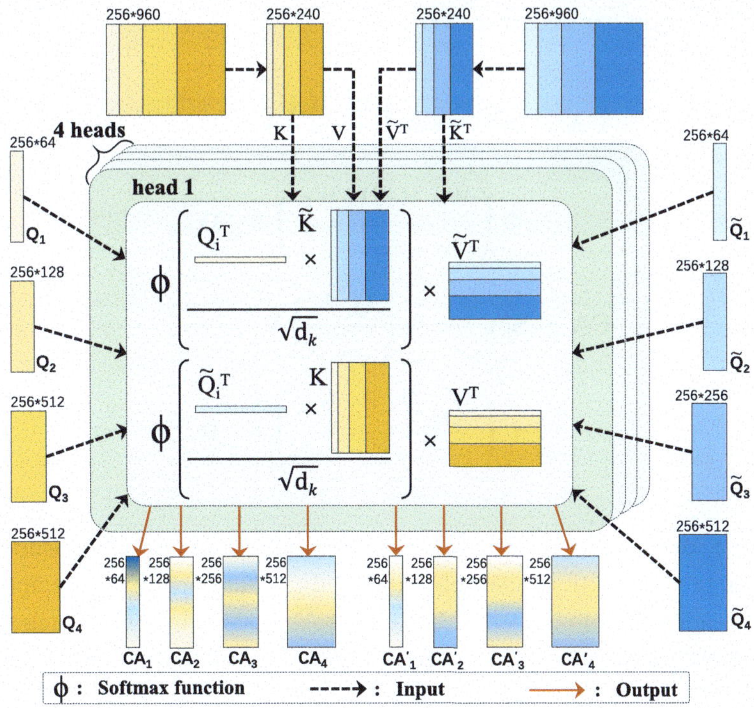

Figure 3.8:
Illustration of the CMA Layer [34].

as $\mathbf{Q}i, \tilde{\mathbf{Q}}i \in \mathbb{R}^{P \times \frac{C_i}{4}}$ and $\mathbf{K}, \tilde{\mathbf{K}}, \mathbf{V}, \tilde{\mathbf{V}} \in \mathbb{R}^{P \times \frac{C_\Sigma}{4}}$, respectively. The update equations for these variables are then given by:

$$\mathbf{Q}_i = \mathbf{R}_i W_{\mathbf{Q}_i}, \mathbf{K} = \mathbf{R}_\Sigma W_{\mathbf{K}}, \mathbf{V} = \mathbf{R}_\Sigma W_{\mathbf{V}}, \tag{3.1}$$

$$\tilde{\mathbf{Q}}_i = \mathbf{D}_i W_{\tilde{\mathbf{Q}}_i}, \tilde{\mathbf{K}} = \mathbf{D}_\Sigma W_{\tilde{\mathbf{K}}}, \tilde{\mathbf{V}} = \mathbf{D}_\Sigma W_{\tilde{\mathbf{V}}}, \tag{3.2}$$

where $W_{\mathbf{Q}_i}$, $W_{\tilde{\mathbf{Q}}_i}$, $W_{\mathbf{K}}$, $W_{\mathbf{V}}$, $W_{\tilde{\mathbf{K}}}$, $W_{\tilde{\mathbf{V}}}$ represent different weights. It is important to note that different heads utilize separate query weights, specifically $W_{\mathbf{Q}_i}$ and $W_{\tilde{\mathbf{Q}}_i}$, while sharing the same weights for keys and values, namely $W_{\mathbf{K}}$, $W_{\mathbf{V}}$, $W_{\tilde{\mathbf{K}}}$ and $W_{\tilde{\mathbf{V}}}$. The update process is illustrated in Fig. 3.8.

Given $\mathbf{Q}_i$, $\tilde{\mathbf{Q}}_i$, $\mathbf{K}$, $\mathbf{V}$, $\tilde{\mathbf{K}}$, and $\tilde{\mathbf{V}}$, the similarity matrix $\mathbf{S}_i$ is determined by the query from the RGB modality and the key from the DSM modality. Conversely, the similarity matrix $\mathbf{S}'_i$ is derived from the query of the DSM modality and the key of the RGB modality. Subsequently, the cross-attention value for the RGB modality, CA_i, which is assisted by the DSM modality, is weighted using $\mathbf{S}_i$, while the cross-attention value for the DSM modality, CA'_i, which is assisted by the RGB modality, is weighted using $\mathbf{S}'_i$. The computation of the multiscale cross-attention in the CMA layer can be expressed as:

$$
\begin{aligned}
\mathrm{CA}_i = \mathbf{S}_i\tilde{\mathbf{V}}^\top &= \Phi\left(\frac{\mathbf{Q}_i^\top\tilde{\mathbf{K}}}{\sqrt{C_\Sigma}}\right)\tilde{\mathbf{V}}^\top, \\
&= \Phi\left(\frac{W_{\mathbf{Q}_i}^\top\mathbf{R}_i^\top\mathbf{D}_\Sigma W_{\tilde{\mathbf{K}}}}{\sqrt{C_\Sigma}}\right)W_{\tilde{\mathbf{V}}}^\top\mathbf{D}_\Sigma^\top,
\end{aligned} \tag{3.3}
$$

$$
\begin{aligned}
\mathrm{CA}_i' = \mathbf{S}_i'\mathbf{V}^\top &= \Phi\left(\frac{\tilde{\mathbf{Q}}_i^\top\mathbf{K}}{\sqrt{C_\Sigma}}\right)\mathbf{V}^\top, \\
&= \Phi\left(\frac{W_{\tilde{\mathbf{Q}}_i}^\top\mathbf{D}_i^\top\mathbf{R}_\Sigma W_{\mathbf{K}}}{\sqrt{C_\Sigma}}\right)W_{\mathbf{V}}^\top\mathbf{R}_\Sigma^\top,
\end{aligned} \tag{3.4}
$$

where $\Phi(\cdot)$ and $(\cdot)^\top$ denote the softmax function and matrix transpose, respectively. After the computation of the cross-attention, a simple MLP and residual operation are applied. In our experiments, cross-attention modules are stacked for N times. During this stage, CA_i and CA_i' computed in the previous round would be used for the next calculation. The resulting output can be expressed as follows:

$$\mathrm{CA}_i^n = \mathrm{CA}_i + \mathrm{MLP}\left(\mathbf{Q}_i + \mathrm{CA}_i^{n-1}\right), \tag{3.5}$$

$$\mathrm{CA}_i'^n = \mathrm{CA}_i' + \mathrm{MLP}\left(\tilde{\mathbf{Q}}_i + \mathrm{CA}_i'^{n-1}\right), \tag{3.6}$$

where the superscript n indicates the CMA layer index with $n = 1, 2, 3, 4$. For simplicity in notation, the layer normalization (LN) is omitted from the equations but is depicted in Fig. 3.8 as the "Normalization" module. The procedures from Eq. (3.3) to Eq. (3.6) are repeated four times to construct a four-layer transformer. Finally, the four outputs of the main modality $\mathbf{CA}_i$ with the size of $P \times C_i$ are processed through the reconstruction module to produce $\mathbf{F}_i$ with the size of $C_i \times \frac{256}{2^{i-1}} \times \frac{256}{2^{i-1}}$ for $i = 1, 2, 3, 4$. In essence, reconstruction reverses the process of feature embedding.

MCATrans: The output $\mathbf{F}_i$ from CMTrans represents the fused feature across four scales after applying cross-attention. These fused features are further enhanced to provide more discriminative and distinguishable integrated information for semantic segmentation. To achieve this goal, a self-attention-based transformer, named MCATrans, is used to capture the global context from the fusion feature space. MCATrans shares a similar structure with CMTrans, but with the CMA layer in CMTrans replaced by the CSA layer, as shown in Fig. 3.9. The correlation information in CSA is then calculated using $\mathbf{Q}_i$, $\mathbf{K}$, $\mathbf{V}$ from the main

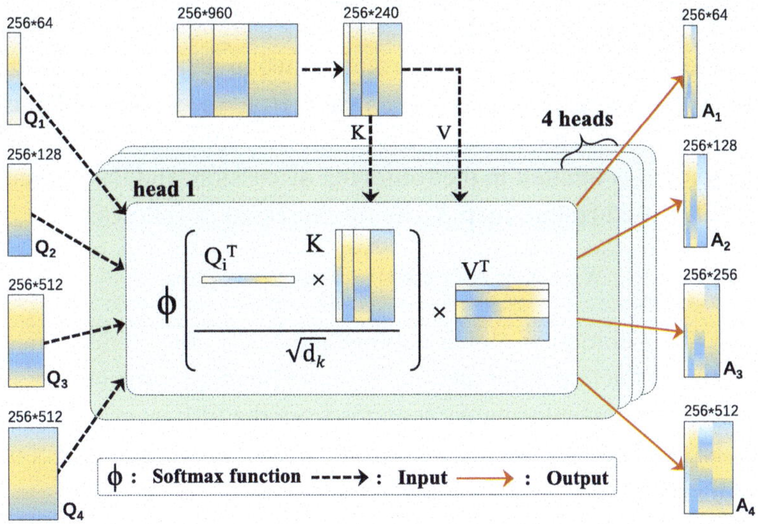

Figure 3.9:
Illustration of the CSA Layer [34].

modality as follows:

$$A_i = \mathbf{S}_i \mathbf{V}^\top = \Phi\left(\frac{\mathbf{Q}_i^\top \mathbf{K}}{\sqrt{C_\Sigma}}\right)\mathbf{V}^\top, \tag{3.7}$$

where $\mathbf{Q}_i$, $\mathbf{K}$, $\mathbf{V}$ are computed from $\mathbf{F}_i$ using the same approach shown in Eq. (3.1) for $i = 1, 2, 3, 4$.

After the self-attention computation, an MLP and residual operation are applied in a manner similar to Eq. (3.5), except that CA_i is replaced by A_i in Eq. (3.7). Subsequently, the result is restored to its original size by the reconstruction module. Finally, the four outputs are combined with the corresponding decoder features from decoder layers D1 to D4 using element-wise summation, resulting in the final full-resolution prediction.

3.3.1.3 Combined Squeeze-and-Excitation (CSE)

To efficiently fuse high-level features by enhancing the most informative channels, a Combined Squeeze-and-Excitation (CSE) structure is used, which incorporates two Squeeze-and-Excitation (SE) modules as detailed in [23]. The SE module is designed to retain and amplify critical feature information due to its global receptive capability. Specifically, the process begins with global average pooling, which serves as a channel descriptor based on the channel attention mechanism. This is followed by a 1×1 convolutional layer with the same number of channels as the input. A Sigmoid activation function is then applied to normalize the weight vector to the range [0, 1]. After obtaining the weight vector, the outer products of the weight vector and the input feature maps from the two branches are computed. The final fused feature map X is derived by summing the results from the RGB and DSM branches. Since the CSE module is applied only at the fifth scale, its inputs include X_5 and Y_5. For simplicity, the subscript $_5$ is omitted in the following discussion, and the notation X, Y, and C is used instead. The fused feature map X can be expressed as follows:

$$X = X \otimes \sigma(\tau(X)) + Y \otimes \sigma(\tau(Y)), \tag{3.8}$$

where $\tau(\cdot)$ denotes the operation of global pooling and 1×1 convolution whereas $\otimes$ and $\sigma(\cdot)$ denote the outer product operator and the Sigmoid function, respectively. This fusion mechanism enhances the most informative features from the fifth layer, which has the largest number of channels, and experiments have shown that it helps in extracting complementary information from the auxiliary DSM modality.

3.3.2 Multilevel multimodal fusion

As summarized in Table 3.1, existing CNN and Transformer-based methods can be grouped into three categories: Class 1, which includes TransFuser [43], SwinFusion [33], CMFNet [34], and MFTransUNet [18], and Class 2, which includes EDFT [63]. Methods in Class 1 and Class 2 typically utilize either CNNs or Transformers for feature extraction or fusion. Building on these works, FTransUNet [36] introduced a more advanced fusion strategy that fully complements existing technologies by combining the strengths of CNNs and Transformers within a multilevel framework. It explored the general network from the perspective of using both CNN and Transformer for feature extraction and fusion, which incurs sufficient modeling of fine-grained local and contextual global information, and subsequently great feature representations.

Table 3.1: Multimodal fusion strategies based on CNN and Transformer [36].

	Feature Extraction	Feature Fusion
Class 1	CNN	Transformer
Class 2	Transformer	CNN or Transformer
FTransUNet	CNN, Transformer	CNN, Transformer

Inspired by the challenges discussed, FTransUNet introduces a multilevel multimodal fusion strategy designed to address the complexities of multimodal remote sensing data fusion for classification. Specifically, FTransUNet first employs a dual-branch model where shallow-level feature maps extracted from each convolutional layer are fused using a shallow-level feature fusion (SFF) module, which incorporates two squeeze-and-excitation (SE) modules [23]. This SE module can be substituted with other advanced versions to achieve shallow-level feature fusion. The CNN-based shallow-level fusion is tailored to capture features of various scales and shapes. After that, deep-level contextual features are extracted and fused using the Fusion Vision Transformer (FVit). Different with TransUNet [6], which focuses on long-range dependencies within a single modality, FVit enhances information within each modality and facilitates information exchange across modalities through a three-stage process. This involves alternating between Adaptively Mutually Boosted Attention (Ada-MBA) layers and self-attention layers. In the first stage, FVit computes self-attention for each modality. In the second stage, the Ada-MBA layer captures long-range relationships between different modalities by performing both cross-attention and self-attention simultaneously. This stage employs two self-attention modules to retain and enhance intramodal information, while two cross-attention modules in the Ada-MBA layer guide feature fusion through mutual-association guidance, fostering cross-modality representations with high interclass separability and low intraclass variation. In the third stage, the fused deep-level features are further enhanced with self-attention. By leveraging this multilevel strategy, FTransUNet effectively encodes multimodality, addressing the complexities inherent in remote sensing data compared to natural images. The final shallow-level and deep-level features are then processed through a cascaded decoder, where features are fused and upsampled using skip connections to reconstruct the input image size. The key contributions of FTransUNet are summarized as follows:

- The FVit utilizes self-attention layers in conjunction with innovative Ada-MBA layers. Specifically, FVit leverages a three-stage architecture that alternates between Ada-MBA layers and self-attention layers, employing a mutual-association guidance mechanism to extract and integrate global contextual information;
- A novel multilevel fusion scheme, namely FTransUNet, addresses the multimodal fusion challenge by integrating FVit with CNN blocks to capture and represent features across various modalities. Specifically, FTransUNet successively fuses detailed shallow-level features and contextual deep-level features in a multilevel manner. This approach enhances the ability to learn semantic-aware regions, thereby improving the performance for multimodal remote sensing data classification.

In contrast to CMFNet, FTransUNet adopts a more hierarchical strategy to processing multimodal information. However, it maintains the fundamental feature-level fusion strategy, as illustrated in Fig. 3.9(b). The following part provides a detailed description of its network structure.

3.3.2.1 Network frame

Fig. 3.10 provides an overview of the FTransUNet. In the CNN backbone, shallow-level features are fused using the SFF module, while the Ada-MBA layer in FVit handles deep-level

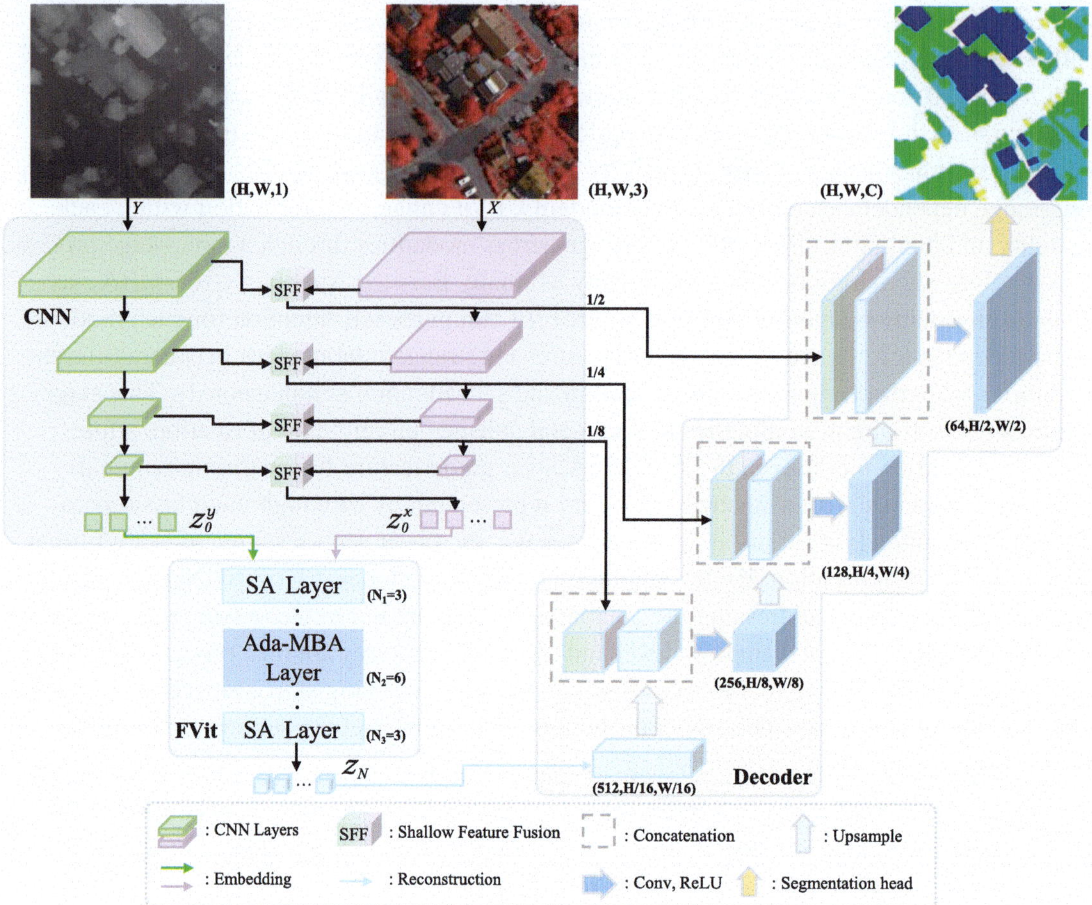

Figure 3.10:
The FTransUNet architecture [36] is illustrated with the encoder on the left and the decoder on the right. The encoder integrates CNN blocks and FVit components. The CNN blocks include two branches with ResNet networks, which extract shallow-level features from multiple modalities and fuse them through four SFF modules. The FVit employs a three-stage process for extracting and fusing deep-level features. On the right, the decoder uses cascaded upsampling techniques [6] to generate segmentation maps from the fused features.

feature fusion. Specifically, FTransUNet processes visible image features (VIS) and DSM features separately through two ResNet branches [16]. The shallow-level features of various scales are enhanced by SFF modules following each CNN block. These fused features are then flattened into sequences and input into FVit, where deep-level features are further fused using Ada-MBA to capture richer contextual information. Finally, the outputs from FVit are combined and reshaped before being fed into a decoder with cascaded upsamplers [6], which refines spatial information with greater precision.

For clarity, dual-modality data are used to explain the FTransUNet. In this section, VIS images are treated as the main modality, while DSM data serve as the assisting modality, as the main modality typically offers more comprehensive information for earth surface classification. It is important to note that the fusion scheme employed in FTransUNet can be easily adapted to other multimodal fusion tasks. The key components of the FTransUNet will be discussed in the following sections.

3.3.2.2 CNN fusion

Denote VIS images and their corresponding DSM data by $X \in \mathbb{R}^{H \times W \times 3}$ and $Y \in \mathbb{R}^{H \times W \times 1}$, respectively, where H and W are the height and width of the inputs. It is important to note that VIS images have three channels, while DSM data have one channel. Following a structure similar to TransUNet [6], FTransUNet employs a dual-branch encoder that initially extracts multiscale features from each modality using separate branches. Each branch within this encoder is composed of four convolutional layers, specifically designed for multiscale feature extraction. The downsampled feature maps of size $C_i \times \frac{H}{2^{i-1}} \times \frac{W}{2^{i-1}}$ are generated by the i-th encoder layer, where i denotes the layer index in the CNN encoder. These shallow-level features, extracted by convolutional operations, are then fused using the SFF modules. Specifically, features obtained from the assisting modality are merged with those from the main modality, i.e., VIS, before the fused features are passed into the next VIS encoder branch. Furthermore, skip connections are employed by directly feeding the outputs of the SFF modules to the corresponding decoder layers, which are designed to restore detailed local and contextual information.

The SFF module begins by aggregating global information using Global Average Pooling (GAP) within both the VIS and DSM branches. For the i-th SFF module, with an input channel size of C_i, the squeeze and excitation process is performed through Global Average Pool (AvgPool) and two convolutional operations with a kernel size of 1×1, followed by ReLU and Sigmoid activation functions. Ultimately, the features from VIS and DSM are weighted and combined through element-wise addition, resulting in the fused shallow-level features.

3.3.2.3 FVit

Let x_I and y_I denote the VIS and DSM feature maps with dimensions $C_I \times \frac{H}{2^{I-1}} \times \frac{W}{2^{I-1}}$, respectively, where I represents the layer index and C_I is the output channel size for the final layer in the CNN backbone. The feature maps x_I and y_I undergo tokenization through two embedding layers followed by a reshape operation. These embedding layers adjust the channel size from C_I to C_{hid}, and the reshape operation flattens the outputs into two sequences of 2D patches, denoted as z_0^x and z_0^y, each of size $C_{\text{hid}} \times L$, where $L = (H \times W)/(2^{I-1} \times 2^{I-1})$ represents the sequence length. Positional embeddings are then added to the vectorized patches z_0^x and z_0^y to retain spatial information. The resulting tokens z_0^x and z_0^y are then input into FVit.

The FVit encoder processes its input through three sequential stages: self-attention layers for deep-level feature enhancement, Ada-MBA layers for deep-level feature fusion, and additional self-attention layers for enhanced fused feature representation, with N_1, N_2, and N_3 layers respectively. Let z_n^x and z_n^y denote the hidden features at the n-th layer in the VIS and DSM branches, respectively, where $n \in 1, 2, \ldots, N_1 + N_2 + N_3$. Throughout the FVit processing, the dimension of the feature maps is maintained at $C_{\text{hid}} \times L$. Specifically, each self-attention layer consists of two self-attention modules, two MultiLayer Perceptron (MLP) modules, and layer normalization (LN) operations, as depicted in Fig. 3.11(a). For multimodal input features z_{n-1}^x and z_{n-1}^y, the self-attention layer is designed to capture global relationships within each modality using the multihead self-attention mechanism [9]. The output of the n-th self-attention layer, for $n = 1, 2, 3$, can be expressed mathematically as follows:

$$\begin{aligned} p_n^x &= SA\left(LN(z_{n-1}^x)\right) + z_{n-1}^x, \\ p_n^y &= SA\left(LN(z_{n-1}^y)\right) + z_{n-1}^y, \\ z_n^x &= MLP\left(LN(p_n^x)\right) + p_n^x, \\ z_n^y &= MLP\left(LN(p_n^y)\right) + p_n^y. \end{aligned} \tag{3.9}$$

Following the deep-level feature enhancement performed by the self-attention layers, FVit employs N_2 Ada-MBA layers to further integrate multimodal features in an abstract semantic space enriched with contextual information, as illustrated in Fig. 3.11(b). During this deep-level feature fusion stage, both cross-attention and self-attention operations are performed simultaneously in the Ada-MBA module to exploit the relationships between the main and assisting modalities. The output of the Ada-MBA layer can be expressed as:

$$\left(g_n^x, g_n^y\right) = Ada\text{-}MBA(LN(z_{n-1}^x), LN(z_{n-1}^y)). \tag{3.10}$$

Let $q_n^x = g_n^x + z_{n-1}^x$ and $q_n^y = g_n^y + z_{n-1}^y$. The fused multimodal features from the n-th layer for $n = 4, 5, \ldots, 9$ can be represented as:

$$\begin{aligned} z_n^x &= MLP\left(LN(q_n^x)\right) + q_n^x, \\ z_n^y &= MLP\left(LN(q_n^y)\right) + q_n^y. \end{aligned} \tag{3.11}$$

Fig. 3.12 depicts the structure of the Ada-MBA module. Leveraging a multihead design [9,53], the Ada-MBA module segments its multimodal feature inputs z_{n-1}^x and z_{n-1}^y into H

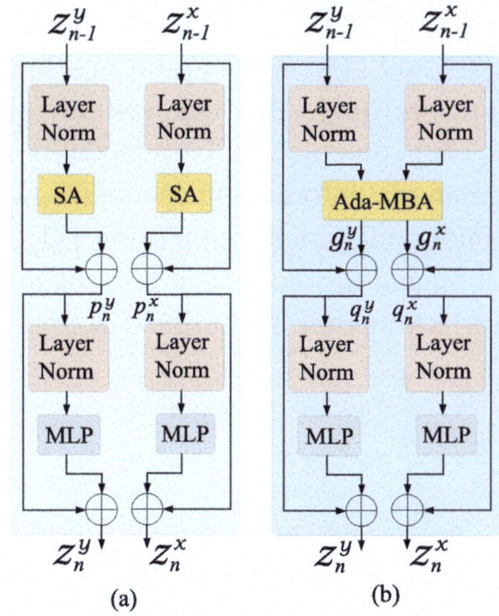

Figure 3.11:
(a) The self-attention layer in FVit. (b) The Ada-MBA layer designed for deep-level feature fusion in FVit. Although both layers share a similar structural framework, they differ in their attention mechanisms. The self-attention layer focuses on extracting multimodal features, while the Ada-MBA layer is responsible for fusing these deep-level semantic features [36].

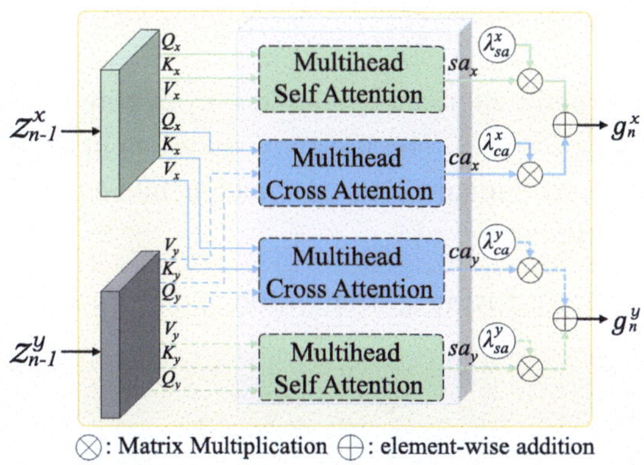

Figure 3.12:
The schematic of the Ada-MBA module. This module incorporates two multihead self-attention for intramodal information extraction and two multihead cross-attention modules for intermodal information fusion. For clarity, separate labels are provided for the self-attention and cross-attention components, although the other matrices remain consistent [36].

equal parts denoted by $z^x_{n-1,h}$ and $z^y_{n-1,h}$ for $h = 1, 2, ..., H$, where H is the number of heads. Since each head performs the same operations, we will omit the index h in subsequent discussions of the fusion attention module within the multihead mechanism. The module computes two sets of matrices, $\{Q_x, K_x, V_x\}$ and $\{Q_y, K_y, V_y\}$, for multimodal information using linear projections U^x_{qkv} and U^y_{qkv}, respectively. After that, self-attention results (sa_x, sa_y) and cross-attention results (ca_x, ca_y) are computed for both modalities simultaneously. Specifically, self-attention extracts intramodal information using $\{Q_x, K_x, V_x\}$ and $\{Q_y, K_y, V_y\}$ while cross-attention captures intermodal information using $\{Q_x, K_y, V_y\}$ and $\{Q_y, K_x, V_x\}$. This process enables the extraction and fusion of deep-level features within a single module. The operations within the Ada-MBA module are outlined as follows:

$$\begin{aligned}
[Q_x, K_x, V_x] &= z^x_{n-1} U^x_{qkv}, \\
[Q_y, K_y, V_y] &= z^y_{n-1} U^y_{qkv}, \\
sa_x &= \varphi\left(\frac{Q_x K_x^\top}{\sqrt{d}}\right) V_x, \\
sa_y &= \varphi\left(\frac{Q_y K_y^\top}{\sqrt{d}}\right) V_y, \\
ca_x &= \varphi\left(\frac{Q_x K_y^\top}{\sqrt{d}}\right) V_y, \\
ca_y &= \varphi\left(\frac{Q_y K_x^\top}{\sqrt{d}}\right) V_x,
\end{aligned} \tag{3.12}$$

where d is the normalization parameter, and $\varphi(\cdot)$ and $(\cdot)^\top$ denote the softmax function and matrix transpose operator, respectively. The last two equations in Eq. (3.12) implement the mutual-association guidance mechanism. This approach achieves guided fusion of the two modalities by exchanging partial matrices.

Next, the adaptive mechanism for fusing self-attention and cross-attention is as follows:

$$\begin{aligned}
g^x_n &= \lambda^x_{sa} sa_x + \lambda^x_{ca} ca_x, \\
g^y_n &= \lambda^y_{sa} sa_y + \lambda^y_{ca} ca_y,
\end{aligned} \tag{3.13}$$

where λ^x_{sa}, λ^x_{sa}, λ^x_{sa}, and λ^x_{sa} denote the learnable weights used to balance the contributions from self-attention and CA.

Finally, the fused feature maps are refined using N_3 self-attention layers. The procedures described in Eqs. (3.9) are applied N_3 times to enhance the fused feature maps for both the VIS and DSM branches. The final output of FVit, denoted as $z_N \in \mathbb{R}^{C_{\text{hid}} \times L}$, represents the feature maps obtained from the last self-attention layer. This output contains the rich contextual information from the multimodal data, which is then fed into the cascaded decoder for further processing.

Comparing the two feature-level fusion methods based on CNN and Transformer architectures, namely CMFNet and FTransUNet, reveals diverse research directions in this field. CMFNet introduces a simultaneous fusion mechanism leveraging channel-wise self-attention, whereas FTransUNet takes a different approach by refining and fusing multiscale features separately. Another key distinction lies in how the fused features are utilized: CMFNet and FTransUNet both employ them as residual connections, while FuseNet [15] and vFuseNet [1] only input them into the next encoder layers. Generally, different strategies come with their own strengths and weaknesses depending on the specific scenario, making it difficult to conclude that any one strategy is entirely ineffective. These methods pave the way for more advanced and interesting multimodal fusion strategies at the feature-level fusion.

3.4 Decision-level fusion methods

3.4.1 Advantages and limitations

Decision-level fusion is a critical approach in remote sensing that integrates the outputs or decisions from multiple classifiers or algorithms to improve the overall accuracy and reliability. Compared with pixel-level fusion or feature-level fusion, decision-level fusion operates at a higher abstraction level, only combining the final decisions from various sources. Decision-level fusion enhances robustness by integrating outputs from multiple classifiers or sensors. This approach reduces the impact of errors or biases in individual data sources, leading to more reliable results. It is flexible and can accommodate a wide range of data types and sources, including data from different sensors, algorithms, or even temporal sequences. This makes it highly adaptable to various applications. It also scales well with the addition of new data sources or classifiers. As more classifiers are added, decision-level fusion can integrate them without requiring significant changes to the fusion process but only the modification of the final decision modules. Since decision-level fusion operates on the final decisions or outputs rather than raw data, it is less sensitive to preprocessing steps which is quite different from pixel-level fusion. This can simplify the fusion process, as different data sources do not need to be prealigned or normalized.

On the other hand, the shortcomings of this technique are also obvious. Decision-level fusion may result in the loss of detailed information because it operates on the decisions or outputs rather than raw data or features. This can be a drawback in applications where fine-grained information is critical. The effectiveness of decision-level fusion heavily depends on the quality and diversity of the individual classifiers. If all classifiers are biased or perform poorly, the fusion result will also be degraded. Meanwhile, since feature extractors and classifiers need to be constructed for different modalities, it would require significant computational resources. Finally, designing appropriate rules or criteria for combining decisions can be complex, especially when integrating outputs from classifiers with different confidence levels or varying performance.

3.4.2 Decision criteria

Here are some examples of how pixel-level fusion is used with specific remote sensing data:

- Majority Voting: In majority voting, each classifier or algorithm casts a vote for a particular class. The class with the most votes is chosen as the final decision. This method is simple and robust, especially when dealing with a large number of classifiers that perform comparably.
- Weighted Voting: It is similar to majority voting, but each classifier's vote is weighted based on its confidence level or past performance. This method gives more influence to classifiers that are more reliable.
- Bayesian Decision Fusion: This approach uses Bayesian probability to combine decisions, where each classifier's output is treated as a conditional probability. The final decision is made based on the maximum posterior probability.
- Dempster-Shafer Theory: This method extends Bayesian fusion by allowing for the representation and combination of uncertainty and partial belief. It does not require prior probabilities and can combine evidence from different sources with varying levels of confidence.
- Rule-Based Fusion: Decisions are combined based on a set of predefined rules or logical conditions. This method is particularly useful when domain knowledge can be encoded into the decision-making process. It also means that it will be difficult to apply without expert knowledge.
- Neural Network-Based Fusion: In this approach, the outputs of multiple classifiers are fed into a neural network, which is trained to make the final decision. This method leverages the neural network's ability to learn complex relationships between the classifier outputs. It allows the construction of a complete end-to-end framework using deep learning models. Therefore, different fusion strategies, such as pixel-level fusion and feature-level fusion, can be seamlessly incorporated into this unified framework, which is the main concept of mixed fusion technique.

In conclusion, decision-level fusion offers significant advantages in enhancing the robustness, flexibility, and accuracy of remote sensing applications by combining the strengths of multiple classifiers or data sources. However, it also comes with limitations, such as the potential loss of detailed information, complexity in rule design, and challenges related to classifier quality and computational demands. Understanding these trade-offs is crucial for effectively implementing decision-level fusion in various remote sensing tasks.

3.5 Conclusion

In the field of remote sensing, pixel-level fusion, feature-level fusion, and decision-level fusion each offer distinct approaches for fusing multimodal data, with unique strengths and limitations.

Pixel-level fusion operates at the most granular level, combining raw data from multiple sources into a single set of fused pixels. This method excels in maintaining spatial resolution and is particularly effective when the input data are highly complementary and accurately aligned. However, it often struggles with noise sensitivity and can result in information loss when the modalities differ significantly in resolution or quality.

Feature-level fusion occurs at a higher abstraction level, where features extracted from different modalities are merged. This approach benefits from the ability to capture more meaningful and complementary information by combining the strengths of various features. It typically offers better robustness to noise compared to pixel-level fusion and is more adaptable to data from different sensors. The main complexity lies in the designation of feature fusion mechanism. However, recent advancements in deep learning have significantly accelerated the progress of this approach.

Decision-level fusion integrates the outcomes of multiple classifiers or decision-making processes, rather than the data or features themselves. This method is advantageous when dealing with heterogeneous data sources, as it allows for flexibility in combining decisions from different algorithms. Decision-level fusion is generally easier to implement with preexisting models and is less sensitive to the alignment and resolution issues that can plague pixel-level fusion. However, it may miss some details that can be captured by earlier fusion strategies, as it relies on the decisions made by individual classifiers.

In conclusion, each fusion technique serves different purposes: pixel-level fusion is ideal for high-resolution, aligned data; feature-level fusion offers a balance between complexity and performance, suitable for diverse sensor inputs; and decision-level fusion provides a versatile, though potentially less integrated, approach to combining decisions from multiple models. The selection of a fusion strategy, or the consideration of a hybrid approach, should be guided by the specific characteristics of the data and the goals of the remote sensing tasks.

References

[1] Nicolas Audebert, Bertrand Le Saux, Sébastien Lefèvre, Beyond RGB: very high resolution urban remote sensing with multimodal deep networks, ISPRS Journal of Photogrammetry and Remote Sensing 140 (2018) 20–32.

[2] Vijay Badrinarayanan, Alex Kendall, Roberto Cipolla, SegNet: a deep convolutional encoder-decoder architecture for image segmentation, IEEE Transactions on Pattern Analysis and Machine Intelligence 39 (12) (2017) 2481–2495.
[3] Xuyang Bai, Zeyu Hu, Xinge Zhu, Qingqiu Huang, Yilun Chen, Hongbo Fu, Chiew-Lan Tai, Transfusion: robust LiDAR-camera fusion for 3D object detection with transformers, in: Proceedings of the IEEE/CVF Conference on Computer Vision and Pattern Recognition, 2022, pp. 1090–1099.
[4] Tadas Baltrušaitis, Chaitanya Ahuja, Louis-Philippe Morency, Multimodal machine learning: a survey and taxonomy, IEEE Transactions on Pattern Analysis and Machine Intelligence 41 (2) (2018) 423–443.
[5] Hu Cao, Yueyue Wang, Joy Chen, Dongsheng Jiang, Xiaopeng Zhang, Qi Tian, Manning Wang, Swin-Unet: Unet-like pure transformer for medical image segmentation, in: Proceedings of European Conference on Computer Vision (ECCV) Workshops, 2022, pp. 205–218.
[6] Jieneng Chen, Yongyi Lu, Qihang Yu, Xiangde Luo, Ehsan Adeli, Yan Wang, Le Lu, Alan L. Yuille, Yuyin Zhou, TransUNet: transformers make strong encoders for medical image segmentation, arXiv preprint, arXiv:2102.04306, 2021.
[7] Mauro Dalla Mura, Saurabh Prasad, Fabio Pacifici, Paulo Gamba, Jocelyn Chanussot, Jón Atli Benediktsson, Challenges and opportunities of multimodality and data fusion in remote sensing, Proceedings of the IEEE 103 (9) (2015) 1585–1601.
[8] Foivos I. Diakogiannis, François Waldner, Peter Caccetta, Chen Wu, ResUNet-a: a deep learning framework for semantic segmentation of remotely sensed data, ISPRS Journal of Photogrammetry and Remote Sensing 162 (2020) 94–114.
[9] Alexey Dosovitskiy, Lucas Beyer, Alexander Kolesnikov, Dirk Weissenborn, Xiaohua Zhai, Thomas Unterthiner, Mostafa Dehghani, Matthias Minderer, Georg Heigold, Sylvain Gelly, et al., An image is worth 16×16 words: transformers for image recognition at scale, in: International Conference on Learning Representations, 2021, pp. 1–22.
[10] Liang Gao, Hui Liu, Minhang Yang, Long Chen, Yaling Wan, Zhengqing Xiao, Yurong Qian, STransFuse: fusing Swin Transformer and convolutional neural network for remote sensing image semantic segmentation, IEEE Journal of Selected Topics in Applied Earth Observations and Remote Sensing 14 (2021) 10990–11003.
[11] Lianru Gao, Jun Li, Mahdi Khodadadzadeh, Antonio Plaza, Bing Zhang, Zhijian He, Huiming Yan, Subspace-based support vector machines for hyperspectral image classification, IEEE Geoscience and Remote Sensing Letters 12 (2) (2014) 349–353.
[12] Pedram Ghamisi, Behnood Rasti, Naoto Yokoya, Qunming Wang, Bernhard Hofle, Lorenzo Bruzzone, Francesca Bovolo, Mingmin Chi, Katharina Anders, Richard Gloaguen, et al., Multisource and multitemporal data fusion in remote sensing: a comprehensive review of the state of the art, IEEE Geoscience and Remote Sensing Magazine 7 (1) (2019) 6–39.
[13] Pall Oskar Gislason, Jon Atli Benediktsson, Johannes R. Sveinsson, Random forests for land cover classification, Pattern Recognition Letters 27 (4) (2006) 294–300.
[14] Luis Gómez-Chova, Devis Tuia, Gabriele Moser, Gustau Camps-Valls, Multimodal classification of remote sensing images: a review and future directions, Proceedings of the IEEE 103 (9) (2015) 1560–1584.
[15] Caner Hazirbas, Lingni Ma, Csaba Domokos, Daniel Cremers, FuseNet: incorporating depth into semantic segmentation via fusion-based CNN architecture, in: Asian Conference on Computer Vision, 2016, pp. 213–228.
[16] Kaiming He, Xiangyu Zhang, Shaoqing Ren, Jian Sun, Deep residual learning for image recognition, in: Proceedings of the IEEE Conference on Computer Vision and Pattern Recognition, 2016, pp. 770–778.
[17] Qibin He, Xian Sun, Wenhui Diao, Zhiyuan Yan, Fanglong Yao, Kun Fu, Multimodal remote sensing image segmentation with intuition-inspired hypergraph modeling, IEEE Transactions on Image Processing 32 (2023) 1474–1487.
[18] Shumeng He, Houqun Yang, Xiaoying Zhang, Xuanyu Li, MFTransNet: a multi-modal fusion with CNN-transformer network for semantic segmentation of HSR remote sensing images, Mathematics 11 (3) (2023) 722.

[19] Xin He, Yong Zhou, Jiaqi Zhao, Di Zhang, Rui Yao, Yong Xue, Swin transformer embedding UNet for remote sensing image semantic segmentation, IEEE Transactions on Geoscience and Remote Sensing 60 (2022) 1–15.
[20] Danfeng Hong, Lianru Gao, Naoto Yokoya, Jing Yao, Jocelyn Chanussot, Qian Du, Bing Zhang, More diverse means better: multimodal deep learning meets remote-sensing imagery classification, IEEE Transactions on Geoscience and Remote Sensing 59 (5) (2020) 4340–4354.
[21] Danfeng Hong, Jingliang Hu, Jing Yao, Jocelyn Chanussot, Xiao Xiang Zhu, Multimodal remote sensing benchmark datasets for land cover classification with a shared and specific feature learning model, ISPRS Journal of Photogrammetry and Remote Sensing 178 (2021) 68–80.
[22] Hamidreza Hosseinpour, Farhad Samadzadegan, Farzaneh Dadrass Javan, CMGFNet: a deep cross-modal gated fusion network for building extraction from very high-resolution remote sensing images, ISPRS Journal of Photogrammetry and Remote Sensing 184 (2022) 96–115.
[23] Jie Hu, Li Shen, Gang Sun, Squeeze-and-excitation networks, in: Proceedings of the IEEE Conference on Computer Vision and Pattern Recognition, 2018, pp. 7132–7141.
[24] Alexander Kirillov, Eric Mintun, Nikhila Ravi, Hanzi Mao, Chloe Rolland, Laura Gustafson, Tete Xiao, Spencer Whitehead, Alexander C. Berg, Wan-Yen Lo, et al., Segment anything, arXiv preprint, arXiv: 2304.02643, 2023.
[25] Philipp Krähenbühl, Vladlen Koltun, Efficient inference in fully connected CRFs with Gaussian edge potentials, Advances in Neural Information Processing Systems 24 (2011) 1–9.
[26] Jiaxin Li, Danfeng Hong, Lianru Gao, Jing Yao, Ke Zheng, Bing Zhang, Jocelyn Chanussot, Deep learning in multimodal remote sensing data fusion: a comprehensive review, International Journal of Applied Earth Observation and Geoinformation 112 (2022) 102926.
[27] Yansheng Li, Yuhan Zhou, Yongjun Zhang, Liheng Zhong, Jian Wang, Jingdong Chen, DKDFN: domain knowledge-guided deep collaborative fusion network for multimodal unitemporal remote sensing land cover classification, ISPRS Journal of Photogrammetry and Remote Sensing 186 (2022) 170–189.
[28] Ze Liu, Yutong Lin, Yue Cao, Han Hu, Yixuan Wei, Zheng Zhang, Stephen Lin, Baining Guo, Swin transformer: hierarchical vision transformer using shifted windows, in: Proceedings of the IEEE/CVF International Conference on Computer Vision, 2021, pp. 10012–10022.
[29] Jonathan Long, Evan Shelhamer, Trevor Darrell, Fully convolutional networks for semantic segmentation, in: Proceedings of the IEEE Conference on Computer Vision and Pattern Recognition, 2015, pp. 3431–3440.
[30] Xiaochen Lu, Junping Zhang, Tong Li, Guilu Zhang, Synergetic classification of long-wave infrared hyperspectral and visible images, IEEE Journal of Selected Topics in Applied Earth Observations and Remote Sensing 8 (7) (2015) 3546–3557.
[31] Luigi Tommaso Luppino, Mads Adrian Hansen, Michael Kampffmeyer, Filippo Maria Bianchi, Gabriele Moser, Robert Jenssen, Stian Normann Anfinsen, Code-aligned autoencoders for unsupervised change detection in multimodal remote sensing images, IEEE Transactions on Neural Networks and Learning Systems 35 (1) (2024) 60–72.
[32] Jiabao Ma, Wujie Zhou, Jingsheng Lei, Lu Yu, Adjacent bi-hierarchical network for scene parsing of remote sensing images, IEEE Geoscience and Remote Sensing Letters 20 (2023) 1–5.
[33] Jiayi Ma, Linfeng Tang, Fan Fan, Jun Huang, Xiaoguang Mei, Yong Ma, SwinFusion: cross-domain long-range learning for general image fusion via Swin Transformer, IEEE/CAA Journal of Automatica Sinica 9 (7) (2022) 1200–1217.
[34] Xianping Ma, Xiaokang Zhang, Man-On Pun, A crossmodal multiscale fusion network for semantic segmentation of remote sensing data, IEEE Journal of Selected Topics in Applied Earth Observations and Remote Sensing 15 (2022) 3463–3474.
[35] Xianping Ma, Xiaokang Zhang, Zhiguo Wang, Man-On Pun, Unsupervised domain adaptation augmented by mutually boosted attention for semantic segmentation of VHR remote sensing images, IEEE Transactions on Geoscience and Remote Sensing 61 (2023) 1–15.
[36] Xianping Ma, Xiaokang Zhang, Man-On Pun, Ming Liu, A multilevel multimodal fusion transformer for remote sensing semantic segmentation, IEEE Transactions on Geoscience and Remote Sensing (2024).

[37] Dimitrios Marmanis, Konrad Schindler, Jan Dirk Wegner, Silvano Galliani, Mihai Datcu, Uwe Stilla, Classification with an edge: improving semantic image segmentation with boundary detection, ISPRS Journal of Photogrammetry and Remote Sensing 135 (2018) 158–172.

[38] Yan Meng, Shanxiong Chen, Yuxuan Liu, Li Li, Zemin Zhang, Tao Ke, Xiangyun Hu, Unsupervised building extraction from multimodal aerial data based on accurate vegetation removal and image feature consistency constraint, Remote Sensing 14 (8) (2022) 1912.

[39] Jiquan Ngiam, Aditya Khosla, Mingyu Kim, Juhan Nam, Honglak Lee, Andrew Y. Ng, Multimodal deep learning, in: ICML, 2011, pp. 689–696.

[40] Keiller Nogueira, Mauro Dalla Mura, Jocelyn Chanussot, William Robson Schwartz, Jefersson Alex Dos Santos, Dynamic multicontext segmentation of remote sensing images based on convolutional networks, IEEE Transactions on Geoscience and Remote Sensing 57 (10) (2019) 7503–7520.

[41] Frosti Palsson, Johannes R. Sveinsson, Magnus Orn Ulfarsson, Jon Atli Benediktsson, Model-based fusion of multi- and hyperspectral images using PCA and wavelets, IEEE Transactions on Geoscience and Remote Sensing 53 (5) (2014) 2652–2663.

[42] Sourabh Paul, Umesh C. Pati, A comprehensive review on remote sensing image registration, International Journal of Remote Sensing 42 (14) (2021) 5396–5432.

[43] Aditya Prakash, Kashyap Chitta, Andreas Geiger, Multi-modal fusion transformer for end-to-end autonomous driving, in: Proceedings of the IEEE/CVF Conference on Computer Vision and Pattern Recognition, 2021, pp. 7077–7087.

[44] Alec Radford, Karthik Narasimhan, Improving language understanding by generative pre-training. Available at: https://api.semanticscholar.org/CorpusID:49313245, 2018.

[45] Olaf Ronneberger, Philipp Fischer, Thomas Brox, U-net: convolutional networks for biomedical image segmentation, in: International Conference on Medical Image Computing and Computer-Assisted Intervention, 2015, pp. 234–241.

[46] Swalpa Kumar Roy, Ankur Deria, Danfeng Hong, Behnood Rasti, Antonio Plaza, Jocelyn Chanussot, Multimodal fusion transformer for remote sensing image classification, IEEE Transactions on Geoscience and Remote Sensing 61 (2023) 1–20.

[47] Daniel Seichter, Mona Köhler, Benjamin Lewandowski, Tim Wengefeld, Horst-Michael Gross, Efficient RGB-D semantic segmentation for indoor scene analysis, in: 2021 IEEE International Conference on Robotics and Automation (ICRA), 2021, pp. 13525–13531.

[48] Yu Shen, Jianyu Chen, Liang Xiao, Delu Pan, Optimizing multiscale segmentation with local spectral heterogeneity measure for high resolution remote sensing images, ISPRS Journal of Photogrammetry and Remote Sensing 157 (2019) 13–25.

[49] Karen Simonyan, Andrew Zisserman, Very deep convolutional networks for large-scale image recognition, in: 3rd International Conference on Learning Representations, 2015, pp. 1–14.

[50] Robin Strudel, Ricardo Garcia, Ivan Laptev, Cordelia Schmid, Segmenter: transformer for semantic segmentation, in: Proceedings of the IEEE/CVF International Conference on Computer Vision, 2021, pp. 7262–7272.

[51] Redha Touati, Max Mignotte, Mohamed Dahmane, Multimodal change detection in remote sensing images using an unsupervised pixel pairwise-based Markov random field model, IEEE Transactions on Image Processing 29 (2019) 757–767.

[52] Yao-Hung Hubert Tsai, Shaojie Bai, Paul Pu Liang, J. Zico Kolter, Louis-Philippe Morency, Ruslan Salakhutdinov, Multimodal transformer for unaligned multimodal language sequences, in: Proceedings of the 57th Annual Meeting of the Association for Computational Linguistics, 2019, pp. 6558–6569.

[53] Ashish Vaswani, Noam Shazeer, Niki Parmar, Jakob Uszkoreit, Llion Jones, Aidan N. Gomez, Łukasz Kaiser, Illia Polosukhin, Attention is all you need, Advances in Neural Information Processing Systems 30 (2017) 1–11.

[54] Libo Wang, Rui Li, Chenxi Duan, Ce Zhang, Xiaoliang Meng, Shenghui Fang, A novel transformer based semantic segmentation scheme for fine-resolution remote sensing images, IEEE Geoscience and Remote Sensing Letters 19 (2022) 1–5.

[55] Libo Wang, Rui Li, Ce Zhang, Shenghui Fang, Chenxi Duan, Xiaoliang Meng, Peter M. Atkinson, UNetFormer: a UNet-like transformer for efficient semantic segmentation of remote sensing urban scene imagery, ISPRS Journal of Photogrammetry and Remote Sensing 190 (2022) 196–214.
[56] Shuang Wang, Dou Quan, Xuefeng Liang, Mengdan Ning, Yanhe Guo, Licheng Jiao, A deep learning framework for remote sensing image registration, ISPRS Journal of Photogrammetry and Remote Sensing 145 (2018) 148–164.
[57] Qi Wei, José Bioucas-Dias, Nicolas Dobigeon, Jean-Yves Tourneret, Hyperspectral and multispectral image fusion based on a sparse representation, IEEE Transactions on Geoscience and Remote Sensing 53 (7) (2015) 3658–3668.
[58] Xin Wu, Danfeng Hong, Jocelyn Chanussot, Convolutional neural networks for multimodal remote sensing data classification, IEEE Transactions on Geoscience and Remote Sensing 60 (2021) 1–10.
[59] Enze Xie, Wenhai Wang, Zhiding Yu, Anima Anandkumar, Jose M. Alvarez, Ping Luo, SegFormer: simple and efficient design for semantic segmentation with transformers, Advances in Neural Information Processing Systems 34 (2021) 12077–12090.
[60] Xiaodong Xu, Wei Li, Qiong Ran, Qian Du, Lianru Gao, Bing Zhang, Multisource remote sensing data classification based on convolutional neural network, IEEE Transactions on Geoscience and Remote Sensing 56 (2) (2017) 937–949.
[61] Zeyu Xu, Zhanfeng Shen, Yang Li, Liegang Xia, Haoyu Wang, Shuo Li, Shuhui Jiao, Yating Lei, Road extraction in mountainous regions from high-resolution images based on DSDNet and terrain optimization, Remote Sensing 13 (1) (2020) 90.
[62] Zhiyong Xu, Weicun Zhang, Tianxiang Zhang, Zhifang Yang, Jiangyun Li, Efficient transformer for remote sensing image segmentation, Remote Sensing 13 (18) (2021) 3585.
[63] Li Yan, Jianming Huang, Hong Xie, Pengcheng Wei, Zhao Gao, Efficient depth fusion transformer for aerial image semantic segmentation, Remote Sensing 14 (2022) 1294.
[64] Xiaokang Zhang, Weikang Yu, Man-On Pun, Wenzhong Shi, Cross-domain landslide mapping from large-scale remote sensing images using prototype-guided domain-aware progressive representation learning, ISPRS Journal of Photogrammetry and Remote Sensing 197 (2023) 1–17.
[65] Xiaokang Zhang, Boning Zhang, Weikang Yu, Xudong Kang, Federated deep learning with prototype matching for object extraction from very-high-resolution remote sensing images, IEEE Transactions on Geoscience and Remote Sensing 61 (2023) 1–16.
[66] Hengshuang Zhao, Jianping Shi, Xiaojuan Qi, Xiaogang Wang, Jiaya Jia, Pyramid scene parsing network, in: Proceedings of the IEEE Conference on Computer Vision and Pattern Recognition, 2017, pp. 2881–2890.
[67] Wujie Zhou, Jianhui Jin, Jingsheng Lei, Jenq-Neng Hwang, CEGFNet: common extraction and gate fusion network for scene parsing of remote sensing images, IEEE Transactions on Geoscience and Remote Sensing 60 (2021) 1–10.
[68] Wujie Zhou, Jianhui Jin, Jingsheng Lei, Lu Yu, CIMFNet: cross-layer interaction and multiscale fusion network for semantic segmentation of high-resolution remote sensing images, IEEE Journal of Selected Topics in Signal Processing 16 (2022) 666–676.
[69] Jingru Zhu, Ya Guo, Geng Sun, Libo Yang, Min Deng, Jie Chen, Unsupervised domain adaptation semantic segmentation of high-resolution remote sensing imagery with invariant domain-level prototype memory, IEEE Transactions on Geoscience and Remote Sensing 61 (2023) 1–18.

CHAPTER 4

Multisensor fusion

Yuxuan Liu

State Key Laboratory of Spatial Datum, Chinese Academy of Surveying and Mapping (CASM), Beijing, China

4.1 General RS data fusion process

In general, the fusion of multimodal remote sensing (RS) data involves two core steps: data alignment and attribute/identity estimation (Fig. 4.1) [1]. Data alignment and data/object correlation together are commonly referred to as matching and alignment in the remote sensing context [2]. Their primary objective is to ensure that measurements are accurately linked to one another and to the corresponding objects of interest. Attribute or identity estimation constitutes the actual fusion step, which involves the combined exploitation of aligned and correlated measurement data within an estimation framework.

After alignment, attribute and identity estimation can be achieved at three levels: pixel level, feature level, and decision level (Fig. 4.2) [3]. Pixel-level fusion involves integration at the stage where the data is collected. The primary goal is to generate enhanced data for various applications, such as image sharpening, superresolution, and 3D model reconstruction

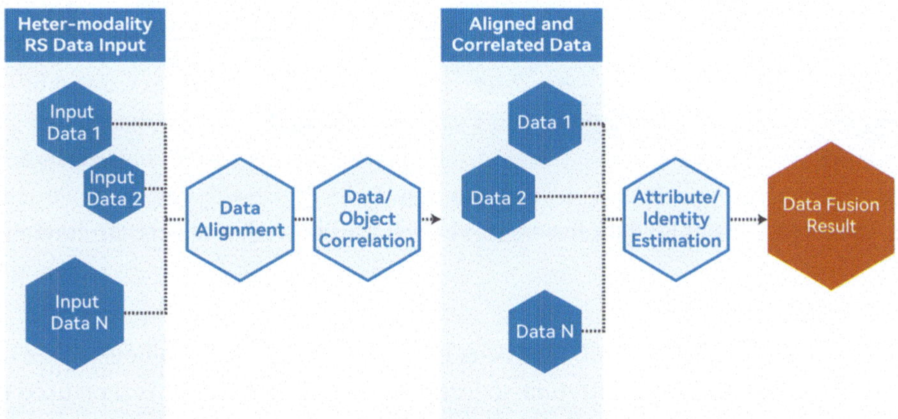

Figure 4.1:
The general process of data fusion for multimodal remote sensing data input.

Multimodal Remote Sensing Fusion and Classification
https://doi.org/10.1016/B978-0-44-329152-4.00011-8

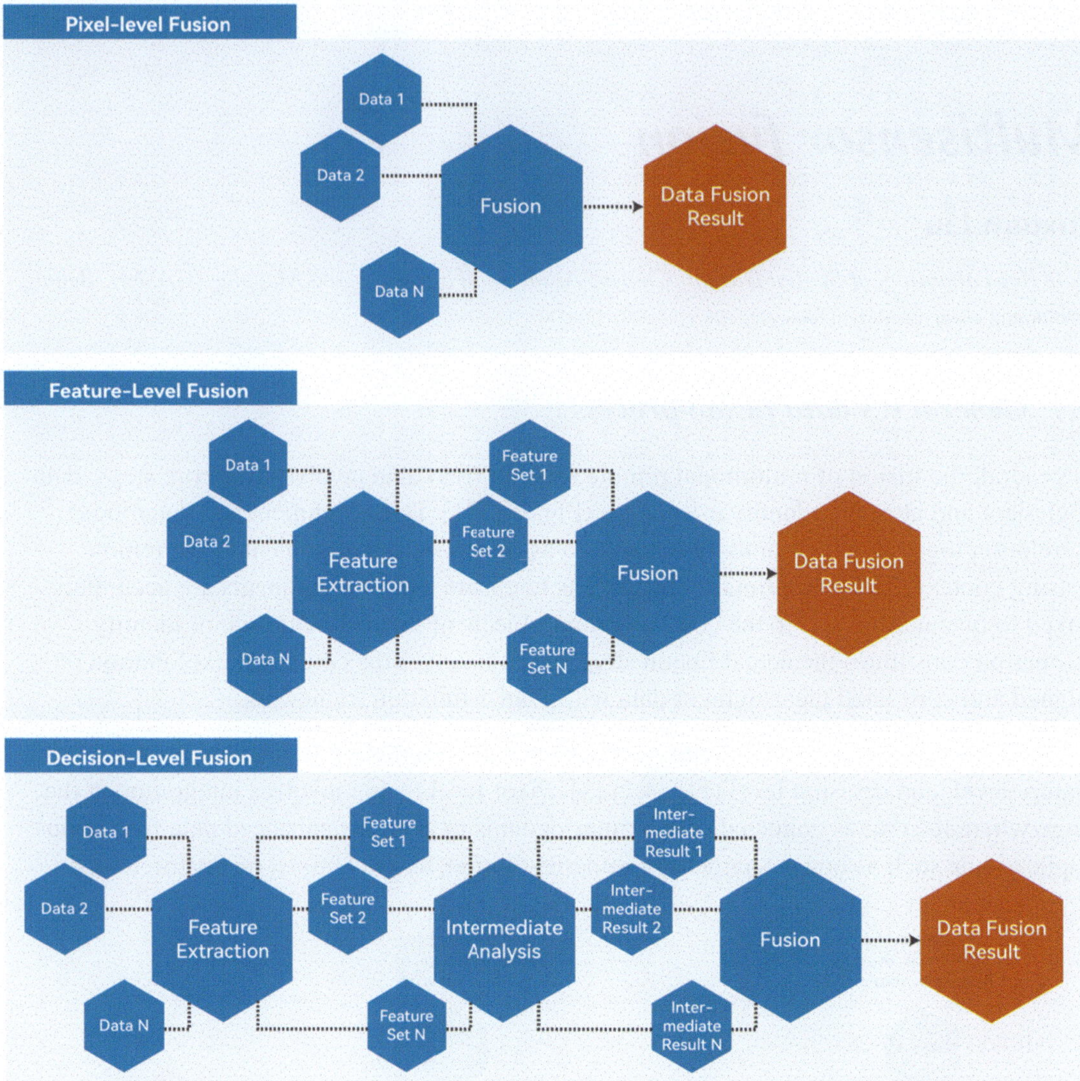

Figure 4.2:
The three levels of RS data fusion: pixel-level, feature-level, and decision-level.

from 2D images. At the feature level, the focus is on generating a comprehensive set of features by combining data from different sources. This enriched dataset is then used for further analysis or decision-making. In land cover classification, for instance, various data types can be merged and fed into a classifier. Alternatively, features like image primitives or spatial characteristics can be extracted from different sources and integrated to reduce

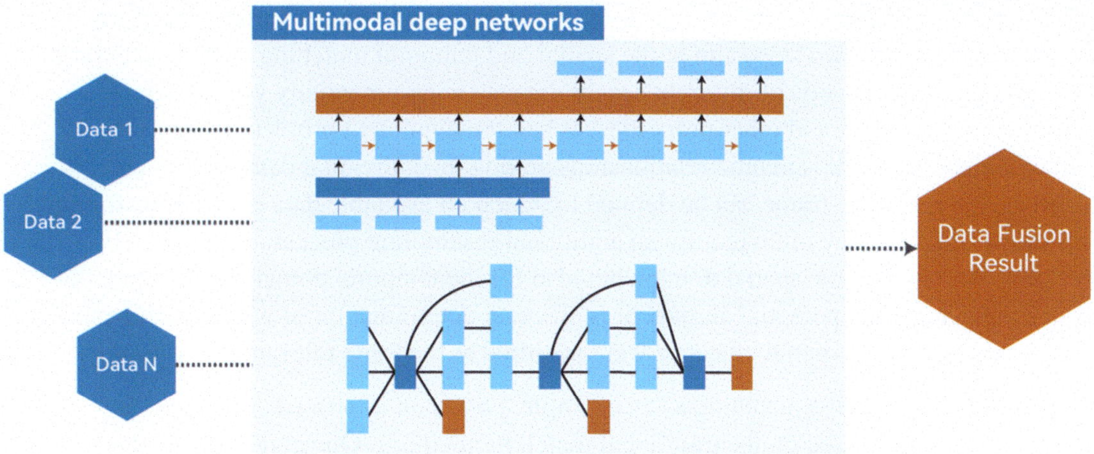

Figure 4.3:
Deep learning-based multimodal RS data fusion.

uncertainty or provide a more detailed description. In decision-level fusion, the integration occurs after the independent analysis of each data source. Information from different sources is combined at the decision-making stage, which enhances the robustness of the final decision. For example, if multiple data sources agree on a decision, this consensus is reinforced; if they differ, methods such as majority voting can be employed to reach a final agreement, minimizing errors. This approach is analogous to ensemble learning in pattern recognition, where combining multiple models improves overall decision accuracy.

In addition to handcrafted methods based on domain-specific knowledge, deep learning has shown significant potential in multimodal remote sensing (RS) fusion. This potential is harnessed through the automatic modeling of complex relationships between input and output data, efficiently managing data alignment and fusion adaptively at the pixel, feature, and decision levels. Although the approach may vary depending on the modalities being fused and the specific tasks involved, these methods can generally be consolidated into a unified framework, as illustrated in Fig. 4.3.

4.1.1 Data matching and alignment

In general terminology, the data alignment step is about coordinate transformations and unit adjustments aimed at achieving a common representation of the heterogeneous measurements and observations describing the object of interest. In analogy, the data/object correlation stage is mainly about associating several multisensor measurements about the object of interest with each other. In other words, it is about letting the data fusion process know which particular measurements are supposed to be fused in the final estimation step. In remote sensing, these two steps aim at the spatial and temporal matching and, if necessary, the alignment, respectively, of different sensor data showing potentially very different radiometric, geometric, and other properties. When the alignment problem is solved and spatial, temporal, and/or semantic relationships among the individual data sources are established, a reference frame can be defined to which all available data can be transformed. This transformation may often require an additional resampling process, which may be necessary not only for the spatial domain but also for the temporal domain. In the end, the result of matching and potential alignment is an exact determination of which measurements belong to the same geospatial object or were acquired at the same relevant point in time.

Taking multimodal image matching as an example, two main improvement strategies have been proposed based on the traditional matching methods. One approach involves converting images of different modalities into a unified intermediate modality and then applying traditional matching algorithms, with the core aim of eliminating modality differences. However, modality conversion relies on specific datasets and transformation models, and its effectiveness is limited in complex scenarios. When modality conversion is inaccurate, matching precision can significantly decline. Another approach introduces strategies resistant to modality variations in feature or template matching, enhancing the robustness of features against modality changes and allowing matching directly in the original modality, which offers greater flexibility and adaptability. Recent research has explored the application of multiscale and multidirectional information, finding that local structural information recovered from multiscale and multidirectional filtering is highly robust to nonlinear radiation deformations (NRD), especially in multimodal images with only translational transformations, showing better matching results. These methods improve matching accuracy and robustness by analyzing local structures and capturing key information that remains invariant across modalities. In practical applications, scale differences and rotations are common geometric changes that impose higher demands on matching algorithms. Theoretically, scale changes can be handled through an image pyramid strategy, but the handling of image rotation is more limited. Traditional methods rely on image gradient or intensity information to calculate the principal direction of features, but NRD makes it difficult to accurately obtain this information, thereby affecting matching accuracy, especially in scenarios with high modality differences and complex deformations, which further exacerbate uncertainties.

With the rapid development of artificial intelligence technology, deep learning methods such as Convolutional Neural Networks (CNN) [4], Generative Adversarial Networks (GAN) [5], Siamese Networks [6], and Transformers [7] are gradually being applied to multimodal image matching. One approach uses GANs for style transfer to achieve modality unification of images from different sources, reducing the impact of NRD, and combines traditional region or feature registration methods to complete the matching. Another approach directly applies deep learning techniques to part or all of the traditional matching process, subdivided into learning-based template matching, feature matching, and end-to-end matching. Template matching involves training neural networks to obtain deep features, replacing intensity information or shallow features in traditional registration, and then achieving matching through neural networks or similarity measures. Feature matching improves registration performance by generating feature detectors or descriptors through deep architectures. End-to-end matching replaces the entire matching process with a complete neural network structure, which, although capable of finding global optimal solutions, consumes significant training resources, has high model complexity, and poor interpretability. While these methods perform well in experiments, they still face challenges in engineering practice. Firstly, there is currently a lack of a large-scale multimodal dataset that covers all image modalities and is fully annotated, leading to a decline in algorithm performance when encountering untrained modalities. Secondly, mainstream learning frameworks still struggle with handling complex spectral data such as remote sensing or medical images, as well as significant scale differences and geometric deformations. Finally, deep learning methods require substantial computational resources during training, limiting their widespread practical application. Fig. 4.4 presents the matching results of several representative methods under different multimodal remote sensing scenarios.

4.1.2 Attribute estimation

Attribute and identity estimation are the very core of the data fusion domain and are mainly driven by different developments in statistical estimation theory and machine learning, respectively. Although most data acquired during space-borne earth-observation missions or well-planned airborne flight campaigns will always be well documented and contain sufficient metadata, most of the potentially available crowd-sourced data (e.g., GPS tracks or smartphone images) will not provide any additional information. It will therefore be necessary to create sophisticated fusion to deliver statistically optimal estimates including stochastically sound quality and reliability measures.

Various fusion approaches have been extensively reviewed in a broad, general sense. Classical probabilistic fusion relies on probability distribution functions and Bayesian estimation, including least-squares, Kalman filter, particle filter, and Markov chain Monte Carlo algorithms [19]. In contrast, evidential belief reasoning, based on the Dempster-Shafer theory, assigns beliefs and plausibility to measurement hypotheses rather than probability distributions, offering a generalization of Bayesian theory with probability mass functions. Fuzzy set theory [20], another approach for imperfect data, introduces partial set membership, allowing imprecise reasoning and better modeling of objects in ill-defined classes. It has

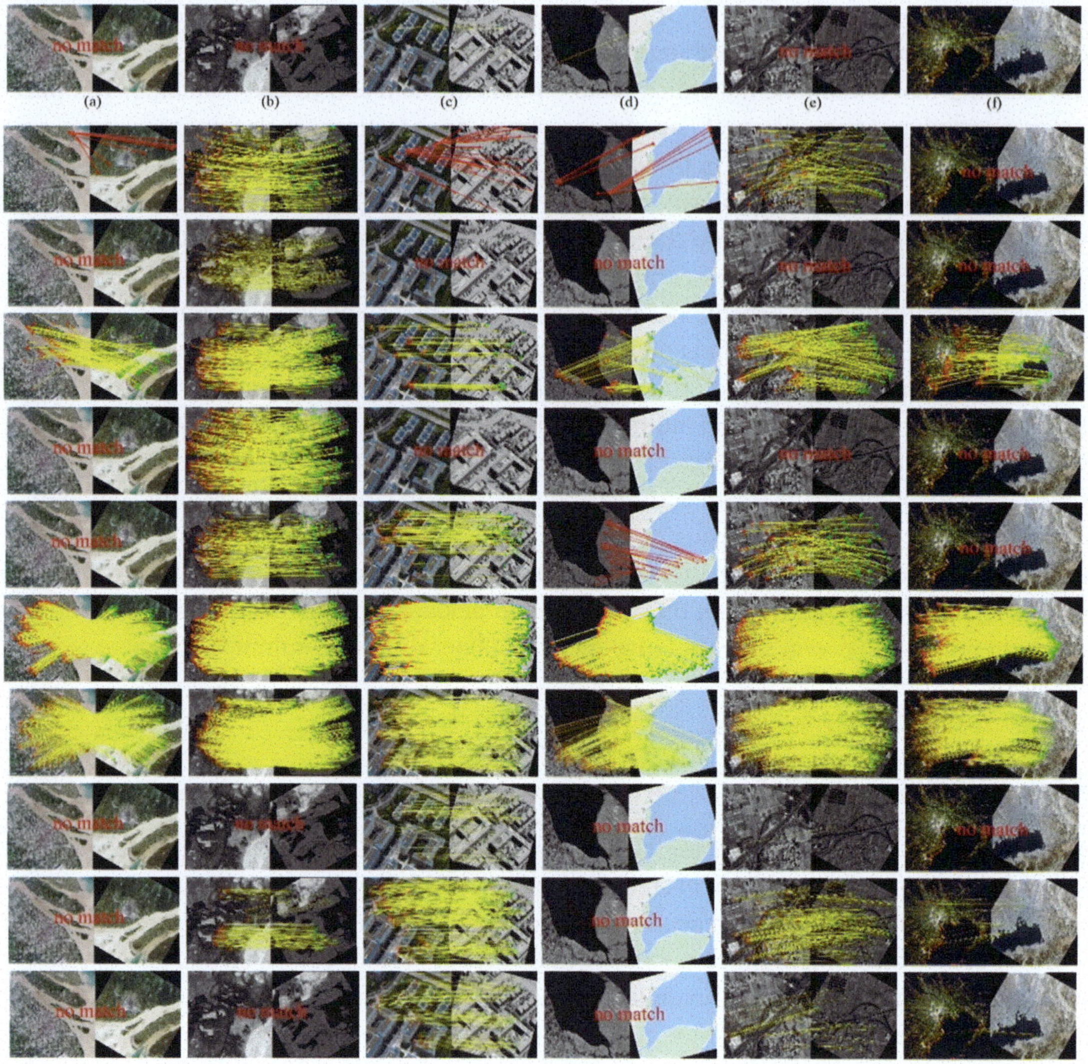

Figure 4.4:
Comparative matching results (a) optical-optical (different season); (b) optical-infrared; (c) optical-LiDAR depth; (d) optical-map; (e) optical-SAR; (f) night-day. From top to bottom, each row corresponds to the matching results of LGHD [8], OS-SIFT [9], PSO-SIFT [10], RIFT [11], CoFSM [12], LNIFT [13], NISR [14], POS-GIFT [15], RedFeat [16], MatchFormer [17], and SemLA [18].

been applied in remote sensing for fusion of displacement measurements from amplitude image correlation and differential SAR interferometry. Fuzzy theory can also complement probabilistic and Dempster-Shafer fusion algorithms. Possibilistic fusion, built on fuzzy set theory, is designed to represent incomplete data, similar to Dempster-Shafer but with a different quantification method. Rough set theory models imprecise data while ignoring uncertainty at different granularity levels, while random set theory uses finite random sets to model target states and measurements instead of conventional vectors. Though most estimation frameworks focus on attribute estimation, machine learning methods like random forests or neural networks offer promising solutions for identity estimation by reducing variance and avoiding overfitting, as seen in remote sensing for improving multisensor classification results.

Machine learning has gained increasing interest across scientific fields, with each machine-learning procedure potentially viewed as an instance of data fusion when training or prediction data come from different sources, and the learned or predicted parameters represent attribute or identity estimation [21]. Beyond this interpretation, machine learning can be an invaluable tool in large remote sensing data fusion frameworks, helping solve complex problems that may be intractable in closed form but become solvable with sufficient training data. Neural networks and ensemble learning methods offer promising solutions for fusion at the decision level, and for remote sensing, machine learning is particularly promising at the feature level, especially in identity estimation tasks like object recognition or land-cover classification. Data availability is crucial in any machine learning setup, particularly with the rise of deep learning. While input data is typically abundant, reference data for training and evaluating models is often limited due to the high costs associated with its acquisition. This challenge is not unique to remote sensing but is especially significant in earth observation applications. Semisupervised learning offers a promising solution to this problem by enabling the use of large amounts of available input data while relying on only a small annotated training set.

4.2 Typical RS data fusion techniques

In order to promote the development of multimodal remote sensing data fusion technology and the application of related technologies, IEEE GRSS has been organizing the Multi-Source Remote Sensing Data Fusion Competition annually since 2015. The contests have covered a wide range of topics, providing insights into the novel use of multimodal data fusion for: (1) Applications where similar modalities (e.g., passive optical) are used to create enhanced data products—specifically, pansharpened images; (2) Applications where disparate modalities (e.g., optical, SAR, and LiDAR) are used simultaneously to better characterize the scene—such as problems related to traditional mapping through classification using multiple modalities, change detection between multitemporal image sequences, etc. Additionally, these contests have attracted novel contributions that fall “outside” the scope identified by the main challenges targeted by each contest. Below, we describe some representative and state-of-the-

Table 4.1: Topics of the IEEE GRSS Data Fusion Contest in the past ten years.

Year	Topic	Data
2015	multiresolution and multisensor fusion at extremely high spatial resolution	Airborne LiDAR data and airborne RGB imagery (Belgium port area)
2016	Urban Mapping and Change Detection	Deimos-2 Panchromatic Image, Deimos-2 Multispectral Image; Iris Video Data
2017	global land use mapping	Landsat 8 Multispectral Image, Sentinel-2 SAR Image, OpenStreetMap Vector Data
2018	Hyperspectral and LiDAR Data Fusion	Airborne Multispectral-LiDAR Data, Airborne Hyperspectral Image, Airborne Visible Image (Houston Area)
2019	Semantic 3D Reconstruction and Stereo	Satellite Visible Image, Satellite Multispectral Image, Airborne LiDAR Data
2020	Global Land Cover Mapping with Weak Supervision	Sentinel-1 SAR Image, Sentinel-2 Multispectral Image, MODIS Image Classification Results
2021	Geospatial AI for Social Good: (1) Detection of Settlements without Electricity; (2) Multitemporal Semantic Change Detection	(1) Sentinel-1 polarimetric SAR dataset, Sentinel-2 multispectral dataset, Landsat 8 multispectral dataset, The Suomi Visible Infrared Imaging Radiometer Suite (VIIRS) night time dataset; (2) USDA National Agriculture Imagery Program (NAIP) data, Landsat-8 (Landsat) data, USGS National Land Cover Database (NLCD) data
2022	Semisupervised Learning for Land Cover Classification	very-high-resolution imagery, DEM information and semantic maps
2023	Large-Scale Fine-Grained Building Reconstruction	SuperView-1, Gaofen-2 and Gaofen-3 satellite images, Gaofen-7 and WorldView with a ground sampling distance (GSD) of roughly 2 m.
2024	Flood Rapid Mapping: Flood rapid mapping with SAR data; Flood rapid mapping with optical data	(1) Sentinel-1 SAR imagery; (2) Sentinel-2 and Landsat optical imageries

art contributions in these areas, and the topics of the IEEE GRSS Data Fusion Contest in the past ten years are summarized in Table 4.1.

4.2.1 High-precision geometric positioning of optical satellite images assisted by LiDAR data

Satellite images are widely used in applications such as environmental monitoring, resource management, urban planning, and disaster monitoring, with high-precision geometric positioning being crucial. Traditional methods rely on manual control points and regional bundle adjustment, which are resource-intensive and time-consuming, especially in complex terrains. LiDAR, with its fast data acquisition and high positioning accuracy, is unaffected by weather conditions, making it ideal for this purpose. The fusion of multisensor data, such as LiDAR and optical satellite images, offers great potential for high-precision geometric orientation [22].

By matching optical satellite images with LiDAR depth maps, numerous correspondences can be established, and these, along with geographic coordinates from the LiDAR map, serve as control information for image orientation. However, differences in physical properties and geometric models between LiDAR and optical images make matching challenging. To address this, we propose using LiDAR as a reference for automatic and accurate orientation of optical satellite images. The AWOG descriptor [23] is applied to generate virtual control points, which are then used to optimize RPC parameters via the RFM model, achieving the required accuracy for 1:50,000 mapping.

4.2.1.1 Methodology

To achieve automatic high-precision geometric orientation of satellite images using LiDAR data, the original LiDAR point cloud is first transformed into a depth map. Then, the RPC parameters from the optical imagery are used to achieve an initial alignment between the LiDAR depth map and the satellite image through the RFM. To optimize matching efficiency and minimize image resolution loss, a block-based strategy divides the image into $n \times n$ regions. Feature values are extracted from each block using the Sobel operator, and template matching with AWOG descriptors is applied to refine the matching points, yielding accurate control points. These control points are then used to recover geographic coordinates. Finally, an iterative least squares adjustment based on the Rational Function Model (RFM) is performed, using both the LiDAR point cloud and optical satellite images as inputs to obtain optimal orientation parameters, achieving precise image orientation for the optical satellite imagery.

4.2.1.2 Data preprocessing

First, geographic coordinates from optical satellite image metadata and corresponding LiDAR data are used to stitch the LiDAR data into a complete dataset. A depth image is then generated, where the grayscale value of each pixel represents the distance from the camera to a point in the scene, capturing the geometric shape and height of the object.

Due to scale and rotation differences between the LiDAR depth map and optical imagery, the original orientation information from the optical image is extracted as RPC parameters. These are used to transform the LiDAR depth map pixel coordinates into ground points, and the RFM is employed to reproject these points onto the optical image, creating an affine transformation model. Resampling of the satellite imagery based on this model roughly mitigates the scale and rotation discrepancies between the LiDAR and optical data.

4.2.1.3 Control points acquisition

1) Feature Extraction and Description

To balance matching efficiency and accuracy, a block-based strategy is employed, dividing the image into $n \times n$ regions. For each block, feature values are extracted using the Sobel operator, and the points with the largest feature values are selected as feature points.

Due to modal differences between the LiDAR depth map and optical satellite imagery, significant nonlinear radiometric distortions exist between the images. To address this, an AWOG descriptor is used, effectively mitigating these grayscale differences. AWOG leverages image gradients to represent image structures, which efficiently counteract variations in illumination and contrast. Furthermore, gradients can be quickly extracted, improving the rate of correct matches while minimizing the impact on matching efficiency.

Firstly, the Sobel operator is used to obtain the gradients $g_x(x, y)$ and $g_y(x, y)$ of the image in the x and y directions. The amplitude $\mathrm{IG}(x, y)$ and the direction $\mathrm{OG}(x, y)$ of the image gradient are then calculated as follows:

$$\mathrm{IG}(x, y) = \sqrt{g_x(x, y)^2 + g_y(x, y)^2} \tag{4.1}$$

$$\mathrm{OG}(x, y) = \arctan^{-1}\left(\frac{g_y(x, y)^2}{g_x(x, y)^2}\right) \tag{4.2}$$

Next, the gradient directions are normalized to the range of 0° to 180° to enhance robustness. The orientation gradient is divided into 8 intervals with 22.5-degree angular spacing, resulting in 9 distinct directions. The angle-weighted features are then obtained using the following formula:

$$\Delta_1(x, y) = \mathrm{OG}(x, y) - \left\lfloor \frac{\mathrm{OG}(x, y)}{22.5} \right\rfloor \times 22.5 \tag{4.3}$$

$$\Delta_2(x, y) = 22.5 - \left\lfloor \frac{\mathrm{OG}(x, y)}{22.5} \right\rfloor \times 22.5 \tag{4.4}$$

$$\mathrm{WIG}_1 = \frac{\Delta_1(x, y)}{22.5} \times \mathrm{IG}(x, y) \tag{4.5}$$

$$\mathrm{WIG}_2 = \frac{\Delta_2(x, y)}{22.5} \times \mathrm{IG}(x, y) \tag{4.6}$$

Finally, for each pixel, a window with 3x3 pixels is used on WIG_1 and WIG_2 to compute weighted gradient magnitudes in the nine directions. The statistical results are then output in vector form.

2) Similarity Measure

Due to its multilevel structure, the AWOG descriptor requires considerable time for spatial domain matching. To enhance computational efficiency, the descriptor is converted to the frequency domain, significantly accelerating the matching process. The similarity between descriptors is measured using squared Euclidean distance, and the image matching function in the frequency domain is derived via the fast Fourier transform. The corrected optical image is then matched with the point cloud depth map, and the RANSAC algorithm is applied to eliminate unreliable matches.

4.2.1.4 Geometric orientation

The obtained matching points are used as control points, and the RFM-based affine transformation compensation model is used to improve the orientation accuracy of the original RPC [24]. The affine transformation model can be expressed as follows:

$$\begin{cases} x_o' = a_0 + a_1 x_o + a_2 y_o \\ y_o' = b_0 + b_1 x_o + b_2 y_o \end{cases} \tag{4.7}$$

where x_o and y_o are the column and row coordinates of the matching points on the target optical satellite image. The parameters a_i and b_i (where $i = 0, 1, 2$) are the coefficients to be determined by solving the affine transformation model. Specifically, a_0 and b_0 compensate the errors in the column and row directions of the image, while a_1 and b_2 account for image errors caused by internal orientation parameter inaccuracies.

To establish the initial corresponding position of the feature point on the optical satellite image relative to the reference LiDAR image, it is necessary to calculate the geographical coordinates of the feature points in terms of latitude and longitude. These coordinates can be obtained using the following formula:

$$\begin{cases} (\lambda^p, \phi^p) = \text{proj2geo}(x_{proj}^p, y_{proj}^p) \\ x_{proj}^p = x_{proj}^{ul} + \text{Re}_l^x \cdot x_l^p \\ y_{proj}^p = y_{proj}^{ul} + \text{Re}_l^y \cdot y_l^p \end{cases} \tag{4.8}$$

where x_l^p and y_l^p are the column and row coordinates of point p on the LiDAR depth map, respectively, and the index l represents the LiDAR scene. Re_l^x and Re_l^y represent the resolution of the reference LiDAR image in the column and row directions, respectively; $(x_{proj}^{ul}, y_{proj}^{ul})$ represents the projection coordinates of the upper-left corner point on the LiDAR reference image; (x_{proj}^p, y_{proj}^p) is the projection coordinate corresponding to point p;

$\text{proj2geo}(x^p_{proj}, y^p_{proj})$ represents a coordinate transformation function that maps point coordinates from the projection coordinate system to the reference geographic coordinate system; (λ^p, ϕ^p) is the latitude and longitude corresponding to point p.

Once the corresponding geographic coordinates of the feature points on the LiDAR depth map are obtained, the laser altimetry data from the LiDAR dataset can be utilized as the elevation for the corresponding points. Using the RPC of the optical image, these points are projected onto the optical image to determine the corresponding image coordinates. The projection process can be expressed using the following formula:

$$\begin{cases} \overline{x^p_{oc}} = RPC_x\left(\overline{\lambda^p}, \overline{\phi^p}, \overline{h^p}\right) \\ \overline{y^p_{oc}} = RPC_y\left(\overline{\lambda^p}, \overline{\phi^p}, \overline{h^p}\right) \end{cases} \tag{4.9}$$

where $\left(\overline{x^p_{oc}}, \overline{y^p_{oc}}\right)$ is the normalized pixel coordinates of p calculated by the RFM model, where the index o denotes the optical image scene and the index c denotes the calculated result. $\left(\overline{\lambda^p}, \overline{\phi^p}, \overline{h^p}\right)$ are the normalized object coordinates of p points.

By combining (4.8) and (4.9), the following regional network adjustment model can be obtained.

$$\begin{cases} v^p_x = -x^p_{om} + x^p_{oc} + \Delta x \\ v^p_y = -y^p_{om} + y^p_{oc} + \Delta y \end{cases} \tag{4.10}$$

where x^p_{om} and y^p_{om} are the column and row coordinates of the point p obtained by image matching on the optical image, and the index m is obtained by matching. x^p_{oc} and y^p_{oc} are the nonnormalized column and row coordinates of point p on the optical image calculated forward by the RFM model.

Finally, the optimal estimation of the affine transformation parameters a_i and b_i are obtained through iterative least squares adjustment. This process minimizes the residual errors between the transformed optical image coordinates and the reference LiDAR coordinates, ensuring the best fit. Once the parameters are determined, the affine transformation matrix is constructed.

The optical satellite image to be positioned is then processed using the affine transformation. After applying the transformation, the RPC parameters are refined to achieve precise orientation. With these updated RPC parameters, high-precision positioning of the LiDAR-assisted satellite images is successfully completed. This approach ensures that the optical satellite image is accurately aligned with the LiDAR data, enabling reliable integration for further analysis. The detailed workflow is given in Fig. 4.5.

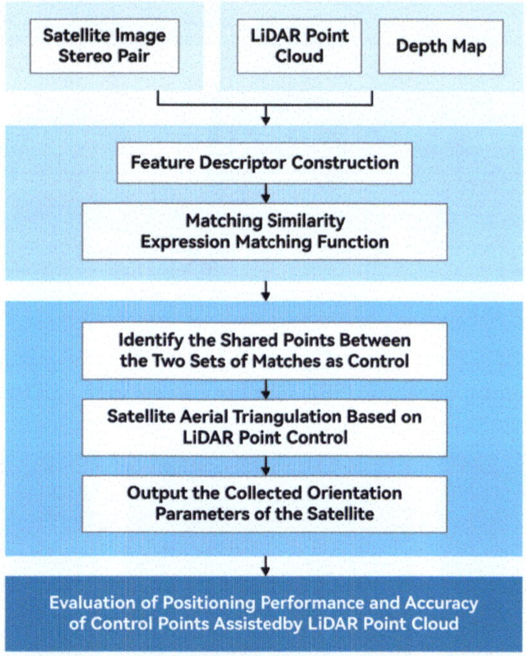

Figure 4.5:
Workflow of Automatic Geometric Positioning of Optical Satellite Images Using LiDAR Data.

4.2.1.5 Experiments and analysis

4.2.1.5.1 Experimental area and data description

Experiments were conducted in the eastern coastal area of Nantong City, Jiangsu Province, to validate the effectiveness of the automatic orientation method for LiDAR data and optical satellite images. The study area, shown in Fig. 4.6(a), is flat, with a low elevation and a height difference of about 10 meters. The experiments used 0.8-meter resolution panchromatic images from the GF7 satellite (captured at 11:00:36 on April 21, 2022), with a strip-type airborne laser point cloud as the reference image. The point density is 20 cm, and the depth map's spatial resolution is 1 meter. The LiDAR data rendering is shown in Fig. 4.6(b).

4.2.1.5.2 Results and analysis

Several parameters were set for the experiment. For the AWOG method, the values of m, n, and the template window size were set to 3, 8, and 71 × 71 pixels, respectively. The number of RANSAC iterations and the total error threshold were set to 2000 and 3 pixels. After matching the satellite image with the LiDAR depth image, over 90 virtual control points were obtained. The distribution of these points on the LiDAR depth map, forward-looking GF-7

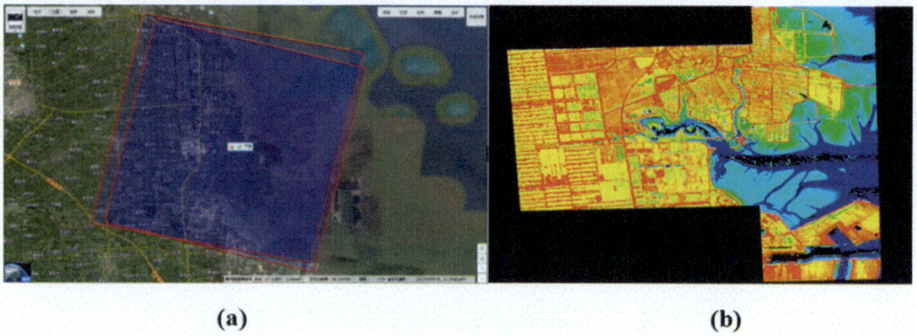

Figure 4.6:
Experimental data. (a) GF7 satellite imagery; (b) LiDAR data.

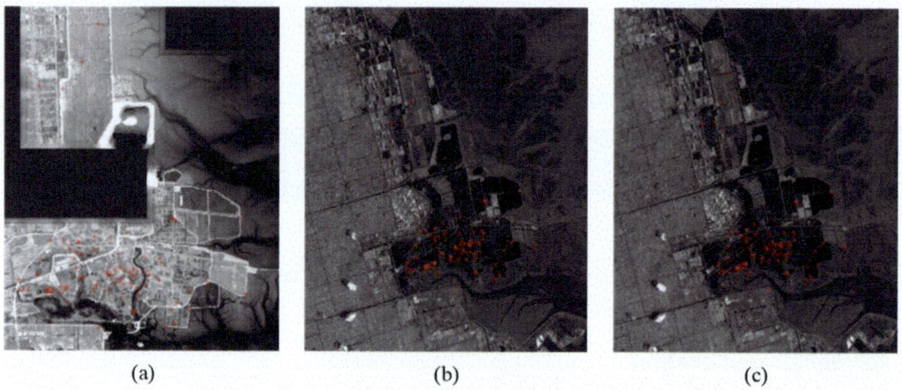

Figure 4.7:
The distribution of obtained corresponding points. (a) LiDAR depth map; (b) the forward-looking GF-7 image; (c) the backward-looking GF-7 image.

image, and backward-looking GF-7 image is shown in Fig. 4.7. A zoomed-in view of three points from Fig. 4.7 is presented in Fig. 4.8.

After the automatic matching calculation and point selection of LiDAR images, a large number of matching points were obtained, and they were evenly distributed across the image. The wide coverage of these points makes them effective control points, ensuring the accuracy and reliability of image orientation. The mean square error after orientation is 1.79 pixels. Based on the statistical results from the orientation iterations, 12 ground control points were selected for adjustment. Table 4.2 shows the errors of the ground control points. Most of the errors are within 1 meter, with the X and Y direction errors ranging from −1 to 1 meter. The RMSE in the X direction is 0.75 meters, and in the Y direction, it is 0.7 meters, both of which are relatively small. The error range in the Z direction is between −4 and 4 meters,

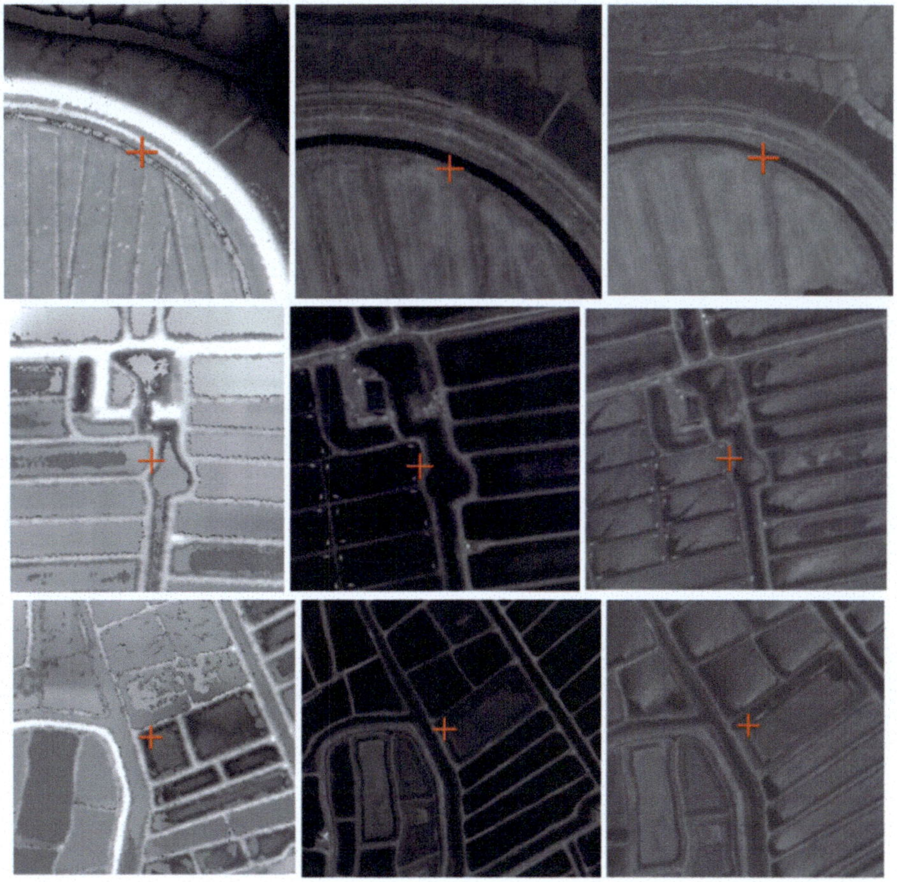

Figure 4.8:
The Zoom-in of three points of Fig. 4.7. The first, second, and third lines are the LiDAR depth image, forward-looking GF-7 image, and backward-looking GF-7 image, respectively.

with an RMSE of 2.6 meters. Although the Z-direction error is slightly larger than the X and Y directions, it remains below 5 meters, meeting the accuracy requirements for 1:50,000 scale topographic mapping. The errors of each ground control point are relatively consistent, and the high accuracy indicates that the results of auxiliary positioning are stable and reliable.

Table 4.3 presents the orientation errors of the GF7 optical satellite image after adjustment. The RMSE in the X and Y directions for both the forward-looking and backward-looking images is approximately 1.2 meters, with the maximum error in both directions being less than 3 meters. Due to the availability of accurate ground control point data, the positioning error of the optical satellite images is relatively small, resulting in high positioning accuracy.

Table 4.2: The Residual Errors of Selected Points.

GCP	X	Y	Z
1	0.80	−0.65	−3.86
2	−0.58	1.54	1.52
3	0.69	−0.79	0.31
4	−0.63	0.49	2.88
5	−0.86	0.47	2.97
6	−0.54	0.31	4.29
7	0.20	0.13	−0.09
8	−0.99	−1.26	−1.39
9	1.16	0.29	0.91
10	−0.64	0.10	0.17
11	0.90	−0.46	−4.08
12	0.48	−0.19	−3.01
RMSE	0.75	0.7	2.60

Table 4.3: The Mean Error of Optical Satellite Image Orientation.

IMAGE	RMSE		MAX	
	X	Y	X	Y
FWD	1.37	1.31	2.90	2.56
BWD	1.21	1.14	2.71	2.43

4.2.2 Superresolution reconstruction of sequential remote sensing images

The spatial resolution of remote sensing images is crucial for evaluating image quality and is mainly determined by sensor precision. However, enhancing sensor performance often incurs high costs due to limitations in sensor production, system expenses, and launch payload capacity. Superresolution reconstruction offers a solution by processing low-resolution images to generate high-resolution results. Despite advancements, adding high-frequency details remains difficult due to wide coverage, complex terrain, and surface features.

Superresolution methods for image sequences are categorized into frequency-domain and spatial-domain approaches. Frequency-domain methods, such as Fourier and Wavelet Transforms, improve resolution by eliminating spectrum aliasing and capturing high-frequency details. While fast and computationally efficient, they are limited by linearity and lack of prior knowledge. Spatial-domain methods, including interpolation, regularization, and learning-based approaches, offer greater flexibility by incorporating factors like motion and spatial priors. However, interpolation can cause distortions, while regularization stabilizes the process. Probabilistic methods are effective but complex, and learning-based methods, though improving, are usually limited to single images.

To address these challenges, a multiscale detail-enhancement superresolution model for remote sensing image sequences is introduced. This model aims to maximize texture detail and spatial resolution while balancing both microscopic and macroscopic information

4.2.2.1 Methodology

Given that the image sequence has already been registered, the process begins with applying a least square filtering method for multiscale decomposition. A reference image is selected based on subjective assessment. The maximum scale decomposition results of the sequence images are used as redundant information for superresolution reconstruction, capturing the common details shared across all images. The remaining observed images are processed by differencing them with the corresponding scale decomposition results of the reference image to extract nonredundant information, which highlights the detail differences between the images in the sequence.

Interpolation is then used to merge the redundant and nonredundant information to generate high-resolution image data. A detail enhancement function is constructed to emphasize medium and small-sized details and integrate this enhanced detail with smoothing information, producing the initial superresolution reconstruction result. To prevent undesired changes in grayscale values due to excessive enhancement, the result is further refined using a local optimization model. The overall technical process of the superresolution reconstruction method is illustrated in Fig. 4.9.

4.2.3 Multiscale decomposition and detail extraction

Multiscale decomposition involves breaking an image down into a smooth layer that captures large-scale edges and one or more detail layers that highlight medium to small-scale textures, by applying different filtering parameters. For this purpose, the least squares filtering is used to decompose the sequence of remote sensing images. The least squares filtering algorithm employs optimization techniques, with the objective function defined as:

$$\min_u \sum_p ((u_p - g_p)^2 + \lambda(\alpha_{x,p}\left(\frac{\partial u}{\partial x}\right)_p^2 + \alpha_{y,p}\left(\frac{\partial u}{\partial y}\right)_p^2)) \tag{4.11}$$

where u represents the input image, g denotes the output image, and p refers to a pixel point on the image. The term $(u_p - g_p)^2$ measures the similarity between the output image and the input image. The parameters $\alpha_{x,p}(g)$ and $\alpha_{y,p}(g)$ are the smoothing weights, while λ is

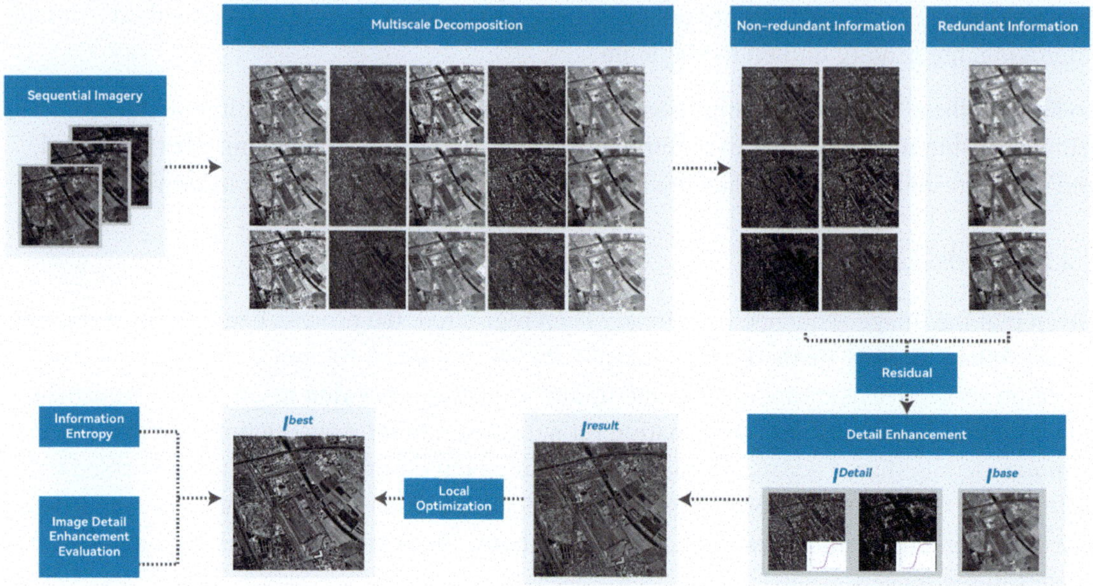

Figure 4.9:
Flowchart of Superresolution Reconstruction Method with Multiscale Detail Enhancement.

a parameter that balances the two data terms. As the value of λ increases, the image becomes smoother. Finding the optimal solution for the objective function involves determining the output image g that minimizes this function. Eq. (4.11) can be transformed into matrix form as follows:

$$\min_{u}(u-g)^T(u-g)+\lambda(u^T\mathbf{D}_x^T\mathbf{A}_x\mathbf{D}_x u+u^T\mathbf{D}_y^T\mathbf{A}_y\mathbf{D}_y u) \tag{4.12}$$

In this equation, $\mathbf{D}_x$ and $\mathbf{D}_y$ are diagonal matrices composed of differential operators, while $\mathbf{A}_x$ and $\mathbf{A}_y$ are diagonal matrices containing $\alpha_{x,p}(g)$ and $\alpha_{y,p}(g)$ respectively. To find the minimum value of Eq. (4.12), we obtain the linear system of equations as shown in Eq. (4.13):

$$\left(I+\lambda\mathbf{L}_g\right)u=g \tag{4.13}$$

where $\mathbf{L}_g=\mathbf{D}_x^T\mathbf{D}_x+\mathbf{D}_y^T\mathbf{D}_y$ represents the standard Laplacian matrix. This indicates that by setting different filter parameters, various scale decomposition results can be obtained. To achieve smoother results at higher scales, filter parameters that enhance image smoothness need to be applied.

Therefore, by applying the least squares filtering method to the sequence of remote sensing images $Image_1, Image_2, \ldots, Image_i$ $(i=1,2,\ldots,n)$, a series of progressively smoothed image information is obtained. To achieve m progressively smoothed images, the least squares filtering method must be applied m times. For example, to construct a multiscale decomposition result at level $j+1$ using $Image_1$ as the reference image, the filtering process produces a series of progressively smoothed images $u^1, \ldots, u^j$ $(j \geq 1)$. Here, u^j represents the smoothing information at the maximum scale, and concurrently, a set of detail information can be extracted.

$$d^j = u^{j-1} - u^j \tag{4.14}$$

where $i = 0, 1, \ldots, m$, $u^0 = Image_1$.

Using the sequence of images as experimental data, let $u^{(i,j)}$ represent the different scales of smoothing information obtained through multiple applications of the least squares filtering method. Let $d^{(i,j)}$ denote the detail information derived from differencing after multiscale decomposition. To fully leverage the nonredundant information among the sequence of images, differencing is performed between the decomposition results at corresponding scales of the observed images and the reference image, yielding nonredundant information at various scales, i.e., $d^{(i,j)} = u^{(i,j-1)} - u^{(1,j)}$. Interpolation methods are then used to obtain a series of high-resolution smoothing information $U^{(i,j)}$ and detail information $D^{(i,j)}$. For simplicity, the reference image $Image_1$ will be used as an example in the following explanations.

4.2.3.1 Detail enhancement

To prevent texture loss in the reconstructed images and to address texture detail issues, a detail enhancement function $S(\beta, D^j)$ is developed based on multiscale decomposition. The function $S(\beta, D^j)$ aims to enhance the medium and small-scale details extracted during the multiscale decomposition process.

$$S\left(\beta, D^j\right) = (2/(1 + exp(-\beta * D^j))) - 1 \tag{4.15}$$

In Eq. (4.15), the texture detail enhancement function $S\left(\beta, D^j\right)$ is defined as a hyperbolic tangent function, where $\beta * D^j$ represents a straightforward scalar multiplication. In this function, the parameter β is required to be a positive number, and D^j denotes the high-resolution detail information obtained through interpolation after multiscale decomposition. Taking the partial derivative of Eq. (4.15) yields:

$$\frac{\partial}{\partial D^j} S\left(\beta, D^j\right) = \frac{1}{2}\beta(1 - S^2(\beta, D^j)) \tag{4.16}$$

It can be demonstrated that when $\beta > 0$, $S(\beta, D^j)$ is an increasing function with respect to D^j. In other words, a larger β enhances the medium and small-scale details in the image more significantly. However, this also increases the risk of image distortion. Therefore, during the experimental process, a detail enhancement parameter β should be carefully defined by balancing it with objective evaluation metrics such as image entropy.

4.2.3.2 Integration and fusion

Image information derived from multiscale decomposition can be easily reconstructed by combining large-scale smoothing information with medium and small-scale detail information. However, even with the edge-preserving least squares filtering algorithm used during multiscale decomposition, residual strong edge information might still be present across different scales. If these strong edges are amplified equally during the detail enhancement process, the high-frequency regions of the image may become oversaturated during subsequent fusion. A simple cutoff approach would lead to significant information loss, while straightforward normalization might distort the image's grayscale value distribution. To effectively mitigate the overamplification of large-scale edges, a sigmoid weighting function f is used to minimize the sharpening effect of these edges.

$$f = \frac{1}{e^{-\theta D^j} + 1} \tag{4.17}$$

The function f is used to suppress the overamplification of large-scale edges, with θ being a constant that controls the extent to which strong edge boundaries are amplified. Based on the information enhancement, the multiscale decomposed image data is combined and fused using Eq. (4.14) to obtain the initial superresolution reconstruction result X_0

$$X_0 = I^{base} + I^{detail} = I^{base} + \sum_{j=1}^{m} D^j \tag{4.18}$$

where I^{base} represents the redundant information from the sequence of images, I^{detail} refers to the nonredundant information, and D^j denotes the high-resolution detail information obtained through interpolation after multiscale detail enhancement.

4.2.3.3 Local detail optimization

Since global constraints may not effectively preserve texture details and sharpened edges in the image, and to prevent significant deviations in grayscale and brightness values after detail enhancement, a local optimization model is proposed. Assuming the reference image is $l \times h$ times $l \times h$, the reconstructed image size is $2l \times 2h$ times $2l \times 2h$. A local optimization window is selected, and downsampling is performed on this window. The downsampled window is then compared to the corresponding window in the reference image, and the

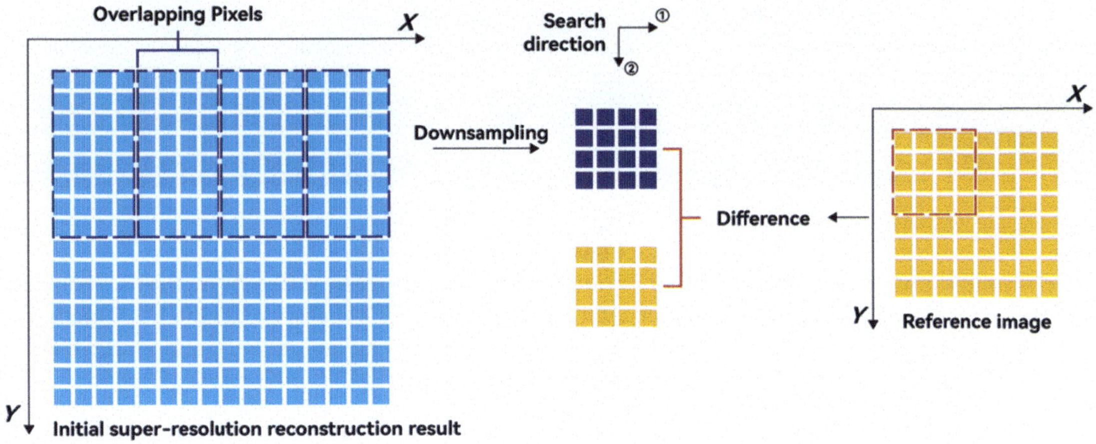

Figure 4.10:
Local Optimization.

absolute sum of the grayscale differences is calculated. If this sum exceeds a preset threshold, the local optimization model is applied iteratively. Gradient descent is used to address the minimization problem defined in Eq. (4.19), and the initial superresolution reconstruction result is updated according to Eq. (4.19). The goal, as illustrated in Fig. 4.10, is to ensure the final result closely matches the reference image.

$$\widehat{X}^{sub} = \arg\min_{X^{sub}} \left\| sX^{sub} - Y^{sub} \right\|_2^2 + \mu \left\| X^{sub} - X_0^{sub} \right\|_2^2 \tag{4.19}$$

where Y^{sub} represents the local optimization subwindow from the reference image, X^{sub} denotes the local optimization subwindow of the candidate high-resolution image, and X_0^{sub} is the local optimization subwindow of the initial superresolution reconstruction. The parameter s indicates the downsampling factor, while μ represents the smoothing scale parameter.

4.2.3.4 Experiments and analysis

To validate the effectiveness of the proposed superresolution reconstruction method, experiments were conducted using both simultaneous and multitemporal datasets. The multiscale detail enhancement superresolution reconstruction framework was applied to achieve 2x image upscaling. The results were then compared with those obtained using bicubic interpolation [25], the Total Variation (TV) method [26], and the Maximum A Posteriori (MAP) method [27].

Experiment 1: Multitemporal panchromatic remote sensing images were utilized for the superresolution reconstruction. The objective was to compare the effectiveness of various

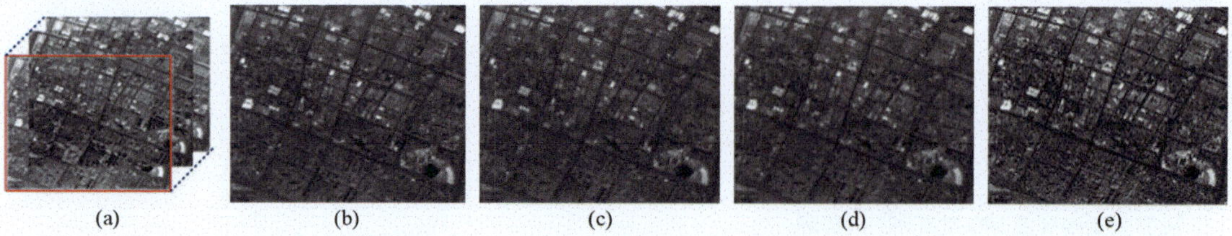

Figure 4.11:
Superresolution reconstruction results of remote sensing images using different methods. From left to right: (a) sequence images; (b) bicubic interpolation; (c) the TV method; (d) the MAP method; (e) the superresolution reconstruction method for remote sensing images with multiscale detail enhancement.

superresolution methods in enhancing the resolution of remote sensing images. The results, shown in Fig. 4.11, demonstrate a clear distinction in performance across the different methods.

In Fig. 4.11(a), the reference image is outlined in red, providing a baseline for comparison. Fig. 4.11(b) shows the result of bicubic interpolation, a common method for image upscaling. Although this method offers a reasonable result, it fails to capture finer details and textures, leading to relatively smooth and less detailed output. Fig. 4.11(c) illustrates the result using the TV method, which is known for preserving edges while reducing noise. This method achieves better results than bicubic interpolation, but still struggles with finer details, particularly in areas where textures are complex. The MAP method, presented in Fig. 4.11(d), uses a probabilistic approach to estimate the most likely high-resolution image given the observed low-resolution data. While this method outperforms bicubic interpolation and TV in terms of preserving details, it still falls short in terms of texture sharpness and overall visual quality when compared to the multiscale detail enhancement method.

Finally, Fig. 4.11(e) demonstrates the result from the multiscale detail enhancement superresolution reconstruction method. This method stands out by providing superior image quality, especially in terms of texture enhancement and fine detail recovery. The method effectively integrates both macroscopic and microscopic information across multiple scales, offering a significant improvement in visual quality compared to the other methods. It enhances small details without sacrificing the overall image coherence, providing a more natural and accurate representation of the remote sensing data. Fig. 4.12 further confirms the superiority of the multiscale detail enhancement method, showing composite images of all four methods. The method's ability to preserve and enhance both global and local features is evident, making it the most effective approach for superresolution reconstruction in remote sensing applications. Overall, the multiscale detail enhancement method not only outperforms other techniques in terms of image quality and texture enhancement but also proves to be

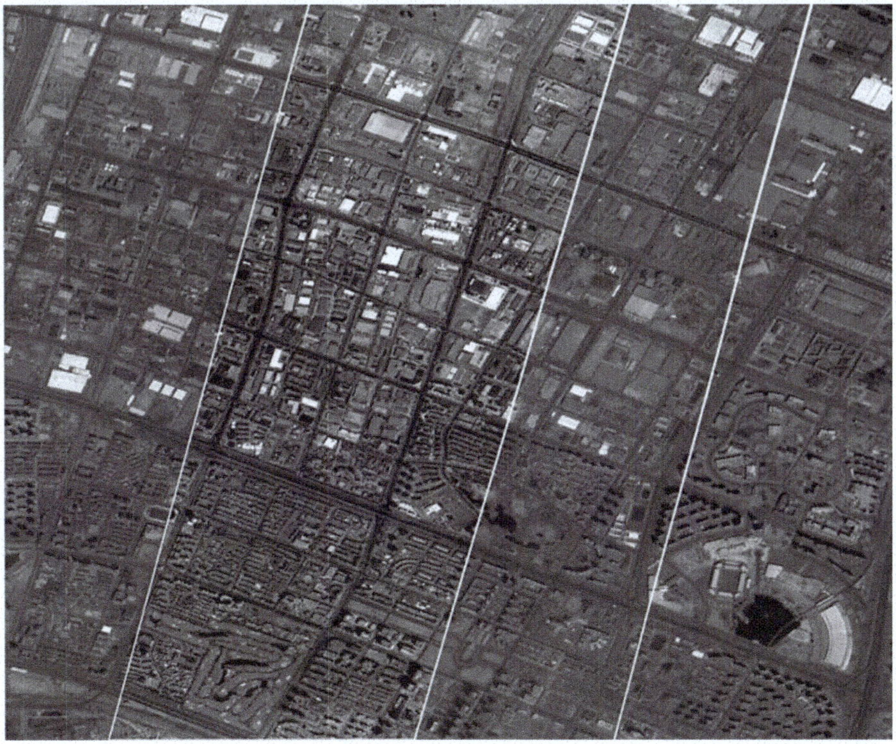

Figure 4.12:
Synthetic images of different reconstruction methods. From left to right: bicubic interpolation; the superresolution reconstruction method for remote sensing images with multiscale detail enhancement; the TV method; the MAP method.

highly effective in handling complex terrain and varying surface features commonly found in remote sensing images.

Experiment 2: In this experiment, simultaneous panchromatic three-line array remote sensing images were used for superresolution reconstruction. The results are presented in Fig. 4.13. Fig. 4.13(a) is the reference image outlined in red. Fig. 4.13(b) to Fig. 4.13(e) show the superresolution reconstruction results using different methods. Fig. 4.14 provides a local zoom of the yellow box area from Fig. 4.13, displaying the reconstruction at a 1:1 resolution.

The multiscale detail enhancement method significantly improves texture detail, making land features more pronounced and better preserving edge structures compared to other methods. From a subjective visual perspective, Fig. 4.13(e) is noticeably clearer than Fig. 4.13(b) to Fig. 4.13(d). This demonstrates that the multiscale detail enhancement method not only

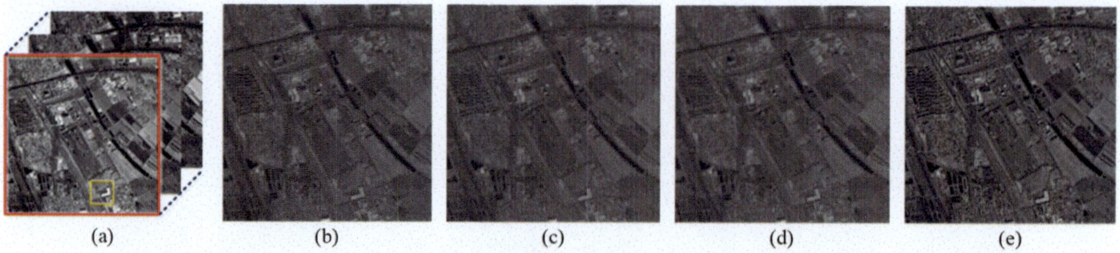

Figure 4.13:
Superresolution reconstruction results of remote-sensing images in the same phase (using different algorithms). From left to right: (a) sequence images; (b) bicubic interpolation; (c) the TV method; (d) the MAP method; (e) the superresolution reconstruction method for remote sensing images with multiscale detail enhancement.

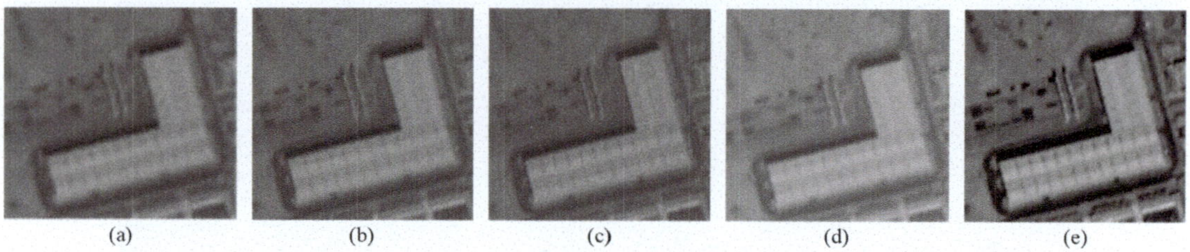

Figure 4.14:
Local magnified display. From left to right: (a) reference image; (b) bicubic interpolation; (c) the TV method; (d) the MAP method; (e) the superresolution reconstruction method for remote sensing images with multiscale detail enhancement.

enhances the spatial resolution of the image but also better preserves edge structures and improves the fine details of the features within the image.

In terms of subjective visual effects, Fig. 4.14(e) is significantly clearer than Fig. 4.14(b) to Fig. 4.14(d), indicating that while improving the spatial resolution of the image, it not only preserves the edge structure information of the image but also improves the fineness of the target scenes in the image.

Compared to the results from Experiment 1, those from Experiment 2 are significantly better. This indicates that simultaneous superresolution reconstruction performs better than multitemporal reconstruction, demonstrating greater robustness with temporally consistent image data. Consequently, for sequence image data, it's preferable to select images from the same time period. When using multitemporal images, choose those from similar time points to minimize changes in land features, ensuring that the reconstructed images are both meaningful and valuable. Compared to bicubic interpolation, TV method, and MAP method, this approach not only provides clearer visual results but also better preserves sharp edges and enriches texture information.

4.2.4 Building extraction based on RS images and LiDAR point clouds

Automatic and accurate building extraction from remotely sensed data is a key challenge in remote sensing, with applications in energy estimation, urban planning, and disaster management. Traditional methods rely on spectral and texture features but struggle with the geometric complexity and heterogeneity in high-resolution images. LiDAR point clouds, offering height and shape information, are more effective for building extraction but require accurate vegetation recognition and postprocessing to improve accuracy.

This section introduces a building extraction method that combines orthorectified remote sensing images with LiDAR data, assuming precise alignment. The method leverages LiDAR height data and image-based vegetation recognition to improve automation and accuracy. Additionally, image features are used to refine the initial building mask, focusing on flat terrains, which are common in urban and rural areas.

4.2.4.1 Methodology

Given that the LiDAR and image data have been precisely reregistered, the approach utilizes the height information from LiDAR data along with vegetation details identified from remote sensing images. This integration enhances both the automation and accuracy of building extraction processes, particularly in relatively flat terrains, which are common in urban and rural areas. The introduced method consists of two main stages: the generation of the initial building mask and the subsequent refinement of the building mask. The performance is demonstrated using the first test area of the Vaihingen dataset from the ISPRS.

4.2.4.2 Initial building mask generation based on vegetation recognition

This stage involves two key steps: generating a normalized digital surface model (nDSM) from LiDAR point clouds and removing vegetation points identified in remote sensing images. The nDSM is created by extracting nonground points from LiDAR data using tools like LAStools, followed by interpolating LiDAR points to generate a digital surface model (DSM) and a digital elevation model (DEM). Subtracting the DEM from the DSM results in the nDSM. A moment-preserving method is applied to binarize the nDSM, filtering out remaining ground points and small objects.

Vegetation removal is critical for accurate building extraction. Spectral information from the near-infrared (NIR) and red bands of remote sensing images is used to compute the

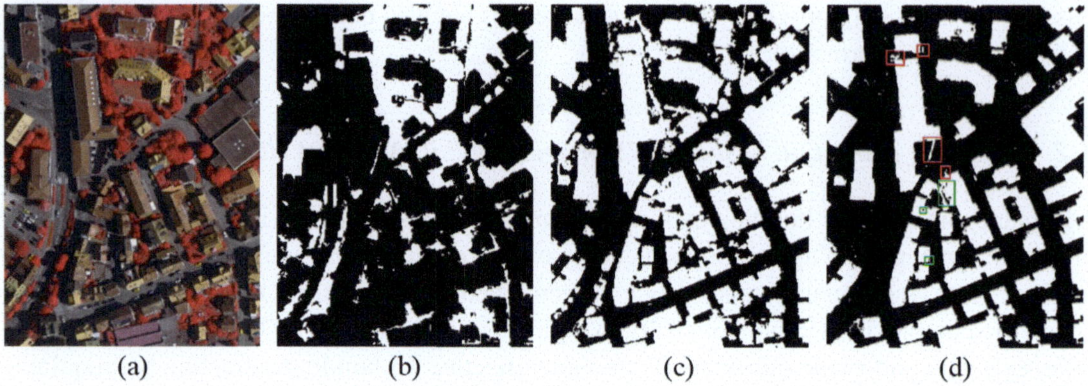

Figure 4.15:
Illustration of the generation of the initial building mask (using the first test area of the Vaihingen dataset from the ISPRS as an example). (a) The remote sensing image. (b) Vegetation recognition result. (c) The nDSM. (d) The initial building mask (the nDSM with recognized vegetation removed). The detected vegetation regions in (b), building regions in (d), and nonground points of the nDSM in (c) are marked in white.

Normalized Difference Vegetation Index (NDVI). The nDSM is resampled to match the spatial resolution of the remote sensing image. Modified histogram subtraction (SH) methods, such as eSH or iSH introduced in [28], are employed, using NDVI and NDGI pairs for vegetation detection in sunlit areas, with additional criteria for shadowed regions.

The resulting initial building mask (as shown in Fig. 4.15), after vegetation removal, preserves the main building structures but may include inaccuracies due to misclassified vegetation or imperfect LiDAR data. These issues are addressed in the refinement stage.

4.2.4.3 Building mask refinement based on consistency constraints by image features

Existing refinement methods often rely on LiDAR-derived features, which may introduce inaccuracies. In contrast, the presented method uses image-derived boundaries through region matching, offering greater stability and accuracy than boundary matching alone.

The refinement process is based on the principle that consistent image features (spectral and texture) should characterize each building or part of a building. The image is segmented using three algorithms: graph-based image segmentation (GS) [29], simple linear iterative clustering (SLIC) [30], and entropy rate superpixel segmentation (ERS) [31]. These segments are then matched with the initial building mask, replacing inaccurate LiDAR boundaries with more precise image-derived ones.

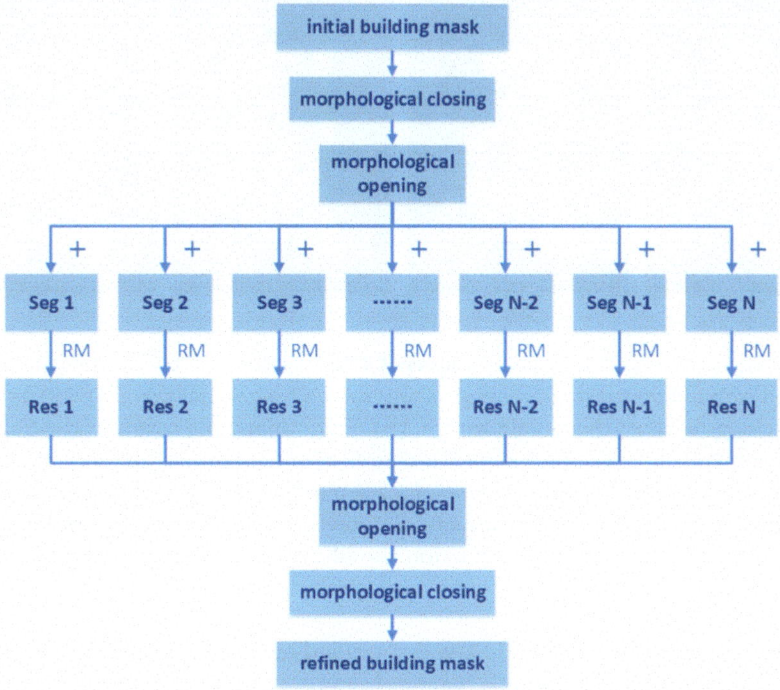

Figure 4.16:
The workflow of building mask refinement based on consistency constraints using image features is as follows: Seg refers to image segmentation obtained by an algorithm, Res denotes the region matching result, and RM stands for region matching. This process refines the initial building mask by leveraging consistent image features for more accurate boundaries.

Fig. 4.16 presents the workflow of the introduced building-mask refinement method. Specifically, the workflow involves performing a closing operation on the initial mask to recover missing building points, segmenting the image, and then conducting region matching. Multiple segmentation algorithms are used to ensure comprehensive coverage and accuracy. The final refined mask is obtained by combining the matching results, followed by morphological operations and size filtering to remove remaining nonbuilding points. A typical building mask refinement result on the Vaihingen dataset (ISPRS) is demonstrated in Fig. 4.17.

4.2.4.4 Experiments and analysis

To validate the effectiveness of the developed method, it was compared with three state-of-the-art methods, DeepLabv3+ [32], U-Net [33], and HOA [34], both qualitatively and quantitatively. DeepLabv3+ and U-Net were selected due to their popularity in deep learning-

Figure 4.17:
The illustration of building mask refinement using the Vaihingen dataset (ISPRS) as an example: (a) region matching result of the GS algorithm, (b) region matching result of the SLIC algorithm, (c) region matching result of the ERS algorithm, and (d) the union of the three results after open operation and region size filtering. The results are overlaid with ground truth, where yellow indicates correct detection, red represents false detection (overdetection), and green denotes missing detection (underdetection).

based semantic segmentation, while HOA was included because it is a recently developed unsupervised approach that also removes vegetation points from nDSM and refines the initial building mask. For convenience, the introduced methods using eSH or iSH will be named biSH+CCIF and iSH+CCIF, respectively.

The Vaihingen dataset from the ISPRS [35] which includes orthorectified remote sensing images and corresponding LiDAR point clouds, was used. The images have three bands: green, red, and near-infrared, from which NDVI and NDGI can be computed. The test area (Fig. 4.18(a)) is predominantly composed of historic buildings with complex shapes.

The comparison results are presented in Fig. 4.18, with the nDSM included for better analysis, which also applies to the other two test areas. From the results, it can be seen that all compared methods (except the nDSM) produce good results, but they also show some overdetected and underdetected regions. Deciding which method performs better is difficult when only considering overall results, as each method performs better in some regions and worse in others. To clarify the differences, Fig. 4.19 highlights the ultra overdetected regions with magenta rectangles, ultra missing regions with white rectangles, and areas where a method outperforms the biSH+CCIF method with blue rectangles.

In Fig. 4.19(a), for overdetection, compared to the biSH+CCIF method, DeepLabv3+ misclassifies a large vegetated region (red region in the first magenta rectangle) and a car (red region in the second magenta rectangle) as buildings. Additionally, overdetection primarily occurs at the building boundaries. For underdetection, DeepLabv3+ misses two large building

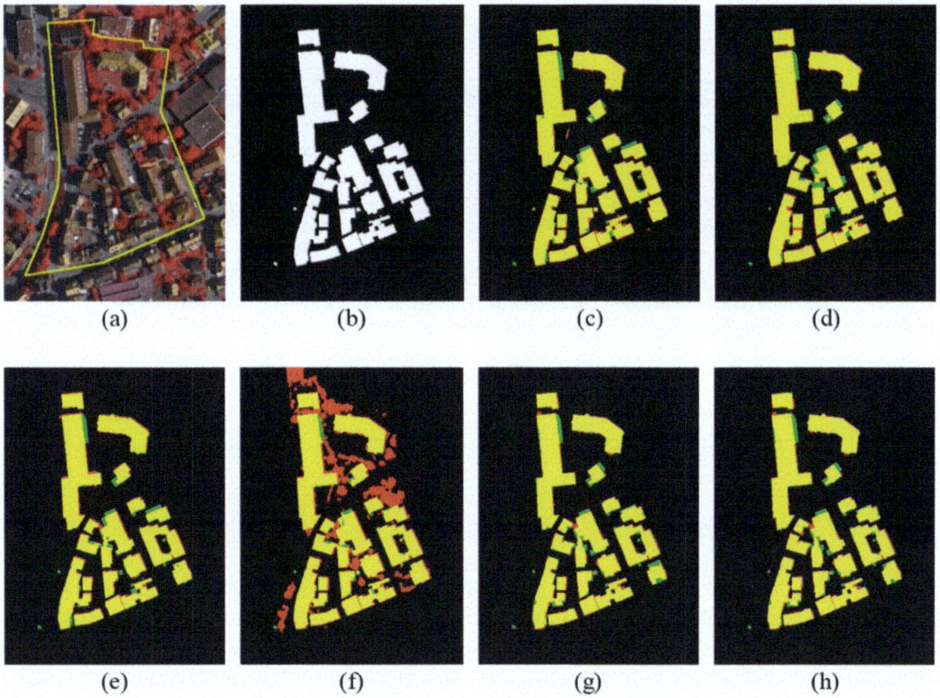

Figure 4.18:
The building extraction results of the compared methods on the Vaihingen dataset. (**a**) The corresponding orthorectified image (range of the test area has been marked by the yellow polyline). (**b**) The ground truth released by the ISPRS (white regions denote the buildings). (**c**) The result of DeepLabv3+. (**d**) The result of U-Net. (**e**) The result of HOA. (**f**) nDSM (temporarily supposing it's a building extraction result). (**g**) The result of the proposed iSH+CCIF method. (**h**) The result of the proposed biSH+CCIF method. In (**c**)-(**h**), yellow regions denote correct detection, red regions denote false detection (overdetection) and green regions denote missing detection (underdetection).

parts (green regions in the first two white rectangles) and smaller building parts (green regions in the third and fourth white rectangles). However, DeepLabv3+ performs better than the biSH+CCIF method in some regions, marked by blue rectangles. The comparison with U-Net and HOA shows similar results, with both methods also having overdetection primarily at building boundaries, which is more significant than DeepLabv3+'s errors. As these methods produce larger error areas, the biSH+CCIF method is considered to perform slightly better. Comparisons between biSH+CCIF and iSH+CCIF methods show very similar results. Therefore, the comparison of DeepLabv3+, U-Net, and HOA with the iSH+CCIF method mirrors the results seen with the biSH+CCIF method, which is why only the comparison results with the biSH+CCIF method are presented for clarity.

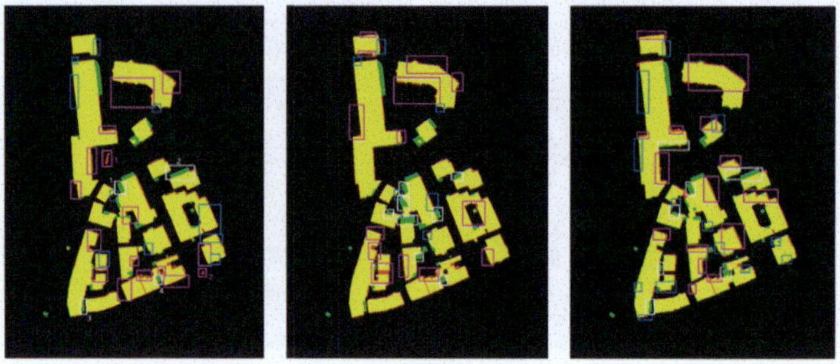

(a) (b) (c)

Figure 4.19:
Rectangle marked comparison between the biSH+CCIF method with **(a)** DeepLabv3+, **(b)** U-Net, and **(c)** HOA. The comparison results were marked on the results of the three compared methods.

References

[1] J. Li, D. Hong, L. Gao, et al., Deep learning in multimodal remote sensing data fusion: a comprehensive review, International Journal of Applied Earth Observation and Geoinformation 112 (2022) 102926.

[2] Z. Li, Y. Liu, Y. Sun, et al., A review of developments in the theory and technology of three-dimensional reconstruction in digital aerial photogrammetry, Acta Geodaetica et Cartographica Sinica 51 (7) (2022) 1437.

[3] M. Dalla Mura, S. Prasad, F. Pacifici, et al., Challenges and opportunities of multimodality and data fusion in remote sensing, Proceedings of the IEEE 103 (9) (2015) 1585–1601.

[4] Y. LeCun, Y. Bengio, G. Hinton, Deep learning, Nature 521 (7553) (2015) 436–444.

[5] I. Goodfellow, J. Pouget-Abadie, M. Mirza, et al., Generative adversarial networks, Communications of the ACM 63 (11) (2020) 139–144.

[6] L. Bertinetto, J. Valmadre, J.F. Henriques, et al., Fully-convolutional Siamese networks for object tracking, in: Computer Vision – ECCV 2016 Workshops, Amsterdam, the Netherlands, October 8-10 and 15-16, 2016, Proceedings, Part II 14, Springer International Publishing, 2016, pp. 850–865.

[7] K. Han, A. Xiao, E. Wu, et al., Transformer in transformer, Advances in Neural Information Processing Systems 34 (2021) 15908–15919.

[8] C.A. Aguilera, A.D. Sappa, R. Toledo, LGHD: a feature descriptor for matching across non-linear intensity variations, in: 2015 IEEE International Conference on Image Processing (ICIP), Quebec City, QC, Canada, 2015, pp. 178–181.

[9] W. Ma, Z. Wen, Y. Wu, et al., Remote sensing image registration with modified SIFT and enhanced feature matching, IEEE Geoscience and Remote Sensing Letters 14 (1) (2017) 3–7.

[10] Y. Xiang, F. Wang, H. You, OS-SIFT: a robust SIFT-like algorithm for high-resolution optical-to-SAR image registration in suburban areas, IEEE Transactions on Geoscience and Remote Sensing 56 (6) (2018) 3078–3090.

[11] J. Li, Q. Hu, M. Ai, RIFT: multi-modal image matching based on radiation-variation insensitive feature transform, IEEE Transactions on Image Processing 29 (2020) 3296–3310.

[12] Y. Yao, Y. Zhang, Y. Wan, et al., Multi-modal remote sensing image matching considering co-occurrence filter, IEEE Transactions on Image Processing 31 (2022) 2584–2597.

[13] J. Li, W. Xu, P. Shi, et al., LNIFT: locally normalized image for rotation invariant multimodal feature matching, IEEE Transactions on Geoscience and Remote Sensing 60 (2022) 1–14.

[14] Z. Fan, L. Zhang, Y. Liu, Nonlinear intensity, scale and rotation invariant matching for multimodal images, arXiv preprint, arXiv:2302.14239, 2023.

[15] Z. Hou, Y. Liu, L. Zhang, POS-GIFT: a geometric and intensity-invariant feature transformation for multimodal images, Information Fusion 102 (2024) 102027.

[16] Y. Deng, J. Ma, ReDFeat: recoupling detection and description for multimodal feature learning, IEEE Transactions on Image Processing 32 (2023) 591–602.

[17] Q. Wang, J. Zhang, K. Yang, et al., Matchformer: interleaving attention in transformers for feature matching, in: Proceedings of the Asian Conference on Computer Vision, 2022, pp. 2746–2762.

[18] H. Xie, Y. Zhang, J. Qiu, et al., Semantics lead all: towards unified image registration and fusion from a semantic perspective, Information Fusion 98 (2023) 101835.

[19] A. Zellner, Bayesian and non-Bayesian estimation using balanced loss functions, in: Statistical Decision Theory and Related Topics V, Springer New York, New York, NY, 1994, pp. 377–390.

[20] H.J. Zimmermann, Fuzzy set theory, Wiley Interdisciplinary Reviews. Computational Statistics 2 (3) (2010) 317–332.

[21] T. Meng, X. Jing, Z. Yan, et al., A survey on machine learning for data fusion, Information Fusion 57 (2020) 115–129.

[22] M. Wang, Y. Wei, Y. Pi, Geometric positioning integrating optical satellite stereo imagery and a global database of ICESat-2 laser control points: a framework and key technologies, Geo-Spatial Information Science 26 (2) (2023) 206–217.

[23] Z. Fan, L. Zhang, Q. Wang, et al., A fast matching method of SAR and optical images using angular weighted orientated gradients, Acta Geodaetica et Cartographica Sinica 50 (10) (2021) 1390.

[24] C.V. Tao, Y. Hu, A comprehensive study of the rational function model for photogrammetric processing, Photogrammetric Engineering and Remote Sensing 67 (12) (2001) 1347–1358.

[25] D. Zhou, X. Shen, Edge-directed bicubic color image interpolation, Acta Automatica Sinica 38 (4) (2012) 525–530.

[26] J.M. Fadili, G. Peyré, Total variation projection with first order schemes, IEEE Transactions on Image Processing 20 (3) (2010) 657–669.

[27] S. Villena, M. Vega, S.D. Babacan, et al., Bayesian combination of sparse and non-sparse priors in image super resolution, Digital Signal Processing 23 (2) (2013) 530–541.

[28] Y. Meng, Z. Hu, X. Chen, et al., Subtracted histogram: utilizing mutual relation between features for thresholding, IEEE Transactions on Geoscience and Remote Sensing 56 (12) (2018) 7415–7435.

[29] P.F. Felzenszwalb, D.P. Huttenlocher, Efficient graph-based image segmentation, International Journal of Computer Vision 59 (2004) 167–181.

[30] R. Achanta, A. Shaji, K. Smith, et al., SLIC superpixels compared to state-of-the-art superpixel methods, IEEE Transactions on Pattern Analysis and Machine Intelligence 34 (11) (2012) 2274–2282.

[31] M.Y. Liu, O. Tuzel, S. Ramalingam, et al., Entropy rate superpixel segmentation, in: CVPR 2011, IEEE, 2011, pp. 2097–2104.

[32] L.C. Chen, Y. Zhu, G. Papandreou, et al., Encoder-decoder with atrous separable convolution for semantic image segmentation, in: Proceedings of the European Conference on Computer Vision (ECCV), Springer, 2018, pp. 801–818.

[33] O. Ronneberger, P. Fischer, T. Brox, U-net: convolutional networks for biomedical image segmentation, in: Medical Image Computing and Computer-Assisted Intervention – MICCAI 2015: 18th International Conference, Munich, Germany, October 5–9, 2015, Proceedings, part III 18, Springer International Publishing, 2015, pp. 234–241.

[34] S. Chen, W. Shi, M. Zhou, et al., Automatic building extraction via adaptive iterative segmentation with LiDAR data and high spatial resolution imagery fusion, IEEE Journal of Selected Topics in Applied Earth Observations and Remote Sensing 13 (2020) 2081–2095.

[35] M. Cramer, The DGPF-test on digital airborne camera evaluation overview and test design, Photogrammetrie, Fernerkundung, Geoinformation (2010) 73–82.

CHAPTER 5

Classification algorithms for multimodal remote sensing

Haoyu Wang[a] and Xiaokang Zhang[b]

[a]School of Information and Control Engineering, China University of Mining and Technology, Xuzhou, China [b]School of Artificial Intelligence, Wuhan University, Wuhan, China

5.1 Reinforcement learning based Markov edge decoupled fusion network for fusion classification of hyperspectral and LiDAR

Hyperspectral image (HSI) can reveal both the spatial distribution of surface objects and their spectral reflection information, thus laying a solid foundation for fine-grained ground object identification [1–3]. LiDAR can provide elevation information, which enables models to distinguish ground objects with identical spectral characteristics but different elevations, such as concrete-built roads and roofs [4,5]. Complementary information can be acquired by fusing HSI and LiDAR data with distinct imaging mechanisms [6–8]. Therefore, multimodal data processing has become a prominent research topic [9,10], as it can effectively and rationally fuse such heterogeneous signals to enhance classification performance.

The acquisition of HSI relies on spectral imaging technology, which captures the continuous spectral response curve of ground objects. In contrast, LiDAR acquisition depends on laser pulse technology that measures the distance between ground objects and the sensor. This discrepancy in imaging principles results in the modal heterogeneity of multimodal data, posing notable challenges for multimodal data mining and fusion.

Challenge 1: Fully Mining Multimodal Data

Ground objects in the real-world exhibit irregular spatial topology structures, but existing fusion classification methods normally use regular feature extraction operators to mine multimodal data, leading to model performance limitations. Graph neural networks [11] provide the possibility to extract features of ground objects with irregular spatial topology structures, but their graph construction strategies are designed for a single modality. These graph construction strategies are unable to establish the information interaction between multimodal data [12], resulting in the potential relevance of multimodal data not being fully explored and integrated. In addition, these graph construction strategies are not customized

Multimodal Remote Sensing Fusion and Classification
https://doi.org/10.1016/B978-0-44-329152-4.00012-X

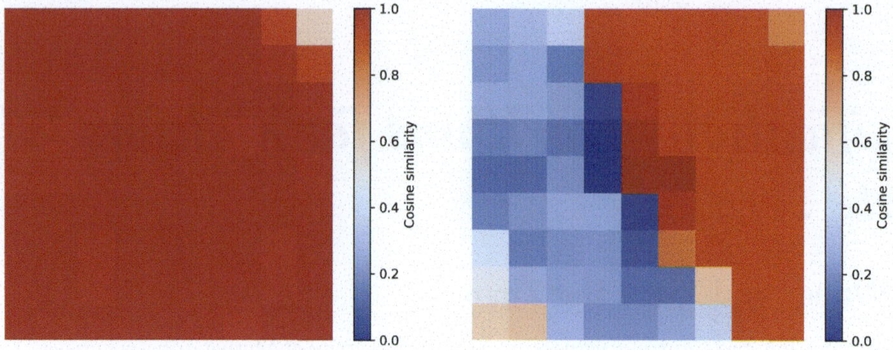

(a) Spatial topology of LiDAR **(b)** Spatial topology of HSI

Figure 5.1:
Inconsistent spatial topologies caused by modal heterogeneity. The higher the similarity, the stronger the class-discrimination information of the pixels in this region. Due to spectral noise in the signal, the left region of HSI exhibits low class discrimination.

for fusion classification tasks, making it difficult to ensure the task adaptability of the obtained graph.

Challenge 2: Reasonably Fusing Multimodal Data

Most existing fusion classification methods accomplish multimodal data fusion under the ideal assumption that the spatial topologies of HSI and LiDAR are fully consistent [13,14]. By calculating the cosine similarity between each pixel and the center pixel (pixel to be classified), we visualize the spatial topologies of two modalities in the same geographic region of a real dataset. As shown in Fig. 5.1, due to the existence of modal heterogeneity, HSI and LiDAR usually present inconsistent spatial topologies in the same geographic region. Therefore, the ideal assumption leads to interference with multimodal data fusion caused by pixels with low class discrimination in a certain modality (e.g., the left region in Fig. 5.1(b)).

To solve Challenge 1, reinforcement learning is considered. Reinforcement learning has agent communication mechanism and reward feedback mechanism, which can capture the potential relevance of multimodal data through collaborative cooperation among agents, providing a solution for graph construction in multimodal scenarios. However, reinforcement learning has not been used in multimodal data processing tasks, and this paper innovatively introduced the idea of reinforcement learning into multimodal graph construction. On the one hand, the multimodal agent communication module is used to establish an agent communication mechanism, capturing the potential relevance of modalities, and providing a new paradigm for the multimodal graph construction. On the other hand, state, action, and reward have been customized for multimodal graph construction tasks. By establishing a task oriented reward feedback mechanism, a multimodal graph that is beneficial for multimodal fusion classification tasks is learned, providing guidance for information exchange and integration between modalities.

To solve Challenge 2, the idea of "divide and conquer" is adopted. First, specific spatial graphs of two modalities are captured by agents separately. Then, the shared and unshared parts of two modalities are clearly defined. Finally, the differentiated fusion strategy for each part is designed, which reasonably utilizes the multimodal data while ensuring its purity. The main contributions are as follows:

1. To our knowledge, graph learning and reinforcement learning, are embedded in a unified multimodal data fusion framework for the first time. Through intelligent graph construction and relationship modeling, multimodal data is mined, and spatial decoupling fusion is used to integrate multimodal information.
2. A reinforcement learning graph network (RLGN) is proposed, which models multimodal graph construction tasks as a parametric partially observable Markov processes. The reward feedback mechanism and agent communication mechanism are customized for this process, providing a new paradigm for multimodal graph construction.
3. Breaking the ideal assumption of multimodal spatial topology sharing in existing fusion classification methods, a decoupled multimodal fusion module (DMFM) is proposed. The multimodal spatial graphs were coupled into shared and unshared parts, and differentiated fusion strategies were designed for each part, providing a new perspective for multimodal data fusion.
4. A multimodal edge attention module (MEAM) is proposed. Based on the characteristics of HSI and LiDAR, two different edge feature extraction methods were designed to capture the dependency relationships between nodes.

5.1.1 Related work

5.1.1.1 Multimodal data mining for multimodal remote sensing classification

Feature extraction operators used for multimodal data mining are generally manually designed [15] or based on deep learning [16]. Manually designed methods aim to help the model accomplish multimodal data fusion by designing feature extraction operators manually based on the inherent data characteristics. Roy et al. [17] used spatial morphology operators to capture information related to the height or shape of different land cover areas from LiDAR, which suppressed noise signal interference in spatial neighborhoods. Pedergnana et al. [18] effectively improved the accuracy of fusion classification by superimposing spatial morphological features from HSI and LiDAR. Deep learning-based methods aim to utilize the powerful feature extraction ability of neural networks to automatically learn high-level semantic features. Zhang et al. [19] used two-way automatic encoders to simultaneously reconstruct HSI and LiDAR, which integrated multimodal data and completed the fusion classification. Yang et al. [20] used the pixel-wise affinity branch to capture spatial topological relationships between different features from LiDAR data, achieving discriminative feature extraction. Xiu et al. [21] proposed a multisource attention network with discriminant graphs and information entities, which emphasized the correlation between pixels and aggregated spatial neighborhood context information, providing the high-quality feature representation for remote sensing classification tasks.

5.1.1.2 Multimodal data fusion for multimodal remote sensing classification

The multimodal data fusion strategy can effectively improve the remote sensing classification performance by utilizing complementary information. The multimodal data fusion strategy has three main forms: Pixel-level fusion [22], decision-level fusion [23], and feature-level fusion [24].

The pixel-level fusion methods aim to fuse multimodal data at the input stage, which integrates multimodal information at the pixel level. Morchhale et al. [25] constructed a new pixel set based on the HSI spectral response and LiDAR elevation information, then used CNN to achieve remote sensing classification. Zhao et al. [22] used Gram-Schmidt for preliminary pixel-level fusion of HSI and LiDAR. They utilized Schmidt forward and inverse transformations to promote interactions between spectral and spatial information. Due to the direct processing of high-dimensional original data, the pixel-level fusion method has a high demand on computational resources.

The decision-level fusion methods primarily integrate weighted integration of classification predictions from various modalities, thereby adjusting the contribution and importance of each modality. Jia et al. [26] extracted Gabor features, amplitude, and phase features from HSIs and LiDAR, and then fused the three features using a weighted majority voting decision fusion strategy. Zhao et al. [27] designed an adaptive weighted multikernel learning method that adaptively integrated the results by combining linear features and multikernel features. However, decision-fusion methods commonly ignore the potential correlations between different modalities.

The feature-level fusion methods aim to fuse extracted multimodal features. This method not only establishes information exchange between different modal data, but also can utilize advanced feature extraction strategies to improve the classification performance of the model. Ding et al. [13] used convolutional operators to capture local dependencies of multimodal data, and exploited Transformer to capture complex local and global spatial correlations of multimodal data. Dong et al. [28] proposed an effective feature fusion network, which utilized a cross-attention mechanism to assign weights and fully utilized complementary information from multiple modalities, obtaining discriminative fusion features. Du et al. [29] aggregated multimodal feature information by constructing a multimodal fusion graph with competitive performance in multimodal remote sensing classification tasks. Hang et al. [30] proposed the CNN-DF-S, which used two partially parameter-shared coupled CNNs to extract two modal features, and simultaneously exploited feature-level and decision-level fusion to integrate these heterogeneous features. Li et al. [31] proposed the maximum correntropy criterion-based robust 3-D-CapsNet (MCC-RCN), which is a multimodal fusion classification method with the robust and strong generalization capability. MCC-RCN is not only able to fuse the multimodal features efficiently, but also can deal with the outliers and noise that exist in the multimodal fusion process.

5.1.2 MEDFN for fusion classification

5.1.2.1 Framework of MEDFN for fusion classification

Fig. 5.2 depicts the training process of MEDFN, which mainly consists of the following stages:

Stage I. Construction of multimodal spatial graph. HSI and LiDAR data are input into the RLGN. First, the agent traverses all pixels of the HSI and LiDAR patches and obtains the observation state in each pixel in turn. The agent then decides whether to select the pixel based on the observation state and historical observation information. Second, pixel selection actions are performed and the next state and corresponding rewards are observed. Third, the above interaction process is repeated to learn the optimal graph construction strategy and maximize the reward. Finally, the agent constructs a two-modal spatial graph based on the learned graph construction strategy.

Stage II. Spatial graph learning. The two-modal spatial graphs are input into MEAM. First, edge feature extractors are used to capture edge features of HSI and LiDAR spatial graphs. Then, edge graph attention (EGA) is used to learn the importance weights of each neighborhood node, and the node features of the two-modal spatial graphs are updated by node feature aggregation.

Stage III. Ground object classification. The updated two-modal spatial graphs are input into the DMFM. First, the two-modal spatial graphs are decoupled into shared and unshared parts. Second, the cross-modal information aggregation mechanism (CIAM) is used to capture the modal-shared features. Third, the knowledge propagation mechanism (KPM) is used to capture the modal-specific features of two modalities, respectively. Finally, multimodal fusion features are obtained by concatenating modal-shared features and modal-specific features, and the classification of ground objects is predicted based on multimodal fusion features.

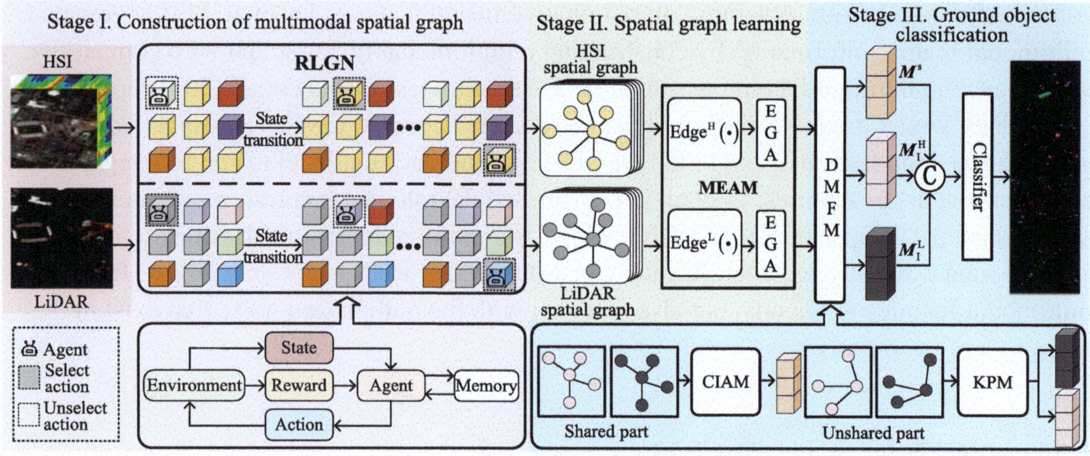

Figure 5.2:
Framework of MEDFN. Stage I: Data is input into RLGN to obtain two-modal spatial graphs. Stage II: The two-modal spatial graphs are input into MEAM for spatial graph learning, resulting in updated two-modal spatial graphs. Stage III: The updated two-modal spatial graphs are decoupled into shared and unshared parts, and modal-shared features and modal-specific features are obtained via CIAM and KPM. Finally, these features are concatenated to generate multimodal fusion features for ground object classification.

5.1.2.2 Multimodal spatial graph construction based on the RLGN

Graph learning can capture spatial dependencies between features in irregular regions. Currently, remote sensing classification methods based on graph learning often use graph construction methods based on spatial prior information or pixel similarity. The above methods not only fail to consider modal characteristics but also lack attention to task. Therefore, the RLGN, which learns the optimal graph construction strategy by optimizing the reward function directly related to the task, is designed. Since the RLGN focuses on structural information closely related to the task, it is able to more accurately capture interpixel dependencies. Next, two aspects of problem formulation and model framework are described.

1) Problem Formulation. The edge relationships between center nodes and neighboring nodes are established through pixel selection, thereby completing spatial graph construction. Specifically, the agent interacts with a customized pixel selection environment to learn the optimal strategy for selecting pixels.

Because agents cannot observe the complete environment during the exploration process, we define the pixel selection problem as a partially observable Markov decision process (POMDP). The five-tuple (S, A, P, R, O) is often used to define this process, where S is a set of states that the agent may be in. A represents a set of actions that can be selected by the agent, which describes the probability of an agent transitioning from one state to the next under a given state and action. P is the state transition probability, which represents the requirements for an agent to transition from one state to the next state under a given state and action. P is usually represented as $P\left(s_{t+1} | s_t, a_t\right)$, where s_t represents the state of the agent at the first time step t, a_t represents the action selected by the agent at that time step, and s_{t+1} represents the next state of the environment feedback after the agent takes the action a_t. R is a reward function that represents the immediate rewards obtained by the agent in each state and action combination. The reward function defines the goal and degree of good or bad actions of an agent, which aims to maximize long-term cumulative rewards by selecting actions. O is a set of partially observable states. Due to this partial observability, the agent cannot directly observe the complete environment but obtains partial information about the environment by obtaining the observation state.

The following further describes how to model pixel selection tasks as parameterized POMDPs.

Observation state (O): At each time step, the observation state of the agent is obtained by concatenating the features of the current pixel and the central pixel (to be classified). This observation state represents the local information that the agent can observe in the environment.

State (S): The POMDP can obtain only local information. The lack of global environmental state information may lead to agent decision biases, which can mislead pixel selection. To solve this problem, long short-term memory (LSTM) is used to model the state of the agent. This allows the agent to use historical observation information to infer and understand global information that it cannot directly obtain. Specifically, we define the state of an agent as the output of LSTM, which includes the current observation state and historical observation information. This enables the agent to clearly distinguish its own state from the above two.

State transition (P): At each time step, the agent first performs a pixel selection action, and then the environment guides the agent to transition to the next adjacent pixel until it has traversed the complete set of pixels. The above operations help agents understand the spatial adjacency relationship between pixels by constructing a serialized state transition process.

Action (A): When processing pixel selection tasks, the agent must decide whether to select the current pixel at each time step. We parameterize the action into a two-dimensional action vector, which encodes the agent's intention to select the current pixel. Once the agent performs a selection action, it triggers environmental interaction. The environment evaluates the current action of the agent, provides immediate rewards and transfers the agent to the next state.

Reward (R): Reward is the behavioral feedback of an agent, which is crucial for the agent's strategy learning process. To enhance the exploration ability of the agent, we customize immediate rewards based on classification tasks. This guides the agent to learn pixel selection strategies that suit the task. The immediate reward r can be expressed as:

$$r = r_n + r_a \tag{5.1}$$

where r_n represents the classification task reward, which is set based on the pixel selection result. Specifically, the set of pixels obtained through pixel selection is first averaged and pooled, and then the k-nearest neighbors are used for classification. If the classification result is correct, all relevant pixel selection actions receive a positive reward, set to $+100$; otherwise, these actions receive a negative reward, set to -100. This design encourages the agents to select the set of pixels that facilitate the correct classification. r_a is the action reward, which can be expressed as follows:

$$r_a = \lambda \left(L_{\text{cls}} \left(\widehat{\boldsymbol{y}}, \boldsymbol{y} \right) - L_{\text{cls}} \left(\tilde{\boldsymbol{y}}, \boldsymbol{y} \right) \right) \tag{5.2}$$

where $L_{\text{cls}}(\cdot)$ denotes the cross-entropy loss, $\tilde{\boldsymbol{y}}$ is the predicted label obtained based on the selected pixels, $\widehat{\boldsymbol{y}}$ is the predicted label obtained based on all pixels, and $\boldsymbol{y}$ is the true label. λ is the reward modulation coefficient, which is used to control the magnitude of the reward signal.

In addition, to establish information exchange among agents to facilitate collaborative cooperation between them, we have designed an agent communication mechanism (ACM). Based on this mechanism, agents from each modality integrate their own states with the states of agents from another modality to obtain a communication state, which encodes the current observation state and historical observation information of the two modal agents. The communication state between two modal agents at time step t can be represented as:

$$\begin{cases} \boldsymbol{s}_t^{\text{H}} = \boldsymbol{W}^{\text{H}} \left(\boldsymbol{\Theta}_t^{\text{L}} \right) + \boldsymbol{\Theta}_t^{\text{H}} \\ \boldsymbol{s}_t^{\text{L}} = \boldsymbol{W}^{\text{L}} \left(\boldsymbol{\Theta}_t^{\text{L}} \right) + \boldsymbol{\Theta}_t^{\text{L}} \end{cases} \tag{5.3}$$

where $\boldsymbol{s}_t^{\text{H}}$ and $\boldsymbol{s}_t^{\text{L}}$ represent the communication states of two modalities, while $\boldsymbol{\Theta}_t^{\text{H}}$ and $\boldsymbol{\Theta}_t^{\text{L}}$ respectively denote the state vectors of these two modalities. $\boldsymbol{W}^{\text{H}}$ and $\boldsymbol{W}^{\text{L}}$ are the learnable communication perception parameters for the two modalities.

2) Model Framework. The agent in RLGN consists of perception component, communication component, and decision component: 1) The perception component is the eye of the agent, responsible for state perception, which consists of multilayer perceptron (MLP) and LSTM. MLP is used to obtain the current observation state based on the input pixel, and LSTM is used to integrate the historical pixel to obtain the state vector that characterizes the current state of the agent. 2) The communication component is the mouth of the agent, responsible for state communication, which consists of ACM, used to establish information exchange between multimodal data. 3) The decision component is the brain of the agent, responsible for action decision-making, which consists of evaluation network, target network, and action selection module (ASM). The evaluation network and target network are used to evaluate the value of the pixel selection action based on the current state of the agent. The ASM can select actions according to the evaluation results.

Fig. 5.3 depicts the framework of the RLGN. First, during the exploration phase, the features of the pixels where the agent is located are input into a MLP to obtain the observation state. This observation state is then input into the LSTM, and the state vector $\boldsymbol{\Theta}_t$ is obtained by integrating the historical observation information and the observation state.

Second, the state vectors of multimodal are input into the ACM. In the ACM, the agent senses the state of another modal agent and combines its own state to obtain the communication state $\boldsymbol{s}_t$.

Third, according to $\boldsymbol{s}_t$, use the evaluation network $Q_{\mathrm{e}}(\cdot)$ to evaluate the value of pixel selection actions. Based on the evaluation results, ASM uses the ε-greedy strategy to execute action $\boldsymbol{a}_t$ and determine whether to select the current pixel.

Then, the agent obtains the reward signal $\boldsymbol{r}_t$ and the next communication state $\boldsymbol{s}_{t+1}$ from the environment and stores the experience data containing $(\boldsymbol{s}_t, \boldsymbol{a}_t, \boldsymbol{r}_t, \boldsymbol{s}_{t+1})$ in the replay buffer.

Next, during the training process, the agent randomly samples a small batch of experience data from the replay buffer. It then evaluates the future expected discount rewards corresponding to the state and action in the experience data based on the current strategy to obtain a Q. In the RLGN, Q includes the current Q and the target Q. The current Q is directly predicted by the evaluation network, and the target Q can be calculated by the Bellman equation based on the target network prediction. The current pixel selection strategy is optimized by minimizing the mean squared error between the current Q and the target Q:

$$L = E_{(\boldsymbol{s}_t, \boldsymbol{a}_t, \boldsymbol{r}_t, \boldsymbol{s}_{t+1})}\left[(\boldsymbol{\psi} - Q_{\mathrm{e}}(\boldsymbol{s}_t, \boldsymbol{a}_t; \boldsymbol{w}_{\mathrm{e}}))^2\right] \tag{5.4}$$

where $\boldsymbol{w}_{\mathrm{e}}$ is the evaluation network parameter, $\boldsymbol{\psi}$ is the target Q and can be represented as:

$$\boldsymbol{\psi} = r_t + \gamma \max_{\boldsymbol{a}} Q_{\mathrm{t}}(s_{t+1}, \boldsymbol{a}; \boldsymbol{w}_{\mathrm{t}}) \tag{5.5}$$

where $Q_{\mathrm{t}}(\cdot)$ is the target network, which is mainly used to stabilize the learning process during the training phase, and $\boldsymbol{w}_{\mathrm{t}}$ is the target network parameter. During the training phase, the parameters of the evaluation network are constantly updated. Directly using the

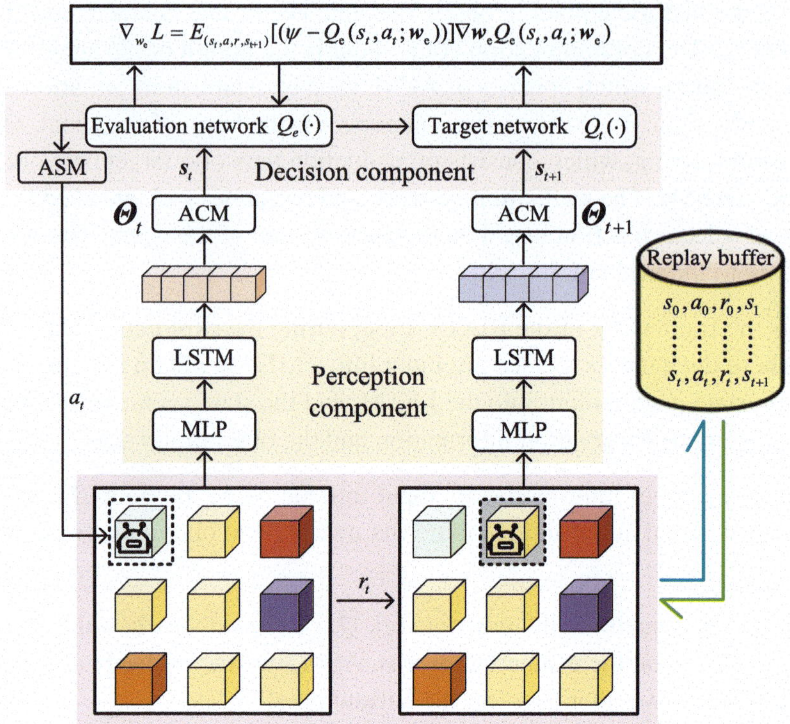

Figure 5.3:
Framework of the RLGN. In the exploration phase, first, the state vector is captured by the perception component. Second, the state vector is input into the ACM, and the communication state s_t is obtained through information communication between agents. Then, s_t is input into the decision component to obtain pixel selection action decision $\boldsymbol{a}_t$. After completing the action, the agent receives the corresponding reward r_t and obtains the communication state s_{t+1} for the next time step. Afterwards, s_t, $\boldsymbol{a}_t$, r_t, s_{t+1} are stored in the replay buffer. During the training phase, the agent samples a batch of experience from the replay buffer to evaluate future expected rewards, and updates the model by optimizing the Bellman equation.

parameters of the evaluation network to calculate $\boldsymbol{\psi}$ may lead to frequent oscillations of $\boldsymbol{\psi}$ and affect the learning stability. Therefore, the target network is introduced, and its parameters are obtained by periodically copying from the evaluation network. Since the target network has a low parameter update frequency, it can provide a more stable $\boldsymbol{\psi}$ for training. $\boldsymbol{\gamma}$ is the reward discount factor, which is used to regulate the effect of immediate rewards r_t and future rewards expected by the target network on the agents. A gradient descent step (Bellman equation) is carried out on L with respect to $\boldsymbol{w}_\mathrm{e}$ according to:

$$\nabla_{\boldsymbol{w}_\mathrm{e}} L = E_{(s_t, \boldsymbol{a}_t, r_t, s_{t+1})}[(\boldsymbol{\psi} - Q_\mathrm{e}(s_t, \boldsymbol{a}_t; \boldsymbol{w}_\mathrm{e}))]\nabla \boldsymbol{w}_\mathrm{e} Q_\mathrm{e}(s_t, \boldsymbol{a}_t; \boldsymbol{w}_\mathrm{e}) \tag{5.6}$$

Finally, the trained agents select each pixel in the HSI and LiDAR data, and the regular multimodal data are converted into irregular graph data by establishing edge relationships between the selected pixels.

5.1.2.3 Spatial graph learning based on the MEAM

The edge relationships between pixels contain abundant spatial topology information for fusion classification of HSI and LiDAR data. This information is very important for revealing dependency relationships between pixels. Therefore, the MEAM, which can effectively capture spatial topology information in HSI and LiDAR data by explicitly considering the edge relationships, is proposed. The framework of the MEAM is shown in Fig. 5.4. Define Graph $G = (V, E)$, where V is the set of nodes and E is the set of edges. First, the edge features of HSI and LiDAR are captured using the edge feature extractor. For each edge (i, j), the edge feature $\boldsymbol{e}_{ij}$ can be expressed as:

$$\begin{cases} \boldsymbol{e}_{ij}^{\mathrm{H}} = \mathrm{Edge}^{\mathrm{H}}\left(\boldsymbol{n}_i, \boldsymbol{n}_j\right) = \mathrm{SLM}\left(\boldsymbol{n}_i + \boldsymbol{n}_j\right) \\ \boldsymbol{e}_{ij}^{\mathrm{L}} = \mathrm{Edge}^{\mathrm{L}}\left(\boldsymbol{n}_i, \boldsymbol{n}_j\right) = \mathrm{CONCAT}\left(\boldsymbol{n}_i, \boldsymbol{n}_j\right) \boldsymbol{W}^{\mathrm{e}} \end{cases} \tag{5.7}$$

where $\boldsymbol{e}_{ij}^{\mathrm{H}}$ and $\boldsymbol{e}_{ij}^{\mathrm{L}}$ represent the edge features of HSI and LiDAR, respectively, $\mathrm{Edge}^{\mathrm{H}}$ and $\mathrm{Edge}^{\mathrm{L}}$ represent the edge feature extractors, $\mathrm{CONCAT}(\cdot)$ is a connected operation, $\boldsymbol{W}^{\mathrm{e}}$ is the edge feature learning parameter, $\boldsymbol{n} = \{\boldsymbol{b}_1, \boldsymbol{b}_2, ..., \boldsymbol{b}_L\}$ is the input feature, and $\boldsymbol{b}$ is the band. SLM denotes the spectral learning module, which uses gated recurrent unit to capture the dependencies between bands in HSI. The hidden state $\boldsymbol{h}_l$ of the gated recurrent unit is updated by the following:

$$\begin{cases} \boldsymbol{r}_l = \mathrm{Sigmoid}\left(\boldsymbol{W}_r \boldsymbol{b}_l + \boldsymbol{U}_r \boldsymbol{b}_{l-1} + \boldsymbol{m}_r\right) \\ z_l = \mathrm{Sigmoid}\left(\boldsymbol{W}_z \boldsymbol{b}_l + \boldsymbol{U}_z \boldsymbol{b}_{l-1} + \boldsymbol{m}_z\right) \\ \tilde{\boldsymbol{h}}_l = \mathrm{Tanh}\left(\boldsymbol{W}_h \boldsymbol{b}_l + \boldsymbol{U}_h \boldsymbol{b}_{l-1} + \boldsymbol{m}_h\right) \\ \boldsymbol{h}_l = (1 - z_{l-1}) \odot \boldsymbol{h}_{l-1} + z_{l-1} \odot \tilde{\boldsymbol{h}}_{l-1} \end{cases} \tag{5.8}$$

where $\odot$ is the Hadamard product, $\mathbf{r}_l$ and $\mathbf{z}_l$ are reset and update gates, $\mathbf{W}_r$, $\mathbf{W}_z$, $\mathbf{W}_h$, $\mathbf{U}_r$, $\mathbf{U}_z$, $\mathbf{U}_h$ are the learnable weight matrices, and $\mathbf{m}_r$, $\mathbf{m}_z$, $\mathbf{m}_h$ are the bias terms.

Second, the attention score α between nodes is obtained using the EGA.

$$\alpha_{ij} = \mathrm{Softmax}(\boldsymbol{e}_{ij}) = \frac{\exp(LeakyReLU(\boldsymbol{W}_{\mathrm{a}} \boldsymbol{e}_{ij}))}{\sum_{j \in \mathcal{N}_i} \exp(LeakyReLU(\boldsymbol{W}_{\mathrm{a}} \boldsymbol{e}_{ij}))} \tag{5.9}$$

where $\boldsymbol{W}_{\mathrm{a}}$ is a learnable attention matrix.

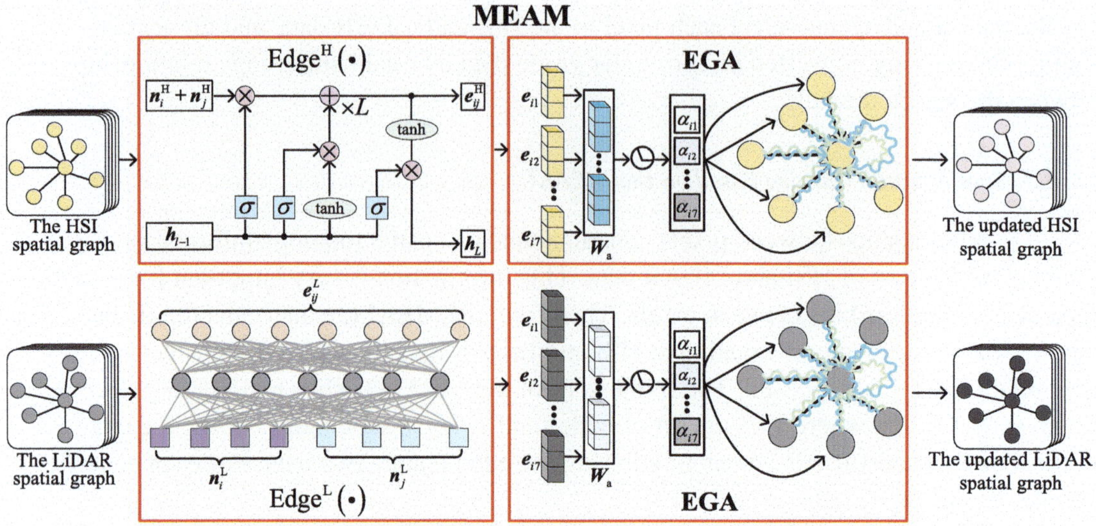

Figure 5.4:
Framework of MEAM. Firstly, the two-modal spatial graphs are input into the edge feature learning network to obtain edge features that reflect the dependency relationships between nodes. Then, the edge graph attention (EGA) is used to learn the attention weights between nodes. Finally, the node features are updated with attention weights to obtain the updated two-modal spatial graph.

Finally, corresponding importance weights are assigned to each node based on α, and rich contextual information is captured using the multihead attention mechanism to aggregate node features as follows:

$$\boldsymbol{F}_i = \sigma\left(\frac{1}{Q}\sum_{q=1}^{Q}\sum_{j\in\mathcal{N}_i}\alpha_{ij}\boldsymbol{W}^q\boldsymbol{h}_j\right) \tag{5.10}$$

where $\boldsymbol{W}^q$ is the learnable node aggregation matrix and Q is the number of attention heads.

5.1.2.4 Ground object classification based on the DMFM

DMFM mainly consists of four parts: Spatial graph decoupling, modal-shared feature learning, modal-specific feature learning, and multimodal data fusion. The framework of the DMFM is shown in Fig. 5.5. First, for the shared part, the CIAM is used to integrate the multimodal complementary information to obtain modal-shared features. For the unshared part, the KPM is used to capture the class-discrimination information to obtain the modal-specific features of the two modalities. Then, the multimodal fusion features are obtained by concatenating the modal-shared features and the modal-specific features of the two modalities.

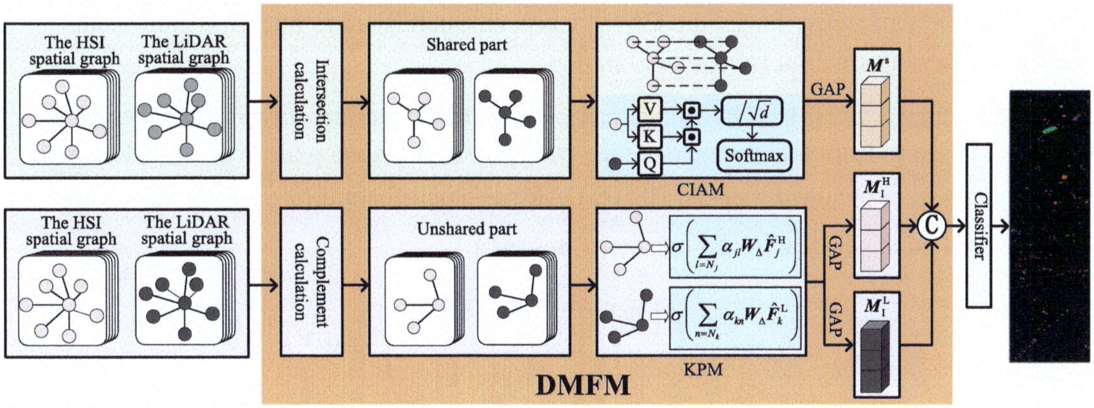

Figure 5.5:
Framework of DMFM. First, shared part and unshared part of two-modal spatial graphs are obtained by intersection calculation and complement calculation. Then, CIAM and KPM are utilized to capture class-discrimination information in both parts respectively. Next, modal-shared features and modal-specific features are obtained by global average pooling (GAP), and they are spliced to obtain fusion features. Finally, the multimodal fusion features are input into the classifier to complete the classification task.

1) Spatial graph decoupling. The two-modal spatial graphs are decoupled into shared and unshared parts as follows:

$$\begin{cases} V_{\mathrm{S}} = V^{\mathrm{H}} \cap V^{\mathrm{L}} \\ V_{\mathrm{I}}^{\mathrm{H}} = C_{V^{\mathrm{H}}} V_{\mathrm{S}} \\ V_{\mathrm{I}}^{\mathrm{L}} = C_{V^{\mathrm{L}}} V_{\mathrm{S}} \end{cases} \tag{5.11}$$

where $V^{\mathrm{H}} = \left\{ \tilde{F}_1^{\mathrm{H}}, ..., \tilde{F}_i^{\mathrm{H}}, ..., \tilde{F}_{E+J}^{\mathrm{H}} \right\}$ and $V^{\mathrm{L}} = \left\{ \tilde{F}_1^{\mathrm{L}}, ..., \tilde{F}_i^{\mathrm{L}}, ..., \tilde{F}_{E+K}^{\mathrm{L}} \right\}$ represent the node feature sets of the two modalities respectively. $\cap$ denotes the intersection operation, and C denotes the complement operation.

$V_{\mathrm{S}} = \left\{ \left(\tilde{F}_1^{\mathrm{H}}, \tilde{F}_1^{\mathrm{L}} \right), ..., \left(\tilde{F}_i^{\mathrm{H}}, \tilde{F}_i^{\mathrm{L}} \right), ..., \left(\tilde{F}_E^{\mathrm{H}}, \tilde{F}_E^{\mathrm{L}} \right) \right\}$ is the shared part. $V_{\mathrm{I}}^{\mathrm{H}} = \left\{ \hat{F}_1^{\mathrm{H}}, ..., \hat{F}_j^{\mathrm{H}}, ..., \hat{F}_J^{\mathrm{H}} \right\}$ and $V_{\mathrm{I}}^{\mathrm{L}} = \left\{ \hat{F}_1^{\mathrm{L}}, ..., \hat{F}_k^{\mathrm{L}}, ..., \hat{F}_K^{\mathrm{L}} \right\}$ are the unshared parts of the two modalities.

2) Modal-shared feature learning. There is a one-to-one correspondence between node pairs belonging to the shared part of two modalities, which can reveal different attributes of the same geographic region. Deeply exploring the potential complementary attributes between modalities can effectively assist the model in completing ground object classification. Inspired by the cognitive mechanism of humans aggregating information between different senses to comprehensively perceive things, the CIAM has been established to integrate multimodal complementary information from shared part.

Specifically, firstly, all HSI-LiDAR node pairs in the shared part are traversed, and cross attention is used to learn the correspondence weights of each node pair; thus, modal-shared node features were obtained. For the HSI-LiDAR node pair feature $\left(\tilde{\boldsymbol{F}}_i^{\mathrm{H}}, \tilde{\boldsymbol{F}}_i^{\mathrm{L}}\right) \in V_{\mathrm{S}}$, the modal-shared node pair feature $\tilde{\boldsymbol{F}}_i^{\mathrm{S}}$ can be expressed as:

$$\tilde{\boldsymbol{F}}_i^{\mathrm{S}} = \mathrm{Softmax}\left(\frac{\left(\boldsymbol{W}_{\mathrm{S}}^{\mathrm{K}}\tilde{\boldsymbol{F}}_i^{\mathrm{L}}\right)\cdot\left(\boldsymbol{W}_{\mathrm{S}}^{\mathrm{Q}}\tilde{\boldsymbol{F}}_i^{\mathrm{H}}\right)}{\sqrt{d}}\right)\cdot\left(\boldsymbol{W}^{\mathrm{V}}\tilde{\boldsymbol{F}}_i^{\mathrm{H}}\right) \tag{5.12}$$

where $\tilde{\boldsymbol{F}}_i^{\mathrm{L}}$ represents the LiDAR node feature, $\tilde{\boldsymbol{F}}_i^{\mathrm{H}}$ represents the HSI node feature, d is the feature dimension, $\boldsymbol{W}^{\mathrm{Q}}$, $\boldsymbol{W}^{\mathrm{K}}$, and $\boldsymbol{W}^{\mathrm{V}}$ represent learnable query vectors, key vectors, and value vectors, respectively.

Secondly, global average pooling is used to aggregate all modal-shared node pair features, obtaining the modal-shared feature $\boldsymbol{M}^{\mathrm{S}}$:

$$\boldsymbol{M}^{\mathrm{S}} = \sum_{i=1}^{E}\tilde{\boldsymbol{F}}_i^{\mathrm{S}} \tag{5.13}$$

where E denotes the number of node pairs in the shared part.

3) Modal-specific feature learning. Thirdly, unshared parts contain node feature with class-discrimination information in a certain modality, so ignoring these unshared parts during multimodal data fusion can lead to the loss of key class-discrimination information. KPM is established to learn modal-specific features, which contain key class-discrimination information from unshared parts. HSI node feature of unshared part $\hat{\boldsymbol{F}}_j^{\mathrm{H}} \in V_{\mathrm{I}}^{\mathrm{H}}$ and LiDAR node feature of unshared part $\hat{\boldsymbol{F}}_k^{\mathrm{L}} \in V_{\mathrm{I}}^{\mathrm{L}}$ can be updated as:

$$\hat{\boldsymbol{F}}_j^{\mathrm{H}} = \sigma\left(\sum_{l=N_j}\alpha_{jl}\boldsymbol{W}_{\Delta}\hat{\boldsymbol{F}}_j^{\mathrm{H}}\right), \hat{\boldsymbol{F}}_k^{\mathrm{L}} = \sigma\left(\sum_{n=N_k}\alpha_{kn}\boldsymbol{W}_{\Delta}\hat{\boldsymbol{F}}_k^{\mathrm{L}}\right) \tag{5.14}$$

where $\sigma(\cdot)$ represents the activation function, $\boldsymbol{W}_{\Delta}$ represents the learnable weight matrix, N_j and N_k represent the neighbor nodes of $\hat{\boldsymbol{F}}_j^{\mathrm{H}}$ and $\hat{\boldsymbol{F}}_k^{\mathrm{L}}$, respectively. α_{jl} and α_{kn} represent attention weights, which can be calculated as:

$$\alpha_{jl} = \boldsymbol{W}^{\mathrm{A}}\left[\boldsymbol{W}_{\Gamma}\hat{\boldsymbol{F}}_j^{\mathrm{H}}||\boldsymbol{W}_{\Gamma}\hat{\boldsymbol{F}}_l^{\mathrm{H}}\right], \alpha_{kn} = \boldsymbol{W}^{\mathrm{A}}\left[\boldsymbol{W}_{\Gamma}\hat{\boldsymbol{F}}_k^{\mathrm{L}}||\boldsymbol{W}_{\Gamma}\hat{\boldsymbol{F}}_n^{\mathrm{L}}\right] \tag{5.15}$$

where $\boldsymbol{W}^{\mathrm{A}}$ represents the attention learning parameter, $\boldsymbol{W}_{\Gamma}$ represents the mapping parameter, and || represents the concatenation operation.

Fourthly, using global average pooling to aggregate two-modal node features of unshared part, obtaining HSI modal-specific feature $\boldsymbol{M}_{\mathrm{I}}^{\mathrm{H}}$ and LiDAR modal-specific feature $\boldsymbol{M}_{\mathrm{I}}^{\mathrm{L}}$ as:

$$\boldsymbol{M}_{\mathrm{I}}^{\mathrm{H}} = \sum_{j=1}^{J} \hat{\boldsymbol{F}}_{j}^{\mathrm{H}}, \boldsymbol{M}_{\mathrm{I}}^{\mathrm{L}} = \sum_{k=1}^{K} \hat{\boldsymbol{F}}_{k}^{\mathrm{L}} \tag{5.16}$$

4) Multimodal data fusion. Finally, the multimodal fusion feature obtained by concatenating $\boldsymbol{M}^{\mathrm{s}}$, $\boldsymbol{M}_{\mathrm{I}}^{\mathrm{H}}$, and $\boldsymbol{M}_{\mathrm{I}}^{\mathrm{L}}$ are input into the classifier to obtain the prediction label $\hat{\boldsymbol{Y}}$:

$$\hat{\boldsymbol{Y}} = \mathrm{CONCAT}\left(\boldsymbol{M}^{\mathrm{s}}, \boldsymbol{M}_{\mathrm{I}}^{\mathrm{H}}, \boldsymbol{M}_{\mathrm{I}}^{\mathrm{L}}\right) \boldsymbol{W}_{\mathrm{cls}} \tag{5.17}$$

where $\boldsymbol{W}_{\mathrm{cls}}$ is a learnable classifier parameter.

5.1.3 Experiments

5.1.3.1 Dataset introduction

Table 5.1 lists the number of training and testing samples used in the experiment. Figs. 5.6–5.8 show the pseudocolor image of the HSI, LiDAR, and ground truth maps of the three datasets.

Houston Dataset[1]: The Houston dataset was collected by the Airborne Laser Surveying Center at Houston University and nearby urban areas, which consists of HSI and LiDAR. Both HSI and LiDAR have undergone careful preprocessing for easy analysis. This multimodal dataset contains 3491905 pixels and 15 classes of ground objects, and HSI has 144 spectral bands.

Trento Dataset[2]: The Trento dataset consists of HSI and LiDAR, which was acquired from the rural area south of Trento, Italy. The dataset includes HSI captured by the Airborne Hyperspectral Imager and LiDAR acquired by the Spaceborne Laser Scanning System. Both HSI and LiDAR have 166600 pixels and 6 classes of ground objects. HSI has 63 bands and a spatial resolution of 1 meter.

MUUFL Dataset[3]: The MUUFL dataset consists of HSI and LiDAR, which was collected by the international Optech company at the University of Gulfport in South Mississippi. This

1 http://www.grss-ieee.org/community/technical-committees/data-fusion/2013-ieee-grss-data-fusion-contest/.
2 https://github.com/wzr78998/Trento.
3 https://github.com/GatorSense/MUUFLGulfport.

Table 5.1: The number of training and testing samples for each dataset.

Houston 2013					Trento					MUUFL				
Ground object class			Sample number		Ground object class			Sample number		Ground object class			Sample number	
Class No.	Color	Name	Training	Testing	Class No.	Color	Name	Training	Testing	Class No.	Color	Name	Training	Testing
H1		Healthy grass	20	1231	T1		Apple trees	20	4014	M1		Trees	20	23226
H2		Stressed grass	20	1234	T2		Buildings	20	2883	M2		Mostly grass	20	4250
H3		Synthetic grass	20	677	T3		Ground	20	459	M3		Mixed ground surface	20	6862
H4		Trees	20	1224	T4		Woods	20	9103	M4		Dirt and sand	20	1806
H5		Soil	20	1222	T5		Vineyard	20	10481	M5		Road	20	6667
H6		Water	20	305	T6		Roads	20	3154	M6		Water	20	446
H7		Residential	20	1248						M7		Building shadow	20	2213
H8		Commercial	20	1224						M8		Building	20	6220
H9		Road	20	1232						M9		Sidewalk	20	1365
H10		Highway	20	1207						M10		Yellow curb	20	163
H11		Railway	20	1215						M11		Cloth panels	20	249
H12		Parking lot 1	20	1213										
H13		Parking lot 2	20	449										
H14		Tennis court	20	408										
H15		Running track	20	640										

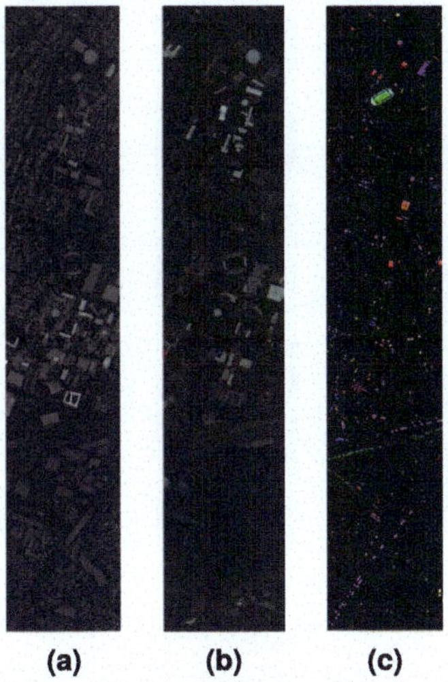

Figure 5.6:
Houston dataset. (a) Pseudocolor image for HSI, (b) LiDAR image, (c) ground truth map.

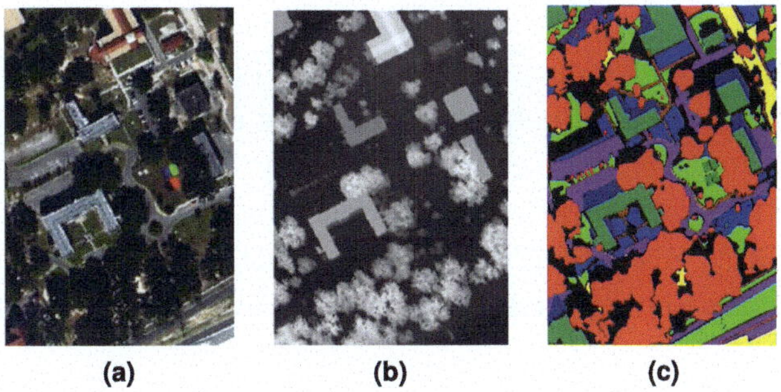

Figure 5.7:
MUUFL dataset. (a) Pseudocolor image for HSI, (b) LiDAR image, (c) ground truth map.

dataset was obtained by synchronously collecting HSI and LiDAR by Gemini LiDAR and CASI-1500 on the same flight platform. Both HSI and LiDAR have 325320 pixels and 11 classes of ground objects, and HSI has 64 spectral bands and a spatial resolution of 1 meter.

5.1.3.2 Experimental setting

To demonstrate the effectiveness of the proposed method based on the real datasets introduced in the previous section, CNN, hierarchical random walk network (HRWN) [22], deep encoder-decoder network (EndNet) [32], cross-channel reconstruction module (CCR-Net) [14], fractional Gabor convolutional network (FGCN) [33], CNN-DF-S [30], and MCC-RCN [31] were selected as comparative methods. Four evaluation indicators were selected

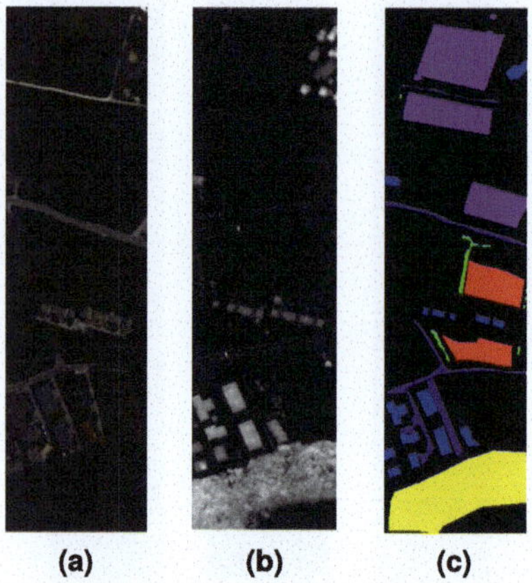

Figure 5.8:
Trento dataset. (a) Pseudocolor image for HSI, (b) LiDAR image, (c) ground truth map.

for the experiment: The accuracy of each class of ground object, overall accuracy (OA), the accuracy of each category (AA), and the kappa coefficient. To ensure as fair a comparison as possible, all experiments were conducted on an Intel i7-12700k CPU, an RTX 2080ti GPU, and a 64 GB memory server. The experimental results of all methods were averaged 10 times. The learning rate is set to 0.001, the weight decal is set to 0.0005, and the batch size is set to 100. For the CNN, 20 HSI samples of each class were used for training, and for the HRWN, EndNet, CCR-Net, FGCN, and MEDFN, 20 HSI and LiDAR samples of each class were used for training. The parameter settings of each comparison method followed the original paper settings.

Tables 5.2–5.4 show the classification performance of different methods on 3 datasets, and Figs. 5.9–5.11 visibly show the classification maps on 3 datasets. The original features and the features extracted by different methods were visualized to obtain Fig. 5.12.

5.1.3.3 Comparative experiments

From Tables 5.2–5.4 and Figs. 5.9–5.12, the following can be concluded:

1. Multimodal methods exhibit better classification performance than CNN, which only processes HSI. This is because HSI is susceptible to a variety of factors during acquisition, leading to the phenomenon where different objects have the same spectral response. Even if a single modality lacks discriminative properties, the multimodal method can utilize discriminative information from another modality to achieve accurate classification.

Table 5.2: Classification Performance of Different Methods on Houston.

Class No.	CNN	HRWN	EndNet	CCR-Net	FGCN	CNN-DF-S	MCC-RCN	MEDFN
H1	98.62	89.60	**99.35**	97.48	95.42	92.20	95.70	85.40
H2	94.41	97.08	**99.51**	86.14	92.11	99.35	97.97	88.38
H3	94.39	**99.85**	**99.85**	97.49	98.92	98.23	98.82	94.66
H4	98.20	92.08	90.28	99.10	96.39	98.20	95.10	**99.30**
H5	99.02	99.51	97.71	99.67	99.74	**99.76**	98.12	99.57
H6	82.62	90.49	95.74	96.72	94.18	93.44	**97.05**	83.48
H7	69.39	79.41	83.89	**89.58**	78.90	77.89	77.32	84.41
H8	67.40	79.49	51.63	69.85	80.74	75.49	81.21	**93.31**
H9	74.76	53.90	73.54	**96.40**	76.37	64.69	78.17	90.52
H10	79.79	86.00	87.57	88.57	87.94	90.89	87.82	**98.02**
H11	75.89	64.28	84.77	80.82	83.38	77.04	86.17	**94.64**
H12	62.41	77.41	73.54	81.62	78.87	**85.24**	71.89	76.81
H13	83.30	85.52	85.75	86.86	83.05	**93.32**	85.52	78.24
H14	98.78	99.02	**100**	99.27	98.41	99.76	**100**	99.26
H15	97.50	99.06	99.38	96.41	98.03	**100**	98.44	**100**
OA (%)	83.75	84.22	86.30	87.79	88.32	87.98	88.52	**91.26**
AA (%)	85.09	86.18	88.17	89.27	89.50	89.70	89.95	**91.07**
Kappa (%)	82.45	82.93	86.21	86.71	87.36	87.00	87.58	**90.83**

Table 5.3: Classification Performance of Different Methods on Trento.

Class No.	CNN	HRWN	EndNet	CCR-Net	FGCN	CNN-DF-S	MCC-RCN	MEDFN
T1	95.04	96.01	95.37	96.11	97.09	98.96	**99.45**	95.54
T2	77.77	81.93	94.03	96.53	92.02	92.08	92.42	**98.34**
T3	97.82	**100**	99.78	96.95	97.82	99.47	98.92	93.43
T4	99.78	99.50	99.53	99.96	99.98	99.57	99.79	**100**
T5	98.57	98.51	99.04	99.74	99.95	98.57	98.74	**100**
T6	79.04	86.21	81.20	81.26	87.83	90.05	92.33	**99.68**
OA (%)	94.41	95.62	96.35	97.03	97.51	97.44	97.92	**99.11**
AA (%)	91.34	93.69	94.82	95.09	95.78	96.45	96.94	**97.83**
Kappa (%)	92.56	94.16	95.15	96.04	96.68	96.57	97.20	**98.73**

2. Among the multimodal methods, the GCN-based methods (MEDFN and FGCN) show better classification performance than the CNN-based methods (HRWN, EndNet, and CCR-Net). Compared to CNN, GCN can capture complex spatial dependencies between ground objects, and is well adapted to the irregular spatial topology of real ground objects.
3. Compared to FGCN, MEDFN exhibits better classification performance. This is because, on the one hand, MEDFN designed communication between modalities when

Table 5.4: Classification Performance of Different Methods on MUUFL.

Class No.	CNN	HRWN	EndNet	CCR-Net	FGCN	CNN-DF-S	MCC-RCN	MEDFN
M1	81.82	81.85	80.01	83.16	82.39	83.63	**84.22**	80.94
M2	65.47	61.94	**80.90**	71.52	76.71	76.19	61.20	76.33
M3	55.96	59.40	57.51	58.18	67.20	50.25	57.67	**71.31**
M4	78.78	79.90	79.62	86.54	87.22	72.81	75.86	**90.77**
M5	75.76	71.21	79.49	72.79	76.96	81.87	79.90	**85.63**
M6	98.32	98.08	92.55	98.56	97.84	96.86	**99.33**	91.83
M7	82.50	**87.36**	84.52	78.24	77.74	79.49	79.44	82.68
M8	72.71	77.16	71.65	80.24	76.70	84.57	**95.77**	83.93
M9	45.62	63.52	61.65	67.34	64.49	59.41	53.55	**77.90**
M10	57.14	66.92	77.44	77.44	81.20	64.42	80.98	**88.72**
M11	98.17	96.35	95.43	99.09	**99.54**	90.76	97.59	**99.54**
OA (%)	74.52	75.36	76.25	77.07	78.35	77.57	78.70	**80.77**
AA (%)	73.81	76.70	78.43	79.37	80.73	76.39	78.68	**84.51**
Kappa (%)	67.63	68.63	69.90	70.87	72.39	71.49	72.80	**75.57**

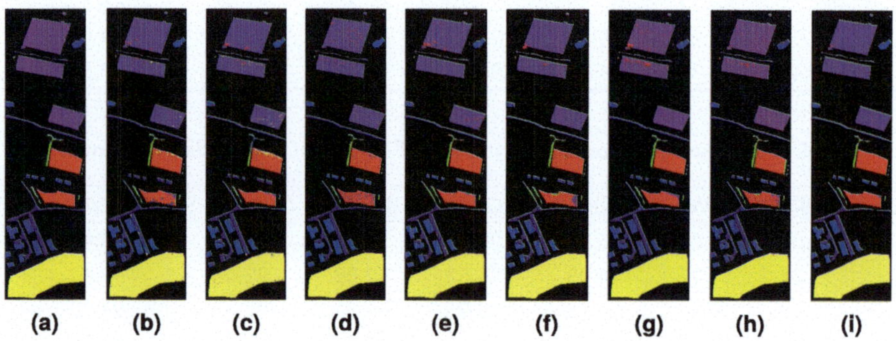

Figure 5.9:
Classification maps obtained on Trento. (a) Truth label, (b) CNN, (c) HRWN, (d) EndNet, (e) CCR-Net, (f) FGCN, (g) CNN-DF-S, (h) MCC-RCN, (i) MEDFN.

constructing two-modal spatial graphs, intelligently capturing the complementary spatial topology between LiDAR and HSI. In addition, the establishment of a reward feedback mechanism ensures that the graph has good task adaptability. On the other hand, by fully considering the edge attributes between nodes, MEDFN can effectively capture the dependency relationships between nodes.

4. Among all multimodal fusion methods, MEDFN has the best classification performance. This is because it breaks the ideal assumption of a shared topology between two modalities and integrates multimodal information based on the idea of divide and

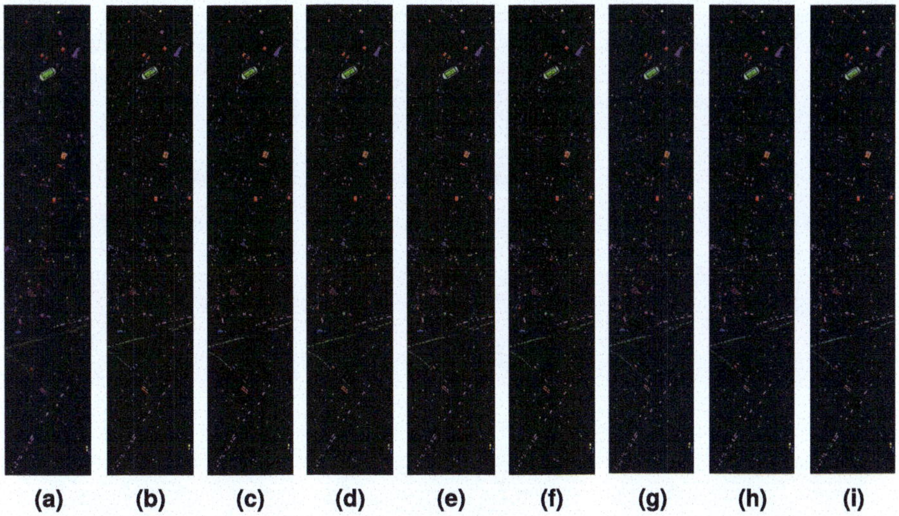

Figure 5.10:
Classification maps obtained on Houston. (a) Truth label, (b) CNN, (c) HRWN, (d) EndNet, (e) CCR-Net, (f) FGCN, (g) CNN-DF-S, (h) MCC-RCN, (i) MEDFN.

conquer While fully utilizing the complementary information, it suppresses intermodal information interference induced by modal heterogeneity.

5. As shown in Figs. 5.9–5.11, each classification map obtained by MEDFN shows the clearest classification boundary. For example, in the Trento dataset, the same plants Apple Trees and Vineyardoften exhibit similar spectral response, making it challenging to clearly distinguish between these two classes of ground object. Compared to other comparative methods, the classification maps show that MEDFN presented the most continuous and accurate classification results.
6. As shown in Fig. 5.12, MEDFN exhibits favorable intraclass consistency and interclass separability. For example, for roads (dark blue) and building (light blue), which are also composed of concrete, MEDFN also shows high interclass variability in feature space.

5.1.3.4 Ablation study

MEDFN contains three key components: RLGN, MEAM, and DMFM. To validate the effectiveness of these components, we performed ablation studies on the above components on three datasets, and the results are shown in Table 5.5.

1) Analysis of RLGN. The RLGN in MEDFN was replaced with the commonly used superpixel graph construction [34] to obtain Baseline-A. It can be seen from Table 5.5 that,

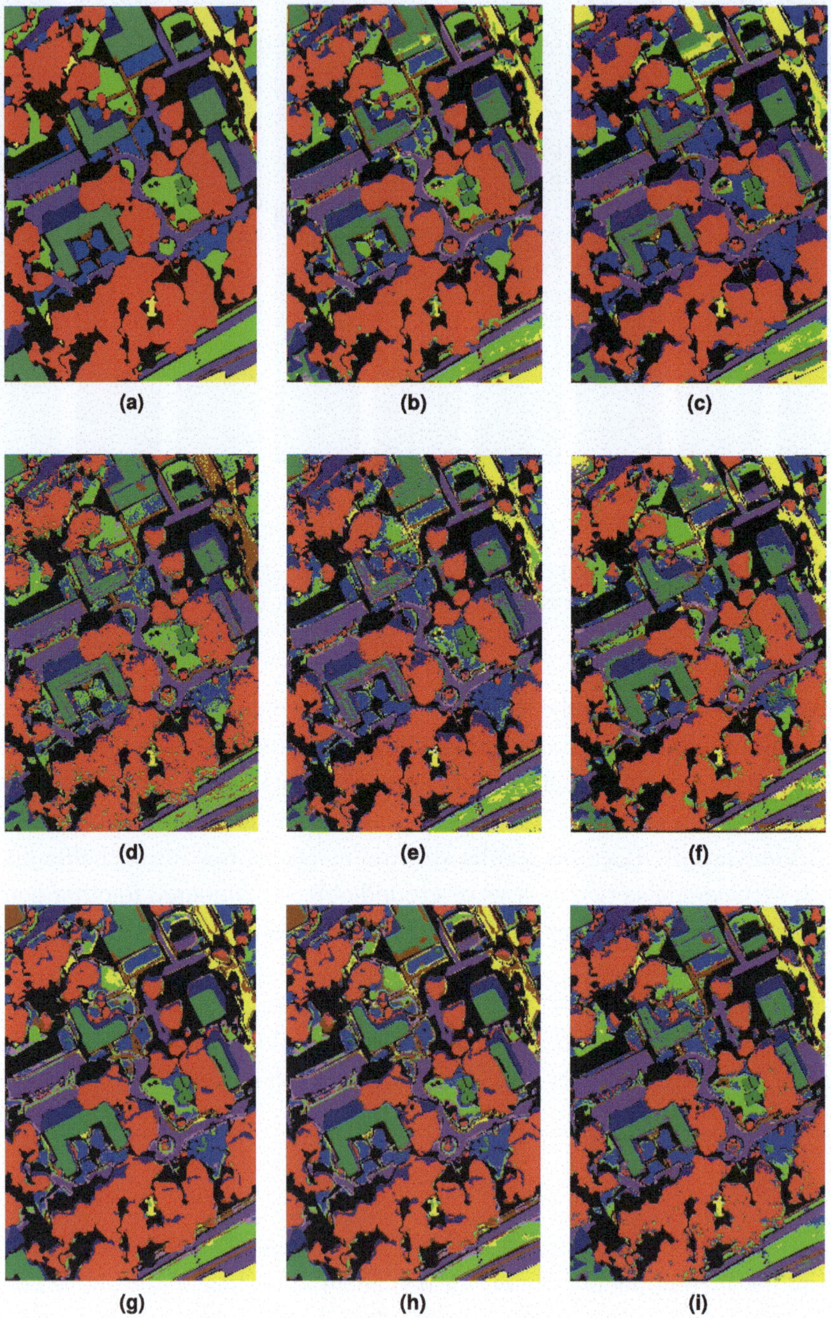

Figure 5.11:
Classification maps obtained on MUFFL. (a) Truth label, (b) CNN, (c) HRWN, (d) EndNet, (e) CCR-Net, (f) FGCN, (g) CNN-DF-S, (h) MCC-RCN, (i) MEDFN.

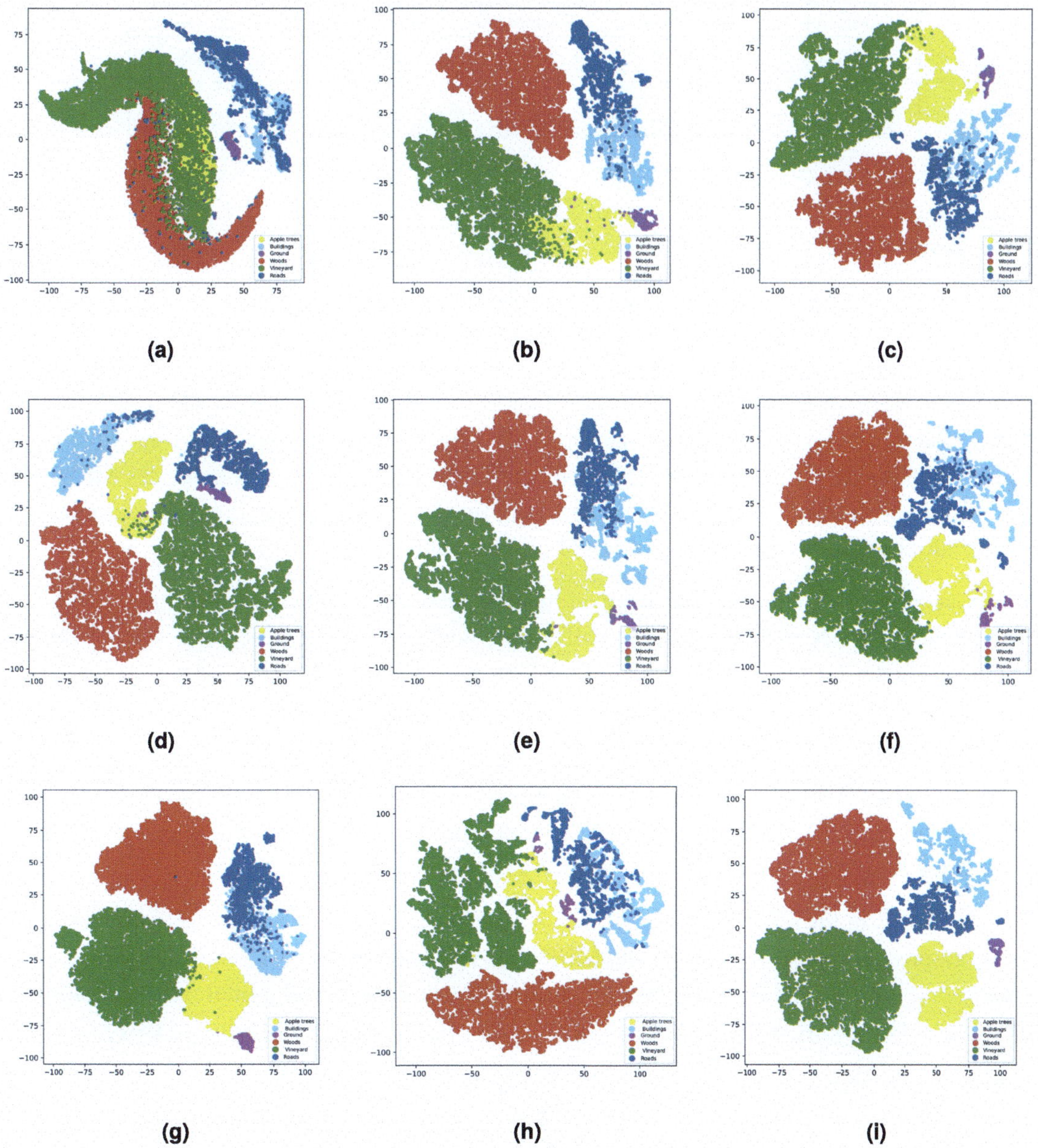

Figure 5.12:
t-distributed stochastic neighbor embedding visualization of features (Trento). (a) Features of original data, (b) HRWN, (c) EndNet, (d) CCR-Net, (e) FGCN, (f) CNN-DF-S, (g) MCC-RCN, (h) Baseline-B, (i) MEDFN.

Table 5.5: OAs of MEDFN and Different Baselines on Three Datasets.

Dataset	Baseline-A	Baseline-B	Baseline-C	Baseline-D
Houston	89.43	90.45	90.54	**91.26**
Trento	97.83	98.47	98.38	**99.11**
MUFFL	78.93	80.05	79.93	**80.77**

compared to Baseline-A, MEDFN improved OA on the three datasets by 1.83%, 1.28%, and 1.84%, respectively.

To further analyze the effect of RLGN, we visualized the adjacency matrix of MEDFN and Baseline-A in a certain region of the Trento datasets. As shown in Fig. 5.13, the orange region indicates that the pixels in this region have been connected to the center pixel during the graph construction process. In contrast, green region indicates that no connection has been established between the two above. Due to the spectral noise signal, different classes of pixels exhibit similar spectral curves. Compared to Baseline-A, RLGN enhances MEDFN's capability to precisely identify different classes of pixels, thereby achieving a more accurate representation of the true spatial topology of ground objects.

The above phenomenon occurs because, on the one hand, the agent communication mechanism of RLGN enables agents to collaborate with each other, which effectively captures the potential relevance of modalities. On the other hand, RLGN can fully explore the spatial topology related to two-modal tasks, and learn the optimal graph construction strategy beneficial to tasks in complex scene changes.

2) Analysis of MEAM. The MEAM in MEDFN was replaced with the commonly used GCN [35] to obtain Baseline-B. From Table 5.5, compared to Baseline-B, MEDFN improved OA on the three datasets by 0.81%, 1.28%, and 0.72%, respectively.

To further analyze the effect of MEAM, we visualized the node features of the two-modal spatial graphs output by Baseline-B and MEDFN. As shown in Fig. 5.12 (h-i): The node features output from MEDFN exhibit higher intraclass consistency and interclass separation, indicating that MEAM is able to improve the ability of class discriminability of two-modal features.

The above phenomenon occurs because, MEAM uses the edge attention mechanism to assign different importance weights to neighbor nodes, allowing the model to focus on high class-discriminant neighbor nodes, thus suppressing neighbor node noise interference.

3) Analysis of DMFM. The DMFM in MEDFN was replaced with the global average pooling and feature concatenation to obtain Baseline-C. From Table 5.5, compared to Baseline-C, MEDFN improved OA on the three datasets by 0.72%, 0.73%, and 0.84%, respectively.

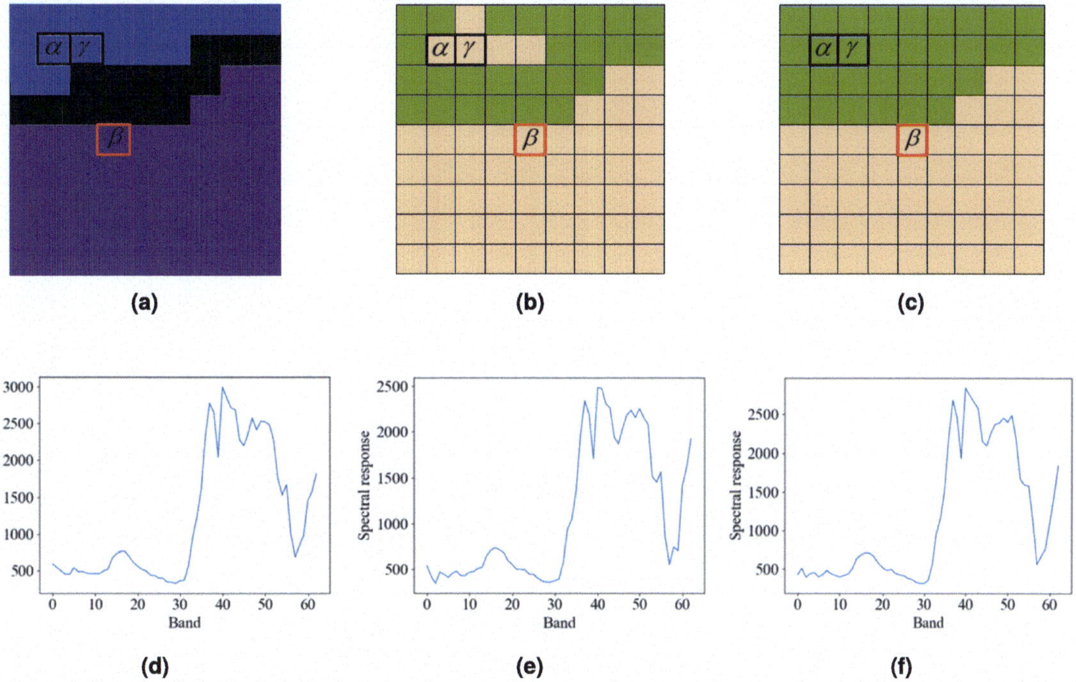

Figure 5.13:
Visualization of adjacency matrices. As shown in (a), a region in the Trento dataset contains three classes including buildings (blue), background (black), and roads (purple). As shown in (d-f), α and γ belonging to the buildings and the center pixel β belonging to the roadshave similar spectral responses. As shown in (b-c), when constructing graphs with RLGN, MEDFN accurately identifies pixels without erroneously establishing connections between these pixels and the central pixels. (a) Ground truth, (b) adjacency matrix of central pixel computed by Baseline-A, (c) adjacency matrix of central pixel learned by MEDFN, (d) spectral response of pixel α, (e) spectral response of pixel β, (f) spectral response of pixel γ.

To further analyze the effect of DMFM, we visualized the mean and variance of the prediction probability for the road using Baseline-C and MEDFN on the Trento dataset. As shown in Fig. 5.14, compared to Baseline-C, MEDFN can deeply mine potential class-discriminant information from unshared regions, presenting higher prediction probability and lower prediction variance for the true class.

The above phenomenon occurs because, during the multimodal data fusion process, DMFM captures the dependency relationships between HSI-LiDAR node pairs in the shared part, enabling the utilization of complementary multimodal information to enhance feature discriminability. This, in turn, improves the model's ability to distinguish ground objects.

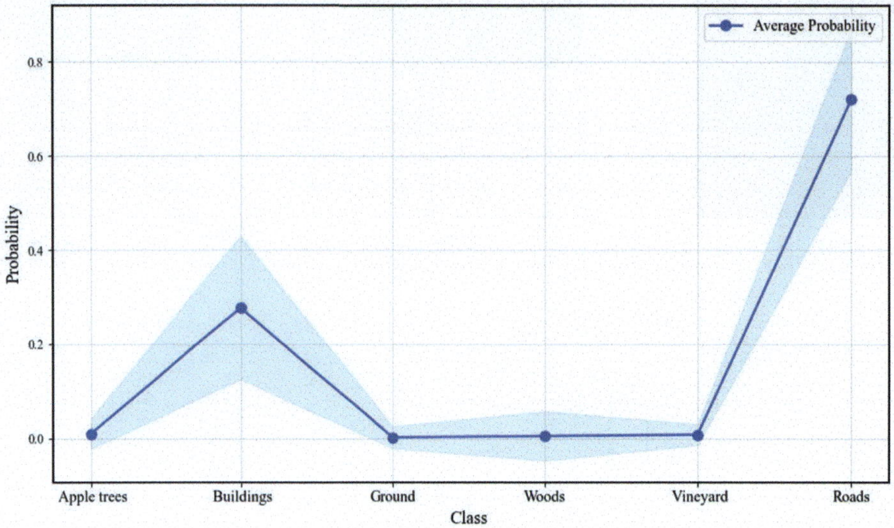

(a)

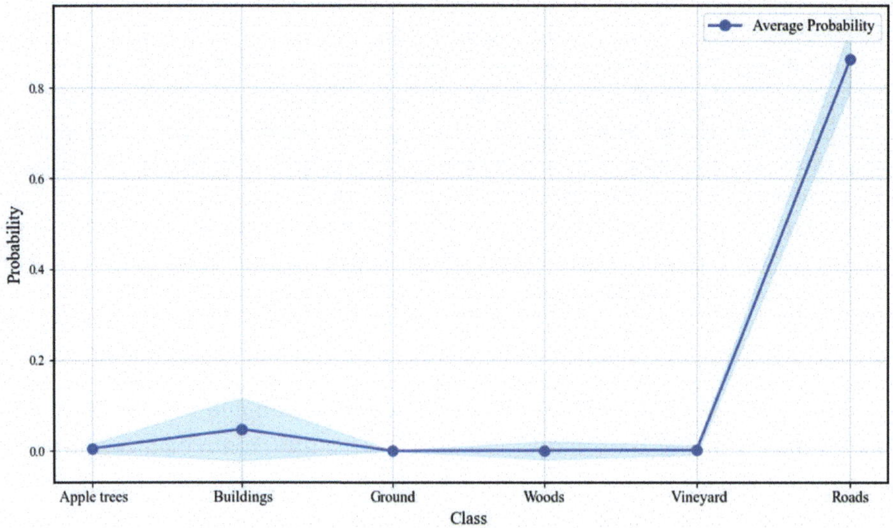

(b)

Figure 5.14:
The prediction probability for road by different models in the Trento dataset. The shaded areas representing variance. (a) Baseline-C, (b) MEDFN.

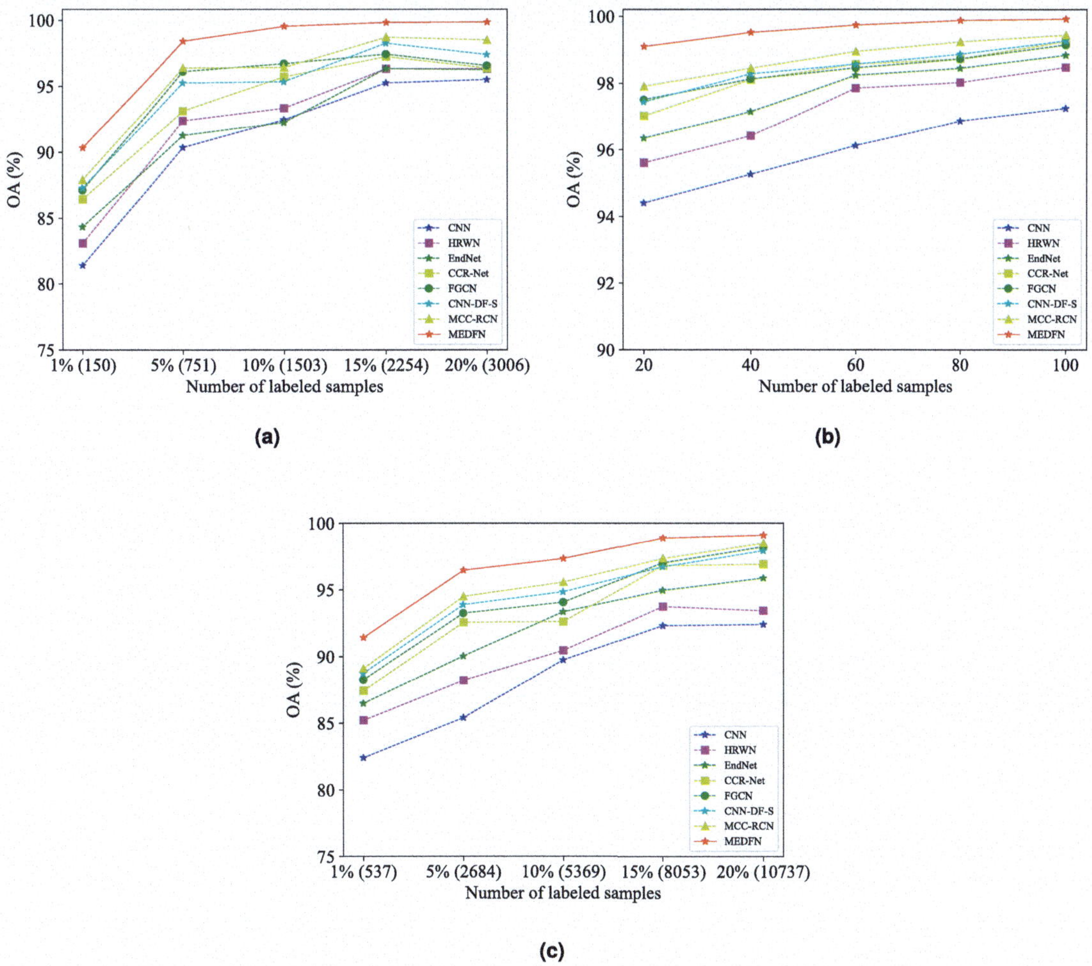

Figure 5.15:
Classification performance of all methods with different training set allocation on the three datasets. (a) Houston dataset, (b) Trento dataset, (c) MUUFL dataset.

5.1.3.5 Training samples

We compared the classification performance of MEDFN and its comparative methods under different training set allocation. For Houston dataset and MUUFL dataset, we gradually increased the number of training samples from 1% to 20%. For Trento dataset, we gradually increased the number of training samples from 20 per class to 100 per class [36]. The experimental results are shown in Fig. 5.15.

1. With the increase of the number of training samples, the OAs of all methods gained improvement. This is because more training samples can better reflect the real data distribution, which helps the model to effectively distinguish different classes of ground objects.
2. With the increase of the number of training samples, the OA rankings of some methods (e.g., CNN-DF-S and FGCN in the MUUFL dataset) changed, indicating that the allocation of training set potentially compromises the comparative performance. MEDFN shows the highest OA under different training set allocation, indicating its good adaptability to changes in the number of samples.

5.1.4 Conclusion

The proposed MEDFN accomplishes the fusion classification task by integrating HSI and LiDAR data. MEDFN is novel due to its use of reinforcement learning to convert regular multimodal data into irregular graph structures and capture multimodal spatial topological information using graph learning, thus significantly improving the accuracy of fusion classification. The main advantages of MEDFN are as follows: 1) Its graph construction strategy of learning is directly related to the classification task and can capture task-adaptive graph structures. 2) It not only considers node features but also learns edge features between ground objects, thus capturing fine spatial topology. 3) By decoupling the shared and unshared parts of multimodal features, the model can deeply explore the correlation and specificity between different modalities. In summary, MEDFN has significant advantages over other methods when facing ground object classification tasks with complex spatial distributions. The experimental results show that MEDFN achieves the most advanced classification performance on three public HSI datasets. In future research, we plan to further apply MEDFN to other types of multimodal data analysis and processing to verify its generality and effectiveness.

5.2 Multimodal remote sensing data classification based on Gaussian mixture variational dynamic fusion network

With the rapid growth of the number of sensors, the properties and applications of various remote sensing data, such as hyperspectral images (HSI) and light detection and ranging (LiDAR), have received extensive attention from researchers. HSI has high spectral resolution [37–39], which can provide key information for distinguishing ground objects that similar in appearance but significantly different in spectral [40]. The use of HSI for ground objects classification has shown promising application in many fields, including environmental monitoring, ground object analysis, agricultural management, and urban planning, etc.

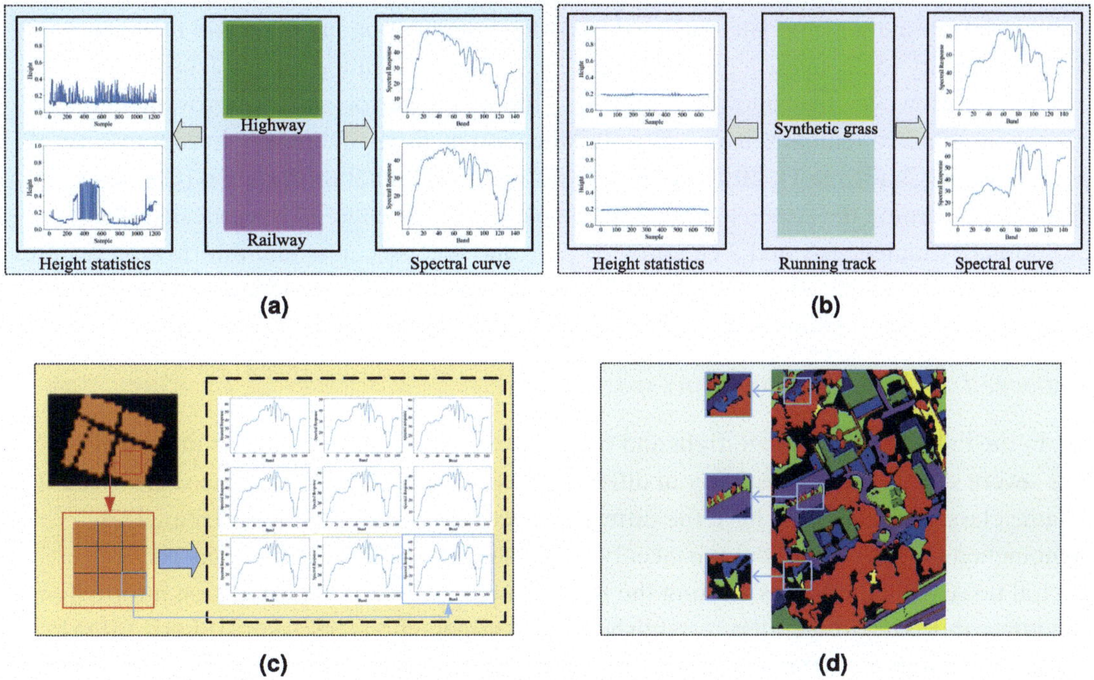

Figure 5.16:
Challenging scenarios for multimodal data fusion classification. (a) LiDAR is beneficial for distinguishing between ground objects that exhibit similar spectral curves. (b) HSI is beneficial for distinguishing between ground objects that exhibit similar heights. (c) Same class ground objects have the different spectral caused by spectral noise. (d) The spatial distribution of ground objects exhibits complex scale changes.

[41], [42]. However, it is difficult to distinguish ground objects with similar spectral using only HSI [43]. As shown in Fig. 5.16(a), highway and railway have similar spectral curves, so it is easy confuse the two classes by using only the spectral information of HSI. The elevation information contained in LiDAR can help the model distinguish the above classes with heights differences. LiDAR provides elevation information for distinguishing ground objects with similar spectral reflectance but different heights, and the quality of its data is rarely affected by weather conditions [13,44,45]. However, for LiDAR data, as shown in Fig. 5.16(b), Synthetic grass and Running track have similar heights, and it is difficult to distinguish them using LiDAR alone, but their spectral change curves are quite different. So in view of the complementary properties of HSI and LiDAR, researchers are committed to using HSI and LiDAR data collaboratively for multimodal remote sensing data classification [19,46].

Multimodal remote sensing data (HSI and LiDAR) classification methods aim to enhance the model ability to distinguish ground object by learning multimodal complementary information. However, the following challenges hinder the performance improvement of existing multimodal remote sensing data classification methods: Difficulty in suppressing spectral noise (Challenge 1), difficulty in fully mining contextual information (Challenge 2), difficulty in learning the strong adaptive fusion pattern (Challenge 3). Specifically, the resolution of Challenges 1 and 2 can provide pure and precise contextual information features for solving Challenge 3. The solution to Challenge 3 can help the model effectively capture complementary information from multimodal data with significant heterogeneity.

Challenge 1: Difficulty in suppressing spectral noise

Due to the influence of weather, light and other factors, some ground objects in the HSI have severe spectral noise, resulting in different class ground objects have the same spectral or same class ground objects have the different spectral, as shown in Fig. 5.16(c). This phenomenon leads to the model mistakenly establishing statistical correlations between spectral noise and labels, thus causing the model to ignore critical identification information. To address this issue, researchers established feature interactions between HSI and LiDAR based on attention mechanisms, guiding the model to suppress noise information. However, these methods only focus on the utilization of LiDAR elevation information, while ignoring its rich spatial prior information.

Considering because of interference of spectral noise, the collected HSI presents an inaccurate spatial topology structure: The spatial continuity of same class ground objects is poor, and the spatial boundary between different class ground objects is not clear. In contrast, LiDAR can accurately reflect the spatial topology structure of ground objects under different weather and lighting conditions [6]: The spatial continuity of same class ground objects is strong, and the spatial boundary between different class ground objects is clear. The spatial topology constraint based on Gaussian mixture (STC-GM) was proposed, using the accurate spatial topology structure of LiDAR as prior information to guide model suppress spectral noise that leads to inaccurate spatial topology structure of HSI.

Challenge 2: Difficulty in fully mining contextual information

In real-world scenarios, the spatial distribution of ground objects often exhibits complex scale changes. As shown in Fig. 5.16(d), for small scale ground objects, attention should be paid to their spectral details in the fusion process. In contrast, when dealing with large scale ground objects, both spectral details and spatial topology structure information should be taken into account. The above phenomenon leads to the difficulty of fully mining the contextual information in multimodal data, which in turn impairs the data fusion effect. Researchers have introduced multiscale techniques to enhance the model contextual mining ability by integrating information from ground objects at multiple scales. However, existing methods still have certain limitations: 1) Ignoring the mutual interference between information of different scale features. 2) Same multiscale information fusion parameters are difficult to adapt to the complex scale changes.

The multiscale dynamic graph aggregation module (MDGAM) was proposed to achieve the fully mining of two-modal contextual information in two perspectives. On the one hand, each scale feature is regarded as graph node, and the class identification information that can be shared among multiscale features is captured. Thus, the information interference between multiscale features is mitigated. On the other hand, MDGAM broke through the limitation of fixed fusion parameters and customized multiscale graph aggregation parameters for different scale.

Challenge 3: Difficulty in learning the strong adaptive fusion pattern

Due to the different imaging mechanisms, there exists modal heterogeneity between the two modalities. This modal heterogeneity leads to the mutual interference of different modal information, which increases the difficulty of the strong adaptive fusion pattern learning. Existing methods learn multimodal data fusion patterns by establishing feature interactions between multimodal data. Although these methods have achieved impressive performance, there are still some limitations: 1) It is difficult to suppress intermodal information interference induced by only establishing feature interaction. 2) The learning of multimodal data fusion pattern depends on the observation of limited samples in batch training, so it is difficult to ensure the adaptability of the model.

To address this issue, the multimodal variational autoencoder (M-VAE) is proposed to learn the strong adaptive fusion pattern. The evidence lower bound of the multimodal joint distribution is derived, and the multimodal joint distribution is learned by optimizing this evidence lower bound. Since the multimodal joint distribution directly models the global complementary fusion pattern of multimodal data through the variational inference mechanism [47], [48], the impact of modal heterogeneity on multimodal data fusion can be effectively suppressed.

The main contributions of this paper are as follows:

1) M-VAE is proposed, which differs from existing fusion methods that directly establish feature interaction between modalities. M-VAE models multimodal joint distributions for multimodal data fusion, avoiding information interference between modalities caused by modal heterogeneity.

2) MDGAM is proposed, which differs from existing methods that use the same multiscale information fusion parameters for different ground objects. MDGAM dynamically customizes

personalized fusion parameters for each ground objects, improving the model adaptability to real-world scenarios with complex scale changes.

3) STC-GM is proposed, which differs from existing methods that often ignore spatial topology information of LiDAR. STC-GM utilizes this rich information to constrain the feature learning process, achieving the spectral noise suppression.

5.2.1 Related work

5.2.1.1 Spectral noise suppression

To suppress the spectral noise in HSI, a widely used method is attention mechanism [49]. The attention mechanism suppresses noise by adjusting the model's attention to different attributes of HSI features. Fan et al. [50] designed a composite attention module containing spectral attention and spatial attention, which assigned weights based on pixel nearest neighbor relationships and spectral band importance. Wang et al. [51] captured second-order feature interactions between multimodal data to obtain accurate attention weight estimation, thereby obtaining pure multimodal features. Li et al. [52] proposed a center spectrum correction module that achieved noise interference suppression by reducing the importance of different adjacent objects.

However, most of these methods have failed to explore the spatial topology structures of LiDAR. This prevents them from further improving their performance in multimodal remote sensing data classification tasks.

5.2.1.2 Contextual information mining

Both HSI and LiDAR contain rich contextual information that reflects the spatial dependence relationship between ground objects, which directly the performance of fusion classification model. CNN, which stacks convolutional layers to expand receptive fields, is widely used for spatial information extraction in multimodal fusion classification task. Zhang et al. [53] designed a three-branch convolution that learns spectral, spatial, and elevation features, establishing the interaction between different levels of features, the model can improve the utilization ability of multimodal spatial information. Xu et al. [54] used CNN and LSTM to extract local and global spatial information of HSI, fully capturing the spatial dependency of ground objects.

The CNN-based fusion classification methods mentioned above are limited by regular convolution kernels, making it difficult to fully capture the spatial dependency relationships between irregular ground objects. Therefore, graph neural network (GCN), which is suitable for irregular structural data, has become a noteworthy method. Xiu et al. [55] explored the graph topological structure hidden in the samples. The information transfer mechanism of GCN is used to capture the irregular spatial neighborhood information. Du et al. [29] constructed a multimodal graph to fuse HSI and LiDAR spatial information, which ensured that the adjacent ground objects of the same spatial topology region are close to each other in the feature space.

The above methods only utilize spatial information from a single scale, which often fail to fully reflect the complex spatial distribution of ground objects [56]. Multiscale feature fusion technology can integrate information from multiple spatial scales, helping the model fully capture the complex spatial dependencies between ground objects. Ding et al. [13] obtained samples at multiple spatial scales by selecting different numbers of neighborhood to be classified. And the multiscale spatial information is aggregated by grouping fusion. Li et al. [57] used convolution kernels of different sizes to obtain multiscale spatial information of two modalities, thereby capturing the dependency between multiscale features.

The performance of above methods is limited by different scales features mutual interference and fixed fusion parameters, and it is difficult to adapt to the scale changes of ground objects.

5.2.1.3 Multimodal data fusion

The most widely used data fusion methods for multimodal fusion classification are decision fusion and feature fusion. Decision fusion improves the accuracy of the final decision by integrating the classification decisions of multimodal data. Yan et al. [58] adopted a dual-branch architecture of HSI and LiDAR, using each branch to independently predict ground object classes. The predictions of the two branches were connected and input into the convolutional layer for fusion predictions. Zhang et al. [59] used a decision fusion strategy based on class label level and posterior probability level to integrate classification decisions from different modalities.

Feature fusion aims to obtain multimodal fusion features by integrating multimodal information in feature space. Hu et al. [60] used the cross-attention mechanism to capture modal sharing information and fused multimodal features. Li et al. [61] first attempted to utilize the representation learning of remote sensing knowledge graphs to generate semantic representations of remote sensing scene categories. By conducting robust cross-modal alignment between visual features and semantic representations, they effectively mitigated the interference between modal information that could be induced by modal heterogeneity. Hang et al. [62] used the cross-reconstruction between HSI and LiDAR features to explore the intrinsic semantic and structural information of training samples. Then the modal sharing feature representation is obtained. Dong et al. [28] proposed a multibranch feature fusion network to obtain multimodal fusion features by simple cat HSI and LiDAR branching features. Li et al. [63] proposed the distribution balancing subconstraint and the variance maximization subconstraint, which promote cross-modal knowledge sharing reducing the multimodal class structure differences between the visual image space and the semantic space. Yao et al. [64] proposed a deep learning-based multimodal remote sensing data classification method, which contains the cross-modal multitask learning mechanism and the cross-modal attention mechanism. This method not only solved the problem of modal incompleteness in multimodal learning, but also realized efficient information exchange between different modalities. Hong et al. [65] integrated the rich spatial information of multimodal remote sensing data by different-resolution feature fusion, and extended the common two-modal fusion to three-modal scenarios. In addition, they presented a complex multimodal remote sensing dataset with challenges. Song et al. [66] designed a novel hashing-based deep metric learning method, which used similarity measure based on Hashing to excavate correlation information. This method made full use of the complementary and related information of HSI and LiDAR data from the perspective of metric learning. Song et al. [67] proposed a deep feature fusion network, which used residual learning to solve the problem of gradient vanishing during multimodal feature learning process, and integrated features of different hierarchical layers to achieve deep multimodal data fusion.

However, most of the above methods ignore the information interference between the modalities, and fail to model the joint distribution between HSI and LiDAR. It makes it difficult to ensure that the key complementary information of both modalities is retained during multimodal data fusion.

5.2.2 GM-VDFN for multimodal remote sensing data classification

5.2.2.1 Framework of GM-VDFN

The training process of GM-VDFN is shown in Fig. 5.17, which can be divided into the following steps:

I. *Multiscale feature learning*. First, the K-nearest neighbor search is used to divide HSI and LiDAR into K-nearest neighbor ground objects clusters. Second, HSI and LiDAR single scale spatial graph are constructed based on the K-nearest ground objects cluster. Third, multiscale spatial graphs of HSI and LiDAR are obtained by adjusting the number of K neighbors and repeating the above process. Fourth, each scale spatial graph of HSI and LiDAR is input into the GCN for node feature aggregation, and the node features of each node are updated. Fifth, calculate the spatial topology matrix of two modalities based on node features, and model the spatial topology matrix as a Gaussian mixture distribution. Finally, the STC-GM is used to align the spatial distributions of two domains, and then force the nodes in the same topological region of two modalities to have similar features, thereby the pure two-modal multiscale feature set is obtained.

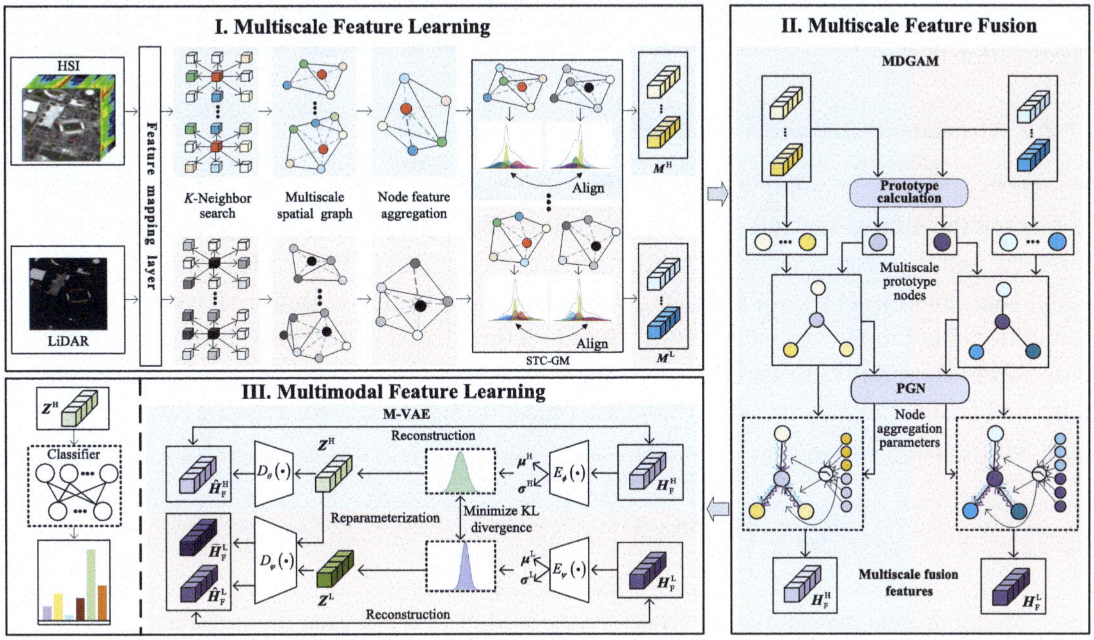

Figure 5.17:
Framework of GM-VDFN. I. The two-modal multiscale spatial graphs were constructed to input GCN for node feature learning, and pure two-modal multiscale features are obtained with the constraint of STC-GM. II. MDGAM is used to fuse the pure two-modal multiscale features separately, obtaining the multiscale fusion features of the two modalities. III. Multiscale fusion features of the two modalities were input into M-VAE to model multimodal joint distributions, and Z^{H} containing multimodal joint distribution information was input into the classifier for class prediction.

II. *Multiscale feature fusion*. First, the pure two-modal multiscale feature set is input into the MDGAM. MDGAM treats the features of each scale as graph nodes, and obtains two-modal multiscale prototype nodes through prototype calculation. Second, the connection relationship between multiscale prototype nodes and each scale node is established, constructing a two-modal multiscale prototype graph. Finally, parameter generated network (PGN) generates node aggregation parameters dynamically based on the edge features between the multiscale prototype node and each scale node. The information transfer mechanism is used to aggregate the multiscale node information, and the multiscale fusion features of two modalities are obtained.

III. *Multimodal feature learning*. First, the multiscale fusion features of the two modalities are input into the M-VAE. By optimizing the evidence lower bound of the M-VAE, the joint distribution of the two modalities is modeled, and the multimodal features are obtained. Then, the multimodal features are entered into the classifier to complete the remote sensing data classification task.

5.2.2.2 Multiscale spatial graph construction

In order to achieve computational acceleration and batch learning similar to miniGCN [35], we first cut the remote sensing data into several patches centered on the target pixels to be classified, and then constructed the spatial graphs of each patch. In both HSI and LiDAR, spatial adjacent objects with high spatial consistency and feature similarity have a high probability of belong to same class [68]. Based on this spatial prior, the K-nearest neighbor composition is used [69] to segment HSI and LiDAR into clusters, so as to make full use of the spatial contextual information around the object (to be classified). When $\boldsymbol{x}_i$ and $\boldsymbol{x}_j$ are K neighbors, an edge connection is created between nodes can be expressed as:

$$A_{i,j} = \begin{cases} D\left(\boldsymbol{x}_i, \boldsymbol{x}_j\right), & \text{if } \boldsymbol{x}_i \in N_{\boldsymbol{x}_j} \text{ and } \boldsymbol{x}_j \in N_{\boldsymbol{x}_i} \\ 0, & \text{otherwise} \end{cases} \tag{5.18}$$

where $\mathrm{N}_{\boldsymbol{x}}$ represents the set of K neighbors of $\boldsymbol{x}$, and $D(\cdot)$ represents the measure of similarity between ground objects, which can be expressed as:

$$D(\boldsymbol{x}_i, \boldsymbol{x}_j) = \frac{\sqrt{\left(\mathrm{cord}(\boldsymbol{x}_i) - \mathrm{cord}(\boldsymbol{x}_j)\right)^{\mathrm{T}} \left(\mathrm{cord}(\boldsymbol{x}_i) - \mathrm{cord}(\boldsymbol{x}_j)\right)}}{\cos\left(\frac{\boldsymbol{x}_i \cdot \boldsymbol{x}_j}{\|\boldsymbol{x}_i\|_2 \|\boldsymbol{x}_j\|_2}\right)} \tag{5.19}$$

where $\mathrm{cord}\left(\boldsymbol{x}_i\right)$ and $\mathrm{cord}\left(\boldsymbol{x}_j\right)$ represent the spatial coordinates of $\boldsymbol{x}_i$ and $\boldsymbol{x}_j$, respectively. The above measure considers the spectral similarity and spatial distance between ground objects, which can comprehensively reflect the dependencies between ground objects.

Then, according to the adjacency matrix $\boldsymbol{A}$, the edge set E of the single-scale spatial graph is obtained. Each ground object is treated as a graph node to obtain the node set V, and the single-scale spatial graph $G = (V, E)$ is obtained.

Finally, by setting different K-nearest neighbor numbers during the construction of HSI and LiDAR spatial graphs, a two-modal multiscale spatial graph set can be obtained:

$$\begin{cases} G^{\mathrm{H}}\left(V^{\mathrm{H}}, E^{\mathrm{H}}\right) = \left\{G^{\mathrm{H}}\left(V_0^{\mathrm{H}}, E_0^{\mathrm{H}}\right), \ldots, G^{\mathrm{H}}\left(V_S^{\mathrm{H}}, E_S^{\mathrm{H}}\right)\right\} \\ G^{\mathrm{L}}\left(V^{\mathrm{L}}, E^{\mathrm{L}}\right) = \left\{G^{\mathrm{L}}\left(V_0^{\mathrm{L}}, E_0^{\mathrm{L}}\right), \ldots, G^{\mathrm{L}}\left(V_S^{\mathrm{L}}, E_S^{\mathrm{L}}\right)\right\} \end{cases} \tag{5.20}$$

where $G^{\mathrm{H}}\left(V^{\mathrm{H}}, E^{\mathrm{H}}\right)$ is the multiscale spatial graph set of HSI, $G^{\mathrm{L}}\left(V^{\mathrm{L}}, E^{\mathrm{L}}\right)$ is the multiscale spatial graph set of LiDAR, and S is the number of scales.

5.2.2.3 Feature extraction module based on GCN

First, given a graph $G = (V, E)$ and its corresponding adjacency matrix $\boldsymbol{A}$, the corresponding Laplacian matrix $\boldsymbol{L} = \boldsymbol{D} - \boldsymbol{A}$ can be obtained, where $\boldsymbol{D}$ is the degree matrix of $\boldsymbol{A}$ and can be expressed as $\boldsymbol{D}_{i,i} = \sum_j \boldsymbol{A}_{i,j}$.

Then, the normalized Laplacian matrix is obtained:

$$\boldsymbol{L}_n = \boldsymbol{I} - \boldsymbol{D}^{-\frac{1}{2}}\boldsymbol{A}\boldsymbol{D}^{-\frac{1}{2}} \tag{5.21}$$

where $\boldsymbol{I}$ is the identity matrix. After spectral decomposition of the normalized Laplace matrix, $\boldsymbol{L}_n = \boldsymbol{U}\boldsymbol{\Sigma}\boldsymbol{U}^{\mathrm{T}}$ can be obtained, where $\boldsymbol{U}$ is the set of eigenvectors in $\boldsymbol{L}_n$, and $\boldsymbol{\Sigma}$ is the diagonal matrix of $\boldsymbol{L}_n$. The convolution operation on the graph can be expressed as:

$$\boldsymbol{v} * \boldsymbol{f}_\theta = \boldsymbol{U}\boldsymbol{f}_\theta(\boldsymbol{\Sigma})\boldsymbol{U}^{\mathrm{T}}\boldsymbol{v} \tag{5.22}$$

where $\boldsymbol{v}$ represents the node feature, $\boldsymbol{f}_\theta$ can be regarded as a function of normalized Laplace matrix eigenvalues $\boldsymbol{\Sigma}$, and θ is a learnable parameter. Since the calculation of the above equation is difficult, it is approximated using the first-order expansion of Chebyshev polynomial [70]:

$$\boldsymbol{v} * \boldsymbol{f}_\theta \approx \theta\left(\boldsymbol{I} + \boldsymbol{D}^{-\frac{1}{2}}\boldsymbol{A}\boldsymbol{D}^{-\frac{1}{2}}\right) \tag{5.23}$$

Finally, the propagation rules for GCN are available:

$$\boldsymbol{H}^{(l+1)} = \sigma\left(\tilde{\boldsymbol{D}}^{-\frac{1}{2}}\tilde{\boldsymbol{A}}\tilde{\boldsymbol{D}}^{-\frac{1}{2}}\boldsymbol{H}^{(l)}\boldsymbol{W}^{(l+1)} + \boldsymbol{b}^{(l+1)}\right) \tag{5.24}$$

where $\tilde{\boldsymbol{A}} = \boldsymbol{A} + \boldsymbol{I}$, $\tilde{\boldsymbol{D}}$ is the degree matrix of $\tilde{\boldsymbol{A}}$, $\boldsymbol{\sigma}$ is the activation function, $\boldsymbol{H}^{(l)}$ is the output of the l-layer graph convolution, $\boldsymbol{W}^{(l+1)}$ and $\boldsymbol{b}^{(l+1)}$ are the learnable weight matrix and bias term in the $(l+1)$-layer graph convolution, respectively.

The input of the feature extraction module is a multiscale spatial graph set of two modalities, which outputs multiscale features sets $\boldsymbol{M}^{\mathrm{H}}$ and $\boldsymbol{M}^{\mathrm{L}}$ of HSI and LiDAR:

$$\begin{cases} \boldsymbol{M}^{\mathrm{H}} = \left[\boldsymbol{H}_1^{\mathrm{H}}, \ldots, \boldsymbol{H}_S^{\mathrm{H}}\right] \\ \boldsymbol{M}^{\mathrm{L}} = \left[\boldsymbol{H}_1^{\mathrm{L}}, \ldots, \boldsymbol{H}_S^{\mathrm{L}}\right] \end{cases} \tag{5.25}$$

5.2.2.4 Spectral noise suppression based on STC-GM

HSI can provide rich spectral information and help distinguish ground objects with different spectral properties. But it is susceptible to spectral noise, making it difficult for model to

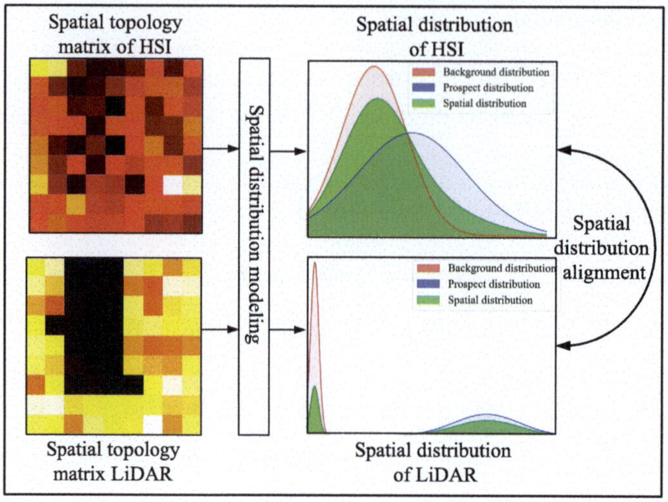

Figure 5.18:
Flowchart of the proposed STC-GM.

distinguish different ground objects. To solve this problem, STC-GM is proposed. The spatial prior that ground objects in the same topology region with a high probability of belonging to the same class, and the precise spatial topology structure of LiDAR are used to constrain the features of the ground objects, realizing the spectral noise suppression. The spatial distribution of HSI and LiDAR is modeled as a Gaussian mixture distribution. By explicitly constraining ground objects in the same topological region to have consistent features, spectral noise is suppressed.

As shown in Fig. 5.18, first, to describe the spatial topology structures of the two modalities (HSI and LiDAR), the similarity between each pixel feature and the central pixel feature in the image patch is calculated to obtain the spatial topology matrixes, and obtain the spatial topology matrices $\boldsymbol{C}^{\mathrm{H}}$ and $\boldsymbol{C}^{\mathrm{L}}$ of HSI and LiDAR:

$$\boldsymbol{C}_i = \frac{\sum_{e=1}^{E} \boldsymbol{v}_i^e \times \boldsymbol{v}_c^e}{\sqrt{\sum_{e=1}^{E} \boldsymbol{v}_i^e} \times \sqrt{\sum_{e=1}^{E} \boldsymbol{v}_c^e}} \tag{5.26}$$

where $\boldsymbol{v}_i^e$ and $\boldsymbol{v}_c^e$ represent the $\boldsymbol{e}$-dimensional features of pixel $\boldsymbol{v}_i$ and central pixel $\boldsymbol{v}_c$ extracted by feature extraction module respectively, and E is the node feature dimension.

Second, based on the spatial topology matrixes, the Gaussian mixture model is used to model the spatial distributions (green) of the two modalities. The spatial distribution includes the prospect distribution (blue) and background distribution (red). The former represents the spatial distribution of pixels of the same class as the central pixel, and the latter represents the spatial distribution of pixels of the different class as the central pixel. Due to the influence of spectral noise, the spatial continuity of the prospect in HSI is poor, and the spatial boundaries between the prospect and background are difficult to distinguish. The former leads to a high variance in the prospect distribution, while the latter results in a unimodal spatial distribution due to the mixing of prospect distribution and background distribution. In contrast, the spatial continuity of the prospect in LiDAR pixels is high, and the spatial boundary between the prospect and background is clear. At this point, the prospect distribution shows low variance, while the spatial distribution shows a bimodal pattern due to the separation of the prospect and background distributions.

We randomly initialize a set of Gaussian distribution mean matrix, variance matrix and their mixture weight of prospect distribution and background distribution. Then, the expectation maximization is used to optimize the parameters of the Gaussian mixture distribution.

In the Expectation step, parameters of the Gaussian distribution are fixed, and the posterior probability that each data point in the spatial topology matrix comes from each Gaussian distribution is calculated. For data point C_b and Gaussian distribution $\mathrm{N}\left(\boldsymbol{\mu}_j, \boldsymbol{\Sigma}_j\right)$, the posterior probability is calculated as:

$$\gamma_{bj} = \frac{\boldsymbol{\Psi}_j \mathrm{N}\left(\boldsymbol{C}_b|\boldsymbol{\mu}_j, \boldsymbol{\Sigma}_j\right)}{\sum_{m=1}^{M} \boldsymbol{\Psi}_m \mathrm{N}\left(\boldsymbol{C}_b|\boldsymbol{\mu}_m, \boldsymbol{\Sigma}_m\right)} \tag{5.27}$$

where $\boldsymbol{\Psi}_j$ is the mixed weight of the Gaussian distribution $\mathrm{N}\left(\boldsymbol{\mu}_j, \boldsymbol{\Sigma}_j\right)$, and $\mathrm{N}\left(\boldsymbol{C}_b|\boldsymbol{\mu}_j, \boldsymbol{\Sigma}_j\right)$ is the probability density of $\mathrm{N}\left(\boldsymbol{\mu}_j, \boldsymbol{\Sigma}_j\right)$ at $\boldsymbol{C}_b$.

In the Maximization step, with a fixed posterior probability, update the parameters of each Gaussian distribution as follows:

$$\begin{cases} \boldsymbol{\mu}_j = \dfrac{\sum_{b=1}^{B} \gamma_{bj} \boldsymbol{C}_b}{\sum_{b=1}^{B} \gamma_{bj}} \\ \boldsymbol{\Sigma}_j = \dfrac{\sum_{b=1}^{B} \gamma_{bj} \left(\boldsymbol{C}_b - \boldsymbol{\mu}_j^{\text{new}}\right)\left(\boldsymbol{C}_b - \boldsymbol{\mu}_j^{\text{new}}\right)^{\mathrm{T}}}{\sum_{b=1}^{B} \gamma_{bj}} \\ \boldsymbol{\Psi}_j = \dfrac{1}{B} \sum_{b=1}^{B} \gamma_{bj} \end{cases} \tag{5.28}$$

where B is the number of data points in the spatial topology matrix.

The Expectation step and Maximization step are repeated until the method converges, and then the Gaussian mixture distribution of the spatial topology matrix is obtained:

$$p\left(C_b\right)=\sum\nolimits_{m=1}^{M}\boldsymbol{\Psi}_m\frac{1}{(2\pi)^{\frac{1}{2}}\left|\boldsymbol{\Sigma}_m\right|^{\frac{1}{2}}}e^{-\frac{1}{2}\left(C_b-\boldsymbol{\mu}_m\right)^T\boldsymbol{\Sigma}_m^{-1}\left(C_b-\boldsymbol{\mu}_m\right)} \tag{5.29}$$

Finally, using the spatial distribution of LiDAR as prior information, the variance of the HSI prospect distribution is reduced through spatial distribution alignment, while promoting the separation of its prospect and background distributions. Correspondingly, STC-GM constrains the output features of the feature extraction module during model training, guiding the model to suppress spectral noise through backpropagation. This operation enhances the spatial continuity of same class ground objects and enhances the distinguishability of spatial boundaries between different class ground objects in HSI. Specifically, spatial distribution alignment is achieved by minimizing the KL divergence between $p^{\mathrm{H}}=\sum\limits_{m=1}^{M}\boldsymbol{\Psi}_m^{\mathrm{H}}N\left(\boldsymbol{C}^{\mathrm{H}}|\boldsymbol{\mu}_m^{\mathrm{H}},\boldsymbol{\Sigma}_m^{\mathrm{H}}\right)$ and $p^{\mathrm{L}}=\sum\limits_{m=1}^{M}\boldsymbol{\Psi}_m^{\mathrm{L}}N\left(\boldsymbol{C}^{\mathrm{L}}|\boldsymbol{\mu}_m^{\mathrm{L}},\boldsymbol{\Sigma}_m^{\mathrm{L}}\right)$. We use Monte Carlo integrals to approximate the KL divergence between p^{H} and p^{L}, and the spatial topology constraint L_{s} can be expressed as:

$$L_{\mathrm{s}}=D_{\mathrm{KL}}\left(p^{\mathrm{L}}||p^{\mathrm{H}}\right)\approx\frac{1}{B}\sum_{b=1}^{B}p^{\mathrm{L}}\left(\boldsymbol{C}_b^{\mathrm{L}}\right)\log\frac{p^{\mathrm{L}}\left(\boldsymbol{C}_b^{\mathrm{L}}\right)}{p^{\mathrm{H}}\left(\boldsymbol{C}_b^{\mathrm{H}}\right)} \tag{5.30}$$

5.2.2.5 Contextual information fully mining based on MDGAM

MDGAM aggregates multiscale features to capture multilevel spatial structure and contextual information from rough to fine, which realizes accurate descriptions of ground objects at different scales.

First, each scale feature in the multiscale feature set $\boldsymbol{M}=\{\boldsymbol{H}_1,\ldots,\boldsymbol{H}_K\}$ is regarded as graph nodes. The edge connections are established between these nodes and multiscale prototype node $\boldsymbol{P}=\frac{1}{K}\sum\limits_{k=1}^{K}\boldsymbol{H}_k$, constructing a two-modal multiscale prototype graph.

Second, the attention coefficient between $\boldsymbol{P}$ and each scale node is calculated, and the attention coefficient $\boldsymbol{\alpha}_k$ of the scale k node can be expressed as:

$$\boldsymbol{\alpha}_k=\frac{\exp\left(\mathrm{LeakyReLU}\left(\boldsymbol{e}_k\right)\right)}{\sum\limits_{i=1}^{K}\exp\left(\mathrm{LeakyReLU}\left(\boldsymbol{e}_i\right)\right)},\ \boldsymbol{e}_k=\boldsymbol{W}^{\mathrm{A}}\left([\boldsymbol{W}\boldsymbol{H}_k||\boldsymbol{W}\boldsymbol{P}]\right) \tag{5.31}$$

where $\boldsymbol{W}^{\mathrm{A}}$ is the attention matrix, and $\boldsymbol{W}$ is the node mapping matrix.

Third, based on the edge feature $\boldsymbol{e}_k$, the node aggregation parameters are dynamically generated to aggregate the multiscale features, obtaining the multiscale fusion feature $\boldsymbol{H}_{\mathrm{F}}$.

$$\boldsymbol{H}_{\mathrm{F}} = \sigma\left(\sum_{k=1}^{K} \alpha_k \widehat{\boldsymbol{W}}_k \boldsymbol{H}_k\right), \ \widehat{\boldsymbol{W}}_k = \mathrm{PGN}(\boldsymbol{e}_k) \tag{5.32}$$

where $\sigma(\cdot)$ is the activation function and $\mathrm{PGN}(\cdot)$ is the parameter generation network, which is composed of multilayer perception. The input of PGN is an edge feature that can reflect the dependency between multiscale prototype nodes and multiscale nodes. The output is the aggregation parameter corresponding to the multiscale node.

The input of MDGAM is a multiscale feature set of two modalities, and the output is the multiscale fusion feature $\boldsymbol{H}_{\mathrm{F}}^{\mathrm{H}}$ of HSI and the multiscale fusion feature $\boldsymbol{H}_{\mathrm{F}}^{\mathrm{L}}$ of LiDAR.

5.2.2.6 Multimodal feature fusion based on M-VAE

To overcome the heterogeneity between HSI and LiDAR and obtain high quality multimodal features, M-VAE is proposed. The multimodal data fusion is transformed into a multimodal joint distribution modeling. Thus, the powerful distribution modeling ability of VAE is utilized to model the multimodal joint distribution of HSI and LiDAR. Specifically, M-VAE derives the evidence lower bound for the multimodal joint distribution of HSI and LiDAR, and learns the multimodal joint distribution by optimizing the evidence lower bound. Furthermore, high-quality fusion features that can capture multimodal joint distribution are obtained in the hidden space.

The evidence lower bound of multimodal joint distribution can be expressed as:

$$\begin{aligned}\log p_{\theta,\varphi}\left(\boldsymbol{H}_{\mathrm{F}}^{\mathrm{L}}, \boldsymbol{H}_{\mathrm{F}}^{\mathrm{H}}\right) &= \log \int_{\boldsymbol{Z}} p_{\theta,\varphi}\left(\boldsymbol{H}_{\mathrm{F}}^{\mathrm{L}}, \boldsymbol{H}_{\mathrm{F}}^{\mathrm{H}}, \boldsymbol{Z}\right) d\boldsymbol{Z} \\ &\geq \int_{\boldsymbol{Z}} \frac{p_{\theta,\psi}\left(\boldsymbol{H}_{\mathrm{F}}^{\mathrm{L}}, \boldsymbol{H}_{\mathrm{F}}^{\mathrm{H}}, \boldsymbol{Z}\right)}{q_{\phi,\varphi}\left(\boldsymbol{Z}|\boldsymbol{H}_{\mathrm{F}}^{\mathrm{L}}, \boldsymbol{H}_{\mathrm{F}}^{\mathrm{H}}\right)} q_{\phi,\varphi}\left(\boldsymbol{Z}|\boldsymbol{H}_{\mathrm{F}}^{\mathrm{L}}, \boldsymbol{H}_{\mathrm{F}}^{\mathrm{H}}\right) d\boldsymbol{Z}\end{aligned} \tag{5.33}$$

Use conditional independence $\boldsymbol{H}_{\mathrm{F}}^{\mathrm{L}} \perp \boldsymbol{H}_{\mathrm{F}}^{\mathrm{H}} | \boldsymbol{Z}$ [71] and Bayer rule to model $q_{\phi,\varphi}\left(\boldsymbol{Z}|\boldsymbol{H}_{\mathrm{F}}^{\mathrm{L}}, \boldsymbol{H}_{\mathrm{F}}^{\mathrm{H}}\right)$:

$$\begin{cases} q_{\phi,\varphi}\left(\boldsymbol{Z}|\boldsymbol{H}_{\mathrm{F}}^{\mathrm{L}}, \boldsymbol{H}_{\mathrm{F}}^{\mathrm{H}}\right) = \dfrac{q_{\phi}\left(\boldsymbol{Z}|\boldsymbol{H}_{\mathrm{F}}^{\mathrm{H}}\right) q_{\varphi}\left(\boldsymbol{H}_{\mathrm{F}}^{\mathrm{L}}|\boldsymbol{Z}\right)}{q_{\phi,\varphi}\left(\boldsymbol{H}_{\mathrm{F}}^{\mathrm{L}}|\boldsymbol{H}_{\mathrm{F}}^{\mathrm{H}}\right)} \\ q_{\phi,\varphi}\left(\boldsymbol{H}_{\mathrm{F}}^{\mathrm{L}}|\boldsymbol{H}_{\mathrm{F}}^{\mathrm{H}}\right) = \displaystyle\int q_{\phi}\left(\boldsymbol{Z}|\boldsymbol{H}_{\mathrm{F}}^{\mathrm{H}}\right) q_{\varphi}\left(\boldsymbol{H}_{\mathrm{F}}^{\mathrm{L}}|\boldsymbol{Z}\right) d\boldsymbol{Z} \end{cases} \tag{5.34}$$

Thus, the new evidence lower bound can be expressed as:

$$\begin{aligned} \log p_{\theta,\psi}\left(\boldsymbol{H}_{\mathrm{F}}^{\mathrm{L}}, \boldsymbol{H}_{\mathrm{F}}^{\mathrm{H}}\right) &\geq L_{\mathrm{F}\{\Theta,\Psi\}}\left(\boldsymbol{H}_{\mathrm{F}}^{\mathrm{L}}, \boldsymbol{H}_{\mathrm{F}}^{\mathrm{H}}\right) \\ &= \int_{Z} \frac{p_{\theta,\psi}\left(\boldsymbol{H}_{\mathrm{F}}^{\mathrm{H}}, \boldsymbol{H}_{\mathrm{F}}^{\mathrm{L}}, \mathbf{Z}\right) q_{\phi,\varphi}\left(\boldsymbol{H}_{\mathrm{F}}^{\mathrm{L}} | \boldsymbol{H}_{\mathrm{F}}^{\mathrm{H}}\right)}{q_{\phi}\left(\mathbf{Z} | \boldsymbol{H}_{\mathrm{F}}^{\mathrm{H}}\right) q_{\varphi}\left(\boldsymbol{H}_{\mathrm{F}}^{\mathrm{L}} | \mathbf{Z}\right)} \frac{q_{\phi}\left(\mathbf{Z} | \boldsymbol{H}_{\mathrm{F}}^{\mathrm{H}}\right) q_{\varphi}\left(\boldsymbol{H}_{\mathrm{F}}^{\mathrm{L}} | \mathbf{Z}\right)}{q_{\phi,\varphi}\left(\boldsymbol{H}_{\mathrm{F}}^{\mathrm{L}} | \boldsymbol{H}_{\mathrm{F}}^{\mathrm{H}}\right)} d\mathbf{Z} \\ &= \mathbb{E}_{q_{\phi}\left(\mathbf{Z}|\boldsymbol{H}_{\mathrm{F}}^{\mathrm{H}}\right)}\left[\frac{q_{\varphi}\left(\boldsymbol{H}_{\mathrm{F}}^{\mathrm{L}} | \mathbf{Z}\right)}{q_{\phi,\varphi}\left(\boldsymbol{H}_{\mathrm{F}}^{\mathrm{L}} | \boldsymbol{H}_{\mathrm{F}}^{\mathrm{H}}\right)} \log \frac{p_{\theta}\left(\boldsymbol{H}_{\mathrm{F}}^{\mathrm{H}} | \mathbf{Z}\right) p_{\psi}\left(\mathbf{Z} | \boldsymbol{H}_{\mathrm{F}}^{\mathrm{L}}\right)}{q_{\phi}\left(\mathbf{Z} | \boldsymbol{H}_{\mathrm{F}}^{\mathrm{H}}\right) q_{\varphi}\left(\boldsymbol{H}_{\mathrm{F}}^{\mathrm{L}} | \mathbf{Z}\right)}\right] \\ &+ \log q_{\phi,\varphi}\left(\boldsymbol{H}_{\mathrm{F}}^{\mathrm{L}} | \boldsymbol{H}_{\mathrm{F}}^{\mathrm{H}}\right) + \log p\left(\boldsymbol{H}_{\mathrm{F}}^{\mathrm{L}}\right) \end{aligned} \tag{5.35}$$

where $\Theta = \{\theta, \psi\}$, $\Psi = \{\phi, \varphi\}$. ϕ and θ are the encoder and decoder parameters for $\boldsymbol{H}_{\mathrm{F}}^{\mathrm{H}}$, ψ and φ are the encoder and decoder parameters for $\boldsymbol{H}_{\mathrm{F}}^{\mathrm{L}}$.

$$\mathbb{E}_{q_{\phi}\left(\mathbf{Z}|\boldsymbol{H}_{\mathrm{F}}^{\mathrm{L}}\right)}\left[\frac{q_{\varphi}\left(\boldsymbol{H}_{\mathrm{F}}^{\mathrm{L}} | \mathbf{Z}\right)}{q_{\phi,\varphi}\left(\boldsymbol{H}_{\mathrm{F}}^{\mathrm{L}} | \boldsymbol{H}_{\mathrm{F}}^{\mathrm{H}}\right)} \log \frac{p_{\theta}\left(\boldsymbol{H}_{\mathrm{F}}^{\mathrm{H}} | \mathbf{Z}\right) p_{\psi}\left(\mathbf{Z} | \boldsymbol{H}_{\mathrm{F}}^{\mathrm{L}}\right)}{q_{\phi}\left(\mathbf{Z} | \boldsymbol{H}_{\mathrm{F}}^{\mathrm{H}}\right) q_{\varphi}\left(\boldsymbol{H}_{\mathrm{F}}^{\mathrm{L}} | \mathbf{Z}\right)}\right]$$

is the weighted likelihood error of two-modal reconstruction. On the one hand, the encoder and decoder are trained by two-modal feature reconstruction. On the other hand, the KL divergence of the two modal posterior distributions $p_{\psi}\left(\mathbf{Z}|\boldsymbol{H}_{\mathrm{F}}^{\mathrm{L}}\right)$ and $p_{\psi}\left(\mathbf{Z}|\boldsymbol{H}_{\mathrm{F}}^{\mathrm{L}}\right)$ is minimized, so that the model learns the modal shareable features. $\log q_{\phi,\varphi}\left(\boldsymbol{H}_{\mathrm{F}}^{\mathrm{L}}|\boldsymbol{H}_{\mathrm{F}}^{\mathrm{H}}\right)$ uses $\boldsymbol{H}_{\mathrm{F}}^{\mathrm{H}}$ to pseudoreconstruct $\boldsymbol{H}_{\mathrm{F}}^{\mathrm{L}}$, which establishes information interaction between two modalities through modal transformation. $\log p\left(\boldsymbol{H}_{\mathrm{F}}^{\mathrm{L}}\right)$ is a constant and can be ignored during optimization. Based on this evidence lower bound, we construct the M-VAE that contains the two-modal encoder, two-modal decoder, and sample weighted module.

First, $\boldsymbol{H}_{\mathrm{F}}^{\mathrm{H}}$ and $\boldsymbol{H}_{\mathrm{F}}^{\mathrm{L}}$ are entered into encoders E_{ϕ} and E_{ψ}, respectively, obtaining the mean and variance of the posterior distribution:

$$\begin{cases} \boldsymbol{\mu}^{\mathrm{H}}, \boldsymbol{\sigma}^{\mathrm{H}} = E_{\phi}\left(\boldsymbol{H}_{\mathrm{F}}^{\mathrm{H}}\right) \\ \boldsymbol{\mu}^{\mathrm{L}}, \boldsymbol{\sigma}^{\mathrm{L}} = E_{\psi}\left(\boldsymbol{H}_{\mathrm{F}}^{\mathrm{L}}\right) \end{cases} \tag{5.36}$$

Second, the reparameterization [72] technique is used to obtain the two-modal latent variables $\mathbf{Z}^{\mathrm{H}}$ and $\mathbf{Z}^{\mathrm{L}}$:

$$\begin{cases} \mathbf{Z}^{\mathrm{H}} = \boldsymbol{\mu}^{\mathrm{H}} + \boldsymbol{\sigma}^{\mathrm{H}} * \boldsymbol{\varepsilon}^{\mathrm{H}} \\ \mathbf{Z}^{\mathrm{L}} = \boldsymbol{\mu}^{\mathrm{L}} + \boldsymbol{\sigma}^{\mathrm{L}} * \boldsymbol{\varepsilon}^{\mathrm{L}} \end{cases} \tag{5.37}$$

where $\boldsymbol{\varepsilon}^{\mathrm{H}}$ and $\boldsymbol{\varepsilon}^{\mathrm{L}}$ are sampled from the standard multivariate Gaussian distribution.

Third, the two-modal latent variables are input into the decoders $\boldsymbol{D}_\theta$ and $\boldsymbol{D}_\varphi$, thus $\boldsymbol{H}_{\mathrm{F}}^{\mathrm{H}}$ and $\boldsymbol{H}_{\mathrm{F}}^{\mathrm{L}}$ are reconstructed:

$$\begin{cases} \hat{\boldsymbol{H}}_{\mathrm{F}}^{\mathrm{H}} = \boldsymbol{D}_\theta\left(\mathbf{Z}^{\mathrm{H}}\right) \\ \hat{\boldsymbol{H}}_{\mathrm{F}}^{\mathrm{L}} = \boldsymbol{D}_\varphi\left(\mathbf{Z}^{\mathrm{L}}\right) \end{cases} \tag{5.38}$$

In addition, in order to capture the complementary properties of the two modalities, the latent variable corresponding to $\boldsymbol{H}_{\mathrm{F}}^{\mathrm{H}}$ is input into decoder $\boldsymbol{D}_\varphi$ for reconstructing $\boldsymbol{H}_{\mathrm{F}}^{\mathrm{L}}$. The pseudoreconstruction feature $\bar{\boldsymbol{H}}_{\mathrm{F}}^{\mathrm{L}}$ is obtained:

$$\bar{\boldsymbol{H}}_{\mathrm{F}}^{\mathrm{L}} = \boldsymbol{D}_\varphi\left(\mathbf{Z}^{\mathrm{H}}\right) \tag{5.39}$$

Fourth, calculate the difference between pseudoreconstructed feature $\bar{\boldsymbol{H}}_{\mathrm{F}}^{\mathrm{L}}$ and reconstruct feature $\hat{\boldsymbol{H}}_{\mathrm{F}}^{\mathrm{L}}$ for weighting the sample. Thus the posterior distribution $q_\phi\left(\mathbf{Z}^{\mathrm{H}} | \boldsymbol{H}_{\mathrm{F}}^{\mathrm{H}}\right)$ can approximate the joint distribution of the two modalities, and the sample weight $\boldsymbol{w}$ can be expressed as:

$$\boldsymbol{w} = \exp\left(\log q_\varphi\left(\boldsymbol{H}_{\mathrm{F}}^{\mathrm{L}} | \mathbf{Z}^{\mathrm{L}}\right) - \log q_{\phi,\varphi}\left(\boldsymbol{H}_{\mathrm{F}}^{\mathrm{L}} | \boldsymbol{H}_{\mathrm{F}}^{\mathrm{H}}\right)\right) \tag{5.40}$$

Fifth, the loss of M-VAE can be obtained:

$$\begin{aligned} L_{\mathrm{F}} = \boldsymbol{w}\Bigg(&\left\|\hat{\boldsymbol{H}}_{\mathrm{F}}^{\mathrm{H}} - \boldsymbol{H}_{\mathrm{F}}^{\mathrm{H}}\right\|_2^2 + \left\|\hat{\boldsymbol{H}}_{\mathrm{F}}^{\mathrm{L}} - \boldsymbol{H}_{\mathrm{F}}^{\mathrm{L}}\right\|_2^2 \\ &-\mathrm{KL}\left(p_\psi\left(\mathbf{Z}^{\mathrm{L}} | \boldsymbol{H}_{\mathrm{F}}^{\mathrm{L}}\right) || q_\phi\left(\mathbf{Z}^{\mathrm{H}} | \boldsymbol{H}_{\mathrm{F}}^{\mathrm{H}}\right)\right)\Bigg) + \left\|\bar{\boldsymbol{H}}_{\mathrm{F}}^{\mathrm{L}} - \boldsymbol{H}_{\mathrm{F}}^{\mathrm{L}}\right) \|_2^2 \end{aligned} \tag{5.41}$$

Finally, $\mathbf{Z}^{\mathrm{H}}$ containing the multimodal joint distribution information is input to the classifier, obtaining the predicted class of the sample, and the classification loss L_{c} is calculated:

$$L_{\mathrm{c}} = \xi\left(\hat{Y}, Y\right), \hat{Y} = \mathrm{Cls}\left(\mathbf{Z}^{\mathrm{H}}\right) \tag{5.42}$$

where $\xi(\cdot)$ represents cross-entropy, $\hat{\boldsymbol{Y}}$ and $\boldsymbol{Y}$ represent predictive labels and true labels, respectively. $\mathrm{Cls}(\cdot)$ represents a classifier network.

5.2.2.7 Total loss

The total loss of GM-VDFN can be expressed as:

$$L = L_{\mathrm{c}} + \lambda_1 L_{\mathrm{s}} + \lambda_2 L_{\mathrm{F}} \tag{5.43}$$

where λ_1 and λ_2 are the balance coefficients that control the loss weights.

5.2.3 Experiment

5.2.3.1 HSI dataset

Four HSI and LiDAR fusion datasets are selected to verify effectiveness of the proposed method. The number of training and test set samples in each dataset is shown in Table 5.6.

1) Houston 2013 dataset. The dataset was collected on the campus of the University of Houston in Texas and nearby areas in 2013, which includes a total of 15029 ground truth samples, covering 15 ground objects classes. The size of the dataset is 3491905 pixels and has a spatial resolution of 2.5 meters. HSI consists of 144 different spectral bands with wavelengths ranging from 0.38 to 1.05 μm. The LiDAR data and HSI have the same spatial resolution, presenting researchers with elevation information of the terrain.

2) Trento dataset. The dataset was collected in a rural area south of Trento, Italy. The dataset has a total of 30214 ground truth samples, covering 6 ground objects classes. The data size of the dataset is 166600 pixels and has a spatial resolution of 1 meter. HSI consists of 63 spectral bands with wavelengths ranging from 0.40 to 0.98 μm. The LiDAR data and HSI have the same spatial resolution, presenting researchers with elevation information of the terrain.

3) MUUFL dataset. The dataset was collected at the University of Southern Mississippi in Long Beach, Mississippi. The dataset has a total of 53687 ground truth samples, covering 11 ground objects classes. The size of this dataset is 325220 pixels. The spatial resolution of HSI is 0.54×1.0 m, while the LiDAR data and HSI have the same spatial resolution, presenting researchers with elevation information of the terrain.

4) The Houston 2018 dataset was collected on the campus of the University of Houston and neighboring urban areas. It comprises 20 classes ground objects with spatial dimensions of 12024768, encompassing a total of 2037660 labeled samples. This dataset includes HSI and LiDAR data spatially aligned. The HSI consists of 48 spectral bands within the wavelength range of 380-1050 nm. The LiDAR provides precise elevation data reflecting the characteristics of the terrain.

5.2.3.2 Experimental setting

To evaluate the effectiveness of the proposed method, we compare it with multiple methods, including CNN [73], hierarchical random walk network (HRWN) [22], network for endmember extraction and hyperspectral unmixing (EndNet) [74], consistent contrastive representation network for multiview clustering (CCR-Net) [14], and fractional Gabor convolutional network (FGCN) [22]. All experiments were conducted on a server equipped with an Intel i7-12700k CPU, RTX 2080ti GPU, and 64 GB of RAM, in the PyTorch framework.

Table 5.6: The Number of Training and Test Sets for Each Dataset.

Houston 2013				Houston 2018			
No.	**Class**	**Train**	**Test**	**No.**	**Class**	**Train**	**Test**
1	Healthy grass	20	1231	1	Health grass	200	39996
2	Stressed grass	20	1234	2	Stress grass	200	130808
3	Synthetic grass	20	677	3	Artificial turf	200	3536
4	Trees	20	1224	4	Evergreen trees	200	55122
5	Soil	20	1222	5	Deciduous trees	200	20972
6	Water	20	305	6	Bare earth	200	18864
7	Residential	20	1248	7	Water	200	1864
8	Commercial	20	1224	8	Res. buildings	200	16699
9	Road	20	1232	9	Nonres. bldg	200	895569
10	Highway	20	1207	10	Roads	200	184083
11	Railway	20	1215	11	Sidewalks	200	136835
12	Parking lot 1	20	1213	12	Crosswalks	200	6859
13	Parking lot 2	20	449	13	Major thoroughfares	200	186238
14	Tennis court	20	408	14	Highways	200	40238
15	Running track	20	640	15	Railways	200	28548
				16	Paved parking	200	46732
				17	Unpaved parking	200	637
				18	Cars	200	27089
				19	Trains	200	22279
				20	Stadium seats	200	28096
Total		300	14729	Total		4000	2033660
Total (T+T)		15029		Total (T+T)		2037660	
Trento				**MUUFL**			
No.	**Class**	**Train**	**Test**	**No.**	**Class**	**Train**	**Test**
1	Apple trees	20	4014	1	Trees	20	23226
2	Buildings	20	2883	2	Mostly grass	20	4250
3	Ground	20	459	3	Mixed surface	20	6862
4	Woods	20	9103	4	Dirt/sand	20	1806
5	Vineyard	20	10481	5	Road	20	6667
6	Roads	20	3154	6	Water	20	446
				7	Bldg shadow	20	2213
				8	Building	20	6220
				9	Sidewalk	20	1365
				10	Yellow curb	20	163
				11	Cloth panels	20	249
Total		120	30094	Total		220	53467
Total (T+T)		30214		Total (T+T)		53687	

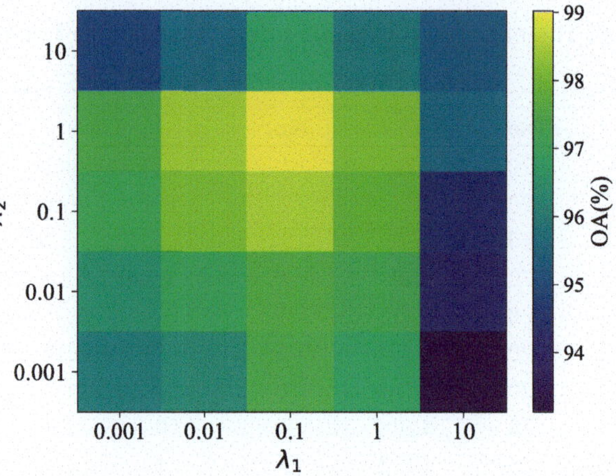

Figure 5.19:
The variation of OA with different λ_1 and λ_2.

In order to ensure the fairness of the experiment, the following strategies on different methods were adopted. For CNN, 20 HSI samples were used for training in each class. For HRWN, EndNet, CCR-Net, FGCN, and GM-VDFN, each class was trained using 20 HSI and LiDAR samples. For each method, Adam was used as the optimizer, the learning rate was set to 0.001, weight decay was 0.0005, and batchsize was 100. In the comparison experiment, the parameter settings of each comparison method follow the original settings. To reduce the influence of randomness on the experimental results, we took the average of all the experimental results of 10 experiments. 3 evaluation indexes were selected in the experiment, including overall accuracy (OA), accuracy of various classes (AA) and Kappa.

5.2.3.3 Parameter analysis

C1. The Influence of Balance Coefficients

For simplicity, we take the Trento dataset as an example, and set balance coefficients λ_1 and λ_2 to $\{0.001, 0.01, 0.1, 1, 10\}$ respectively. Fig. 5.19 demonstrates the variation of OA with different balance coefficient settings, from which it can be seen that:

1) With the increase of λ_1, OA shows an increasing and then decreasing trend. This is because: The small λ_1 limits the effectiveness of STC-GM, which makes it difficult to effectively constrain the model to suppress spectral noise during the learning process, impairing the model performance. The large λ_2 leads to over-alignment of the spatial topology structure of HSI and LiDAR, which prevents the model from capturing complementary information.

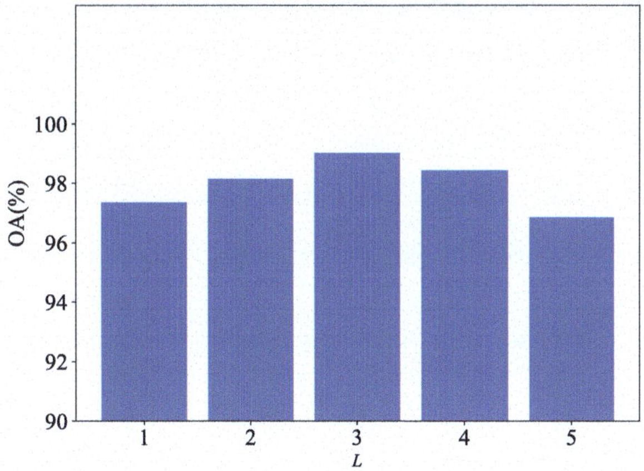

Figure 5.20:
The variation of OA with different L.

2) With the increase of λ_2, OA shows an increasing and then decreasing trend, which is because: The small λ_2 limits the modeling ability of M-VAE on the multimodal joint distribution, resulting in the difficulty of suppressing the modal heterogeneity. The large λ_2 lets the model to focus too much on the task of modeling the joint multimodal distribution and neglect the key classification task, resulting in limited model performance.

3) The highest OA was achieved by the model when $\lambda_1 = 0.1$ and $\lambda_2 = 1$, so we set λ_1 to 0.1 and λ_2 to 1.

C2. The Influence of GCN Layers Number

For simplicity, we take the Trento dataset as an example, and set GCN layers number L to $\{1, 2, 3, 4, 5\}$, Fig. 5.20 demonstrates the variation of OA with different L, from which it can be seen that: With the increase of L, OA shows an increasing and then decreasing trend. This is because the small L limits the learning ability of the model, while large L leads to the problem of over-smoothing, where different classes of ground object have similar feature representations. Therefore, we set L to 3 to strike a balance between model learning ability and over-smoothing.

5.2.3.4 Comparative experiments

It can be observed from Tables 5.7–5.10:

1) Compared to CNN using only HSI data, the multimodal methods achieved higher classification accuracy. This is because LiDAR provides elevation information of ground objects, which can accurately describe the different heights of ground objects.

Table 5.7: Classification Performance on Houston 2013.

Class	CNN	HRWN	EndNet	CCR-Net	FGCN	GLT-Net	GM-VDFN
Healthy grass	87.81	90.09	93.58	95.94	88.79	97.24	**97.64**
Stressed grass	87.28	87.84	88.25	92.95	89.79	96.60	**97.89**
Synthetic grass	71.20	97.34	99.11	99.70	**100**	**100**	98.67
Trees	93.87	93.30	98.04	95.18	93.14	**99.35**	94.20
Soil	99.10	99.59	**99.67**	99.18	99.10	99.26	98.45
Water	**98.03**	96.72	91.48	78.36	**98.03**	97.71	95.08
Residential	88.30	75.08	90.54	88.14	88.54	90.63	**92.23**
Commercial	78.59	63.32	81.45	72.88	**89.05**	83.42	77.94
Road	88.31	85.15	74.19	62.50	**89.29**	80.11	72.97
Highway	81.52	85.75	83.60	77.88	78.79	75.81	**91.30**
Railway	51.11	77.53	88.15	77.61	84.86	87.41	**97.78**
Parking lot 1	64.72	75.27	66.12	79.72	50.29	90.44	**91.10**
Parking lot 2	**95.32**	92.20	63.25	78.40	91.98	87.08	86.41
Tennis court	99.02	99.75	99.26	92.16	97.79	**100**	**100**
Running track	**100**	**100**	98.28	95.94	**100**	**100**	99.69
OA (%)	83.60	85.67	87.24	85.36	87.36	91.28	**92.04**
AA (%)	85.61	84.52	87.67	85.77	89.30	92.34	**92.76**
Kappa (%)	82.29	87.93	86.21	84.16	86.34	90.58	**91.40**

Table 5.8: Classification Performance on Trento.

Class	CNN	HRWN	EndNet	CCR-Net	FGCN	GLT-Net	GM-VDFN
Apple trees	94.29	98.68	98.75	97.56	91.31	97.91	**99.93**
Buildings	96.08	77.49	70.93	94.24	**99.55**	98.89	96.57
Ground	**99.56**	97.17	97.60	80.83	97.17	94.12	92.81
Woods	99.53	99.55	**100**	99.89	**100**	99.90	**100**
Vineyard	90.81	96.94	99.04	99.14	98.98	98.73	**100**
Roads	89.63	93.15	97.84	89.51	**98.18**	95.91	94.93
OA (%)	94.42	95.70	96.45	97.40	97.89	98.62	**99.02**
AA (%)	94.98	94.28	94.03	93.53	97.53	**97.58**	97.37
Kappa (%)	92.64	93.83	95.26	96.53	97.18	98.17	**98.69**

2) Compared to HRWN, EndNet, and CCR-Net, FGCN demonstrates better classification performance. This is attributed to FGCN utilization of graph convolution for feature extraction, which enables it to adapt to the irregular spatial topology structure of ground objects. By suppressing information interference from background and different ground objects in spatial neighborhood, FGCN is able to accurately classify different classes of ground objects.

Table 5.9: Classification Performance on Muufl.

Class	CNN	HRWN	EndNet	CCR-Net	FGCN	GLT-Net	GM-VDFN
Trees	80.01	90.63	76.68	76.63	**92.60**	84.64	87.01
Mostly grass	70.66	77.60	**79.32**	63.22	68.66	70.78	71.85
Mixed ground surface	51.85	56.34	54.58	**74.10**	58.64	64.09	62.20
Dirt and sand	78.68	32.34	84.94	74.03	84.77	82.00	**86.16**
Road	68.86	44.53	**86.16**	81.51	65.71	76.15	76.76
Water	**99.78**	83.86	98.21	97.76	87.44	98.88	98.42
Building shadow	69.36	64.26	72.80	71.62	83.42	**93.18**	83.54
Building	88.75	86.67	86.50	87.57	69.18	86.72	**92.42**
Sidewalk	61.25	**73.48**	46.37	64.03	52.82	57.44	58.79
Yellow curb	61.35	**93.25**	74.23	87.12	69.94	80.37	81.82
Cloth panels	89.56	95.18	93.57	93.17	97.19	**98.39**	95.80
OA (%)	74.47	75.46	75.97	76.79	78.51	79.82	**81.06**
AA (%)	74.56	68.20	77.58	79.16	75.49	81.15	**81.34**
Kappa (%)	67.64	72.56	69.82	70.80	71.92	74.32	**75.47**

3) The GM-VDFN exhibits the best classification performance. This is because: First, the STC-GM makes the features of HSI and LiDAR have topological consistency in space by aligning the spatial distribution of the two modalities, which improves the intraclass consistency of features. Second, MDGAM dynamically explores personalized fusion pattern for different samples, which adapts to various spatial scales of ground objects. Third, M-VAE models the multimodal joint distribution of HSI and LiDAR, which effectively solves the challenge brought by heterogeneity to multimodal data fusion, and realizes the effective integration of multimodal information.

5.2.3.5 Comparative experiments under the standard settings

Considering that Houston 2013 and Trento datasets provide the standard training and test sets, we have used the standard training set and test sets of Houston 2013 and Trento datasets as shown in Table 5.11 to compare the classification performance of the proposed method and its comparison methods, and the results are shown in Tables 5.12–5.13. It can be seen that GM-VDFN achieved the highest classification accuracy under the standard training set and test set settings of Houston 2013 and Trento datasets, verifying its effectiveness in the multimodal remote sensing data classification task.

5.2.3.6 Visual analysis

F1. Visual Analysis of Classification Maps

Table 5.10: Classification Performance on Houston 2018.

Class	CNN	HRWN	EndNet	CCR-Net	FGCN	GLT-Net	GM-VDFN
Health grass	88.99	**96.25**	92.23	93.58	90.94	94.63	90.93
Stress grass	83.86	88.28	86.94	**89.39**	88.71	88.12	87.67
Artificial turf	**100**	**100**	**100**	**100**	**100**	**100**	**100**
Evergreen trees	91.63	95.23	94.81	94.97	95.66	95.50	**96.50**
Deciduous trees	87.41	**97.00**	91.59	92.93	95.65	94.03	96.41
Bare earth	97.83	95.62	98.39	99.16	99.63	98.99	**99.82**
Water	**100**	98.96	**100**	**100**	**100**	**100**	98.44
Residential buildings	86.32	88.79	89.31	88.61	91.04	91.34	**93.39**
Nonresidential buildings	85.75	85.26	88.94	88.49	86.91	**91.23**	89.75
Roads	37.12	61.68	52.97	57.82	59.02	57.70	**61.95**
Sidewalks	43.70	60.22	47.30	48.45	62.71	57.72	**65.30**
Crosswalks	71.91	60.37	82.01	83.59	87.15	**88.60**	88.14
Major thoroughfares	45.69	49.43	56.26	62.68	64.84	64.68	**67.50**
Highways	85.88	94.08	90.86	93.44	96.78	96.22	**96.94**
Railways	97.48	99.66	99.53	99.25	99.58	**99.75**	99.74
Paved parking lots	80.81	**95.03**	92.05	93.45	94.28	94.76	94.02
Unpaved parking lots	**100**	**100**	**100**	**100**	**100**	**100**	**100**
Cars	85.41	92.56	91.72	90.04	93.38	92.30	**95.82**
Trains	94.22	98.27	99.22	99.54	99.44	99.62	**99.68**
Stadium seats	99.31	98.78	99.79	99.80	99.81	99.73	**99.84**
OA (%)	75.33	80.25	80.58	81.71	82.56	84.04	**84.68**
AA (%)	83.17	87.77	87.70	88.14	90.28	90.25	**91.08**
Kappa (%)	68.96	75.20	75.34	76.77	77.92	79.61	**80.48**

Figs. 5.21–5.23 present the classification maps of all methods on three datasets, from which it can be seen that the classification maps of GM-VDFN demonstrate the clearest classification boundaries. This is because: 1) GM-VDFN can capture the dependence relationship between ground objects in irregular spatial region, effectively suppressing the interference of background and different-class ground objects. 2) GM-VDFN mines modal information from multiple spatial scales, enabling the model to fully learn the spatial topology structure of ground objects. By forcing ground objects in the same spatial topology region to have similar feature representation, the spatial consistency of the classification result is improved.

F2. Visual Analysis

Fig. 5.24 displays the t-distributed stochastic neighbor embedding (t-SNE) of multimodal fusion classification methods on the Trento. It can be observed that GM-VDFN exhibits the best intraclass consistency and interclass separation. This is because: On one hand, STC-GM

Table 5.11: Standard Training and Test Sets of Houston 2013 and Trento Datasets.

Houston 2013				Trento			
Class		Sample number		Class		Sample number	
No	Name	Training	Test	No	Name	Training	Test
1	Healthy grass	198	1053	1	Apple trees	129	3905
2	Stressed grass	190	1064	2	Buildings	125	2778
3	Synthetic grass	192	505	3	Ground	105	374
4	Trees	188	1056	4	Woods	154	8969
5	Soil	186	1056	5	Vineyard	184	10317
6	Water	182	143	6	Roads	122	3052
7	Residential	196	1072				
8	Commercial	191	1053				
9	Road	193	4059				
10	Highway	191	1036				
11	Railway	184	1054				
12	Parking lot 1	192	1042				
13	Parking lot 2	184	285				
14	Tennis court	181	247				
15	Running track	187	473				
	Total	2832	12197		Total	819	29395
Total (Training + Test)		15029		Total (Training + Test)		30214	

Table 5.12: OAs of Houston 2013 Dataset and Trento Dataset.

Dataset	CNN	HRWN	EndNet	CCR-Net	FGCN	GLT-Net	GM-VDFN
Houston 2013	85.13	87.93	88.75	89.83	90.71	93.82	**95.43**
Trento	92.11	96.80	97.43	97.95	98.13	98.95	**99.43**

Table 5.13: Kappas of Houston 2013 Dataset and Trento Dataset.

Dataset	CNN	HRWN	EndNet	CCR-Net	FGCN	GLT-Net	GM-VDFN
Houston 2013	84.67	86.71	87.26	88.73	89.86	94.23	**94.62**
Trento	89.51	95.73	97.09	97.43	97.84	98.36	**98.92**

can effectively suppress spectral noise by using spatial prior and accurate spatial topology information of LiDAR, which reduces the intraclass variance of the feature. On the other hand, GM-VDFN fully excavates multiscale modal information, and uses the complementary properties of HSI and LiDAR to effectively separate different ground objects.

(a) (b) (c) (d) (e) (f) (g) (h)

Figure 5.21:
(a) Label map (b) CNN (c) HRWN (d) EndNet (e) CCR-Net (f) FGCN (g) GLT-Net (h) GM-VDFN.

(a) (b) (c) (d) (e) (f) (g) (h)

Figure 5.22:
(a) Label map (b) CNN (c) HRWN (d) EndNet (e) CCR-Net (f) FGCN (g) GLT-Net (h) GM-VDFN.

5.2.3.7 Ablation study

To verify contribution of different techniques of GM-VDFN to the ultimate classification result, we ablate MDGAM, STC-GM, M-VAE, as shown in Table 5.14, and the OAs of these ablation models are shown in Table 5.15.

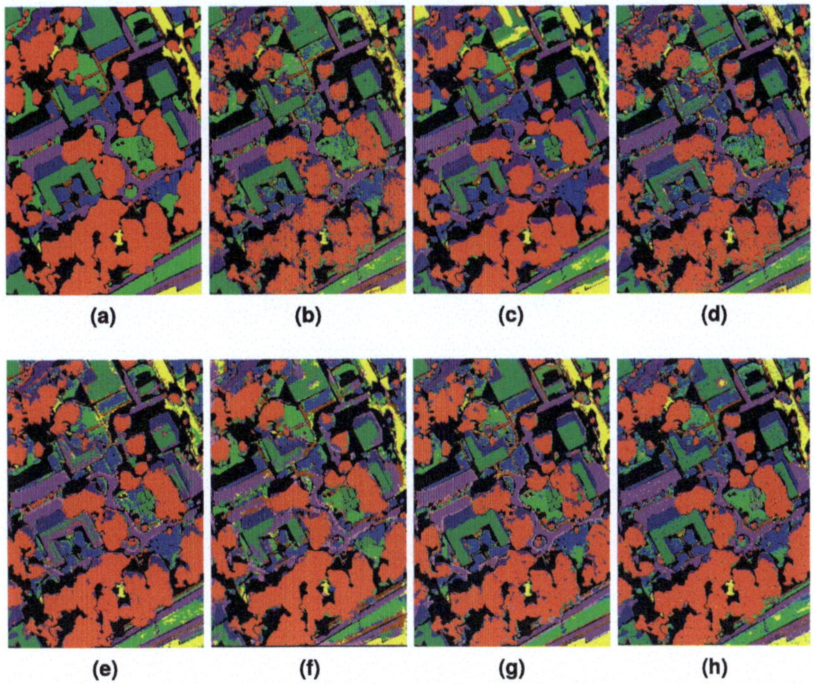

Figure 5.23:
(a) Label map (b) CNN (c) HRWN (d) EndNet (e) CCR-Net (f) FGCN (g) GLT-Net (h) GM-VDFN.

Table 5.14: Ablation Experiments.

	GM-VDFN1	GM-VDFN2	GM-VDFN3	GM-VDFN
MDGAM	✘	✔	✔	✔
STC-GM	✘	✘	✔	✔
M-VAE	✘	✘	✘	✔

Table 5.15: Impact of Different Techniques on OAs (%).

Dataset	GM-VDFN1	GM-VDFN2	GM-VDFN3	GM-VDFN
Houston 2013	88.15	89.57	91.02	**92.04**
Trento	95.56	97.12	98.21	**99.02**
MUUFL	76.13	77.52	78.45	**80.65**

1) Effectiveness of MDGAM

Compared to GM-VDFN2, the OAs of GM-VDFN1 decreased by 1.42%, 1.56%, and 1.39% on three datasets. This indicates that MDDGM can fully exploit the modal information by exploring multiscale feature fusion patterns that are suitable for different ground objects.

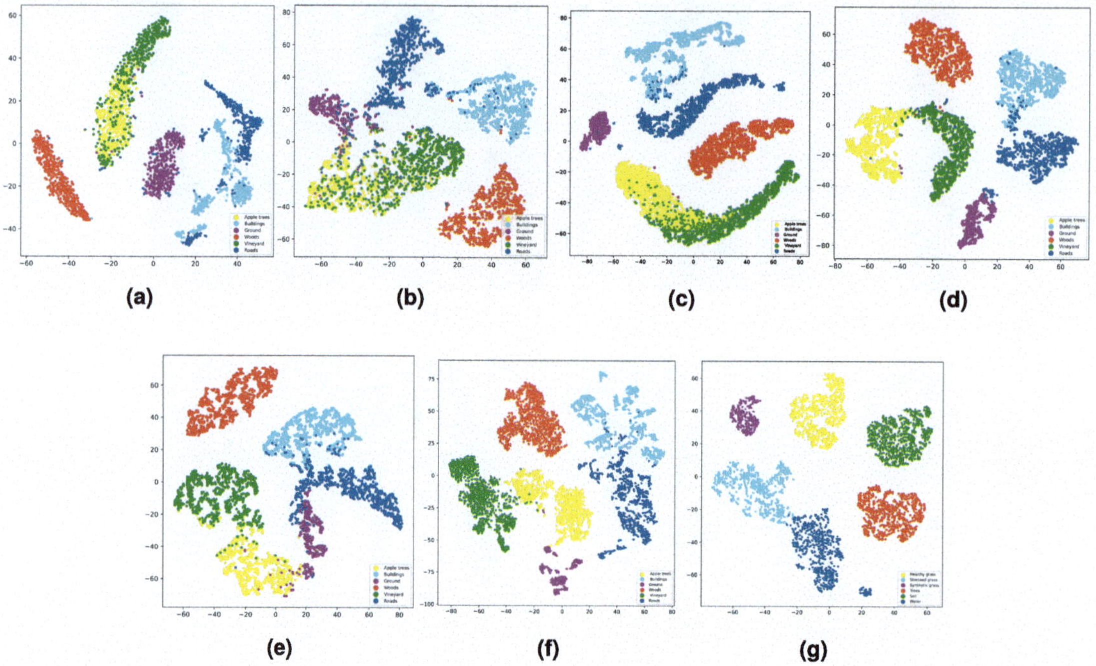

Figure 5.24:
(a) CNN (b) HRWN (c) EndNet (d) CCR-Net (e) FGCN (f) GLT-Net (g) GM-VDFN.

2) Effectiveness of STC-GM

Compared with GM-VDFN3, which uses STC-GM to align the spatial distributions of two modalities, the OAs of GM-VDFN2 decreased by 1.45%, 1.09%, and 0.93% on three datasets respectively. This indicates that STC-GM can effectively suppress spectral noise and enhance the discriminability of the model.

To show the effectiveness of STC-GM more clearly, we visualize the constraint effect of STC-GM, as shown in Fig. 5.25. From Fig. 5.25 (a), in the absence of STC-GM, the ground objects in the same spatial topology region are affected by spectral noise and show poor consistency. From Fig. 5.25 (b-c), the spatial distribution of HSI and LiDAR is aligned with STC-GM, which results in a consistent feature representation of ground objects in the same spatial topology region of HSI.

3) Effectiveness of M-VAE

Compared to GM-VDFN, which obtains multimodal fusion features using M-VAE, GM-VDFN3 has a decrease in OAs of 1.02%, 0.81%, and 2.20% on the three datasets, respectively. It is shown that the multimodal fusion features obtained by learning the

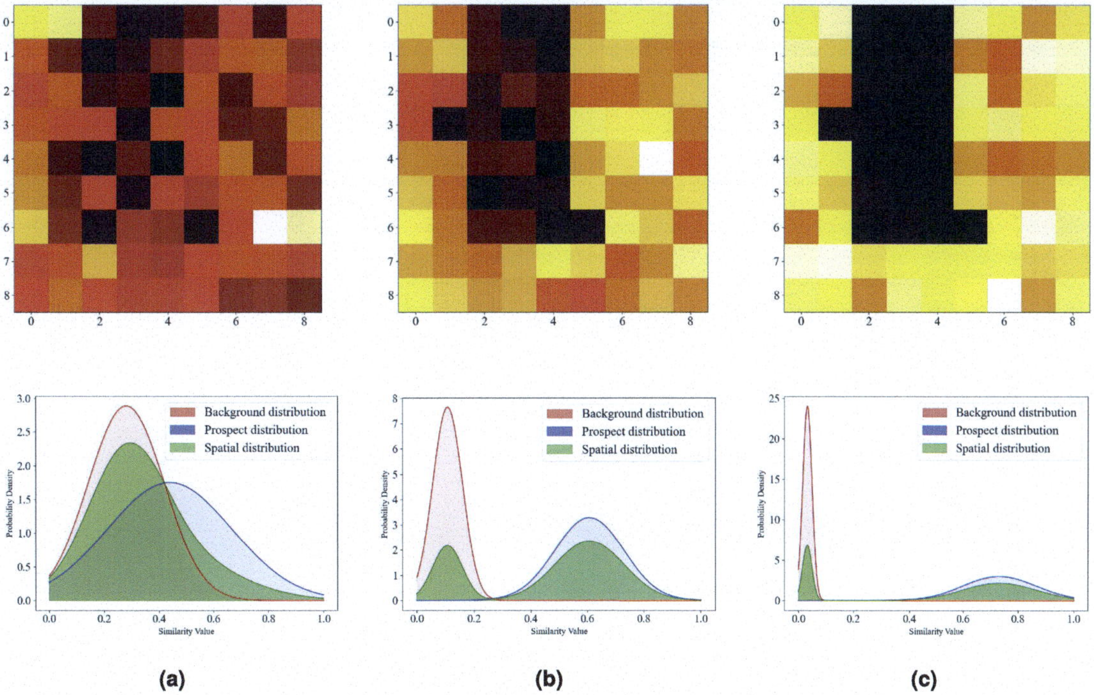

Figure 5.25:
Visualization of STC-GM. (a) In the absence of STC-GM, the spatial topology matrix (top) and spatial distribution (bottom) calculated based on HSI features. (b) With STC-GM, the spatial topology matrix (top) and spatial distribution (bottom) calculated based on HSI features are calculated. (c) Spatial topology matrix (top) and spatial distribution (bottom) calculated based on LiDAR features.

joint distribution of GM-VDFN3 can explore the complex nonlinear relationship between the modalities. This can better preserve the complementary features and provide more discriminative features for the ground object classification.

5.2.3.8 The influence of training set allocation settings

We compared the performance of GM-VDFN and its comparison methods under different training set allocation settings. For the Houston 2013 dataset and MUUFL dataset, we gradually increased the number of training samples from 1% to 20%. For the Trento dataset, we gradually increased the number of training samples from 20 per class to 100 per class. The experimental results are shown in Fig. 5.26.

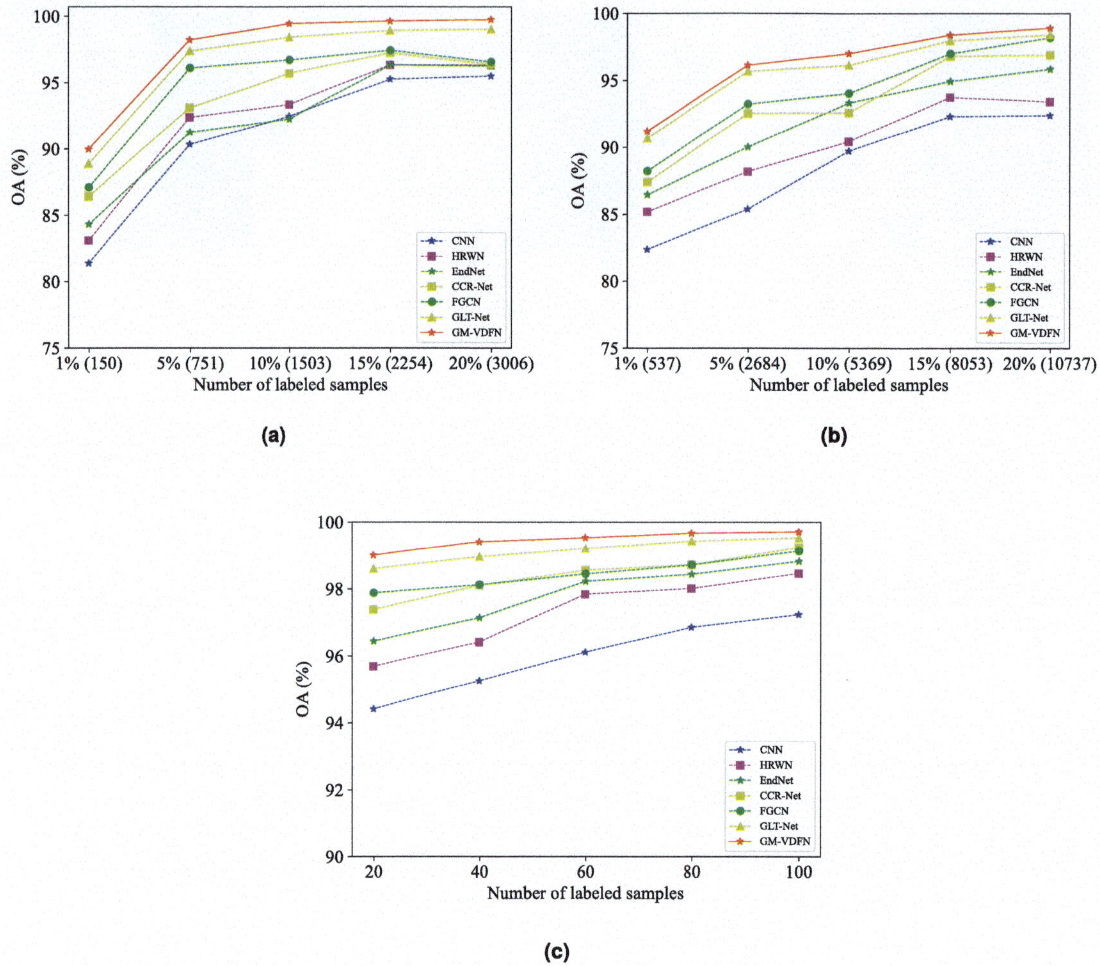

Figure 5.26:
Classification performance of all methods with different training set allocation settings on the three datasets. (a) Houston 2013 dataset. (b) MUUFL dataset. (c) Trento dataset.

1) As the number of training samples increases, the OAs of all methods increase, which is due to the fact that more training samples can better reflect the real data distribution and can help the model to better distinguish the classes of ground objects.

2) GM-VDFN exhibits the highest classification performance under different training set allocation settings, indicating its good adaptability to changes in sample number.

Table 5.16: Running Time, FLOPs, and Parameters of Different Methods.

Methods	CNN	HRWN	EndNet	CCR-Net	FGCN	GLT-Net	GM-VDFN
Running time (s)	384.25	165.31	92.42	1351.32	12135.12	743.24	658.25
FLOPs (M)	1.83	1.53	0.89	2.46	13.46	312.28	8.46
Parameters (M)	1.62	1.42	0.93	1.35	1.13	0.55	1.24

5.2.3.9 Computational cost and running time

To show computational costs and running times of all methods, we took the Houston 2013 dataset as an example and counted their running time, floating point operations (FLOPs), and number of parameters, as shown in Table 5.16. Among the all methods, GM-VDFN ranks fourth in terms of running time, fifth in terms of FLOPs, and fourth in terms of number of parameters, and computational cost is in the middle. This is because the multiscale operation of GM-VDFN increases the scale of model inputs, which results in sacrificing computational cost.

5.2.4 Conclusion

A novel fusion network is proposed, which effectively solves the three major challenges in HSI and LiDAR multimodal fusion classification tasks: Spectral noise suppression, full mining modal information, and multimodal data fusion. GM-VDFN has the following advantages: 1) By aligning the spatial distribution of HSI and LiDAR, it can improve the intraclass consistency of ground features and reduce the performance damage caused by spectral noise to the model. 2) GM-VDFN captures shareable class identification information in multiscale features, and fully mine contextual information by customizing personalized fusion modals for different samples. 3) GM-VDFN can transform the problem of multimodal fusion into a multimodal joint distribution learning problem, overcoming the obstacles caused by heterogeneity between modalities and providing a new paradigm for multimodal data fusion. The experimental results show that GM-VDFN can achieve the most advanced classification performance on four public HSI datasets.

However, this paper only focuses on processing HSI and LiDAR data that have the same spatial resolution and are collected at the same time. It is difficult for our proposed method to use the multisource remote sensing data with different spatial resolutions and multitemporal. The SpectralGPT [75] is able to capture shared knowledge from multisource remote sensing data with different attributes, providing a general foundation model for remote sensing data processing. Inspired by [75], in future work, we will try to combine the self-supervised learning and encoder for visible tokens to establish a collaborative learning mechanism.

Through the above operation, the rich knowledge of multisource remote sensing data is integrated to improve the classification accuracy of the model.

References

[1] B. Du, M. Zhang, L. Zhang, R. Hu, D. Tao, PLTD: patch-based low-rank tensor decomposition for hyperspectral images, IEEE Transactions on Multimedia 19 (1) (Jan. 2017) 67–79.
[2] C. Shi, C.-M. Pun, Multiscale superpixel-based hyperspectral image classification using recurrent neural networks with stacked autoencoders, IEEE Transactions on Multimedia 22 (2) (Feb. 2020) 487–501.
[3] Q. Zhu, W. Deng, Z. Zheng, Y. Zhong, Q. Guan, W. Lin, L. Zhang, D. Li, A spectral-spatial-dependent global learning framework for insufficient and imbalanced hyperspectral image classification, IEEE Transactions on Cybernetics 52 (11) (Nov. 2022) 11709–11723.
[4] T. Lu, K. Ding, W. Fu, S. Li, A. Guo, Coupled adversarial learning for fusion classification of hyperspectral and LiDAR data, Information Fusion 93 (May. 2023) 118–131.
[5] X. Wang, Y. Feng, R. Song, Z. Mu, C. Song, Multi-attentive hierarchical dense fusion net for fusion classification of hyperspectral and LiDAR data, Information Fusion 82 (Jun. 2022) 1–18.
[6] M. Zhang, W. Li, Y. Zhang, R. Tao, Q. Du, Hyperspectral and LiDAR data classification based on structural optimization transmission, IEEE Transactions on Cybernetics 53 (5) (May. 2023) 3153–3164.
[7] R. Dian, S. Li, X. Kang, Regularizing hyperspectral and multispectral image fusion by CNN denoiser, IEEE Transactions on Neural Networks and Learning Systems 32 (3) (Mar. 2021) 1124–1135.
[8] D. Hong, L. Gao, R. Hang, B. Zhang, J. Chanussot, Deep encoder–decoder networks for classification of hyperspectral and LiDAR data, IEEE Geoscience and Remote Sensing Letters 19 (Aug. 2022) 1–5.
[9] X. Zhao, M. Zhang, R. Tao, W. Li, W. Liao, L. Tian, W. Philips, Fractional Fourier image transformer for multimodal remote sensing data classification, IEEE Transactions on Neural Networks and Learning Systems (Jul. 2022).
[10] P. Ghamisi, B. Rasti, N. Yokoya, Q. Wang, B. Hofle, L. Bruzzone, F. Bovol, M. Chi, K. Anders, R. Gloaguen, P. Atkinson, J. Benediktsson, Multisource and multitemporal data fusion in remote sensing: a comprehensive review of the state of the art, IEEE Geoscience and Remote Sensing Magazine 7 (1) (Mar. 2019) 6–39.
[11] S. Wan, S. Pan, P. Zhong, X. Chang, J. Yang, C. Gong, Dual interactive graph convolutional networks for hyperspectral image classification, IEEE Transactions on Geoscience and Remote Sensing 60 (May. 2021) 1–14.
[12] H. Zhang, J. Zou, L. Zhang, EMS-GCN: an end-to-end mixhop superpixel-based graph convolutional network for hyperspectral image classification, IEEE Transactions on Geoscience and Remote Sensing 60 (Mar. 2022) 1–16.
[13] K. Ding, T. Lu, W. Fu, S. Li, F. Ma, Global–local transformer network for HSI and LiDAR data joint classification, IEEE Transactions on Geoscience and Remote Sensing 60 (Oct. 2022) 1–13.
[14] X. Wu, D. Hong, J. Chanussot, Convolutional neural networks for multimodal remote sensing data classification, IEEE Transactions on Geoscience and Remote Sensing 60 (Nov. 2022) 1–10.
[15] M. Zhang, W. Li, Q. Du, L. Gao, B. Zhang, Feature extraction for classification of hyperspectral and LiDAR data using patch-to-patch CNN, IEEE Transactions on Cybernetics 50 (1) (Jan. 2020) 100–111.
[16] J. Li, H. Huo, C. Li, R. Wang, Q. Feng, AttentionFGAN: infrared and visible image fusion using attention-based generative adversarial networks, IEEE Transactions on Multimedia 23 (May. 2021) 1383–1396.
[17] S.K. Roy, A. Deria, D. Hong, M. Ahmad, A. Plaza, J. Chanussot, Hyperspectral and LiDAR data classification using joint CNNs and morphological feature learning, IEEE Transactions on Geoscience and Remote Sensing 60 (May. 2022) 1–16.
[18] M. Pedergnana, P.R. Marpu, M. Dalla Mura, J.A. Benediktsson, L. Bruzzone, Classification of remote sensing optical and LiDAR data using extended attribute profiles, IEEE Journal of Selected Topics in Signal Processing 6 (7) (Nov. 2012) 856–865.

[19] M. Zhang, W. Li, R. Tao, H. Li, Q. Du, Information fusion for classification of hyperspectral and LiDAR data using IP-CNN, IEEE Transactions on Geoscience and Remote Sensing 60 (Jul. 2022) 1–12.

[20] Y. Yang, D. Zhu, T. Qu, Q. Wang, F. Ren, C. Cheng, Single-stream CNN with learnable architecture for multisource remote sensing data, IEEE Transactions on Geoscience and Remote Sensing 60 (Apr. 2022) 1–18.

[21] D. Xiu, Z. Pan, Y. Wu, Y. Hu, MAGE: multisource attention network with discriminative graph and informative entities for classification of hyperspectral and LiDAR data, IEEE Transactions on Geoscience and Remote Sensing 60 (Sep. 2022) 1–14.

[22] X. Zhao, R. Tao, W. Li, H. Li, Q. Du, W. Liao, W. Philips, Joint classification of hyperspectral and LiDAR data using hierarchical random walk and deep CNN architecture, IEEE Transactions on Geoscience and Remote Sensing 58 (10) (Oct. 2020) 7355–7370.

[23] J. Li, Y. Ma, R. Song, B. Xi, D. Hong, Q. Du, A triplet semisupervised deep network for fusion classification of hyperspectral and LiDAR data, IEEE Transactions on Geoscience and Remote Sensing 60 (Oct. 2022) 1–13.

[24] J. Yang, C. Wu, B. Du, L. Zhang, Enhanced multiscale feature fusion network for HSI classification, IEEE Transactions on Geoscience and Remote Sensing 59 (12) (Dec. 2021) 10328–10347.

[25] S. Morchhale, V.P. Pauca, R.J. Plemmons, T.C. Torgersen, Classification of pixel-level fused hyperspectral and LiDAR data using deep convolutional neural networks, in: Proc. Workshop Hyperspectral Image Signal Proces.: Evol. Remote Sens., Oct. 2016, pp. 1–5.

[26] S. Jia, Z. Zhan, M. Xu, Q. Huang, J. Zhou, X. Jia, Multiple feature-based superpixel-level decision fusion for hyperspectral and LiDAR data classification, IEEE Transactions on Geoscience and Remote Sensing 59 (2) (Feb. 2021) 1437–1452.

[27] C. Zhao, X. Gao, Y. Wang, J. Li, Efficient multiple-feature learning-based hyperspectral image classification with limited training samples, IEEE Transactions on Geoscience and Remote Sensing 54 (7) (Jul. 2016) 4052–4062.

[28] W. Dong, T. Zhang, J. Qu, S. Xiao, T. Zhang, Y. Li, Multibranch feature fusion network with self- and cross-guided attention for hyperspectral and LiDAR classification, IEEE Transactions on Geoscience and Remote Sensing 60 (Jun. 2022) 1–12.

[29] X. Du, X. Zheng, X. Lu, A.A. Doudkin, Multisource remote sensing data classification with graph fusion network, IEEE Transactions on Geoscience and Remote Sensing 59 (12) (Dec. 2021) 10062–10072.

[30] R. Hang, Z. Li, P. Ghamisi, D. Hong, G. Xia, Q. Liu, Classification of hyperspectral and LiDAR data using coupled CNNs, IEEE Transactions on Geoscience and Remote Sensing 58 (7) (Feb. 2020) 4939–4950.

[31] H.-C. Li, W.-Y. Wang, L. Pan, W. Li, Q. Du, R. Tao, Robust capsule network based on maximum correntropy criterion for hyperspectral image classification, IEEE Journal of Selected Topics in Applied Earth Observations and Remote Sensing 13 (Feb. 2020) 738–751.

[32] D. Hong, L. Gao, R. Hang, B. Zhang, J. Chanussot, Deep encoder-decoder networks for classification of hyperspectral and LiDAR data, IEEE Transactions on Geoscience and Remote Sensing 19 (Aug. 2022).

[33] X. Zhao, R. Tao, W. Li, W. Philips, W. Liao, Fractional Gabor convolutional network for multisource remote sensing data classification, IEEE Transactions on Geoscience and Remote Sensing 60 (2022) 5503818.

[34] M.S. Kotzagiannidis, C.-B. Schnlieb, Semi-supervised superpixel-based multi-feature graph learning for hyperspectral image data, IEEE Transactions on Geoscience and Remote Sensing 60 (Sep. 2021) 1–12.

[35] D. Hong, L. Gao, J. Yao, B. Zhang, A. Plaza, J. Chanussot, Graph convolutional networks for hyperspectral image classification, IEEE Transactions on Geoscience and Remote Sensing 59 (7) (Jul. 2021) 5966–5978.

[36] Y. Peng, Y. Zhang, B. Tu, C. Zhou, Q. Li, Multiview hierarchical network for hyperspectral and LiDAR data classification, IEEE Journal of Selected Topics in Applied Earth Observations and Remote Sensing 15 (Jan. 2022) 1454–1469.

[37] H. Yan, Y. Zhao, J. Chan, S. Kong, Spectral super-resolution based on dictionary optimization learning via spectral library, IEEE Transactions on Geoscience and Remote Sensing 61 (2022) 5400216.

[38] J. He, J. Li, Q. Yuan, H. Shen, L. Zhang, Spectral response function-guided deep optimization-driven network for spectral super-resolution, IEEE Transactions on Neural Networks and Learning Systems 33 (9) (Sept. 2022) 4213–4227.

[39] X. Zhang, W. Huang, Q. Wang, X. Li, SSR-NET: spatial-spectral reconstruction network for hyperspectral and multispectral image fusion, IEEE Transactions on Geoscience and Remote Sensing 59 (7) (Jul. 2021) 5953–5965.

[40] Y. Kong, X. Wang, Y. Cheng, Spectral-spatial feature extraction for HSI classification based on supervised hypergraph and sample expanded CNN, IEEE Journal of Selected Topics in Applied Earth Observations and Remote Sensing 11 (11) (Nov. 2018) 4128–4140.

[41] W. Gao, F. Liu, J. Liu, L. Xiao, X. Tang, Spatial-spectral adaptive learning with pixelwise filtering for hyperspectral image classification, IEEE Transactions on Geoscience and Remote Sensing 61 (2023) 5515315.

[42] S. Li, W. Song, L. Fang, Y. Chen, P. Ghamisi, J. Benediktsson, Deep learning for hyperspectral image classification: an overview, IEEE Transactions on Geoscience and Remote Sensing 57 (9) (Apr. 2019) 6690–6709.

[43] D. Li, X. Hu, S. Wang, C. Zhang, R. Zhou, H. Zhou, Hyperspectral images ground object recognition based on split attention, in: IEEE Int. Conf. Big Data, Artif. Intell. Internet Things Eng., ICBAIE, Apr. 2021, pp. 324–330.

[44] A. Wang, Collaborative classification of hyperspectral and LiDAR data based on dual-branch convolutional neural network, in: Dig Int. Geosci. Remote Sens. Symp., Sept. 2022, pp. 2738–2741.

[45] F. Guo, Z. Li, Q. Meng, L. Wang, J. Zhang, Dual graph convolution joint dense networks for hyperspectral and LiDAR data classification, in: Dig Int. Geosci. Remote Sens. Symp., Sept. 2022, pp. 1141–1144.

[46] Y. Zhang, Y. Peng, B. Tu, Y. Liu, Local information interaction transformer for hyperspectral and LiDAR data classification, IEEE Journal of Selected Topics in Applied Earth Observations and Remote Sensing 16 (Dec. 2022) 1130–1143.

[47] C. Zhang, J. Btepage, H. Kjellstrm, S. Mandt, Advances in variational inference, IEEE Transactions on Pattern Analysis and Machine Intelligence 41 (8) (Aug. 2019) 2008–2026.

[48] R. Borsoi, T. Imbiriba, J. Bermudez, Deep generative endmember modeling: an application to unsupervised spectral unmixing, IEEE Transactions on Computational Imaging 6 (Oct. 2019) 374–384.

[49] M. Paoletti, X. Tao, L. Han, Z. Wu, S. Moreno-lvarez, J. Haut, Deep attention-driven HSI scene classification based on inverted dot-product, in: Dig Int. Geosci. Remote Sens. Symp., Sept. 2022, pp. 1380–1383.

[50] Y. Fan, MSLAENet: multiscale learning and attention enhancement network for fusion classification of hyperspectral and LiDAR data, IEEE Journal of Selected Topics in Applied Earth Observations and Remote Sensing 15 (Nov. 2022) 10041–10054.

[51] M. Wang, F. Gao, J. Dong, H. Li, Q. Du, Nearest neighbor-based contrastive learning for hyperspectral and LiDAR data classification, IEEE Transactions on Geoscience and Remote Sensing 61 (2023) 5501816.

[52] J. Li, Y. Liu, R. Song, Y. Li, K. Han, Q. Du, Sal^{2}RN: a spatial-spectral salient reinforcement network for hyperspectral and LiDAR data fusion classification, IEEE Transactions on Geoscience and Remote Sensing 61 (2022) 5500114.

[53] T. Zhang, S. Xiao, W. Dong, J. Qu, Y. Yang, A mutual guidance attention-based multi-level fusion network for hyperspectral and LiDAR classification, IEEE Transactions on Geoscience and Remote Sensing 19 (2021) 5509105.

[54] Q. Xu, C. Yang, J. Tang, B. Luo, Grouped bidirectional LSTM network and multistage fusion convolutional transformer for hyperspectral image classification, IEEE Transactions on Geoscience and Remote Sensing 60 (2022) 5541314.

[55] D. Xiu, Z. Pan, Y. Wu, Y. Hu, AGE: multisource attention network with discriminative graph and informative entities for classification of hyperspectral and LiDAR data, IEEE Transactions on Geoscience and Remote Sensing 60 (2022) 5539714.

[56] S. Zeng, Z. Wang, C. Gao, Z. Kang, D. Feng, Hyperspectral image classification with global-local discriminant analysis and spatial-spectral context, IEEE Journal of Selected Topics in Applied Earth Observations and Remote Sensing 11 (12) (Dec. 2018) 5005–5018.

[57] H. Li, W. Hu, W. Li, J. Li, Q. Du, A. Plaza, A^3 CLNN: spatial, spectral and multiscale attention convLSTM neural network for multisource remote sensing data classification, IEEE Transactions on Neural Networks and Learning Systems 33 (2) (Feb. 2022) 747–761.

[58] H. Yan, E. Zhang, J. Wang, C. Leng, J. Peng, MTFFN: multimodal transfer feature fusion network for hyperspectral image classification, IEEE Geoscience and Remote Sensing Letters 19 (2022) 6008005.

[59] Y. Zhang, H. Yang, S. Prasad, E. Pasolli, J. Jung, M. Crawford, Ensemble multiple kernel active learning for classification of multisource remote sensing data, IEEE Journal of Selected Topics in Applied Earth Observations and Remote Sensing 8 (2) (Feb. 2015) 845–858.

[60] Y. Hu, H. He, L. Weng, Hyperspectral and LiDAR data land-use classification using parallel transformers, in: Dig Int Geosci Remote Sens Symp., Sept. 2022, pp. 703–706.

[61] Y. Li, D. Kong, Y. Zhang, Y. Tan, L. Chen, Robust deep alignment network with remote sensing knowledge graph for zero-shot and generalized zero-shot remote sensing image scene classification, ISPRS Journal of Photogrammetry and Remote Sensing 179 (Sep. 2021) 145–158.

[62] R. Hang, X. Qian, Q. Liu, Cross-modality contrastive learning for hyperspectral image classification, IEEE Transactions on Geoscience and Remote Sensing 60 (2022) 5532812.

[63] Y. Li, Z. Zhu, J. Yu, Y. Zhang, Learning deep cross-modal embedding networks for zero-shot remote sensing image scene classification, IEEE Transactions on Geoscience and Remote Sensing 59 (Dec. 2021) 10590–10603.

[64] J. Yao, B. Zhang, C. Li, D. Hong, J. Chanussot, Extended vision transformer (ExViT) for land use and land cover classification: a multimodal deep learning framework, IEEE Transactions on Geoscience and Remote Sensing 61 (2023) 5514415.

[65] D. Hong, B. Zhang, H. Li, Y. Li, J. Yao, C. Li, M. Werner, J. Chanussot, A. Zipf, X. Zhu, Cross-city matters: a multimodal remote sensing benchmark dataset for cross-city semantic segmentation using high-resolution domain adaptation networks, Remote Sensing of Environment 299 (2023) 113856.

[66] W. Song, Y. Dai, Z. Gao, L. Fang, Y. Zhang, Hashing-based deep metric learning for the classification of hyperspectral and LiDAR data, IEEE Transactions on Geoscience and Remote Sensing 61 (2023) 5704513.

[67] W. Song, S. Li, L. Fang, T. Lu, Hyperspectral image classification with deep feature fusion network, IEEE Transactions on Geoscience and Remote Sensing 56 (6) (Feb. 2018) 3173–3184.

[68] C. Wang, L. Zhang, W. Wei, Y. Zhang, Dynamic super-pixel normalization for robust hyperspectral image classification, IEEE Transactions on Geoscience and Remote Sensing 61 (2023) 5505713.

[69] J. Liu, C. Xu, C. Yin, W. Wu, Y. Song, K-core based temporal graph convolutional network for dynamic graphs, IEEE Transactions on Knowledge and Data Engineering 34 (8) (Aug. 2022) 3841–3853.

[70] B. Jiang, Z. Zhang, D. Lin, J. Tang, B. Luo, Semi-supervised learning with graph learning-convolutional networks, in: Proc IEEE Comput Soc Conf Comput Vision Pattern Recognit., Jan. 2020, pp. 11305–11312.

[71] J. Thomas, M. Sebastian, H.S. Philip, N. Siddharth, T. Rainforth, Capturing label characteristics in VAEs, in: Int. Conf. Learn. Represent., ICLR, May 2021.

[72] M. Ye, J. Chen, F. Xiong, Y. Qian, Learning a deep structural subspace across hyperspectral scenes with cross-domain VAE, IEEE Transactions on Geoscience and Remote Sensing 60 (2022) 5522613.

[73] Y. Chen, H. Jiang, C. Li, X. Jia, P. Ghamisi, Deep feature extraction and classification of hyperspectral images based on convolutional neural networks, IEEE Transactions on Geoscience and Remote Sensing 54 (10) (Oct. 2016) 6232–6251.

[74] S. Ozkan, B. Kaya, G. Akar, EndNet: sparse autoEncoder network for endmember extraction and hyperspectral unmixing, IEEE Transactions on Geoscience and Remote Sensing 57 (1) (Jan. 2019) 482–496.

[75] D. Hong, B. Zhang, X. Li, Y. Li, C. Li, J. Yao, N. Yokoya, H. Li, P. Ghamisi, X. Jia, A. Plaza, P. Gamba, J.A. Benediktsson, J. Chanussot, SpectralGPT: spectral remote sensing foundation model, IEEE Transactions on Pattern Analysis and Machine Intelligence (2024), https://doi.org/10.1109/TPAMI.2024.3362475.

CHAPTER 6

Change detection and monitoring

Min Zhang[a], Lukang Wang[b], Wenzhong Shi[c], and Xiaokang Zhang[d]

[a]School of Geosciences and Info-Physics, Central South University, Changsha, China [b]College of Geodesy and Geomatics, Shandong University of Science and Technology, Qingdao, China [c]The Department of Land Surveying and Geo-Informatics, The Hong Kong Polytechnic University, Hong Kong SAR, China [d]School of Artificial Intelligence, Wuhan University, Wuhan, China

6.1 Introduction

Change detection is the process of identifying differences in the state of an object or phenomenon by observing it at different times [176]. It is one of the major problems in earth observation and has been extensively researched in recent decades. Multitemporal RS data, such as satellite imagery and aerial imagery, can provide abundant information to identify land use and land cover (LULC) differences in a specific area across a period of time. This is very crucial in various applications, such as urban planning, environmental monitoring, agriculture investigation, disaster assessment, and map revision.

With the ongoing development of Earth observation techniques, huge amounts of RS data with a high spectral-spatial-temporal resolution are now available, which brings new requirements of change detection techniques and greatly promotes their development. To address the problems brought about by finer spatial and spectral resolution images during the change detection process, many change detection approaches are proposed. Here, they are broadly divided into two categories: traditional and AI-based. Fig. 6.1 presents the general flow of traditional change detection and AI-based change detection.

Existing change detection reviews have focused mainly on the design of change detection techniques in multitemporal hyperspectral images (HSIs) and high-spatial-resolution images [126,128,176,189]. The techniques they reviewed are mainly traditional change detection approaches, which can be generally summarized into the following groups:

- Visual analysis: the change map is obtained by manual interpretation, which can provide highly reliable results based on expert knowledge but is time-consuming and labor-intensive;

Multimodal Remote Sensing Fusion and Classification
https://doi.org/10.1016/B978-0-44-329152-4.00013-1

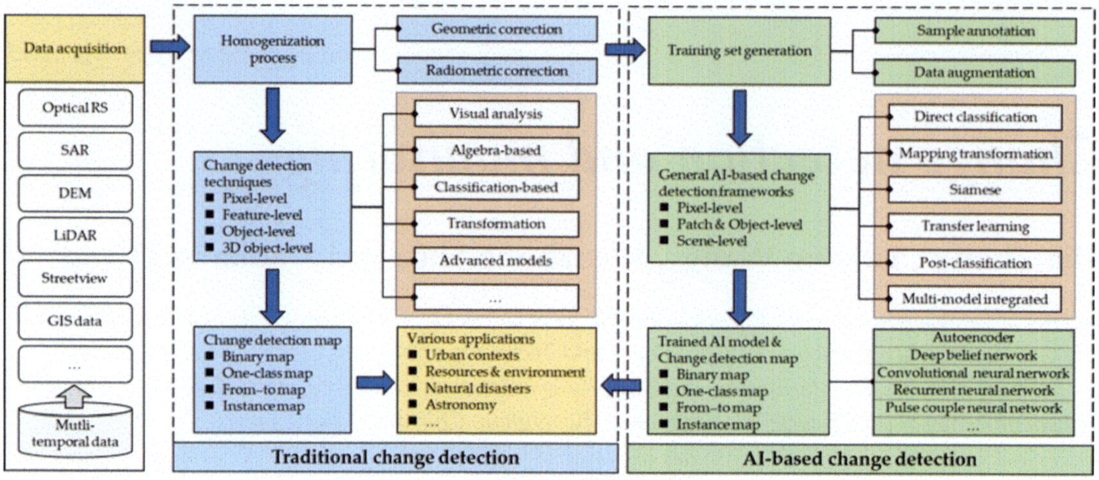

Figure 6.1:
General schematic diagram of change detection.

- Algebra-based methods: the change map is obtained by performing algebraic operations or transformations on multitemporal data, such as image differencing, image regression, image ratioing, and change vector analysis (CVA);
- Transformation: data reduction methods, such as principle component analysis (PCA), Tasseled Cap (KT), multivariate alteration detection (MAD), Gram-Schmidt (GS), and Chi-Square, are used to suppress correlated information and highlight variance in multitemporal data;
- Classification-based methods: changes are identified by comparing multiple classification maps (i.e., postclassification comparison), or using a trained classifier to directly classify data from multiple periods (i.e., multidate classification or direct classification);
- Advanced models: advanced models, such as the Li-Strahler reflectance model, the spectral mixture model, and the biophysical parameter method, are used to convert the spectral reflectance values of multiperiod data into physically based parameters or fractions to perform change analysis, and this is more intuitive and has physical meaning, but it is complicated and time-consuming;
- Others: hybrid approaches and others, such as knowledge-based, spatial-statistics-based, and integrated GIS and RS methods, are used.

According to the detection unit, these methods can also be classified based on pixel-level, feature-level, object-level, and three-dimensions (3D) object-level, and have been systematically reviewed in the literature [20,89,158]. Due to the rapid development of computer technology, the research of traditional change detection approaches has turned to integrating AI techniques. In both traditional change detection flow and AI-based flow, the first step is data acquisition and the aim of change detection is to obtain the change detection map for various applications; after preparing the data, traditional approaches typically consist of two steps, including a homogenization process and a change detection process, while the AI-based approaches generally require an extra training set generation process and an AI model training process for change detection. Obviously, the key components of AI-based approaches are the AI techniques.

AI techniques, also called machine intelligence, can provide a better performance in various data-processing tasks. It can be defined as a system's ability to correctly interpret external data, to learn from such data, and to use those learnings to achieve specific goals and tasks through flexible adaptation [101]. AI techniques in this chapter focus on the recent emergence of deep learning methods, new network structures, and intelligent machine learning methods, which are inspired by biological systems. Traditional machine learning methods, such as support vector machines and decision trees, have not been considered due to their relatively low intelligence and existing reviews [89].

Many new approaches integrating AI techniques have been developed to improve the accuracy and automation of change detection. A wide body of RS research has suggested that AI-based change detection approaches are superior to the traditional in terms of feature extraction [52,227]. Due to the powerful modeling and learning capabilities, AI techniques can model the relationship between the image object and its real-world geographical feature as closely as possible, which enables the detection of more real change information. Generally, they utilize spatial-context information in multitemporal data to learn hierarchical feature representations, and these high-level feature representations are more effective and robust in change detection tasks.

This chapter provides a deep review of the application of AI technologies in RS change detection processing. It focuses on the state-of-the-art methods, applications, and challenges of AI for change detection in multitemporal data.

6.2 Implementation process of AI-based change detection

Fig. 6.1 illustrates the general flow of AI-based change detection. The key is to obtain a high-performance trained AI model. In detail, as presented in Fig. 6.2, the implementation process of AI-based change detection includes the following four main steps:

1. Homogenization: Due to differences in illumination and atmospheric conditions, seasons, and sensor attitudes at the time of acquisition, multiperiod data usually need to be homogenized before change detection. Geometric and radiometric correction are two commonly used methods [23,201]. The former aims to geometrically align two or more

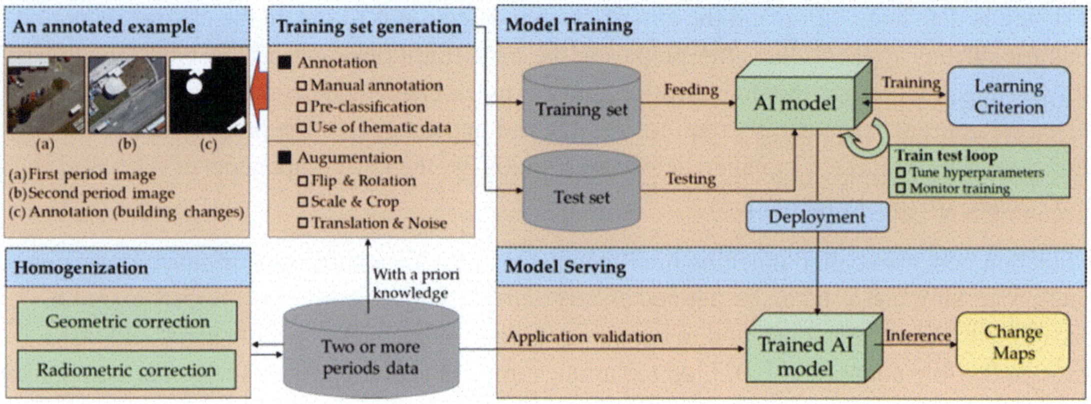

Figure 6.2:
Implementation process of AI-based change detection (black arrows indicate workflow and red arrow indicates an example).

given pieces of data, which can be achieved through registration or coregistration. Given two period data, only when they are overlaid can the comparison between corresponding positions be meaningful [225]. The latter aims to eliminate radiance or reflectance differences caused by the digitalization process of sensors and atmospheric attenuation distortion caused by absorption and scattering in the atmosphere [128], which helps to reduce false alarms caused by these radiation errors in change detection. For heterogeneous data, a special AI model structure can be designed for feature space transformation to achieve change detection (see Section 6.4.1.2);

2. Training set generation: To develop the AI model, a large, high-quality training set is required that can help algorithms to understand that certain patterns or series of outcomes come with a given question. Multiperiod data are labeled or annotated using certain techniques (e.g., manual annotation [232], preclassification [229], use of thematic data [93]) to make it easy for the AI model to learn the characteristics of the changed objects. Fig. 6.2 presents an annotated example for building change detection, which is composed of two-period RS images and a corresponding ground truth labeled with building changes at the pixel level. Based on the ground truth, i.e., prior knowledge, the AI model can be trained in a supervised manner. To alleviate the problem of lack of training data, data augmentation, which is widely used, is a good strategy, such as horizontal or vertical flip, rotation, change in scale, crop, translation, or adding noise, which can significantly increase the diversity of data available for training models, without actually collecting new data;
3. Model training: After the training set is generated, it can usually be divided into two datasets according to the number of samples or the geographic area: a training set for AI

model training and a test set for accuracy evaluation during the training process [108]. The training and testing processes are performed alternately and iteratively. During the training process, the model is optimized according to a learning criterion, which can be a loss function in deep learning (e.g., softmax loss [60], contrastive loss [220], Euclidean loss [15], or cross-entropy loss [73]). By monitoring the training process and test accuracy, the convergence state of the AI model can be obtained, which can help in adjusting its hyperparameters (such as the learning rate), and also in judging whether the model performance has reached the best (i.e., termination) condition;
4. Model serving: By deploying a trained AI model, change maps can be generated more intelligently and automatically for practical applications. Moreover, this can help validate the generalization ability and robustness of the model, which is an important aspect of evaluating the practicality of the AI-based change detection technique.

The above steps provide a general implementation process of AI-based change detection, but the structure of the AI model is diverse and needs to be well designed according to different application situations and the training data, which will be introduced in Sections 6.4 and 6.5. It is worth mentioning that existing mature frameworks such as TensorFlow, Keras, Pytorch, and ONNX, help researchers more easily realize the design, training, and deployment of AI models, and their development documents provide detailed introductions.

6.3 Data sources for change detection

With the development of data acquisition platforms such as satellites, drones, and ground survey vehicles, the massive multisource RS data that they produce bring forth new application requirements for land change monitoring. In particular, multisensor high-spatial and high-temporal-resolution data require more automated and robust change detection methods to reduce the cost of manual interpretation. By summarizing the data types used for change detection, we can deeply analyze the applicability of existing change detection methods to the data. In this chapter, the types of data used for change detection are divided into optical RS images, SAR images, and street view images. It should be noted that street view images are usually not used as RS data but as auxiliary data [67,100,162], so it is not common in the RS community. Still, there are overlapping ideas for change detection. In this chapter, street view images are treated as a kind of RS data in a broad sense and reviewed, because they can provide street-level observation data. Moreover, the methods combining heterogeneous data for change detection are summarized and analyzed. Examples of different data sources are shown in Fig. 6.3. Optical RS and SAR images are gathered with passive and active sensors, respectively, covering different electromagnetic spectral ranges. Other data sources, such as digital elevation models (DEMs), geographic information system (GIS) data, and point cloud data, can provide valuable supplementary attributes. Overhead remote sensing

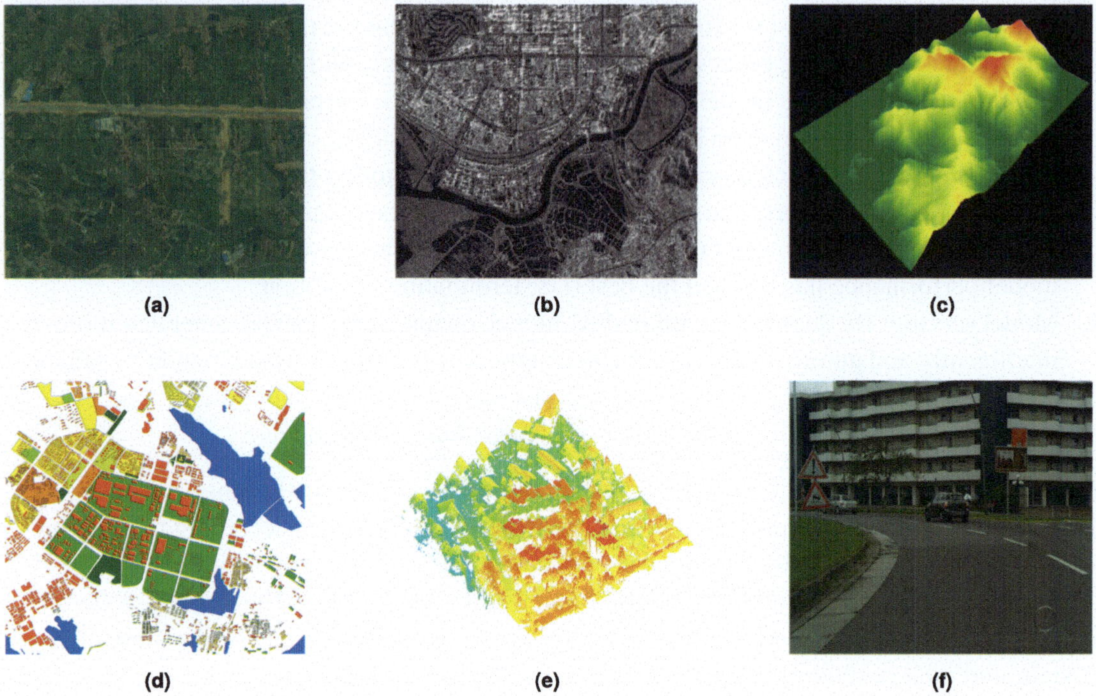

Figure 6.3:
Examples of different data sources for change detection: (a) Optical RS image (obtained by Quickbird); (b) SAR image (obtained by the Advanced Land Observing Satellite (ALOS) Phased Array type L-band Synthetic Aperture Radar (PALSAR)); (c) digital elevation model (DEM); (d) geographic information system (GIS) data (from OpenStreetMap); (e) Point cloud data (from International Society for Photogrammetry and Remote Sensing (ISPRS) benchmarks); (f) Street view image (from Cityscapes datasets [34]).

collects information over large spatial areas, but its time resolution is relatively low. Street view images can provide nearly real-time information at street-level. For detailed descriptions of optical RS images, SAR images, street view images, and combined heterogeneous data, please refer to sections below.

6.3.1 Optical RS images

Optical RS images can be divided into hyperspectral, multispectral, and panchromatic images according to the number of bands. HSIs are volumetric image cubes that consist of hundreds of spectral bands. They have narrow bands over a wide portion of the electromagnetic spectrum; the band range is generally less than 10 nm. Multispectral images typically contain multiple bands but fewer than 15 bands. The spectral resolution of multispectral images is

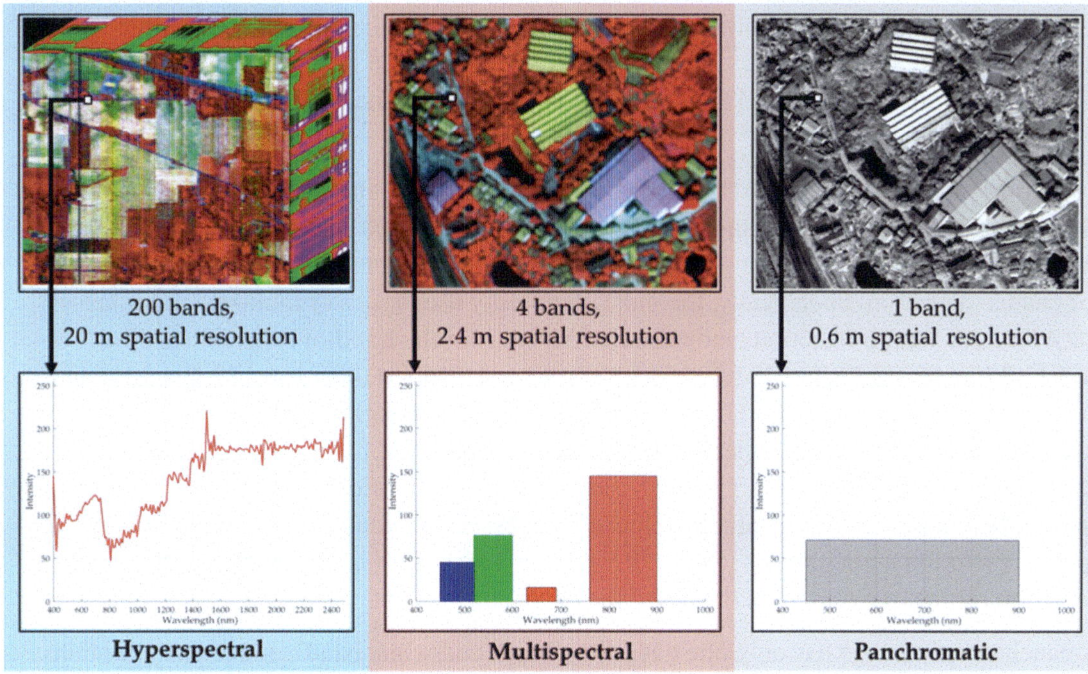

Figure 6.4:
Examples of hyperspectral, multispectral, and panchromatic images, where the hyperspectral image is from Indian Pines; multispectral and panchromatic images are from Quickbird.

in the range of 0.1 times the wavelengths. Panchromatic images have a single band that is formed by using the total light energy in the visible spectrum (instead of partitioning it into different spectra). A side-by-side example of hyperspectral, multispectral, and panchromatic images is shown in Fig. 6.4. HSIs are gathered by the AVIRIS sensor [6]; multispectral and panchromatic images are taken from the Quickbird satellite. The second row of Fig. 6.4 shows the spectral intensity of a selected image pixel and the spectral resolution of the three types of images. Therefore, images with different number of bands, reflecting the spectral resolution, require different methods for change detection.

HSIs have hundreds or even thousands of continuous and narrow bands, which can provide abundant spectral and spatial information. Multitemporal HSIs are of great significance in distinguishing the subtle changes in ground objects through their high-dimensional feature information. The detailed information on spectral changes presents promising change detection performance. However, it increases the redundancy of the data and makes it difficult to interpret. Moreover, due to the generally low spatial resolution of HSIs, the textures around the pixels are vague, and mixed pixels occupy a large proportion. Change detection methods for HSIs must address the problems of high dimensionality, mixed pixels, high computational cost, and a limited training dataset. Effective AI algorithms can be employed to solve these problems and have been proved to achieve satisfactory performance [87,114,177,199].

Multispectral images can be acquired economically and stably, with spatial resolution ranging from low to high. They can provide rich colors, textures, and other properties. Images with high spatial resolution or very high spatial resolution (10 to 100 cm/pixel) can also reflect the structure information of the ground objects [81]. Consequently, they are widely used for change detection. Specifically, the most commonly used types of multispectral images for AI-based change detection methods are derived from the Landsat series of satellites [1,3,11,37,54,63–65,78,113,132,133,137,138,140,141,146,151,164–166,173,194,206] and the Sentinel series of satellites [7,156], due to their low acquisition cost and high time and space coverage. In addition, other satellites, such as Quickbird [4,21,107,125,148,149,169], SPOT series [19,142,143,154], Gaofen series [23,50,226], Worldview series [5,71,72,86,167], provide high and very high spatial resolution images, and various aircrafts provide very high spatial resolution aerial images [51,55,79,94,95,98,108,144,185,228], allowing the change detection results to retain more details of the changes.

A panchromatic image has only one band (i.e., black and white band), and usually contains a couple of hundred nanometer bandwidth. The bandwidth enables it to hold a high signal-noise ratio, making the panchromatic data available at a high and very high spatial resolution. Therefore, panchromatic images are usually fused with multispectral images to obtain richer spectral information and spatial information for change detection with high and very high spatial resolution. In addition, they can be used directly for change detection [205]. Optical RS images are widely utilized for change detection as they provide abundant spectral and spatial information. However, optical sensors rely upon the sun illumination and the used wavelength is close to visible light or 1 mm. Therefore, they are often affected by solar radiation and clouds. SAR, on the other hand, uses a wavelength of 1 cm to 1 m and has its own illumination source. Thus, it can image at both day and night, in almost all weather conditions.

6.3.2 SAR images

SAR is a technique which uses signal processing to improve the resolution beyond the limitation of physical antenna aperture [18]. The sensor is mounted on an aircraft or a satellite, and is used to make a high-resolution image of the earth's surface. SAR is independent of atmospheric and sunlight condition, so it has become a valuable source of information in change detection. With the development of SAR imaging technology, multiplatform, multiband, multipolarization SAR images provide more abundant data sources for change detection tasks. However, SAR images always suffer from the effect of speckle noises, which results in a more difficult process of change detection than optical RS images. Their three key problems include: i) suppressing speckle noise; ii) designing a change metric or a change indicator; and iii) using a threshold or a classifier based on a change metric to generate a final change map. Change detection methods using AI techniques, especially an autoencoder (AE) [24,39,61,70,109,116,130,131,155,180,182] and a convolutional neural network (CNN) [36,44,92,115,120,127,168], to suppress speckle noise and extract features has been proven to be the state of the art. In the overall process and framework of methods, they are similar to the methods based on optical RS images, and the detailed framework and AI model introduction are analyzed in Sections 6.4 and 6.5.

6.3.3 Street view images

Unlike optical RS and SAR images, street view images are captured at eye-level instead of overhead. They provide more detailed information in relatively small areas and at more observation angles, which can be used for dynamic or real-time change detection. Change detection based on street view images focuses on changes in the dynamic urban visual landscape, such as the addition or removal of specific landmarks, pedestrians, vehicles, and other roadside buildings.

A critical challenge is on how to identify noisy changes caused by various illuminations, camera viewpoints, occlusions, and shadows in detecting changes using street view images. These noisy changes are interwoven with semantic changes, making it difficult to define and measure the wanted semantic changes in street view images. Thus, using AI algorithms, mainly CNN [13,77,88,99,171,195], to learn deep features for change detection, requires street view images that have been spatially registered.

6.3.4 Combining heterogeneous data

According to whether the source of multiperiod data is the same, the change detection methods can be divided into homogeneous data change detection and heterogeneous data change detection. Homogeneous data comes from the same type of sensors, and they have the same properties, spectral distribution, and feature space, while heterogeneous data come from different types of sensors, they have different properties and feature spaces, so they cannot be analyzed directly for difference image. Although change detection using heterogeneous data is more challenging, it has fewer restrictions on the type of input data and can be used in more situations. Different sensors can complement each other to provide richer information on ground objects. For example, using optical RS images and SAR images for change detection, the former can provide rich texture information, while the latter can be acquired without atmospheric restrictions. It can be used for emergency change detection in areas where data are insufficient or disasters occur. Much work on this issue has been proposed that uses AI methods to detect changes in SAR and optical RS images [7,68,121,134,145,215,218,219,225]. In addition, GIS maps [221], point cloud data [228], DEMs [29,85], and digital surface models (DSMs) [91] are used in combination with optical RS images or SAR images for change detection. These different data types can satisfy different application requirements and are selected according to the actual situation.

6.4 General AI-based change detection frameworks

The input of the change detection task is multitemporal data, which are homogeneous or heterogeneous data in two or more periods. According to the deep feature extraction or latent feature representation learning process of the bitemporal data, the AI-based change detection frameworks can be summarized into three types: single-stream, double-stream, and multimodel integrated. In addition, we further analyze their unsupervised scheme in these frameworks, which is a very important and challenging research issue in AI.

6.4.1 Single-stream framework

There are two main types of single-stream framework structures for AI-based change detection, as shown in Fig. 6.5, namely a direct classification structure and a mapping transformation-based structure.

They usually only need a core AI model to achieve change detection, so they can be regarded as a single-stream structure. It is worth noting that, in practice, some studies have made improvements based on these structures to meet specific change detection purposes, and a detailed analysis is given below.

6.4.1.1 Direct classification structure

The direct classification methods use various data processing approaches to fuse the two or more periods of data into intermediate data, and a single AI-based classifier is then used to perform feature learning and achieve two or multiple classifications of the fusion data. That is, as shown in Fig. 6.5a, this structure converts the change detection task into a classification task, also known as a two-channel structure in some of the literature [44,217]. Its two key research issues are the choice of data fusion approach and AI-based classifier.

To obtain the fusion data from multiperiod data, the two most common approaches are using change analysis methods and direct concatenation. Change analysis methods, such as CVA [63], differencing by log-ratio operator [122,229] or change measures [2,131], are able to directly provide change intensity information (i.e., the difference data) in multitemporal

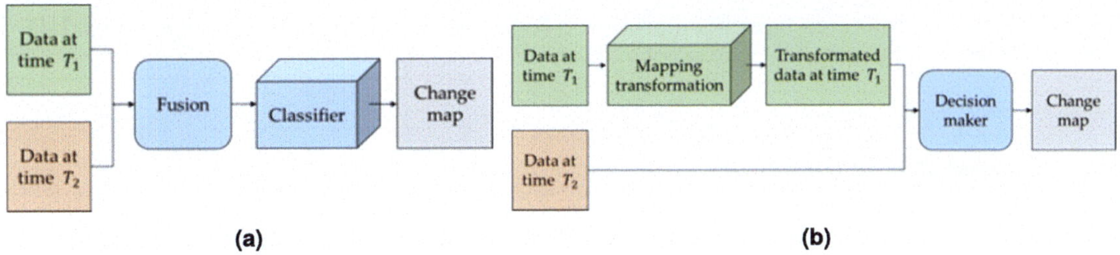

Figure 6.5:
Schematic diagram of single-stream framework structures of AI-based change detection: (a) the direct classification structure; (b) the mapping transformation-based structure.

data, which can highlight change information and facilitate change detection. The direct concatenation method can retain all the information of the multiperiod data, so the change information is extracted by the subsequent classifier. In general, the one-dimensional input data is directly concatenated [37,73,75,109,192,212], while the two-dimensional data is concatenated by channel [58,92,104,115]. Moreover, the fusion of original data and difference data [60,61] is another good strategy, which can keep all the information while highlighting the difference information.

The classifier uses AI techniques to classify the fused data into two types (i.e., changed or unchanged) or multiple types (different types of changes) [142]. Its performance and related training data are the key to finally obtaining satisfactory change maps. More details are reviewed in Section 6.5.

6.4.1.2 Mapping transformation-based structure

The mapping transformation-based framework structure is usually used to detect changes in different domains or heterogeneous data. Its main idea is to use the AI method to learn the feature mapping transformation, and use it to perform feature transformation on one kind of data, as shown in Fig. 6.5b. The transformed features correspond to the features of another kind of data. In short, it transforms data from one feature space to another feature space. Finally, the change map can be obtained by performing decision analysis on the corresponding features of the two kinds of data. In [225], a mapping neural network (MNN) is designed to learn the mapping function between the multispatial-resolution data, and the feature similarity analysis is then implemented to build a change map. This method also achieves change detection between SAR and optical images. In [166], the authors use an ANN to achieve relative radiometric normalization, and then detect changes of the two-period data under the same radiation condition. Moreover, using this idea of mapping transformation, several improved structures have been proposed for detecting changes in heterogeneous data [121,134,218,219] or different domain data [52].

6.4.2 Double-stream framework

Since the change detection task is usually based on two periods of data, that is, two inputs, the double-stream structure is very common for change detection and can be summarized into three types, as shown in Fig. 6.6. They are a siamese structure, a transfer learning-based structure, and a postclassification structure.

6.4.2.1 Siamese structure

The siamese structure processes multitemporal data through two branches with shared or partially shared parameters, so that corresponding features from different periods can be compared directly. It is widely used when the two-period inputs have similar properties and when feature-level similarity learning is beneficial for change detection.

6.4.2.2 Transfer learning-based structure

The transfer learning-based structure usually has two training phases, namely the deep feature learning phase and the fine-tuning phase. In the deep feature learning phase, the AI model is usually supervised, pretrained with sufficient labeled samples in other domain data [48,156,168]. The fine-tuning phase is optional and, in this phase, only a small number of labeled samples are required for fine-tuning [59,103,123,203,215] or additional classifier training [185,223]. Therefore, the change map can be directly obtained by the trained classifier. Without fine-tuning, final change maps can be obtained based on the two-period feature maps using change analysis, such as low rank analysis [84], CVA [169], clustering [107,167], and threshold [47,99]. This means that no more labeled samples are needed for further training. Moreover, based on the idea of transfer learning, the pretrained AI model can also be used to generate training samples or masks to achieve the unsupervised scheme [154], which is a very practical strategy.

6.4.2.3 Postclassification structure

As shown in Fig. 6.6c, the postclassification structure consists of two classifiers, which can usually be converted into classification tasks and trained in a joint or independent way. It provides a classification map for each period data and the change map with change directions can be obtained by comparing classification maps. Nevertheless, the accuracy of the change detection results of these methods depends on the performance of the classifier.

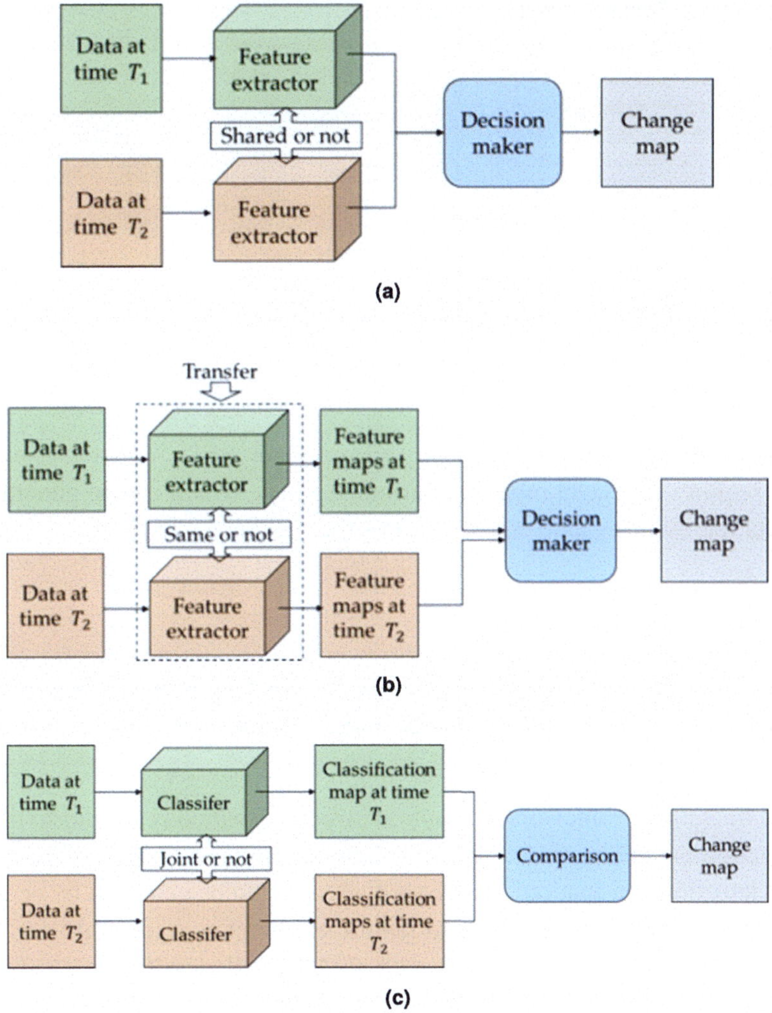

Figure 6.6:
Schematic diagram of double-stream framework structure of AI-based change detection: (a) the Siamese structure; (b) the transfer-learning-based structure; (c) the postclassification structure.

A number of studies [1,11,14,19,91,133,143,146,149,173,209] have proven that AI techniques increase the accuracy of land-cover classification to a notable level and the results can be further used for change detection. By converting the direct geometric or spectral comparison to label changes, the postclassification structure can be regarded as a very general and practical structure, and it provides a type change matrix. Advantageously, it works robustly for data acquired under different acquisition conditions (illumination condition, sensor attitude, season, etc.) or even different sensors [78,85,194]. Supervised training of the AI-

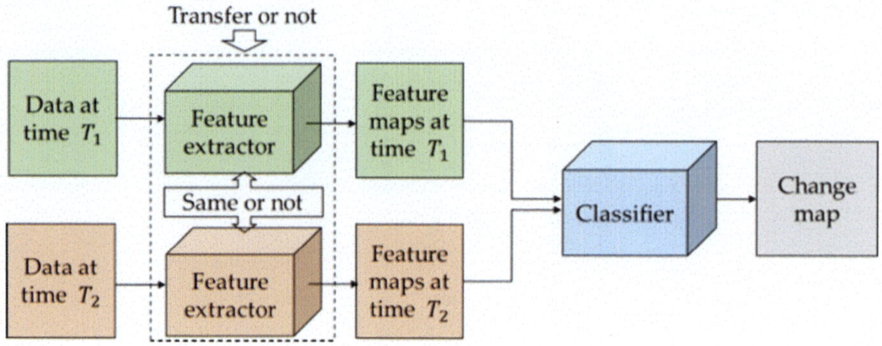

Figure 6.7:
Schematic diagram of multimodel integrated structure of AI-based change detection.

based classifier requires a large number of training samples, which can be generated by existing GIS data representing land cover [221] or thematic maps [62].

6.4.3 Multimodel integrated structure

Many works have integrated multiple AI models to improve the performance of change detection methods. Considering the large number and complex structure, only a representative structure is summarized, as shown in Fig. 6.7.

The multimodel integrated framework is a hybrid structure, which is similar to the double stream structure, but it contains more types of AI models and can also be trained in multiple stages. Change detection is a spatiotemporal analysis and can be achieved by acquiring the spatial-spectral features through an AI-based feature extractor as a spectral-spatial module, and then modeling temporal dependency through an AI-based classifier as a temporal module [23,125,138,177]. Moreover, this hybrid structure is skillfully used for unsupervised change detection [70] and object-level change detection [79]. This makes the whole change detection process more complicated while improving performance.

6.4.4 Unsupervised schemes in change detection frameworks

AI-based change detection frameworks usually include feature extractors or classifiers, which require supervised and unsupervised training. Since obtaining a large number of labeled samples for supervised training is usually time-consuming and labor-intensive, many efforts have been made to achieve AI-based change detection in an unsupervised or semisupervised manner. As introduced in Section 6.4.2.2, transfer learning can reduce or even eliminate the need for training samples, but these are not pure unsupervised schemes, as samples from other

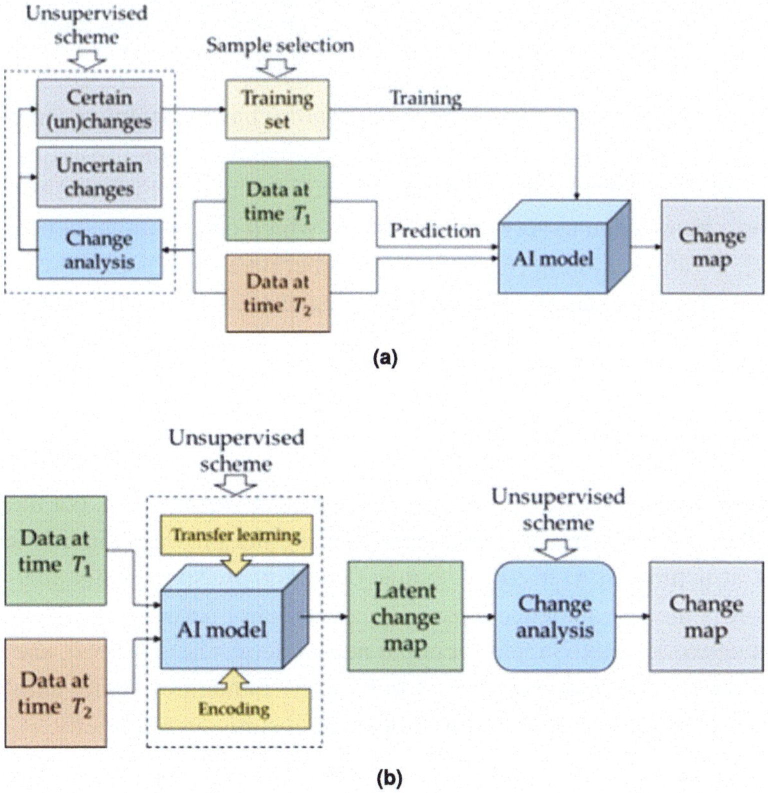

Figure 6.8:
Flow chart of the most commonly used unsupervised schemes.

domains are required. In addition, the most commonly used unsupervised scheme is to use the change analysis method and sample selection strategy to select absolute changed or/and unchanged as training samples for AI models. Its flow chart is shown in Fig. 6.8a.

It can be seen that there are two change detection stages in this scheme. The first stage, i.e., preclassification, is usually simple but worth studying, and most of them are unsupervised methods, which can be implemented with difference analysis and clustering [109], such as K-means [45], fuzzy c-means (FCM) [22,56,57,61,70,111,115,117,181,185,212], spatial FCM [104,116], or hierarchical FCM [36,60]. This stage in some works is implemented by threshold analysis [199,229], saliency analysis [154], or well-designed rules [71,72,122, 134,172,177,230]. After obtaining high-confidence changed or/and unchanged samples, the AI model can be trained in a supervised manner for change detection in the second stage. Moreover, another commonly used unsupervised scheme is based on the latent change map, as shown in Fig. 6.8b. In addition to the pretrained model obtained by transfer learning, it can be generated by an unsupervised AI model (e.g., AEs), and the final change map is then generated by using a clustering algorithm [15,24,31,130,207,226,231].

Although unsupervised change detection does not require labeled training samples, sometimes the lack of prior knowledge makes it unsuitable for change detection involving semantic information. Weakly and semisupervised schemes use inaccurate or insufficient labeled samples as a priori knowledge to solve this problem, which can be implemented with label aggregation [39], iterative learning [38,165], deep generative models [32] (see Section 6.5.2 for a more detailed review), sample generation strategies [82], or novel cost functions [110,114].

6.5 Mainstream networks in AI

Section 6.4 summarizes general AI-based change detection frameworks, but the detailed introduction of their feature extractors and classifiers are not provided. In this section, the various network structures in AI used for change detection are specifically analyzed. At present, they mainly include autoencoders (AEs), deep belief networks (DBNs), CNNs, recurrent neural networks (RNNs), pulse couple neural networks (PCNNs), and generative adversarial networks (GANs), as can be seen in Fig. 6.9. In addition, other networks or methods in AI used for change detection have also been summarized briefly.

Obviously, the input for change detection includes two or more periods of data, so the change detection task can be converted into a process of obtaining change information directly from the multiperiod sequence data. RNN, as a memory network whose input is sequence data, is very suitable for this situation. As shown in Fig. 6.9d, its core part is a directed graph that can be unfold to a chain, with units (i.e., RNN cells) linked in sequence. An RNN cell has two inputs: one is the current time input $\mathbf{x_t}$, which is used to update the state in real time, and the other is the state of the hidden layer $\mathbf{h_{t-1}}$ at the previous time, which is used to remember the state. The network at different times shares the same set of parameters F.

A long short-term memory (LSTM) network, a special RNN that alleviates the problem of gradient disappearance and gradient explosion during long sequence training, has been employed as a temporal module (see Section 6.4.3) in change detection tasks [23,125,137, 138,177]. In [132], the authors used an improved LSTM network to acquire and record the change information of multitemporal RS data. Advantageously, the trained model could be transferred to other data domains; that is, it has a good generalization ability. Further, in [133], based on the same idea, i.e., knowledge transfer, the authors propose an RNN-based framework to detect annual urban dynamics of four cities. These methods can help to address the problem of temporal spectral variance and insufficient samples in the long-term detection of urban change.

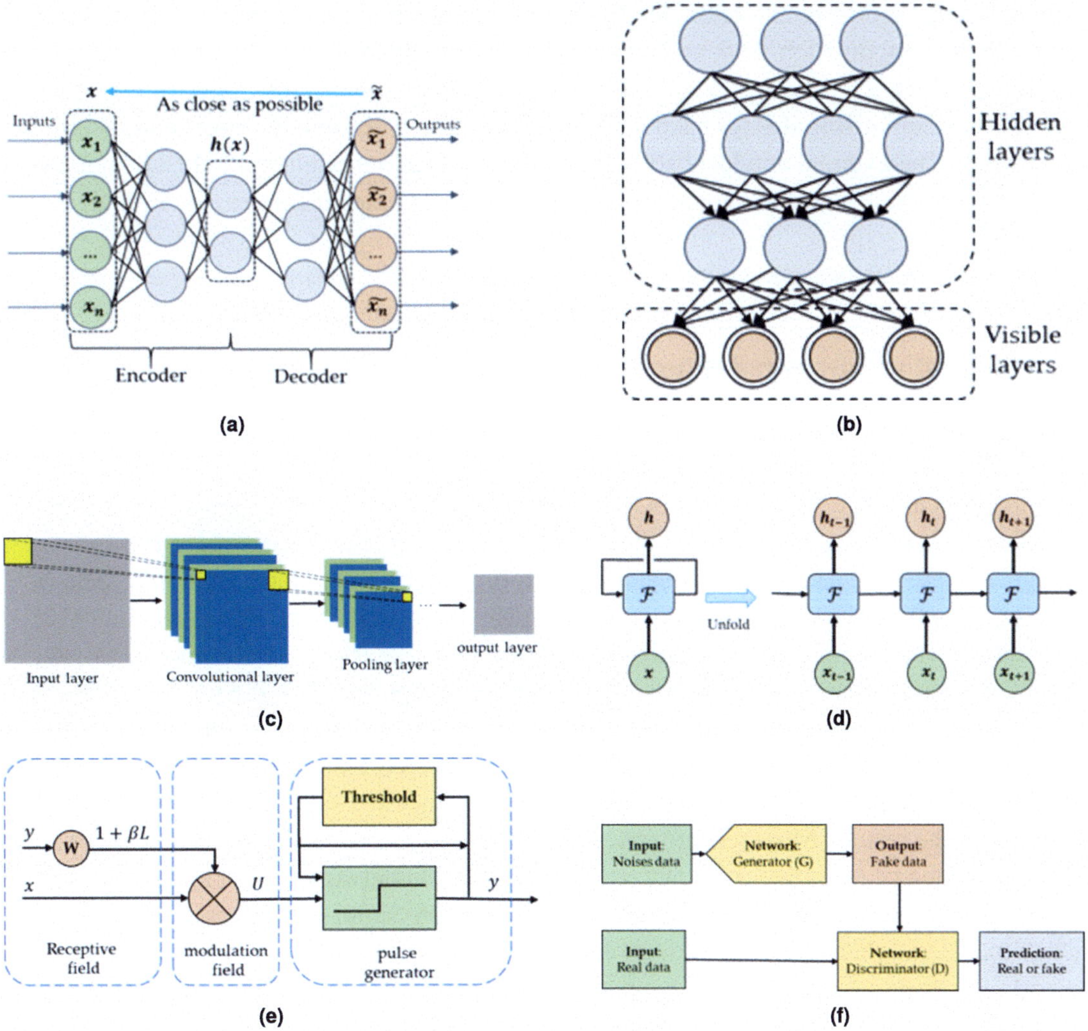

Figure 6.9:
Schematic diagram of general network architectures in AI used for change detection: (a) autoencoder (AE); (b) deep belief network (DBN); (c) convolutional neural network (CNN); (d) recurrent neural network (RNN); (e) pulse couple neural network (PCNN); (f) generative adversarial networks (GANs).

6.5.1 Pulse couple neural network

The PCNN is a bionic neural network inspired by the visual cortex of mammals. Unlike traditional neural networks, it does not require a learning and training stage to extract effective information from very complex backgrounds, which means that it is unsupervised and can easily be used as a feature extractor. As shown in Fig. 6.9e, it mainly consists of three parts: a received field, a modulation field, and a pulse generator. A PCNN receives a two-dimensional input image, and each neuron corresponds to one pixel in the image. The value of each pixel serves as an external stimulus for each neuron, and its connected neighboring neurons provide local stimuli. The external and local stimuli are combined in a modulation field and a pulse generator to produce a pulsed output. As the number of iterations increases, the PCNN generates a pulse sequence that can be used for image segmentation and feature extraction and, similarly, for change detection [7,86,124,134,140,148,157,234].

6.5.2 Generative adversarial networks

GANs are algorithmic architectures that use two neural networks, namely, a generator and a discriminator, to contest with each other in a game to obtain the best generation and discrimination model [74], as shown in Fig. 6.9f. The generator learns to generate plausible data as negative training examples for the discriminator, while the discriminator learns to distinguish these fake data from real data. Therefore, with a small amount of labeled data (i.e., real data), a well-performed discrimination model can be trained for change detection [232]. Different from using random Gaussian noises to generate fake data [71], the authors of [69] used a CNN-based generator to produce fake difference maps based on the joint distribution of two-period data, which can help to weaken the impact of the bad pixels on the network. In [83], the authors designed a W-Net as a generator to decrease the network parameters and make them easy to train, and used many manually annotated samples to improve the reliability of the results. More recent work [145] proposed a conditional GAN to achieve change detection of heterogeneous data, where a translation network and an approximation network were used to transform the heterogeneous data into same feature space, and the change map could then be obtained by direct comparison. Similarly, a coupling translation network was employed in [68]. For mapping landslides, the authors of [51] used a mapping generator to transform the pre- and postlandslide images into the same domain, and a Siamese network was then employed to detect landslide changes.

To obtain a well-functioning GAN, the methods need a well-designed loss function and a good training strategy; otherwise, the model results may be unsatisfactory due to the freedom of the neural network. In addition, real data are needed to ensure the reliability of the network. These are challenges that exist in many practical applications.

6.5.3 Other artificial neural networks and AI methods

There are many types of artificial neural networks in AI and the mainstream network structures used for change detection are described above. In addition, other networks, such as Hopfield networks [26,63,64,113,184,200], back propagation networks [2,35,37,210], multilayer perceptrons (MLPs) [16,40,136,149,152,188], extreme learning machines [25], and self-organizing map (SOM) networks [66,141,150,179,187,211,213], do not require a large number of training samples to learn high-level abstract features as deep neural networks do, but due to their shallow network structure, low sample size requirements, and easy training process, they are also widely used in change detection tasks and can achieve satisfactory results. Since they can be regarded as traditional machine learning techniques, we will not make more detailed comments here due to space limitations and existing reviews [33,89,102].

In addition to the neural network in AI, there are other AI techniques used for implementing change detection. Recently, dictionary learning has been employed, and it focuses on learning internal feature representations from datasets [190,208,209,214], just like AEs. The cellular automata (CA), a spatially and temporally discrete model inspired by cellular behavior, can help to model future changes in LULC [46] and predict urban spatial expansion [49]. The development of these AI techniques has significantly promoted research on change detection, which helps to develop more automatic, intelligent and accurate methods to meet the needs of various applications.

6.6 Applications

In practical applications, according to the change information, the final change maps can be grouped into four types, namely, binary maps, one-class maps, from-to maps, and instance maps, as can be seen in Fig. 6.10.

A binary map uses 1 and 0 to indicate change and no change. It contains any changes and cannot provide an additional type of information for the changed ground object. A one-class map provides single-type change information, which indicates the appearance and disappearance of a specific type of ground object. For example, the result of building change detection is a one-class map indicating the addition and removal of buildings, which can be used for urban management [144]. A from-to map provides change transfer information, indicating that the ground object is changed from one type to another, and these change types are determined by the classifier [221]. A change instance map provides boundaries for each change instance, which can be the result of object-based change detection [94]. These change detection maps can be generated by trained AI models and used in various applications. To demonstrate the future potential application demands and development possibilities of AI-based change detection techniques, the current attempts and work in various application fields are summarized in this section.

The development of AI-based change detection techniques has greatly facilitated many applications and has improved their automation and intelligence. Most AI-based change

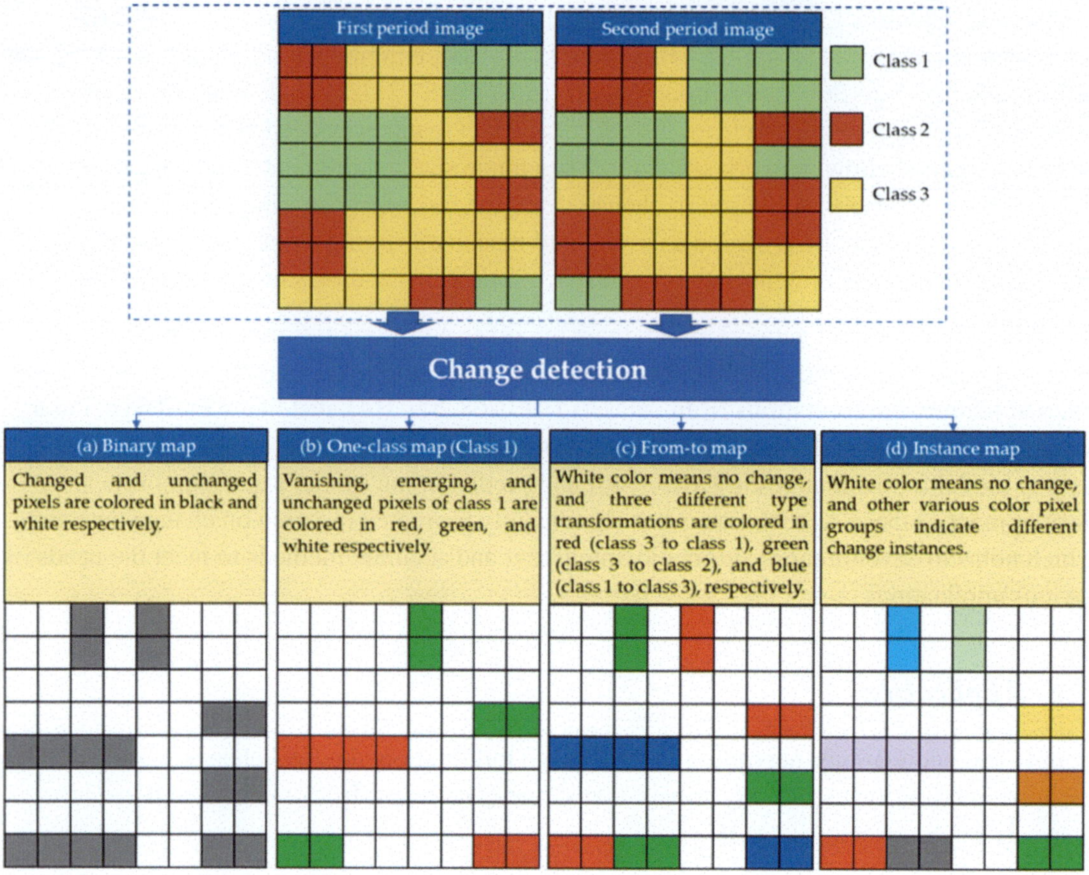

Figure 6.10:
Schematic diagram of four change maps. To meet different application requirements, four different change maps can be obtained by detecting the change of the two-period images, i.e., (a) binary maps, (b) one-class maps, (c) from-to maps, and (d) instance maps.

detection generates binary maps, and these studies only focus on the algorithm itself, without a specific application field. Therefore, it can be considered that they are generally suitable for LULC change detection. In this section, we focus on the techniques that are associated with specific applications, and they can be broadly divided into four categories:

- Urban contexts: urban expansion, public space management, and building change detection;
- Resources and environment: human-driven environmental changes, hydroenvironmental changes, sea ice, surface water, and forest monitoring;

- Natural disasters: landslide mapping and damage assessment;
- Astronomy: planetary surfaces.

We provide an overview of the various change detection techniques in the literature for the different application categories. The works and data types associated with these applications are listed in Table 6.1.

Urbanization is a significant factor causing land surface change. As a result of population growth, the expansion of urbanization plays an important role in transforming natural land cover into urban facilities for people. In [133], the authors proposed a new framework based on transfer learning and an RNN for urban area extraction and change detection. Its overall accuracy of single-year urban maps is approximately 96% among the four target cities (Beijing, New York, Melbourne, and Munich). Using a genetic-algorithm-evolved ANN [191] or a CNN [90], urban changes were obtained by taking the difference in the predicted urban distribution maps. For public space management, change detection based on street view images is a good way to identify the encroachment of public spaces. In [195], a CNN-using Siamese structure and transfer learning achieved 98.3% pixel accuracy in the VL-CMU-CD dataset. In addition, many studies focus on building changes. Due to the small scale of buildings, change detection is usually carried out based on high- or very-high-spatial-resolution RS data, such as aerial photos [94,144] and satellite images from Quickbird [236] or Worldview 2 [86]. However, due to differences in the experimental data, it is difficult to evaluate which one has the best performance. Although some methods are based on the WHU building dataset [95], unfortunately, their experimental data are different with regard to area, which makes it difficult to compare them directly. In [94], the authors proposed two CNN models, a mask R-CNN for object-based instance segmentation and a multiscale CNN for pixel-based semantic segmentation, and their intersection of union (IoU) accuracy of buildings was more than 0.83. In [98], the authors proposed a pyramid feature-based attention-guided Siamese network to detect building changes and the IoU of change map exceeded 0.97. Sometimes, the problem of constant cloud coverage prevents optical RS images from being fully utilized, and SAR data represent a good alternative [92]. On the other hand, to obtain more building information to promote change detection, airborne laser scanning data [228] and building thematic data [195] can provide 3-D information and a priori knowledge of buildings, respectively, and proved to be effective.

Land cover changes typically reflect changes in climatology and hydrology. Thus, AI-based change detection techniques can provide effective methods to monitor changes in resources and the environment. For example, forest monitoring is usually achieved by detecting changes using multiperiod Landsat satellite images [75,105,118,206]. In addition to the Landsat data, the authors of [146] used an ANN to investigate LULC changes and the effect on outlet runoff by detecting LULC change locations. Surface water is essential for humans; using an ANN

Table 6.1: Summary of main applications of AI-based change detection techniques.

Applications		Data Types
Urban contexts	Urban expansion	a. Satellite images [133,191] b. SAR image [90]
	Public space management	Street view images [195]
	Building change detection	a. Aerial images [94,144,185] b. Satellite images [86,236] c. Satellite/Aerial images [95,98] d. Airborne laser scanning data and aerial images [228] e. SAR images [92] f. Satellite images and GIS map [195]
Resources & environment	Human-driven environmental changes	Satellite images [21]
	Hydroenvironmental changes	Satellite images [146]
	Sea ice	SAR images [59]
	Surface water	Satellite images [163,178]
	Forest monitoring	Satellite images [75,105,118,206]
Natural disasters	Landslide mapping	a. Aerial images [51,108] b. Satellite images [29,42,188]
	Damage assessment	a. Satellite images (caused by tsunami [135,183], particular incident [82], flood [153], or earthquake [93]) b. Aerial images (caused by tsunami [55]) c. SAR images (caused by fires [155], or earthquake [168]) d. Street view images (caused by tsunami [99]) e. Street view images and GIS map (caused by tsunami [170])
Astronomy	Planetary surfaces	Satellite images [103]

[163] or CNN [178], its changes can be detected effectively. Obtaining sea ice changes is crucial for navigation safety and climate research in the polar regions. The authors of [59] employed a transfer learning-based framework and a CNN model to detect sea ice changes from two-period SAR images, and obtained results with kappa coefficient exceeding 94%.

Natural disaster is a powerful agent that changes the appearance of the landscape. Therefore, change detection techniques based on RS data are significant measures to monitor natural disasters. For example, more automatic and accurate landslide mapping can be achieved through CNN-based change detection methods [29,42,51,108]. Damage assessment is an important application field of change detection. After a natural disaster, AI-based change techniques can help to identify damaged areas using preevent and postevent data. Existing research mainly includes tsunamis [55,99,135,170,183], floods [153], fires [155], and earthquakes [93,168].

Many change detection techniques focus on applications that are closely related to people's daily lives, but in [103], a new AI-based approach is proposed for planetary surface change detection, employing a transfer learning-based framework and convolutional AE models. The experiments showed that the proposed method was superior to a difference image-based method.

6.7 Challenges and opportunities for AI-based change detection

By combining general AI-based change detection frameworks and the network structure summarized in Sections 6.4 and 6.5, change detection for various applications can be implemented. The design of the AI model usually needs to consider the multiperiod input data type, training set size, and desired change map. Specifically, a mapping transformation-based structure or pseudosiamese structure should be considered first based on heterogeneous data. To obtain from-to change maps, postclassification structure is the best choice. If training samples are insufficient, a transfer learning-based structure can help alleviate this problem, and the use of AEs and GANs can also reduce the dependence on ground truth. Change detection based on long-term sequence data is usually implemented using the multimodel integrated structure and an RNN model. A CNN has strong feature extraction capability, and is the best choice when there are sufficient training samples.

Various applications of AI-based change detection have demonstrated that AI techniques have achieved great success in the field of change detection in RS community. However, there are many challenges in the processes, and they relate to the following:

- With the development of various platforms and sensors, they bring significant challenges such as high-dimensional datasets (the high spatial resolution and hyperspectral features), complex data structures (nonlinear and overlapping distributions), and the nonlinear optimization problem (high computational complexity). The complexity of multisource data greatly contributes to the difficulty of learning robust and discriminative representations from training data with AI techniques. This can be considered a challenge of heterogeneous big data processing;

- The supervised AI methods require massive training samples, which are usually obtained by time-consuming and labor-intensive processes such as human interpretation of RS products and field surveys. It is a big challenge to achieve a robust model of AI-based methods with insufficient training samples. Unsupervised AI techniques need to be developed;
- There are various efficient and accurate AI models and frameworks, as we review in Sections 6.4 and 6.5. At present, researchers constantly propose novel AI-based change detection approaches endlessly. Still, it is also a great challenge to choose an efficient one and ensure its accuracy for different applications. The reliability of AI needs to be considered in practical applications.

Some researchers have explored solutions to these problems and proposed useful strategies. We will discuss them separately and give our views.

6.7.1 Heterogeneous big data processing

Heterogeneity is one of the main characteristics of big data and heterogeneous data, causing problems in the generation and analytics of change detection results [197]. From the data source perspective, RS technology can provide various data types for change detection, such as SAR, GIS data, high-resolution satellite images, and various time and space measured data. These data with high variability of data types and formats are difficult to use due to missing values, high data redundancy, and untruthfulness. Moreover, the generalization ability of existing AI methods needs to be improved in RS data processing, especially in heterogeneous big data processing [95]. Therefore, in our opinion, the following aspects need further study:

- Although some AI-based change detection methods based on heterogeneous data have achieved satisfactory results, as summarized in Section 6.3.4, the types of sensor and data size of these studies are relatively limited. Moreover, they mainly consider change detection between different source data rather than finding the fusion of data in the same period. The full use of multisource data at the same period (e.g., optical RS images and DEM) and data fusion theories (i.e., mutual compensation of various types of data), combined with AI techniques, would help improve the accuracy of change detection sufficiently;
- Since current change detection methods mainly depend on the detection of 2D information, with the development of 3D reconstruction techniques, using 3D data to

detect changes in buildings, etc., is also a direction of future development [158]. Among such techniques, 3D reconstruction based on oblique images or laser point cloud data and 3D information integration based on aerial imagery and ground-level street view imagery (i.e., air-ground integration) are the hot topics of research. There are still no effective AI techniques that implement 3D change detection;

- The processing of RS big data requires a large amount of computing resources, limiting the implementation of the AI model. For example, the processing of large-format data usually needs to be processed in blocks, which easily leads to edge problems. The large amount of data means that large trainable parameters in the AI model are required, resulting in a difficult training process and consuming a high amount of computing resources. Therefore, it is necessary to balance the amount of data and the number of trainable parameters. They pose challenges to the design of AI-based change detection approaches.

In short, heterogeneous big data should be considered when designing AI models for change detection so that it can be practically used for RS big data processing, which is worth pursuing.

6.7.2 Unsupervised AI

Although domain knowledge can be used to help design representations in traditional machine learning methods, the quest for AI is motivating the design of more powerful unsupervised representation-learning algorithms [8]. This is because unsupervised AI possesses the capacity of learning hierarchy features directly from the data itself, and can be used to make data-driven decisions. The research on unsupervised AI can be considered in the following aspects:

- Due to the lack of labeled samples to train efficient AI models in the past few years, many researchers have devoted great efforts to these problems and have consistently produced impressive results. New unsupervised AI techniques are constantly emerging, including GAN, transfer learning, and AEs, as summarized in Section 6.4.4. Although these techniques alleviate the lack of samples to a certain extent, there is still room for improvement;
- Change detection is generally regarded as a low-likelihood problem (i.e., the unchanged in the change map is much larger than the changed), with the uncertainty of the change location and direction. The current unsupervised AI techniques do not easily solve this problem due to the lack of prior knowledge. Excluding supervised AI, weakly- and semisupervised AI techniques are feasible solutions, but further research is needed to improve performance. Nevertheless, pure unsupervised AI technique for change detection should be the ultimate goal;

- One of the reasons for studying unsupervised AI techniques is the lack of training samples, i.e. prior knowledge. Considering that the Internet has entered the Web 2.0 era (emphasizing user-generated content, ease of use, participatory culture and interoperability for end users), using crowd-source data as a priori knowledge is a good alternative solution. For example, OpenStreetMap, a free wiki world map, can provide a large amount of annotation data labeled by volunteers for the training of AI models. Although the label precision of some crowd-sourced data is not high, the AI model can also be trained in a weakly supervised manner to achieve change detection.

On the other hand, given the current trends in CV, unsupervised AI techniques will remain a hot research field and even more popular in change detection as well.

6.7.3 Reliability of AI

Although many change detection frameworks using AI present the model structure, their trainable parameters are opaque, like black boxes, which make it difficult to determine why they do what they do or how they work [76]. The reliability of AI aims to develop techniques to help improve the reliability and interpretability of the change detection methods. Therefore, it is necessary to develop robust AI and interpretable AI for change detection. The relevant theoretical literature can be found in [41,139]. We only discuss strategies that can be used to improve the reliability of change detection results from the following aspects:

- Strategy 1: Reduce errors caused by data sources, such as using preprocessing (e.g., spectral and radiometric correction) to reduce the uncertainty of data caused by geometric errors and spectral differences, or fusing multiple data to improve the reliability of the original data, thereby improving the reliability of the change detection results. To date, there have been some studies considering the impact of registration [175] and algorithm fusion [224];
- Strategy 2: Improve the interpretability of AI models through a submodular model structure, which can help to understand the operation principle of the entire AI model by understanding the function of each submodule. For example, the region-proposals component in R-CNN can be clearly understood as a generator to predict the regions of the objects;
- Strategy 3: Improve the robustness of AI models by integration of multiple algorithms and results. Ensemble learning is a good solution [12,201], which can improve the accuracy of the final result by using the results of multiple models;
- Strategy 4: Improve the practicality of the AI model results by integrating postprocessing algorithms, such as the Markov random field [80], the conditional random field [28], and level set evolution [204], which can help remove noise points and provide accurate boundaries. This is critical for some cartographic applications;

- Strategy 5: Improve the fineness of change maps through refined detection units. According to the detection unit of change detection, it can be divided into scene level, patch or superpixel level, pixel level and subpixel level from coarse to fine. From the aspect of reliability, subpixel level is the best choice because it alleviates the problem of mixed pixels in RS images. However, this easily leads to high computational complexity. Therefore, using different detection units according to different land cover types is the best solution, which requires a well-designed AI model;
- Strategy 6: Improve the representation of change maps by detecting changes in each instance. As we introduced in Section 6.6, the change maps can be grouped into binary maps, one-class maps, from-to maps, and instance maps. The instance change map is more practical but still lacks research. It can provide change information for each instance, and is more reflective of real-world changes. Moreover, it can avoid the limitation of the binary map without semantic information and the restriction of the from-to map by the classification system, thereby improving the reliability of the final result.

When using AI techniques for change detection, factors that affect the reliability of data preprocessing, model training, change feature extraction, and accuracy assessment should be considered. This aims toward the most reasonable AI framework to improve the reliability of change detection results.

6.7.4 Visual foundation models in AI

The concept of "Foundation Models [9]," proposed by the Stanford Institute for Human-Centered Artificial Intelligence's (HAI) Center for Research on Foundation Models (CRFM) in August 2021, is defined as "models trained on broad data (generally using large-scale self-supervised learning) that can be adapted (e.g., via fine-tuning) to a wide range of downstream tasks." This definition highlights the foundation models' characteristics of leveraging extensive data for pretraining and their wide applicability across various scenarios. Initially achieving breakthrough success in natural language processing, particularly through the development of Large Language Models (LLMs), like GPT series [10], PaLM [30], T5 [160], LLaMa [193], and ERNIE [186], these models, with their deep learning of language semantics and syntax from massive textual data, can perform a wide range of complex linguistic tasks, including text generation, translation, sentiment analysis, and question-answering systems, marking a new era in AI research and applications. Exploration of foundation models has also been conducted in the field of vision. Models like CLIP [159] and ALIGN [97] trained on a vast number of image-text pairs, demonstrate the capability to understand and link image content with textual descriptions. They map images and text into a shared representational space, enabling potent cross-modal capabilities. Derived models, such as Florence [216], RegionCLIP [235], CLIP2Video [53], and CLIP-ViL [174], along with integration with modules like DALL-E [161], have shown adaptability to various computer vision tasks, like image classification, object detection, visual question answering, and image generation.

In RS image processing, a related task in computer vision, the application of visual foundation models has shown significant potential and broad prospects [17,112,119,222]. As change detection in RS imagery is essentially a semantic segmentation task, models specifically developed for segmentation, like CLIPSeg [129], SegGPT [202], Segment Anything Model (SAM) [106], and SEEM [237], are closer to change detection tasks and exhibit immense potential in precisely identifying and tracking terrestrial changes. Recent research, [27,96, 147,196] efforts have begun to explore the applicability of these visual foundation models in the field of RS image segmentation, offering innovative perspectives and methodologies. For more specific change detection tasks, Ding et al. [43] integrated FastSAM [233] as an encoder in a supervised learning model for feature extraction in RS imagery, exploring its potential advantages in semisupervised change detection tasks. Wang et al. [198] combined the localization capability of CAM with SAM's zero-shot segmentation ability, establishing a weakly supervised change detection framework that achieves precise pixel-level change detection on VHR RS images using only image-level labels. These explorations demonstrate the opportunities visual foundation models bring to RS image processing, accelerating data processing speed, enhancing task accuracy, and reducing reliance on large-scale annotated datasets, offering new directions for future developments in change detection and related tasks in RS image processing. Nevertheless, applying visual foundation models in RS image processing still faces several challenges that need to be overcome through continuous technological innovation and research depth:

- In incomplete supervision change detection scenarios, visual foundation models can serve as a powerful auxiliary tool. Researchers can generate high-quality pseudolabels using visual foundation models combined with appropriate prompts, reducing reliance on extensive accuracy annotations. However, the potential of visual foundation models extends beyond this; developing effective learning algorithms to leverage foundation models' advantages in incomplete supervision and integrating them more directly into the main process of change detection are crucial areas for further exploration.
- Existing research shows that pretraining datasets for visual foundation models often lack images specific to certain domains, like RS imagery. Further exploration into developing specialized foundation models using large-scale RS datasets could enable models to capture the unique features of remote sensing imagery more accurately, facilitating zero-shot transfer to related tasks. However, processing and analyzing large-scale RS datasets require immense computational resources.
- In scenarios with limited computational resources, fine-tuning visual foundation models through open interfaces is a practical solution. Employing a partial weight-locking

strategy allows researchers to update the model for specific RS image-related tasks selectively. This method not only conserves computational resources but also ensures the model's ability to adapt quickly to new tasks. Developing more effective fine-tuning strategies to maintain the model's generalizability and ensuring its continuous update and maintenance remain significant challenges.

In this section, a summary of the challenges and opportunities for AI-based change detection techniques has been delineated and we have put forward our prospects. The development of AI-based change detection techniques depends on the future endeavor on overcoming these challenges; the efforts and innovations of researchers would push forward further successes of the techniques.

6.8 Conclusion

This chapter presents the latest methods, applications, and challenges of the AI-based change detection techniques. For beginners, the implementation process of AI-based change detection is introduced. Considering that the validity of training data is one of the major challenges, the commonly used data sources and existing datasets used for change detection were fully surveyed. Although the current public datasets have increased significantly, openly labeled datasets for change detection are still scarce and deficient, which requires the joint efforts of the RS community. The systematic analysis of the general network frameworks and commonly used networks in AI adopted for change detection shows that great progress has been made in the combination of AI for change detection, but there are still many challenges in change detection with heterogeneous big data processing, unsupervised AI, the reliability of AI, and visual foundation models in AI. This means that further research needs to be pushed forward. This chapter offers a clearer organization and will help researchers understand this field.

References

[1] Abdelgadir A. Abuelgasim, W.D. Ross, S. Gopal, C.E. Woodcock, Change detection using adaptive fuzzy neural networks: environmental damage assessment after the Gulf War, Remote Sensing of Environment 70 (2) (1999) 208–223.

[2] Hossein Aghababaee, Jalal Amini, Yu-Chang Tzeng, Improving change detection methods of SAR images using fractals, Scientia Iranica 20 (1) (2013) 15–22.

[3] Asim Anees, Jagannath Aryal, Małgorzata M. Oeilly, Timothy J. Gale, Tim Wardlaw, A robust multi-kernel change detection framework for detecting leaf beetle defoliation using Landsat 7 ETM+ data, ISPRS Journal of Photogrammetry and Remote Sensing 122 (2016) 167–178.

[4] Mohammed El Amin Arabi, Moussa Sofiane Karoui, Khelifa Djerriri, Optical remote sensing change detection through deep Siamese network, in: IGARSS 2018-2018 IEEE International Geoscience and Remote Sensing Symposium, IEEE, 2018, pp. 5041–5044.

[5] Ting Bai, Kaimin Sun, Shiquan Deng, Yan Chen, Comparison of four machine learning methods for object-oriented change detection in high-resolution satellite imagery, in: MIPPR 2017: Remote Sensing Image Processing, Geographic Information Systems, and Other Applications, vol. 10611, SPIE, 2018, pp. 99–106.

[6] Marion Baumgardner, Larry Biehl, David Landgrebe, 220 band AVIRIS hyperspectral image data set: June 12, 1992 Indian Pine Test Site 3, https://doi.org/10.4231/R7RX991C, 2015.

[7] Alessia Benedetti, Matto Picchiani, Fabio Del Frate, Sentinel-1 and Sentinel-2 data fusion for urban change detection, in: IGARSS 2018-2018 IEEE International Geoscience and Remote Sensing Symposium, IEEE, 2018, pp. 1962–1965.

[8] Yoshua Bengio, Aaron C. Courville, Pascal Vincent, Unsupervised feature learning and deep learning: a review and new perspectives, CoRR, arXiv:1206.5538, 2012.

[9] Rishi Bommasani, Drew A. Hudson, Ehsan Adeli, Russ Altman, Simran Arora, Sydney von Arx, Michael S. Bernstein, Jeannette Bohg, Antoine Bosselut, Emma Brunskill, et al., On the opportunities and risks of foundation models, arXiv preprint, arXiv:2108.07258, 2021.

[10] Tom Brown, Benjamin Mann, Nick Ryder, Melanie Subbiah, Jared D. Kaplan, Prafulla Dhariwal, Arvind Neelakantan, Pranav Shyam, Girish Sastry, Amanda Askell, et al., Language models are few-shot learners, Advances in Neural Information Processing Systems 33 (2020) 1877–1901.

[11] Lorenzo Bruzzone, Roberto Cossu, RBF neural network approach for detecting land-cover transitions, in: Image and Signal Processing for Remote Sensing VII, vol. 4541, SPIE, 2002, pp. 223–231.

[12] Lorenzo Bruzzone, Roberto Cossu, Gianni Vernazza, Detection of land-cover transitions by combining multidate classifiers, Pattern Recognition Letters 25 (13) (2004) 1491–1500.

[13] Shuhui Bu, Qing Li, Pengcheng Han, Pengyu Leng, Ke Li, Mask-CDNet: a mask based pixel change detection network, Neurocomputing 378 (2020) 166–178.

[14] Cong Cao, Suzana Dragićević, Songnian Li, Land-use change detection with convolutional neural network methods, Environments 6 (2) (2019) 25.

[15] Guo Cao, Bisheng Wang, Haro-Carrión Xavier, Di Yang, Jane Southworth, A new difference image creation method based on deep neural networks for change detection in remote-sensing images, International Journal of Remote Sensing 38 (23) (2017) 7161–7175.

[16] Laura Castellana, Annarita Dddabbo, Guido Pasquariello, A composed supervised/unsupervised approach to improve change detection from remote sensing, Pattern Recognition Letters 28 (4) (2007) 405–413.

[17] Keumgang Cha, Junghoon Seo, Taekyung Lee, A billion-scale foundation model for remote sensing images, arXiv preprint, arXiv:2304.05215, 2023.

[18] Yee Kit Chan, Voon Koo, An introduction to synthetic aperture radar (SAR), Progress in Electromagnetics Research B 2 (2008) 27–60.

[19] Ni-Bin Chang, Min Han, Wei Yao, Liang-Chien Chen, Shiguo Xu, Change detection of land use and land cover in an urban region with SPOT-5 images and partial Lanczos extreme learning machine, Journal of Applied Remote Sensing 4 (1) (2010) 043551.

[20] Gang Chen, Geoffrey J. Hay, Luis M.T. Carvalho, Michael A. Wulder, Object-based change detection, International Journal of Remote Sensing 33 (14) (2012) 4434–4457.

[21] Hao Chen, Yi Hua, Qinglong Ren, Ye Zhang, Comprehensive analysis of regional human-driven environmental change with multitemporal remote sensing images using observed object-specified dynamic Bayesian network, Journal of Applied Remote Sensing 10 (1) (2016) 016021.

[22] Hongruixuan Chen, Chen Wu, Bo Du, Liangpei Zhang, Deep Siamese multi-scale convolutional network for change detection in multi-temporal VHR images, in: 2019 10th International Workshop on the Analysis of Multitemporal Remote Sensing Images (MultiTemp), IEEE, 2019, pp. 1–4.

[23] Hongruixuan Chen, Chen Wu, Bo Du, Liangpei Zhang, Le Wang, Change detection in multisource VHR images via deep Siamese convolutional multiple-layers recurrent neural network, IEEE Transactions on Geoscience and Remote Sensing 58 (4) (2019) 2848–2864.

[24] Huan Chen, Licheng Jiao, Miaomiao Liang, Fang Liu, Shuyuan Yang, Biao Hou, Fast unsupervised deep fusion network for change detection of multitemporal SAR images, Neurocomputing 332 (2019) 56–70.

[25] Jiaoyan Chen, Guozhou Zheng, Cong Fang, Ningyu Zhang, Huajun Chen, Zhaohui Wu, Time-series processing of large scale remote sensing data with extreme learning machine, Neurocomputing 128 (2014) 199–206.

[26] Keming Chen, Chunlei Huo, Zhixin Zhou, Hanqing Lu, Unsupervised change detection in high spatial resolution optical imagery based on modified Hopfield neural network, in: 2008 Fourth International Conference on Natural Computation, vol. 4, IEEE, 2008, pp. 281–285.

[27] Keyan Chen, Chenyang Liu, Hao Chen, Haotian Zhang, Wenyuan Li, Zhengxia Zou, Zhenwei Shi, RSPrompter: learning to prompt for remote sensing instance segmentation based on visual foundation model, IEEE Transactions on Geoscience and Remote Sensing 62 (2024) 1–17.

[28] Liang-Chieh Chen, George Papandreou, Iasonas Kokkinos, Kevin Murphy, Alan L. Yuille, DeepLab: semantic image segmentation with deep convolutional nets, atrous convolution, and fully connected CRFs, IEEE Transactions on Pattern Analysis and Machine Intelligence 40 (4) (2017) 834–848.

[29] Zhong Chen, Yifei Zhang, Chao Ouyang, Feng Zhang, Jie Ma, Automated landslides detection for mountain cities using multi-temporal remote sensing imagery, Sensors 18 (3) (2018) 821.

[30] Aakanksha Chowdhery, Sharan Narang, Jacob Devlin, Maarten Bosma, Gaurav Mishra, Adam Roberts, Paul Barham, Hyung Won Chung, Charles Sutton, Sebastian Gehrmann, et al., Palm: scaling language modeling with pathways, Journal of Machine Learning Research 24 (240) (2023) 1–113.

[31] Yan Chu, Guo Cao, Hassan Hayat, Change detection of remote sensing image based on deep neural networks, in: 2016 2nd International Conference on Artificial Intelligence and Industrial Engineering (AIIE 2016), Atlantis Press, 2016, pp. 262–267.

[32] Clayton Connors, Ranga Raju Vatsavai, Semi-supervised deep generative models for change detection in very high resolution imagery, in: 2017 IEEE International Geoscience and Remote Sensing Symposium (IGARSS), IEEE, 2017, pp. 1063–1066.

[33] Pol Coppin, Inge Jonckheere, Kristiaan Nackaerts, Bart Muys, Eric Lambin, Review ArticleDigital change detection methods in ecosystem monitoring: a review, International Journal of Remote Sensing 25 (9) (2004) 1565–1596.

[34] Marius Cordts, Mohamed Omran, Sebastian Ramos, Timo Rehfeld, Markus Enzweiler, Rodrigo Benenson, Uwe Franke, Stefan Roth, Bernt Schiele, The cityscapes dataset for semantic urban scene understanding, in: Proceedings of the IEEE Conference on Computer Vision and Pattern Recognition, 2016, pp. 3213–3223.

[35] C.P. Dalmiya, N. Santhi, An enhanced back propagation method for change analysis of remote sensing images with adaptive preprocessing, European Journal of Remote Sensing 53 (sup1) (2020) 41–52.

[36] Bin Cui, Yonghong Zhang, Li Yan, Jujie Wei, Hongn Wu, An unsupervised SAR change detection method based on stochastic subspace ensemble learning, Remote Sensing 11 (11) (2019) 1314.

[37] X. Long Dai, Slamak Khorram, Remotely sensed change detection based on artificial neural networks, Photogrammetric Engineering and Remote Sensing 65 (1999) 1187–1194.

[38] Rodrigo Caye Daudt, Bertrand Le Saux, Alexandre Boulch, Yann Gousseau, Guided anisotropic diffusion and iterative learning for weakly supervised change detection, in: Computer Vision and Pattern Recognition Workshops, 2019.

[39] Shaunak De, Davide Pirrone, Francesca Bovolo, Lorenzo Bruzzone, Avik Bhattacharya, A novel change detection framework based on deep learning for the analysis of multi-temporal polarimetric SAR images, in: 2017 IEEE International Geoscience and Remote Sensing Symposium (IGARSS), IEEE, 2017, pp. 5193–5196.

[40] Fabio Del Frate, Fabio Pacifici, Domenico Solimini, Monitoring urban land cover in Rome, Italy, and its changes by single-polarization multitemporal SAR images, IEEE Journal of Selected Topics in Applied Earth Observations and Remote Sensing 1 (2) (2008) 87–97.

[41] Thomas G. Dietterich, Steps toward robust artificial intelligence, AI Magazine 38 (3) (2017) 3–24.

[42] Anzi Ding, Qingyong Zhang, Xinmin Zhou, Bicheng Dai, Automatic recognition of landslide based on CNN and texture change detection, in: 2016 31st Youth Academic Annual Conference of Chinese Association of Automation (YAC), IEEE, 2016, pp. 444–448.

[43] Lei Ding, Kun Zhu, Daifeng Peng, Hao Tang, Kuiwu Yang, Lorenzo Bruzzone, Adapting segment anything model for change detection in VHR remote sensing images, IEEE Transactions on Geoscience and Remote Sensing (2024).

[44] Huihui Dong, Wenping Ma, Yue Wu, Maoguo Gong, Licheng Jiao, Local descriptor learning for change detection in synthetic aperture radar images via convolutional neural networks, IEEE Access 7 (2018) 15389–15403.

[45] Bo Du, Lixiang Ru, Chen Wu, Liangpei Zhang, Unsupervised deep slow feature analysis for change detection in multi-temporal remote sensing images, IEEE Transactions on Geoscience and Remote Sensing 57 (12) (2019) 9976–9992.

[46] Z.O. Durmusoglu, A.A. Tanriover, Modelling land use/cover change in Lake Mogan and surroundings using CA-Markov Chain Analysis, https://doi.org/10.22438/jeb/38/5(si)/gm-15, 2017.

[47] Arabi Mohammed El Amin, Qingjie Liu, Yunhong Wang, Convolutional neural network features based change detection in satellite images, in: First International Workshop on Pattern Recognition, vol. 10011, SPIE, 2016, pp. 181–186.

[48] Arabi Mohammed El Amin, Qingjie Liu, Yunhong Wang, Zoom out CNNs features for optical remote sensing change detection, in: 2017 2nd International Conference on Image, Vision and Computing (ICIVC), IEEE, 2017, pp. 812–817.

[49] Fenglei Fan, Yunpeng Wang, Zhishi Wang, Temporal and spatial change detecting (1998–2003) and predicting of land use and land cover in Core corridor of Pearl River Delta (China) by using TM and ETM+ images, Environmental Monitoring and Assessment 137 (2008) 127–147.

[50] Jianchao Fan, Kai Lin, Min Han, A novel joint change detection approach based on weight-clustering sparse autoencoders, IEEE Journal of Selected Topics in Applied Earth Observations and Remote Sensing 12 (2) (2019) 685–699.

[51] Bo Fang, Gang Chen, Li Pan, Rong Kou, Lizhe Wang, GAN-based Siamese framework for landslide inventory mapping using bi-temporal optical remote sensing images, IEEE Geoscience and Remote Sensing Letters 18 (3) (2020) 391–395.

[52] Bo Fang, Li Pan, Rong Kou, Dual learning-based Siamese framework for change detection using bi-temporal VHR optical remote sensing images, Remote Sensing 11 (11) (2019) 1292.

[53] Han Fang, Pengfei Xiong, Luhui Xu, Yu Chen, CLIP2Video: mastering video-text retrieval via image clip, arXiv preprint, arXiv:2106.11097, 2021.

[54] Idan Feldberg, Nathan S. Netanyahu, Maxim Shoshany, A neural network-based technique for change detection of linear features and its application to a Mediterranean region, in: IEEE International Geoscience and Remote Sensing Symposium, vol. 2, IEEE, 2002, pp. 1195–1197.

[55] Aito Fujita, Ken Sakurada, Tomoyuki Imaizumi, Riho Ito, Shuhei Hikosaka, Ryosuke Nakamura, Damage detection from aerial images via convolutional neural networks, in: 2017 Fifteenth IAPR International Conference on Machine Vision Applications (MVA), IEEE, 2017, pp. 5–8.

[56] Feng Gao, Junyu Dong, Bo Li, Qizhi Xu, Automatic change detection in synthetic aperture radar images based on PCANet, IEEE Geoscience and Remote Sensing Letters 13 (12) (2016) 1792–1796.

[57] Feng Gao, Xiaopeng Liu, Junyu Dong, Guoqiang Zhong, Muwei Jian, Change detection in SAR images based on deep semi-NMF and SVD networks, Remote Sensing 9 (5) (2017) 435.

[58] Yunhao Gao, Feng Gao, Junyu Dong, Shengke Wang, Change detection from synthetic aperture radar images based on channel weighting-based deep cascade network, IEEE Journal of Selected Topics in Applied Earth Observations and Remote Sensing 12 (11) (2019) 4517–4529.

[59] Yunhao Gao, Feng Gao, Junyu Dong, Shengke Wang, Transferred deep learning for sea ice change detection from synthetic-aperture radar images, IEEE Geoscience and Remote Sensing Letters 16 (10) (2019) 1655–1659.

[60] Jie Geng, Xiaorui Ma, Xiaojun Zhou, Hongyu Wang, Saliency-guided deep neural networks for SAR image change detection, IEEE Transactions on Geoscience and Remote Sensing 57 (10) (2019) 7365–7377.

[61] Jie Geng, Hongyu Wang, Jianchao Fan, Xiaorui Ma, Change detection of SAR images based on supervised contractive autoencoders and fuzzy clustering, in: 2017 International Workshop on Remote Sensing with Intelligent Processing (RSIP), IEEE, 2017, pp. 1–3.
[62] Saman Ghaffarian, Norman Kerle, Edoardo Pasolli, Jamal Jokar Arsanjani, Post-disaster building database updating using automated deep learning: an integration of pre-disaster OpenStreetMap and multi-temporal satellite data, Remote Sensing 11 (20) (2019) 2427.
[63] Ashish Ghosh, Badri Narayan Subudhi, Lorenzo Bruzzone, Integration of Gibbs Markov random field and Hopfield-type neural networks for unsupervised change detection in remotely sensed multitemporal images, IEEE Transactions on Image Processing 22 (8) (2013) 3087–3096.
[64] Susmita Ghosh, Lorenzo Bruzzone, Swarnajyoti Patra, Francesca Bovolo, Ashish Ghosh, A context-sensitive technique for unsupervised change detection based on Hopfield-type neural networks, IEEE Transactions on Geoscience and Remote Sensing 45 (3) (2007) 778–789.
[65] Susmita Ghosh, Swarnajyoti Patra, Ashish Ghosh, An unsupervised context-sensitive change detection technique based on modified self-organizing feature map neural network, International Journal of Approximate Reasoning 50 (1) (2009) 37–50.
[66] Susmita Ghosh, Moumita Roy, Ashish Ghosh, Semi-supervised change detection using modified self-organizing feature map neural network, Applied Soft Computing 15 (2014) 1–20.
[67] Nehla Ghouaiel, Sébastien Lefèvre, Coupling ground-level panoramas and aerial imagery for change detection, Geo-Spatial Information Science 19 (3) (2016) 222–232.
[68] Maoguo Gong, Xudong Niu, Tao Zhan, Mingyang Zhang, A coupling translation network for change detection in heterogeneous images, International Journal of Remote Sensing 40 (9) (2019) 3647–3672.
[69] Maoguo Gong, Xudong Niu, Puzhao Zhang, Zhetao Li, Generative adversarial networks for change detection in multispectral imagery, IEEE Geoscience and Remote Sensing Letters 14 (12) (2017) 2310–2314.
[70] Maoguo Gong, Hailun Yang, Puzhao Zhang, Feature learning and change feature classification based on deep learning for ternary change detection in SAR images, ISPRS Journal of Photogrammetry and Remote Sensing 129 (2017) 212–225.
[71] Maoguo Gong, Yuelei Yang, Tao Zhan, Xudong Niu, Shuwei Li, A generative discriminatory classified network for change detection in multispectral imagery, IEEE Journal of Selected Topics in Applied Earth Observations and Remote Sensing 12 (1) (2019) 321–333.
[72] Maoguo Gong, Tao Zhan, Puzhao Zhang, Qiguang Miao, Superpixel-based difference representation learning for change detection in multispectral remote sensing images, IEEE Transactions on Geoscience and Remote Sensing 55 (5) (2017) 2658–2673.
[73] Maoguo Gong, Jiaojiao Zhao, Jia Liu, Qiguang Miao, Licheng Jiao, Change detection in synthetic aperture radar images based on deep neural networks, IEEE Transactions on Neural Networks and Learning Systems 27 (1) (2015) 125–138.
[74] Ian J. Goodfellow, Jean Pouget-Abadie, Mehdi Mirza, Bing Xu, David Warde-Farley, Sherjil Ozair, Aaron Courville, Yoshua Bengio, Generative adversarial nets, Advances in Neural Information Processing Systems 27 (2014).
[75] Sucharita Gopal, Curtis Woodcock, Remote sensing of forest change using artificial neural networks, IEEE Transactions on Geoscience and Remote Sensing 34 (2) (1996) 398–404.
[76] Riccardo Guidotti, Anna Monreale, Salvatore Ruggieri, Franco Turini, Fosca Giannotti, Dino Pedreschi, A survey of methods for explaining black box models, ACM Computing Surveys (CSUR) 51 (5) (2018) 1–42.
[77] Enqiang Guo, Xinsha Fu, Jiawei Zhu, Min Deng, Yu Liu, Qing Zhu, Haifeng Li, Learning to measure change: fully convolutional Siamese metric networks for scene change detection, arXiv preprint, arXiv: 1810.09111, 2018.
[78] Min Han, Chengkun Zhang, Yang Zhou, Object-wise joint-classification change detection for remote sensing images based on entropy query-by fuzzy ARTMAP, GIScience and Remote Sensing 55 (2) (2018) 265–284.

[79] Pengcheng Han, Cunbao Ma, Qing Li, Pengyu Leng, Shuhui Bu, Ke Li, Aerial image change detection using dual regions of interest networks, Neurocomputing 349 (2019) 190–201.

[80] Pengfei He, Wenzhong Shi, Zelang Miao, Hua Zhang, Liping Cai, Advanced Markov random field model based on local uncertainty for unsupervised change detection, Remote Sensing Letters 6 (9) (2015) 667–676.

[81] Yuhong He, Qihao Weng, High Spatial Resolution Remote Sensing: Data, Analysis, and Applications, CRC Press, 2018.

[82] Rachid Hedjam, Abdelhamid Abdesselam, Farid Melgani, Change detection from unlabeled remote sensing images using SIAMESE ANN, in: IGARSS 2019-2019 IEEE International Geoscience and Remote Sensing Symposium, IEEE, 2019, pp. 1530–1533.

[83] Bin Hou, Qingjie Liu, Heng Wang, Yunhong Wang, From W-Net to CDGAN: bitemporal change detection via deep learning techniques, IEEE Transactions on Geoscience and Remote Sensing 58 (3) (2019) 1790–1802.

[84] Bin Hou, Yunhong Wang, Qingjie Liu, Change detection based on deep features and low rank, IEEE Geoscience and Remote Sensing Letters 14 (12) (2017) 2418–2422.

[85] Dong-ming Huang, Chun-tao Wei, Jun-chen Yu, Jian-lin Wang, A method of detecting land use change of remote sensing images based on texture features and DEM, in: International Conference on Intelligent Earth Observing and Applications 2015, vol. 9808, SPIE, 2015, pp. 613–618.

[86] Fenghua Huang, Ying Yu, Tinghao Feng, Automatic building change image quality assessment in high resolution remote sensing based on deep learning, Journal of Visual Communication and Image Representation 63 (2019) 102585.

[87] Fenghua Huang, Ying Yu, Tinghao Feng, Hyperspectral remote sensing image change detection based on tensor and deep learning, Journal of Visual Communication and Image Representation 58 (2019) 233–244.

[88] Carlos Gómez Huélamo, Pablo F. Alcantarilla, Luis Miguel Bergasa, Elena López-Guillén, Change detection tool based on GSV to help DNNs training, in: Advances in Physical Agents: Proceedings of the 19th International Workshop of Physical Agents (WAF 2018), Madrid, Spain, November 22–23, 2018, Springer, 2019, pp. 115–131.

[89] Masroor Hussain, Dongmei Chen, Angela Cheng, Hui Wei, David Stanley, Change detection from remotely sensed images: from pixel-based to object-based approaches, ISPRS Journal of Photogrammetry and Remote Sensing 80 (2013) 91–106.

[90] Shota Iino, Riho Ito, Kento Doi, Tomoyuki Imaizumi, Shuhei Hikosaka, Generating high-accuracy urban distribution map for short-term change monitoring based on convolutional neural network by utilizing SAR imagery, in: Earth Resources and Environmental Remote Sensing/GIS Applications VIII, vol. 10428, SPIE, 2017, pp. 11–21.

[91] Shota Iino, Riho Ito, Kento Doi, Tomoyuki Imaizumi, Shuhei Hikosaka, CNN-based generation of high-accuracy urban distribution maps utilising SAR satellite imagery for short-term change monitoring, International Journal of Image and Data Fusion 9 (4) (2018) 302–318.

[92] Raveerat Jaturapitpornchai, Masashi Matsuoka, Naruo Kanemoto, Shigeki Kuzuoka, Riho Ito, Ryosuke Nakamura, Newly built construction detection in SAR images using deep learning, Remote Sensing 11 (12) (2019) 1444.

[93] Min Ji, Lanfa Liu, Runlin Du, Manfred F. Buchroithner, A comparative study of texture and convolutional neural network features for detecting collapsed buildings after earthquakes using pre- and post-event satellite imagery, Remote Sensing 11 (10) (2019) 1202.

[94] Shunping Ji, Yanyun Shen, Meng Lu, Yongjun Zhang, Building instance change detection from large-scale aerial images using convolutional neural networks and simulated samples, Remote Sensing 11 (11) (2019) 1343.

[95] Shunping Ji, Shiqing Wei, Meng Lu, Fully convolutional networks for multisource building extraction from an open aerial and satellite imagery data set, IEEE Transactions on Geoscience and Remote Sensing 57 (1) (2018) 574–586.

[96] Wei Ji, Jingjing Li, Qi Bi, Tingwei Liu, Wenbo Li, Li Cheng, Segment anything is not always perfect: an investigation of SAM on different real-world applications, https://doi.org/10.1007/s11633-023-1385-0, 2024.

[97] Chao Jia, Yinfei Yang, Ye Xia, Yi-Ting Chen, Zarana Parekh, Hieu Pham, Quoc Le, Yun-Hsuan Sung, Zhen Li, Tom Duerig, Scaling up visual and vision-language representation learning with noisy text supervision, in: International Conference on Machine Learning, PMLR, 2021, pp. 4904–4916.

[98] Huiwei Jiang, Xiangyun Hu, Kun Li, Jinming Zhang, Jinqi Gong, Mi Zhang, PGA-SiamNet: pyramid feature-based attention-guided Siamese network for remote sensing orthoimagery building change detection, Remote Sensing 12 (3) (2020) 484.

[99] Ken Sakurada, Takayuki Okatani, Change detection from a street image pair using CNN features and superpixel segmentation, in: Proceedings of the British Machine Vision Conference, 2015, p. 61.

[100] Jian Kang, Marco Körner, Yuanyuan Wang, Hannes Taubenböck, Xiao Xiang Zhu, Building instance classification using street view images, ISPRS Journal of Photogrammetry and Remote Sensing 145 (2018) 44–59.

[101] Andreas Kaplan, Michael Haenlein, Siri, Siri, in my hand: who the fairest in the land? On the interpretations, illustrations, and implications of artificial intelligence, Business Horizons 62 (1) (2019) 15–25.

[102] Anuj Karpatne, Zhe Jiang, Ranga Raju Vatsavai, Shashi Shekhar, Vipin Kumar, Monitoring land-cover changes: a machine-learning perspective, IEEE Geoscience and Remote Sensing Magazine 4 (2) (2016) 8–21.

[103] Hannah Rae Kerner, Kiri L. Wagstaff, Brian D. Bue, Patrick C. Gray, James F. Bell, Heni Ben Amor, Toward generalized change detection on planetary surfaces with convolutional autoencoders and transfer learning, IEEE Journal of Selected Topics in Applied Earth Observations and Remote Sensing 12 (10) (2019) 3900–3918.

[104] Hatem Magdy Keshk, Xu-Cheng Yin, Change detection in SAR images based on deep learning, International Journal of Aeronautical and Space Sciences 21 (2) (2020) 549–559.

[105] Salman H. Khan, Xuming He, Fatih Porikli, Mohammed Bennamoun, Forest change detection in incomplete satellite images with deep neural networks, IEEE Transactions on Geoscience and Remote Sensing 55 (9) (2017) 5407–5423.

[106] Alexander Kirillov, Eric Mintun, Nikhila Ravi, Hanzi Mao, Chloe Rolland, Laura Gustafson, Tete Xiao, Spencer Whitehead, Alexander C. Berg, Wan-Yen Lo, et al., Segment anything, in: Proceedings of the IEEE/CVF International Conference on Computer Vision, 2023, pp. 4015–4026.

[107] Mohammed El Amin Larabi, Souleyman Chaib, Khadidja Bakhti, Kamel Hasni, Mohammed Amine Bouhlala, High-resolution optical remote sensing imagery change detection through deep transfer learning, Journal of Applied Remote Sensing 13 (4) (2019) 046512.

[108] Tao Lei, Yuxiao Zhang, Zhiyong Lv, Shuying Li, Shigang Liu, Asoke K. Nandi, Landslide inventory mapping from bitemporal images using deep convolutional neural networks, IEEE Geoscience and Remote Sensing Letters 16 (6) (2019) 982–986.

[109] Yu Lei, Xiaodong Liu, Jiao Shi, Chao Lei, Jing Wang, Multiscale superpixel segmentation with deep features for change detection, IEEE Access 7 (2019) 36600–36616.

[110] Heng-Chao Li, Gang Yang, Wen Yang, Qian Du, William J. Emery, Deep nonsmooth nonnegative matrix factorization network with semi-supervised learning for SAR image change detection, ISPRS Journal of Photogrammetry and Remote Sensing 160 (2020) 167–179.

[111] Mengke Li, Ming Li, Peng Zhang, Yan Wu, Wanying Song, Lin An, SAR image change detection using PCANet guided by saliency detection, IEEE Geoscience and Remote Sensing Letters 16 (3) (2018) 402–406.

[112] Xiang Li, Congcong Wen, Yuan Hu, Zhenghang Yuan, Xiao Xiang Zhu, Vision-language models in remote sensing: current progress and future trends, IEEE Geoscience and Remote Sensing Magazine 12 (2) (2024) 32–66.

[113] Xiaodong Li, Feng Ling, Yun Du, Qi Feng, Yihang Zhang, A spatial–temporal Hopfield neural network approach for super-resolution land cover mapping with multi-temporal different resolution remotely sensed images, ISPRS Journal of Photogrammetry and Remote Sensing 93 (2014) 76–87.

[114] Xuelong Li, Zhenghang Yuan, Qi Wang, Unsupervised deep noise modeling for hyperspectral image change detection, Remote Sensing 11 (3) (2019) 258.

[115] Yangyang Li, Cheng Peng, Yanqiao Chen, Licheng Jiao, Linhao Zhou, Ronghua Shang, A deep learning method for change detection in synthetic aperture radar images, IEEE Transactions on Geoscience and Remote Sensing 57 (8) (2019) 5751–5763.

[116] Yangyang Li, Linhao Zhou, Cheng Peng, Licheng Jiao, Spatial fuzzy clustering and deep auto-encoder for unsupervised change detection in synthetic aperture radar images, in: IGARSS 2018-2018 IEEE International Geoscience and Remote Sensing Symposium, IEEE, 2018, pp. 4479–4482.

[117] Frank Liao, Elizabeth Koshelev, Malcolm Milton, Yuanwei Jin, Enyue Lu, Change detection by deep neural networks for synthetic aperture radar images, in: 2017 International Conference on Computing, Networking and Communications (ICNC), IEEE, 2017, pp. 947–951.

[118] Erik J. Lindquist, Rémi Dnnunzio, Assessing global forest land-use change by object-based image analysis, Remote Sensing 8 (8) (2016) 678.

[119] Fan Liu, Delong Chen, Zhangqingyun Guan, Xiaocong Zhou, Jiale Zhu, Qiaolin Ye, Liyong Fu, Jun Zhou, RemoteCLIP: a vision language foundation model for remote sensing, IEEE Transactions on Geoscience and Remote Sensing (2024).

[120] Fang Liu, Licheng Jiao, Xu Tang, Shuyuan Yang, Wenping Ma, Biao Hou, Local restricted convolutional neural network for change detection in polarimetric SAR images, IEEE Transactions on Neural Networks and Learning Systems 30 (3) (2018) 818–833.

[121] Jia Liu, Maoguo Gong, Kai Qin, Puzhao Zhang, A deep convolutional coupling network for change detection based on heterogeneous optical and radar images, IEEE Transactions on Neural Networks and Learning Systems 29 (3) (2016) 545–559.

[122] Jia Liu, Maoguo Gong, Jiaojiao Zhao, Hao Li, Licheng Jiao, Difference representation learning using stacked restricted Boltzmann machines for change detection in SAR images, Soft Computing 20 (2016) 4645–4657.

[123] Junfu Liu, Keming Chen, Guangluan Xu, Xian Sun, Menglong Yan, Wenhui Diao, Hongzhe Han, Convolutional neural network-based transfer learning for optical aerial images change detection, IEEE Geoscience and Remote Sensing Letters 17 (1) (2019) 127–131.

[124] Rui Liu, Zhenhong Jia, Xizhong Qin, Jie Yang, Nikola Kasabov, SAR image change detection method based on pulse-coupled neural network, Journal of the Indian Society of Remote Sensing 44 (2016) 443–450.

[125] Ruochen Liu, Zhihong Cheng, Langlang Zhang, Jianxia Li, Remote sensing image change detection based on information transmission and attention mechanism, IEEE Access 7 (2019) 156349–156359.

[126] Sicong Liu, Daniele Marinelli, Lorenzo Bruzzone, Francesca Bovolo, A review of change detection in multitemporal hyperspectral images: current techniques, applications, and challenges, IEEE Geoscience and Remote Sensing Magazine 7 (2) (2019) 140–158.

[127] Tao Liu, Ying Li, Ying Cao, Qiang Shen, Change detection in multitemporal synthetic aperture radar images using dual-channel convolutional neural network, Journal of Applied Remote Sensing 11 (4) (2017) 042615.

[128] Dengsheng Lu, Paul Mausel, Eduardo Brondizio, Emilio Moran, Change detection techniques, International Journal of Remote Sensing 25 (12) (2004) 2365–2401.

[129] Timo Lüddecke, Alexander Ecker, Image segmentation using text and image prompts, in: Proceedings of the IEEE/CVF Conference on Computer Vision and Pattern Recognition, 2022, pp. 7086–7096.

[130] Bin Luo, Chudi Hu, Xin Su, Yajun Wang, Differentially deep subspace representation for unsupervised change detection of SAR images, Remote Sensing 11 (23) (2019) 2740.

[131] Ning Lv, Chen Chen, Tie Qiu, Arun Kumar Sangaiah, Deep learning and superpixel feature extraction based on contractive autoencoder for change detection in SAR images, IEEE Transactions on Industrial Informatics 14 (12) (2018) 5530–5538.

[132] Haobo Lyu, Hui Lu, Lichao Mou, Learning a transferable change rule from a recurrent neural network for land cover change detection, Remote Sensing 8 (6) (2016) 506.

[133] Haobo Lyu, Hui Lu, Lichao Mou, Wenyu Li, Jonathon Wright, Xuecao Li, Xinlu Li, Xiao Xiang Zhu, Jie Wang, Le Yu, et al., Long-term annual mapping of four cities on different continents by applying a deep information learning method to Landsat data, Remote Sensing 10 (3) (2018) 471.

[134] Wenping Ma, Yunta Xiong, Yue Wu, Hui Yang, Xiangrong Zhang, Licheng Jiao, Change detection in remote sensing images based on image mapping and a deep capsule network, Remote Sensing 11 (6) (2019) 626.

[135] Akansha Mehrotra, Krishna Kant Singh, Madhav Ji Nigam, Kirat Pal, Detection of tsunami-induced changes using generalized improved fuzzy radial basis function neural network, Natural Hazards 77 (2015) 367–381.

[136] M. Ersoy Mirici, S. Berberoglu, A. Akin, Onur Satir, Land use/cover change modelling in a Mediterranean rural landscape using Multi-Layer Perceptron and Markov Chain (MLP-MC), Applied Ecology and Environmental Research 16 (1) (2018).

[137] Lichao Mou, Lorenzo Bruzzone, Xiao Xiang Zhu, Learning spectral-spatial-temporal features via a recurrent convolutional neural network for change detection in multispectral imagery, IEEE Transactions on Geoscience and Remote Sensing 57 (2) (2018) 924–935.

[138] Lichao Mou, Xiao Xiang Zhu, A recurrent convolutional neural network for land cover change detection in multispectral images, in: IGARSS 2018-2018 IEEE International Geoscience and Remote Sensing Symposium, IEEE, 2018, pp. 4363–4366.

[139] Shane T. Mueller, Robert R. Hoffman, William Clancey, Abigail Emrey, Gary Klein, Explanation in human-AI systems: a literature meta-review, synopsis of key ideas and publications, and bibliography for explainable AI, arXiv preprint, arXiv:1902.01876, 2019.

[140] Victor-Emil Neagoe, Adrian-Dumitru Ciotec, Serban-Vasile Carata, A new multispectral pixel change detection approach using pulse-coupled neural networks for change vector analysis, in: 2016 IEEE International Geoscience and Remote Sensing Symposium (IGARSS), IEEE, 2016, pp. 3386–3389.

[141] Victor-Emil Neagoe, Radu-Mihai Stoica, Alexandru-Ioan Ciurea, A modular neural network model for change detection in earth observation imagery, in: 2013 IEEE International Geoscience and Remote Sensing Symposium-IGARSS, IEEE, 2013, pp. 3321–3324.

[142] H. Nemmour, Y. Chibani, Fuzzy neural network architecture for change detection in remotely sensed imagery, International Journal of Remote Sensing 27 (4) (2006) 705–717.

[143] Hassiba Nemmour, Youcef Chibani, Neural network combination by fuzzy integral for robust change detection in remotely sensed imagery, EURASIP Journal on Advances in Signal Processing 2005 (2005) 1–9.

[144] Keisuke Nemoto, Ryuhei Hamaguchi, Masakazu Sato, Aito Fujita, Tomoyuki Imaizumi, Shuhei Hikosaka, Building change detection via a combination of CNNs using only RGB aerial imageries, in: Remote Sensing Technologies and Applications in Urban Environments II, vol. 10431, SPIE, 2017, pp. 107–118.

[145] Xudong Niu, Maoguo Gong, Tao Zhan, Yuelei Yang, A conditional adversarial network for change detection in heterogeneous images, IEEE Geoscience and Remote Sensing Letters 16 (1) (2018) 45–49.

[146] Vahid Nourani, Kiyoumars Roushangar, Gholamreza Andalib, An inverse method for watershed change detection using hybrid conceptual and artificial intelligence approaches, Journal of Hydrology 562 (2018) 371–384.

[147] Lucas Prado Osco, Qiusheng Wu, Eduardo Lopes De Lemos, Wesley Nunes Gonçalves, Ana Paula Marques Ramos, Jonathan Li, José Marcato Junior, The segment anything model (SAM) for remote sensing applications: from zero to one shot, International Journal of Applied Earth Observation and Geoinformation 124 (2023) 103540.

[148] Fabio Pacifici, Fabio Del Frate, Automatic change detection in very high resolution images with pulse-coupled neural networks, IEEE Geoscience and Remote Sensing Letters 7 (1) (2009) 58–62.

[149] Fabio Pacifici, Fabio Del Frate, Chiara Solimini, William J. Emery, An innovative neural-net method to detect temporal changes in high-resolution optical satellite imagery, IEEE Transactions on Geoscience and Remote Sensing 45 (9) (2007) 2940–2952.
[150] Swarnajyoti Patra, Susmita Ghosh, Ashish Ghosh, Unsupervised change detection in remote-sensing images using one-dimensional self-organizing feature map neural network, in: 9th International Conference on Information Technology (ICIT'06), IEEE, 2006, pp. 141–142.
[151] Swarnajyoti Patra, Susmita Ghosh, Ashish Ghosh, Unsupervised change detection in remote-sensing images using modified self-organizing feature map neural network, in: 2007 International Conference on Computing: Theory and Applications (ICCTA'07), IEEE, 2007, pp. 716–720.
[152] Swarnajyoti Patra, Susmita Ghosh, Ashish Ghosh, Change detection of remote sensing images with semi-supervised multilayer perceptron, Fundamenta Informaticae 84 (3–4) (2008) 429–442.
[153] Bo Peng, Zonglin Meng, Qunying Huang, Caixia Wang, Patch similarity convolutional neural network for urban flood extent mapping using bi-temporal satellite multispectral imagery, Remote Sensing 11 (21) (2019) 2492.
[154] Daifeng Peng, Haiyan Guan, Unsupervised change detection method based on saliency analysis and convolutional neural network, Journal of Applied Remote Sensing 13 (2) (2019) 024512.
[155] Peter Planinšič, Dušan Gleich, Temporal change detection in SAR images using log cumulants and stacked autoencoder, IEEE Geoscience and Remote Sensing Letters 15 (2) (2018) 297–301.
[156] Andrea Pomente, Matteo Picchiani, Fabio Del Frate, Sentinel-2 change detection based on deep features, in: IGARSS 2018-2018 IEEE International Geoscience and Remote Sensing Symposium, IEEE, 2018, pp. 6859–6862.
[157] Chiara Pratola, Fabio Del Frate, Giovanni Schiavon, Domenico Solimini, Toward fully automatic detection of changes in suburban areas from VHR SAR images by combining multiple neural-network models, IEEE Transactions on Geoscience and Remote Sensing 51 (4) (2013) 2055–2066.
[158] Rongjun Qin, Jiaojiao Tian, Peter Reinartz, 3D change detection–approaches and applications, ISPRS Journal of Photogrammetry and Remote Sensing 122 (2016) 41–56.
[159] Alec Radford, Jong Wook Kim, Chris Hallacy, Aditya Ramesh, Gabriel Goh, Sandhini Agarwal, Girish Sastry, Amanda Askell, Pamela Mishkin, Jack Clark, et al., Learning transferable visual models from natural language supervision, in: International Conference on Machine Learning, PmLR, 2021, pp. 8748–8763.
[160] Colin Raffel, Noam Shazeer, Adam Roberts, Katherine Lee, Sharan Narang, Michael Matena, Yanqi Zhou, Wei Li, Peter J. Liu, Exploring the limits of transfer learning with a unified text-to-text transformer, Journal of Machine Learning Research 21 (140) (2020) 1–67.
[161] Aditya Ramesh, Prafulla Dhariwal, Alex Nichol, Casey Chu, Mark Chen, Hierarchical text-conditional image generation with clip latents, arXiv preprint, arXiv:2204.06125, 2022.
[162] Krishna Regmi, Mubarak Shah, Bridging the domain gap for ground-to-aerial image matching, in: Proceedings of the IEEE/CVF International Conference on Computer Vision, 2019, pp. 470–479.
[163] Komeil Rokni, Anuar Ahmad, Karim Solaimani, Sharifeh Hazini, A new approach for surface water change detection: integration of pixel level image fusion and image classification techniques, International Journal of Applied Earth Observation and Geoinformation 34 (2015) 226–234.
[164] Moumita Roy, Susmita Ghosh, Ashish Ghosh, A neural approach under active learning mode for change detection in remotely sensed images, IEEE Journal of Selected Topics in Applied Earth Observations and Remote Sensing 7 (4) (2013) 1200–1206.
[165] Moumita Roy, Susmita Ghosh, Ashish Ghosh, A novel approach for change detection of remotely sensed images using semi-supervised multiple classifier system, Information Sciences 269 (2014) 35–47.
[166] Vahid Sadeghi, Farshid Farnood Ahmadi, Hamid Ebadi, A new fuzzy measurement approach for automatic change detection using remotely sensed images, Measurement 127 (2018) 1–14.
[167] Sudipan Saha, Francesca Bovolo, Lorenzo Brurzone, Unsupervised multiple-change detection in VHR optical images using deep features, in: IGARSS 2018-2018 IEEE International Geoscience and Remote Sensing Symposium, IEEE, 2018, pp. 1902–1905.

[168] Sudipan Saha, Francesca Bovolo, Lorenzo Bruzzone, Destroyed-buildings detection from VHR SAR images using deep features, in: Image and Signal Processing for Remote Sensing XXIV, vol. 10789, SPIE, 2018, pp. 336–344.

[169] Sudipan Saha, Francesca Bovolo, Lorenzo Bruzzone, Unsupervised deep change vector analysis for multiple-change detection in VHR images, IEEE Transactions on Geoscience and Remote Sensing 57 (6) (2019) 3677–3693.

[170] Ken Sakurada, Daiki Tetsuka, Takayuki Okatani, Temporal city modeling using street level imagery, Computer Vision and Image Understanding 157 (2017) 55–71.

[171] Ken Sakurada, Weimin Wang, Nobuo Kawaguchi, Ryosuke Nakamura, Dense optical flow based change detection network robust to difference of camera viewpoints, arXiv preprint, arXiv:1712.02941, 2017.

[172] Farnaam Samadi, Gholamreza Akbarizadeh, Hooman Kaabi, Change detection in SAR images using deep belief network: a new training approach based on morphological images, IET Image Processing 13 (12) (2019) 2255–2264.

[173] Karen C. Seto, Weiguo Liu, Comparing ARTMAP neural network with the maximum-likelihood classifier for detecting urban change, Photogrammetric Engineering and Remote Sensing 69 (9) (2003) 981–990.

[174] Sheng Shen, Liunian Harold Li, Hao Tan, Mohit Bansal, Anna Rohrbach, Kai-Wei Chang, Zhewei Yao, Kurt Keutzer, How much can clip benefit vision-and-language tasks?, arXiv preprint, arXiv:2107.06383, 2021.

[175] Wenzhong Shi, Ming Hao, Analysis of spatial distribution pattern of change-detection error caused by misregistration, International Journal of Remote Sensing 34 (19) (2013) 6883–6897.

[176] Ashbindu Singh, Review article digital change detection techniques using remotely-sensed data, International Journal of Remote Sensing 10 (6) (1989) 989–1003.

[177] Ahram Song, Jaewan Choi, Youkyung Han, Yongil Kim, Change detection in hyperspectral images using recurrent 3D fully convolutional networks, Remote Sensing 10 (11) (2018) 1827.

[178] Ahram Song, Yeji Kim, Yongil Kim, Change detection of surface water in remote sensing images based on fully convolutional network, Journal of Coastal Research 91 (SI) (2019) 426–430.

[179] Yan Song, Xiuxiao Yuan, Honggen Xu, Yun Yang, A novel image change detection method based on enhanced growing self-organization feature map, in: Geoinformatics 2006: Remotely Sensed Data and Information, vol. 6419, SPIE, 2006, pp. 340–346.

[180] Linzhi Su, Xin Cao, Fuzzy autoencoder for multiple change detection in remote sensing images, Journal of Applied Remote Sensing 12 (3) (2018) 035014.

[181] Linzhi Su, Maoguo Gong, Puzhao Zhang, Mingyang Zhang, Jia Liu, Hailun Yang, Deep learning and mapping based ternary change detection for information unbalanced images, Pattern Recognition 66 (2017) 213–228.

[182] Linzhi Su, Jiao Shi, Puzhao Zhang, Zhao Wang, Maoguo Gong, Detecting multiple changes from multi-temporal images by using stacked denosing autoencoder based change vector analysis, in: 2016 International Joint Conference on Neural Networks (IJCNN), IEEE, 2016, pp. 1269–1276.

[183] Jérémie Sublime, Ekaterina Kalinicheva, Automatic post-disaster damage mapping using deep-learning techniques for change detection: case study of the Tohoku tsunami, Remote Sensing 11 (9) (2019) 1123.

[184] Badri Narayan Subudhi, Susmita Ghosh, Ashish Ghosh, Spatial constraint Hopfield-type neural networks for detecting changes in remotely sensed multitemporal images, in: 2013 IEEE International Conference on Image Processing, IEEE, 2013, pp. 3815–3819.

[185] Bin Sun, Guo-Zhong Li, Min Han, Qiu-Hua Lin, A deep learning approach to detecting changes in buildings from aerial images, in: Advances in Neural Networks–ISNN 2019: 16th International Symposium on Neural Networks, ISNN 2019, Moscow, Russia, July 10–12, 2019, Proceedings, Part II 16, Springer, 2019, pp. 414–421.

[186] Yu Sun, Shuohuan Wang, Yukun Li, Shikun Feng, Xuyi Chen, Han Zhang, Xin Tian, Danxiang Zhu, Hao Tian, Hua Wu, Ernie: enhanced representation through knowledge integration, arXiv preprint, arXiv:1904.09223, 2019.

[187] Shenghao Tang, Tong Li, Xinghao Cheng, A novel remote sensing image change detection algorithm based on self-organizing feature map neural network model, in: 2016 International Conference on Communication and Electronics Systems (ICCES), IEEE, 2016, pp. 1–7.
[188] Cristina Tarantino, Palma Blonda, Guido Pasquariello, Remote sensed data for automatic detection of land-use changes due to human activity in support to landslide studies, Natural Hazards 41 (2007) 245–267.
[189] Andrew P. Tewkesbury, Alexis J. Comber, Nicholas J. Tate, Alistair Lamb, Peter F. Fisher, A critical synthesis of remotely sensed optical image change detection techniques, Remote Sensing of Environment 160 (2015) 1–14.
[190] Murase Tomoya, Tanaka Kanji, Change detection under global viewpoint uncertainty, arXiv preprint, arXiv:1703.00552, 2017.
[191] Xiaohua Tong, Xue Zhang, Miaolong Liu, Detection of urban sprawl using a genetic algorithm-evolved artificial neural network classification in remote sensing: a case study in Jiading and Putuo districts of Shanghai, China, International Journal of Remote Sensing 31 (6) (2010) 1485–1504.
[192] Azzedine Touazi, Djamel Bouchaffra, A k-nearest neighbor approach to improve change detection from remote sensing: application to optical aerial images, in: 2015 15th International Conference on Intelligent Systems Design and Applications (ISDA), IEEE, 2015, pp. 98–103.
[193] Hugo Touvron, Thibaut Lavril, Gautier Izacard, Xavier Martinet, Marie-Anne Lachaux, Timothée Lacroix, Baptiste Rozière, Naman Goyal, Eric Hambro, Faisal Azhar, et al., LLaMA: open and efficient foundation language models, arXiv preprint, arXiv:2302.13971, 2023.
[194] S. Varamesh, S.M. Hosseini, M. Rahimzadegan, Detection of land use changes in northeastern Iran by Landsat satellite data, Applied Ecology and Environmental Research 15 (3) (2017).
[195] Ashley Varghese, Jayavardhana Gubbi, Akshaya Ramaswamy, P. Balamuralidhar, ChangeNet: a deep learning architecture for visual change detection, in: Proceedings of the European Conference on Computer Vision (ECCV) Workshops, 2018, pp. 1–16.
[196] Di Wang, Jing Zhang, Bo Du, Minqiang Xu, Lin Liu, Dacheng Tao, Liangpei Zhang, SAMRS: scaling-up remote sensing segmentation dataset with segment anything model, Advances in Neural Information Processing Systems 36 (2023) 8815–8827.
[197] Lidong Wang, Heterogeneous data and big data analytics, in: ACIS, vol. 3, 2017, pp. 8–15.
[198] Lukang Wang, Min Zhang, Wenzhong Shi, CS-WSCDNet: class activation mapping and segment anything model-based framework for weakly supervised change detection, IEEE Transactions on Geoscience and Remote Sensing 61 (2023) 1–12.
[199] Qi Wang, Zhenghang Yuan, Qian Du, Xuelong Li, GETNET: a general end-to-end 2-D CNN framework for hyperspectral image change detection, IEEE Transactions on Geoscience and Remote Sensing 57 (1) (2018) 3–13.
[200] Qunming Wang, Wenzhong Shi, Peter M. Atkinson, Zhongbin Li, Land cover change detection at subpixel resolution with a Hopfield neural network, IEEE Journal of Selected Topics in Applied Earth Observations and Remote Sensing 8 (3) (2014) 1339–1352.
[201] Xin Wang, Sicong Liu, Peijun Du, Hao Liang, Junshi Xia, Yunfeng Li, Object-based change detection in urban areas from high spatial resolution images based on multiple features and ensemble learning, Remote Sensing 10 (2) (2018) 276.
[202] Xinlong Wang, Xiaosong Zhang, Yue Cao, Wen Wang, Chunhua Shen, Tiejun Huang, SegGPT: segmenting everything in context, arXiv preprint, arXiv:2304.03284, 2023.
[203] Yong Wang, Bo Du, Lixiang Ru, Chen Wu, Hui Luo, Scene change detection via deep convolution canonical correlation analysis neural network, in: IGARSS 2019-2019 IEEE International Geoscience and Remote Sensing Symposium, IEEE, 2019, pp. 198–201.
[204] Zian Wang, David Acuna, Huan Ling, Amlan Kar, Sanja Fidler, Object instance annotation with deep extreme level set evolution, in: Proceedings of the IEEE/CVF Conference on Computer Vision and Pattern Recognition, 2019, pp. 7500–7508.
[205] Wahyu Wiratama, Donggyu Sim, Fusion network for change detection of high-resolution panchromatic imagery, Applied Sciences 9 (7) (2019) 1441.

[206] Curtis E. Woodcock, Scott A. Macomber, Mary Pax-Lenney, Warren B. Cohen, Monitoring large areas for forest change using Landsat: generalization across space, time and Landsat sensors, Remote Sensing of Environment 78 (1–2) (2001) 194–203.
[207] Chen Wu, Hongruixuan Chen, Bo Du, Liangpei Zhang, Unsupervised change detection in multitemporal VHR images based on deep kernel PCA convolutional mapping network, IEEE Transactions on Cybernetics 52 (11) (2021) 12084–12098.
[208] Chen Wu, Lefei Zhang, Liangpei Zhang, A scene change detection framework for multi-temporal very high resolution remote sensing images, Signal Processing 124 (2016) 184–197.
[209] Chen Wu, Liangpei Zhang, Bo Du, Kernel slow feature analysis for scene change detection, IEEE Transactions on Geoscience and Remote Sensing 55 (4) (2017) 2367–2384.
[210] Ke Wu, Qian Du, Yi Wang, Yetao Yang, Supervised sub-pixel mapping for change detection from remotely sensed images with different resolutions, Remote Sensing 9 (3) (2017) 284.
[211] Ruliang Xiao, Runxi Cui, Mingwei Lin, Lifei Chen, Youcong Ni, Xinhong Lin, SOMDNCD: image change detection based on self-organizing maps and deep neural networks, IEEE Access 6 (2018) 35915–35925.
[212] Junfeng Xu, Baoming Zhang, Haitao Guo, Jun Lu, Yuzhun Lin, Combining iterative slow feature analysis and deep feature learning for change detection in high-resolution remote sensing images, Journal of Applied Remote Sensing 13 (2) (2019) 024506.
[213] Chen Xue, Li Xiaowen, Ma Jianwen, Urban change detection based on self-organizing feature map neural network, in: IGARSS 2004. 2004 IEEE International Geoscience and Remote Sensing Symposium, vol. 5, IEEE, 2004, pp. 3428–3431.
[214] Gang Yang, Heng-Chao Li, Wei-Ye Wang, Wen Yang, William J. Emery, Unsupervised change detection based on a unified framework for weighted collaborative representation with RDDL and fuzzy clustering, IEEE Transactions on Geoscience and Remote Sensing 57 (11) (2019) 8890–8903.
[215] Meijuan Yang, Licheng Jiao, Fang Liu, Biao Hou, Shuyuan Yang, Transferred deep learning-based change detection in remote sensing images, IEEE Transactions on Geoscience and Remote Sensing 57 (9) (2019) 6960–6973.
[216] Lu Yuan, Dongdong Chen, Yi-Ling Chen, Noel Codella, Xiyang Dai, Jianfeng Gao, Houdong Hu, Xuedong Huang, Boxin Li, Chunyuan Li, et al., Florence: a new foundation model for computer vision, arXiv preprint, arXiv:2111.11432, 2021.
[217] Sergey Zagoruyko, Nikos Komodakis, Learning to compare image patches via convolutional neural networks, in: Proceedings of the IEEE Conference on Computer Vision and Pattern Recognition, 2015, pp. 4353–4361.
[218] Tao Zhan, Maoguo Gong, Xiangming Jiang, Shuwei Li, Log-based transformation feature learning for change detection in heterogeneous images, IEEE Geoscience and Remote Sensing Letters 15 (9) (2018) 1352–1356.
[219] Tao Zhan, Maoguo Gong, Jia Liu, Puzhao Zhang, Iterative feature mapping network for detecting multiple changes in multi-source remote sensing images, ISPRS Journal of Photogrammetry and Remote Sensing 146 (2018) 38–51.
[220] Yang Zhan, Kun Fu, Menglong Yan, Xian Sun, Hongqi Wang, Xiaosong Qiu, Change detection based on deep Siamese convolutional network for optical aerial images, IEEE Geoscience and Remote Sensing Letters 14 (10) (2017) 1845–1849.
[221] Chi Zhang, Shiqing Wei, Shunping Ji, Meng Lu, Detecting large-scale urban land cover changes from very high resolution remote sensing images using CNN-based classification, ISPRS International Journal of Geo-Information 8 (4) (2019) 189.
[222] J. Zhang, Z. Zhou, G. Mai, L. Mu, M. Hu, S. Li, Text2Seg: remote sensing image semantic segmentation via text-guided visual foundation models, arXiv preprint, arXiv:2304.10597, 2023.
[223] Min Zhang, Wenzhong Shi, A feature difference convolutional neural network-based change detection method, IEEE Transactions on Geoscience and Remote Sensing 58 (10) (2020) 7232–7246.
[224] Penglin Zhang, Wenzhong Shi, Man Sing Wong, Jiangping Chen, A reliability-based multi-algorithm fusion technique in detecting changes in land cover, Remote Sensing 5 (3) (2013) 1134–1151.

[225] Puzhao Zhang, Maoguo Gong, Linzhi Su, Jia Liu, Zhizhou Li, Change detection based on deep feature representation and mapping transformation for multi-spatial-resolution remote sensing images, ISPRS Journal of Photogrammetry and Remote Sensing 116 (2016) 24–41.

[226] Puzhao Zhang, Maoguo Gong, Hui Zhang, Jia Liu, Yifang Ban, Unsupervised difference representation learning for detecting multiple types of changes in multitemporal remote sensing images, IEEE Transactions on Geoscience and Remote Sensing 57 (4) (2018) 2277–2289.

[227] Wuxia Zhang, Xiaoqiang Lu, The spectral-spatial joint learning for change detection in multispectral imagery, Remote Sensing 11 (3) (2019) 240.

[228] Zhenchao Zhang, George Vosselman, Markus Gerke, Claudio Persello, Devis Tuia, Michael Ying Yang, Detecting building changes between airborne laser scanning and photogrammetric data, Remote Sensing 11 (20) (2019) 2417.

[229] Jiaojiao Zhao, Maoguo Gong, Jia Liu, Licheng Jiao, Deep learning to classify difference image for image change detection, in: 2014 International Joint Conference on Neural Networks (IJCNN), IEEE, 2014, pp. 411–417.

[230] Qiunan Zhao, Jingjing Ma, Maoguo Gong, Hao Li, Tao Zhan, Three-class change detection in synthetic aperture radar images based on deep belief network, Journal of Computational and Theoretical Nanoscience 13 (6) (2016) 3757–3762.

[231] Wei Zhao, Zhirui Wang, Maoguo Gong, Jia Liu, Discriminative feature learning for unsupervised change detection in heterogeneous images based on a coupled neural network, IEEE Transactions on Geoscience and Remote Sensing 55 (12) (2017) 7066–7080.

[232] Wenzhi Zhao, Lichao Mou, Jiage Chen, Yanchen Bo, William J. Emery, Incorporating metric learning and adversarial network for seasonal invariant change detection, IEEE Transactions on Geoscience and Remote Sensing 58 (4) (2019) 2720–2731.

[233] Xu Zhao, Wenchao Ding, Yongqi An, Yinglong Du, Tao Yu, Min Li, Ming Tang, Jinqiao Wang, Fast segment anything, arXiv preprint, arXiv:2306.12156, 2023.

[234] Yanfei Zhong, Wenfeng Liu, Ji Zhao, Liangpei Zhang, Change detection based on pulse-coupled neural networks and the NMI feature for high spatial resolution remote sensing imagery, IEEE Geoscience and Remote Sensing Letters 12 (3) (2014) 537–541.

[235] Yiwu Zhong, Jianwei Yang, Pengchuan Zhang, Chunyuan Li, Noel Codella, Liunian Harold Li, Luowei Zhou, Xiyang Dai, Lu Yuan, Yin Li, et al., RegionCLIP: region-based language-image pretraining, in: Proceedings of the IEEE/CVF Conference on Computer Vision and Pattern Recognition, 2022, pp. 16793–16803.

[236] Bin Zhu, Hongmin Gao, Xin Wang, Mengxi Xu, Xiaobin Zhu, Change detection based on the combination of improved SegNet neural network and morphology, in: 2018 IEEE 3rd International Conference on Image, Vision and Computing (ICIVC), IEEE, 2018, pp. 55–59.

[237] Xueyan Zou, Jianwei Yang, Hao Zhang, Feng Li, Linjie Li, Jianfeng Wang, Lijuan Wang, Jianfeng Gao, Yong Jae Lee, Segment everything everywhere all at once, Advances in Neural Information Processing Systems 36 (2023) 19769–19782.

Applications in carbon neutrality

Ningning Zhu

College of Geography and Planning, Chengdu University of Technology, Chengdu, China

7.1 Research status

Forests constitute the largest carbon pools in terrestrial ecosystems, and elucidating their baseline carbon stock and carbon sink potential is crucial [1,2,17]. Remote sensing, owing to its macroscale, comprehensive, dynamic, rapid, and reproducible nature, has addressed the limitations in carbon accounting for forest vegetation [3,4]. We introduce a novel system for forest vegetation carbon accounting, grounded in the structure and growth equations of individual trees. Specifically, it comprises: (1) Integrating airborne and terrestrial LiDAR data to extract structural parameters, including diameter at breast height (DBH), tree height, and crown width, thereby establishing a carbon stock calculation method for forest plots at the individual tree level; (2) Developing a pixel-level regional forest carbon stock model with physical interpretability, utilizing canopy height and crown closure as key variables, to mitigate the uncertainty of machine/deep learning remote sensing regression inversion; and (3) Accurately estimating regional forest carbon sinks by forecasting future forest canopy height and crown closure, based on the pixel-level forest carbon stock model and individual tree growth equations. We adhere to the overarching theme of forest plot carbon stock – regional forest carbon stock – regional forest carbon sink, expanding from plot to regional spatial scales and extending from carbon stock to carbon sink across temporal scales, thereby establishing a novel remote sensing-based system for forest vegetation carbon accounting [34,53]. The overall technical roadmap is shown in Fig. 7.1.

7.1.1 Research status on forest plot carbon stock calculation using LiDAR remote sensing

Structural parameters of individual trees, such as diameter at breast height (DBH), tree height, and crown width, are key to calculating forest plot carbon stocks. These parameters are extracted by fusing ground-based LiDAR and UAV LiDAR data through air-ground data fusion. Unlike urban scenes, forest point clouds lack traditional geometric features (points, lines, planes, etc.) suitable for registration, posing challenges for the automated

Multimodal Remote Sensing Fusion and Classification
https://doi.org/10.1016/B978-0-44-329152-4.00014-3

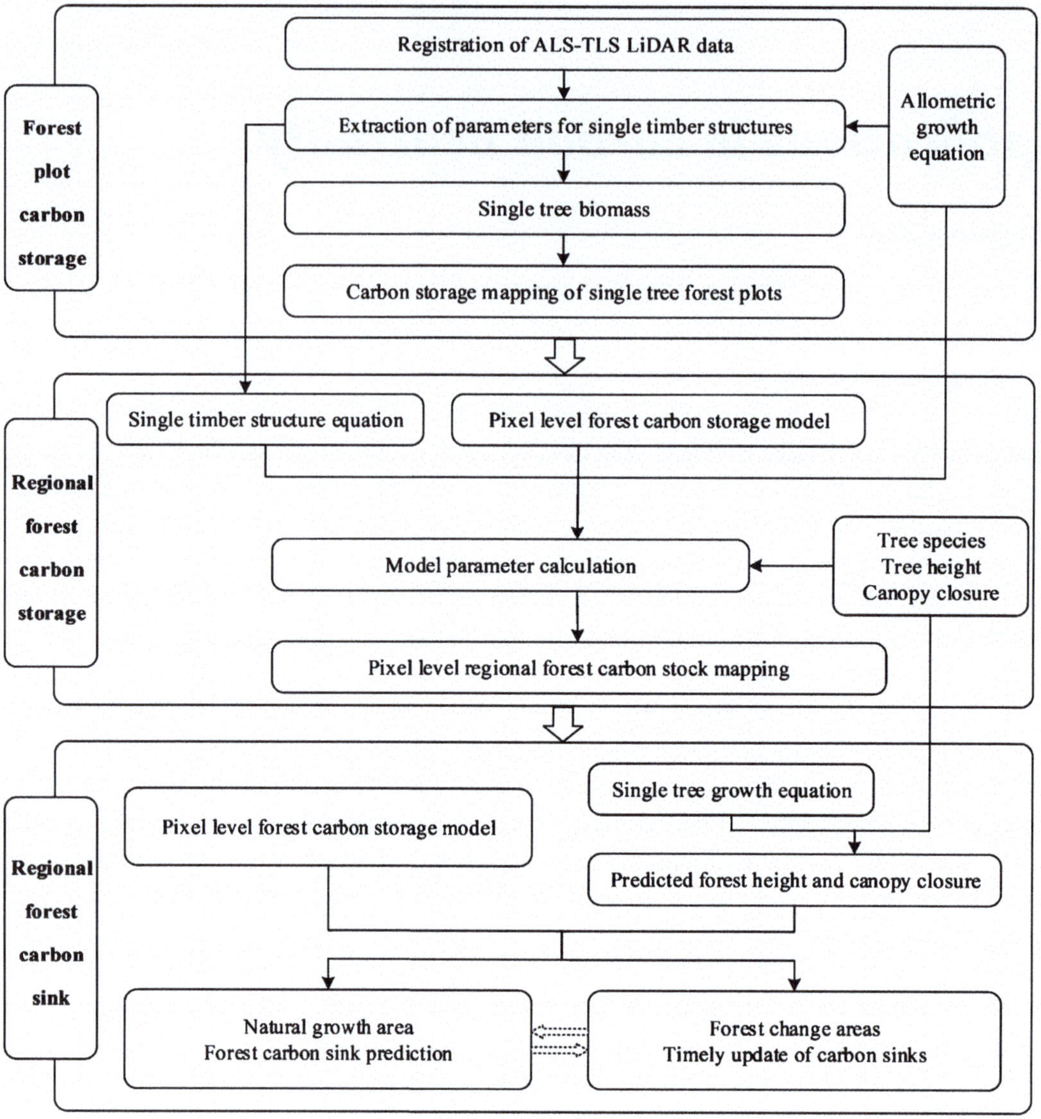

Figure 7.1:
Technology roadmap.

registration of UAV and ground-based point clouds in forests. Hauglin et al. [35] used a normalized feature descriptor based on DBH and tree height to register ground-based and airborne point clouds in a coniferous forest in southeastern Norway, but manual adjustment of different search ranges was required. Polewski et al. [36] used tree positions as registration primitives, calculating horizontal and vertical distance features between each tree and others for similarity measurement, achieving a horizontal registration accuracy of 0.66 m. Paris et al. [37] applied normalized cross-correlation to register Canopy Height Model (CHM) images derived from ground-based and airborne point clouds to obtain planar transformation parameters; however, the vertical registration problem remained unsolved, and the method was only suitable for relatively open forest areas, requiring improvement for dense forests.

Currently, there are two main approaches for individual tree segmentation. The first is based on the CHM model. This method uses local maxima obtained from the CHM as seed points and applies mature image segmentation algorithms (e.g., region growing, watershed algorithm) to segment the generated CHM raster image, enabling individual tree crown detection and segmentation. Yang et al. [38] used a local maximum method to detect individual tree tops in the CHM as markers for the watershed algorithm, avoiding oversegmentation. Geng et al. [39] detected tree tops by finding local maxima and combined this with a marker-controlled watershed algorithm to segment crown boundaries in airborne laser point clouds, showing that the correlation coefficients between estimated and measured tree height and crown width reached 0.786 and 0.9, respectively. Wang et al. [40] introduced a spectral clustering method combined with mean shift voxelization and the Nystrom method, significantly improving the accuracy and efficiency of individual tree segmentation from airborne LiDAR point clouds while maintaining the superior performance of spectral clustering. Qin et al. [41] integrated airborne LiDAR, hyperspectral, and ultrahigh-resolution RGB data, using the Watershed Spectral Texture-controlled Normalized Cut (WST-Ncut) algorithm to segment a subtropical broadleaf forest, achieving an accuracy of 0.86. The second approach is direct segmentation based on 3D point clouds. Jiang et al. [42] used an improved semantic segmentation network combined with cylindrical convolution and dynamic shift segmentation strategies to segment urban street trees, achieving 92% accuracy. Li et al. [43] respectively applied a marker-controlled watershed algorithm and a point cloud distance clustering algorithm, adjusting corresponding parameters to segment individual trees in high-canopy-density plantations, concluding that both algorithms are suitable for relatively accurate individual tree segmentation in such forests. Yang et al. [44] proposed a branch topology-based individual tree segmentation method, increasing the average segmentation accuracy by 70%. Meng et al. [45] combined airborne LiDAR data and spectral image data, selecting the optimal segmentation scale to accurately locate individual tree tops, improving the oversegmentation issue of single-scale methods and increasing individual tree identification accuracy from 0.67 to 0.92.

7.1.2 Research status on regional forest carbon stock calculation

After obtaining individual tree DBH and height parameters, individual-tree-level forest carbon stock models can be established [46,47], which are used to calculate forest carbon stocks at small plot scales. Statistical sampling methods are then applied to estimate forest carbon stocks over large areas [5–7]. This approach is affected by the number and spatial distribution of sample plots. For example, China has 200,000 forest sample plots, leading to enormous workload and difficulty in evaluating calculation results. Remote sensing methods offer advantages of continuous spatial distribution and timeliness, enabling pixel-level acquisition of spectral and vertical structural information over large areas [48,49]. Machine learning or deep learning approaches are used to establish relationships between remote sensing feature variables and forest carbon stocks, with accuracy depending on the selection of independent variables [50–52]. However, forcing a regression relationship between them makes the model lack theoretical interpretability and broad applicability.

Optical remote sensing data are the most commonly used data for forest carbon stock calculation due to their easy accessibility, long time series, and multiscale characteristics [8]. Medium-resolution remote sensing satellites Landsat and Sentinel are commonly used for regional-scale forest carbon stock calculation [9,10], while low-resolution satellites MODIS and AVHRR are often used for global-scale forest carbon stock inversion [11,12]. Medium-resolution SAR data have also been used for regional-scale forest carbon stock calculation [13]. However, the accuracy of optical remote sensing and SAR data significantly decreases in forests with carbon stocks exceeding 100–300 Mg/ha [14,15]. LiDAR has certain penetration capability, can directly observe forest height, and is not affected by saturation, but rapid acquisition of global-scale airborne LiDAR data is a challenging task. Although many studies have used remote sensing data for forest carbon stock mapping, remote sensing inversion methods still have large errors and uncertainties [8]. Feature variables extracted from multisource remote sensing data are used to establish relationships with carbon stocks, including vegetation indices [16,17], texture features [18], spectral features, forest canopy density, and forest height [19,20]. Many algorithms have been used to model the relationship between remote sensing variables and forest carbon stocks, commonly including K-Nearest Neighbor (KNN), artificial neural networks, random forest, and support vector machines. Fu et al. [21] used the KNN method to establish the relationship between MODIS data and forest biomass, estimating that the average forest biomass in Northeast China around 2000 was 101.98 Mg/ha. Purohit et al. [22] used the random forest algorithm with spectral and texture variables extracted from Landsat 8 and Sentinel-1 to predict biomass of different forest types in the Doon Valley, achieving an R^2 of 0.87, RMSE of 62.56 Mg/ha, and rRMSE of 18.7%. Stumer et al. [23] used a neural network method combining forest inventory data and satellite remote sensing data to estimate forest carbon stocks in the Thuringia region of Germany, showing comparable performance to the KNN method. Safari et al. [24] compared the performance of four machine learning algorithms, namely random forest, support vector machine, boosted regression trees, and multivariate adaptive regression splines, for forest carbon stock calculation using multitemporal Landsat 8 imagery, finding that boosted regression trees had lower accuracy, support vector machines had large bias, and random forest performed most robustly.

7.1.3 Research status on regional forest carbon sequestration estimation

According to the Ninth National Forest Inventory, China's forest area has reached 22.96% of its land area, of which 33% are young forests, indicating significant carbon sequestration potential between 2010 and 2050. Currently, growth models are mainly used to predict forest carbon sequestration, while also accounting for changes in forest area and the influence of environmental factors on forest stand growth [28]. Using mature forests as a reference is an important approach to explore carbon sequestration. Liu et al. [30] defined the forest biomass carbon stock when forest age reaches the upper limit of overmature forests as the forest biomass carbon capacity, and the difference between this capacity and the current (or a certain year's) forest biomass carbon stock as the forest biomass carbon sequestration potential. Other studies estimate the carbon sequestration potential of newly added forests based on the planned afforestation areas of various forestry projects and the carbon accumulation rates of different vegetation types. For existing forests, the carbon sequestration potential is estimated using a modified carbon sequestration rate factor method [31], which is suitable for estimating the carbon sequestration potential of forests at a certain developmental stage. International scholars often use simulation methods to assess forest carbon sequestration potential, for example, simulating forest carbon sequestration potential under the evolution of forest management models [26], different forest fire risk levels and fire treatment methods [25], and the impact of human activities on land use and land cover change [27]. Domestic scholars, in addition to using simulation methods, also focus on incorporating development targets set in medium- and long-term plans and policy documents issued by relevant sectors [29], making assumptions about key information such as forest coverage rate and forestland area in the study area, and thereby analyzing the forest carbon sequestration potential by the end of the planning period.

Different methods for estimating forest carbon sequestration have their own advantages and disadvantages. The stock change method requires that two inventory surveys use the same methods and the same biomass models and parameters. Plot-based data are the most accurate and reliable, but the workload is heavy, and accuracy can only be guaranteed at the overall sampling scale. The gain-loss method obtains forest carbon sequestration through empirical statistical models, but it is region-specific, requiring redetermination of empirical parameter values when applied to other areas, limiting model applicability [32]. Additionally, directly using forest inventory data to predict forest carbon sequestration has two problems: (1) the classification of forest age groups does not distinguish between natural and planted forests, yet the age group boundaries differ greatly between them, and the setting of these boundaries directly affects the establishment of forest growth equations; and (2) it does not account for differences in growth rates of the same tree species across different regions, simply replacing regional-scale forest growth rates with plot-scale forest growth rates.

7.2 Calculation of carbon storage in forest plots

The use of ground-based LiDAR (TLS) and unmanned aerial vehicle LiDAR (ULS) to accurately obtain the diameter at breast height and tree height parameters of individual trees has improved the efficiency of traditional plot surveys; Using airborne laser point clouds and hyperspectral satellite images for single tree scale forest carbon storage calculation avoids the tedious work of large-scale measurement [5–7]. Fig. 7.2 shows the technical process of this method.

The automatic registration methods for multisource point clouds in forest scenes using TLS-TLS and TLS-ULS mainly include three steps: keypoint detection, initial correspondence generation, and RANSAC coarse error removal. The keypoints should meet the following requirements: (1) reliable repeatability between different viewpoints in the same forest area; (2) local surfaces should have sufficient descriptive information to enable points to have unique features. The extraction of single tree structural parameters from LiDAR point cloud includes three steps: nontree point removal, single tree segmentation, and single tree structural parameter extraction. The point cloud is divided into tree points and nontree points, and the tree height parameter is measured after single tree segmentation of the tree points. A certain thickness of point cloud is taken from the lowest point of the single tree to fit the diameter at breast height parameter, as shown in Fig. 7.3. The thickness is set to 0.1 m, 0.2 m, and 0.3 m respectively. Circular fitting is used for the cross-sectional point cloud projected onto the horizontal plane. Due to the influence of tree nodes or branches, the diameter parameters of different thicknesses will be different. The mean of fitting with a difference less than 5 cm is taken as the final diameter at breast height parameter. The numbers in the figure are the cross-sectional values of different thicknesses. The chest diameter parameter for point fitting, measured in centimeters.

The data collection efficiency of ground stations and airborne platforms is different, and the coverage area of ground stations is generally smaller than that of airborne platforms. Using the single tree diameter at breast height D, tree height H, and crown width Cw extracted from the overlapping area of forest sample land air ground data, different tree species' single tree structure equations $H = f(D)$ and $Cw = f(D)$ are established. As shown in Fig. 7.4, the blue dots represent the scatter plot of tree height diameter at breast height, and the red dots represent the scatter plot of crown width diameter at breast height. Considering the growth characteristics of individual trees, tree height and crown width increase with the increase

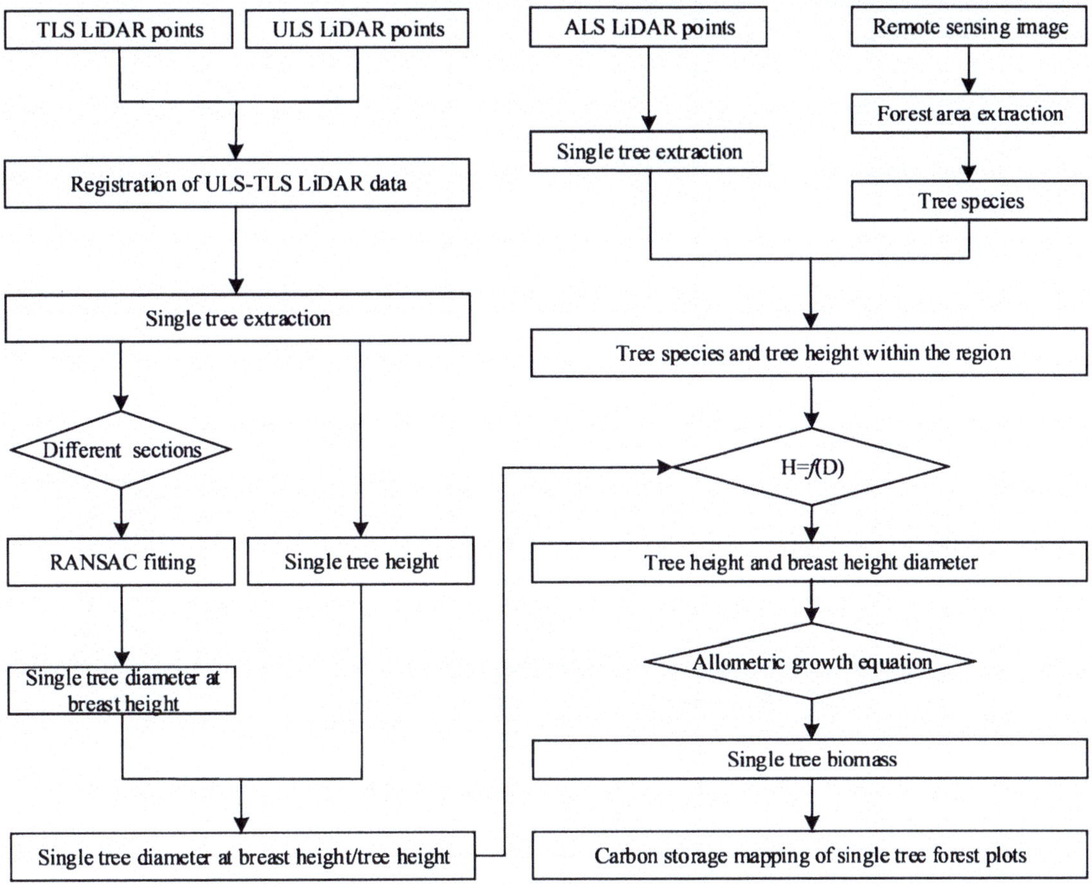

Figure 7.2:
Technical flow chart for calculating carbon stock in forest plot at the single-tree level.

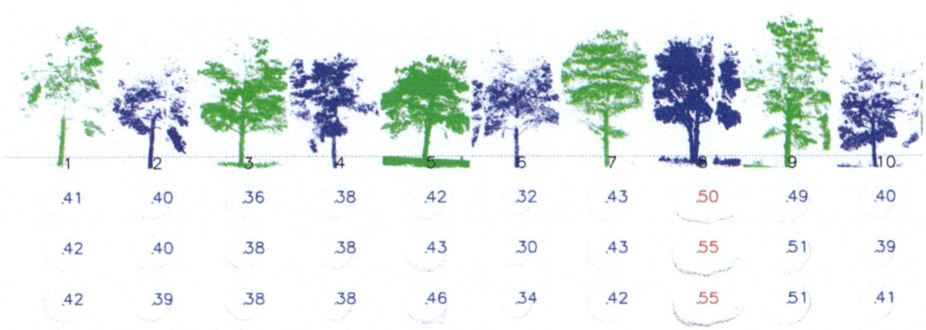

Figure 7.3:
Extraction of single-tree diameter at breast height parameters.

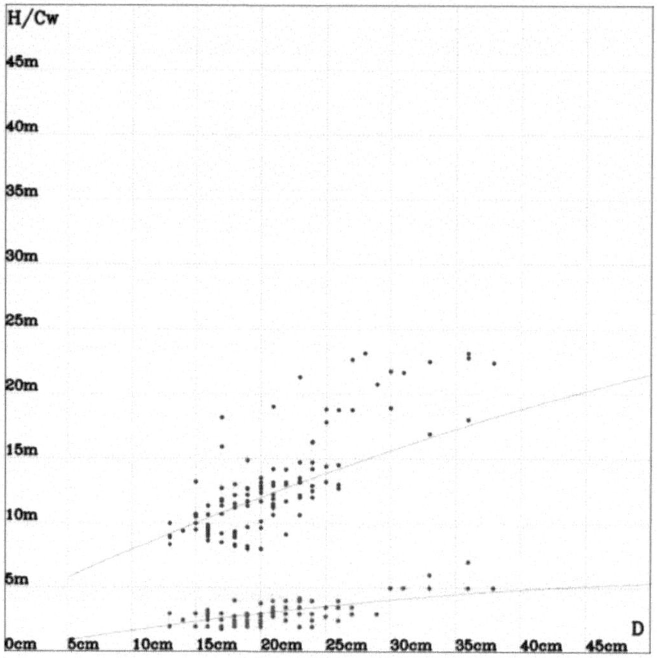

Figure 7.4:
Single timber structure equation (Chinese fir).

of breast height diameter, and the growth rate gradually decreases and tends to stabilize. A quadratic curve $y = ax^2 + bx + c$ can be used for fitting.

Extract all individual trees from the drone point cloud of the forest plot, obtain tree height parameters, use the tree structure equation to calculate the diameter at breast height of individual trees $D = f^{-1}(H)$, and then use the allometric growth equation $W = f(D, H)$ to calculate the biomass of individual trees [33]. Multiply the biomass W by the carbon content coefficient CF of the corresponding tree species to convert it into carbon storage $C = CF \bullet (W_T + W_R)$, as shown in Table 7.1. Accumulate all individual trees in the forest plot to complete the carbon storage mapping of individual tree level forest plots.

Fig. 7.5 shows the carbon storage mapping of forest plots using airborne point clouds within a range of 1 km × 1 km. As shown in Fig. 7.5 (a), there are 3.89 million points in the entire area with a point density of 3.6 points/m^2. The blue border represents the verification area, which includes 4223 eucalyptus trees through manual investigation. The number of automatically extracted single trees is 4314, with an extraction error of 2.15%. 82,003 single trees were automatically extracted in the entire area. According to the statistical chart of tree height distribution below the figure, the height of eucalyptus trees is concentrated between 5-20 m, with an average height of 14.2 m. The diameter at breast height parameter is calculated

Table 7.1: Allometric growth equations of different tree species.

China fir	Pine	Eucalyptus
$W_S = 0.073429(D^2H)^{0.86262}$ $W_P = 0.013775(D^2H)^{0.84463}$ $W_B = 0.000482(D^2H)^{1.23314}$ $W_L = 0.019638(D^2H)^{0.78969}$ $W_T = W_S + W_P + W_B + W_L$ $W_R = 0.04305(D^2H)^{0.7358}$	$W_T = 0.071556(D^2H)^{0.857209}$ $W_R = \frac{W_T}{6.23}$	$W_S = 0.0902526D^{2.44815}$ $W_B = 0.0049163D^{2.81779}$ $W_L = 0.012694D^{2.26839}$ $W_T = W_S + W_B + W_L$ $W_R = \frac{W_T}{7.45}$

Note: D Unit: cm, H unit: m; W Unit: t; W_S for trunk biomass, W_P for bark biomass, W_B for branch biomass, and W_L for leaf biomass; the total biomass of the aboveground part W_T and the biomass of the underground part W_R; the carbon coefficient CF is approximately 0.5.

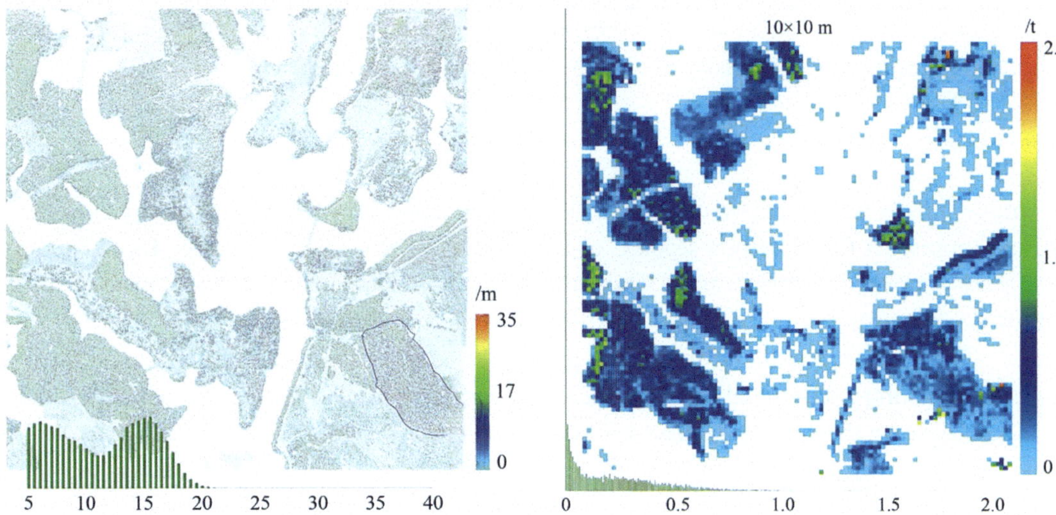

(a) Forest plot carbon stock mapping of single tree level (b) Forest plot carbon stock mapping of pixel level

Figure 7.5:
Forest plot carbon stock mapping.

using the single tree structure equation, and the average diameter at breast height is 12.3 cm. The diameter and height of a single tree are input into the allometric growth equation, multiplied by the carbon content coefficient of eucalyptus trees, and the accumulated carbon storage of forest vegetation in this area is 1195 tons. Divide the entire region into grids and accumulate the carbon storage of individual trees within the grid. Fig. 7.5 (b) shows a pixel level forest plot carbon storage map with a resolution of 10 m, which can be overlaid with 10 m resolution remote sensing images for regression inversion of forest carbon storage on a larger scale.

7.3 Calculation of regional forest carbon storage

Establish an explicit calculation model for pixel level forest carbon storage based on remote sensing using two core variables, forest canopy density and height. Forest carbon storage is represented as the product of tree species density, forest area, and height, and the model parameters are automatically solved by the single tree structure equation. The detailed technical roadmap is shown in Fig. 7.6.

To construct a forest carbon storage calculation method with interpretability and controllable accuracy, the power function multiplication form of diameter at breast height (D) and tree height (H) used in the single tree level carbon storage model is borrowed, that is $C = c_1 \bullet D^{c_2} \bullet H^{c_3}$, D and H respectively represent the plane and elevation information of the single tree. A pixel level carbon storage model is constructed, as shown in Eq. (7.1), where the carbon storage C of each pixel is represented by four variables: remote sensing image resolution P, forest canopy density F, terrain slope, and forest height H.

$$C = a \cdot S^b \cdot H^c \tag{7.1}$$

The model revolves around two core variables, forest height (H) and forest canopy closure (F), and represents forest carbon stock C as the product of tree species density, forest area S ($S = P^2 F / \cos\theta$), and elevation H. This allows for simple and efficient calculation of carbon stock at different times, achieving high-resolution dynamic monitoring of forest carbon sinks on a large scale. Using the single tree structure equation $H = f(D)$ and $Cw = f(D)$ constructing a simulated forest scene, combined with the allometric growth equation, theoretically solve the carbon storage model parameters a, b, and c [34].

By setting single trees with different breast height diameters and spacing to construct a simulated forest scene, the simulated forest pixels with arbitrary saturation can be obtained, as shown in Fig. 7.7 (a). A total of 147 trees are determined by the single tree structural equation; Fig. 7.7 (b) shows the division of a 100 m × 100 m area into a 20 m × 20 m grid, with a total of 25 pixels. Using simulated forest pixels with different spatial scales and saturation levels, calculate canopy closure F, average height H, and carbon storage C using Eqs. (7.2)–(7.4) to solve model parameters a, b, and c.

$$F = \frac{\sum_{i=1}^{n} S_i}{P^2} \tag{7.2}$$

$$H = \frac{\sum_{i=1}^{n} h_i S_i}{\sum_{i=1}^{n} S_i} \tag{7.3}$$

$$C = \sum_{i=1}^{n} \frac{C_i S_i}{\pi (Cw_i/2)^2} \tag{7.4}$$

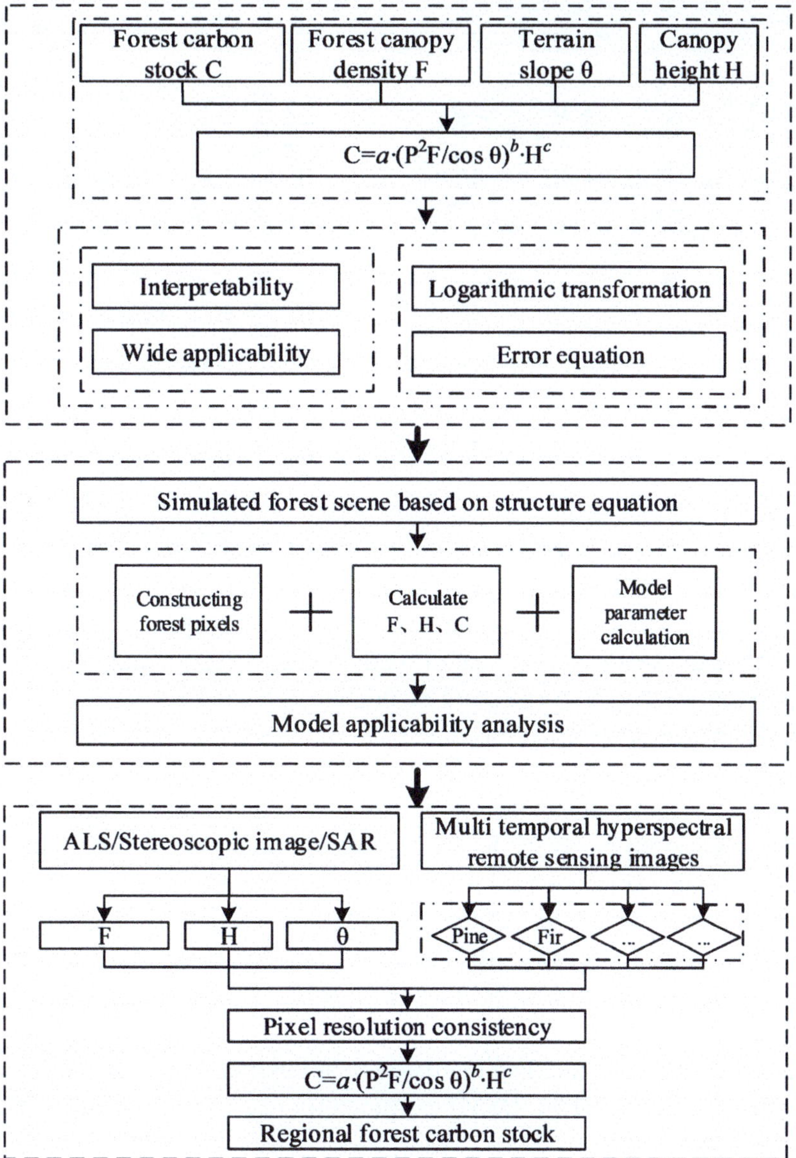

Figure 7.6:
Technical route for building a universal model of forest carbon stock.

where P is the pixel resolution, n is the number of single trees falling into that pixel, Cw_i is the crown width of the i-th tree, S_i represents the area of the crown width of the i-th tree within that pixel, h_i represents the height of the i-th tree, and C_i represents the carbon storage of the i-th tree.

(a) Simulated forest three-dimensional view (b) Simulated forest top view

Figure 7.7:
Build simulated forest pixels.

Based on the pixel level forest carbon storage model, this article completed the world's first 2-meter resolution forest carbon storage mapping on the order of 200,000 square kilometers, as shown in Fig. 7.8 (a). The Guangxi Zhuang Autonomous Region covers an area of 237,600 km^2 and has a forest coverage rate of 62.6%. The forest height H with a 2-meter resolution was obtained using airborne point cloud data collected between 2019 and 2021. The tree species distribution information came from the forest resource second-class survey. At a fine scale of 2 meters, the canopy closure was 0 or 1. By substituting variables H and F into the model for pixel by pixel calculation, it can be concluded that the carbon storage of forest vegetation in Guangxi is 349.29 Tg (2021), while one study [33] calculated the carbon storage of forest vegetation in Guangxi based on the data from the seventh national forest resource continuous inventory (2004-2008) to be 320.22 Tg, which is relatively close to each other; Fig. 7.8 (b) and Fig. 7.8 (c) show the statistical maps of forest vegetation carbon storage and carbon density in 14 prefecture level cities in Guangxi. Compared with existing methods, this model avoids heavy manual survey work and overcomes the problem of traditional remote sensing regression inversion lacking theoretical interpretability. It can fully utilize remote sensing data products (DEM, DSM) to efficiently map high-resolution forest carbon storage on a large scale.

7.4 Regional forest carbon sink prediction

Regional forest carbon sink prediction includes two parts: forest carbon sink prediction based on pixel level forest carbon storage models and forest carbon sink update based on revisit satellite images [28–32]. The detailed technical roadmap is shown in Fig. 7.9.

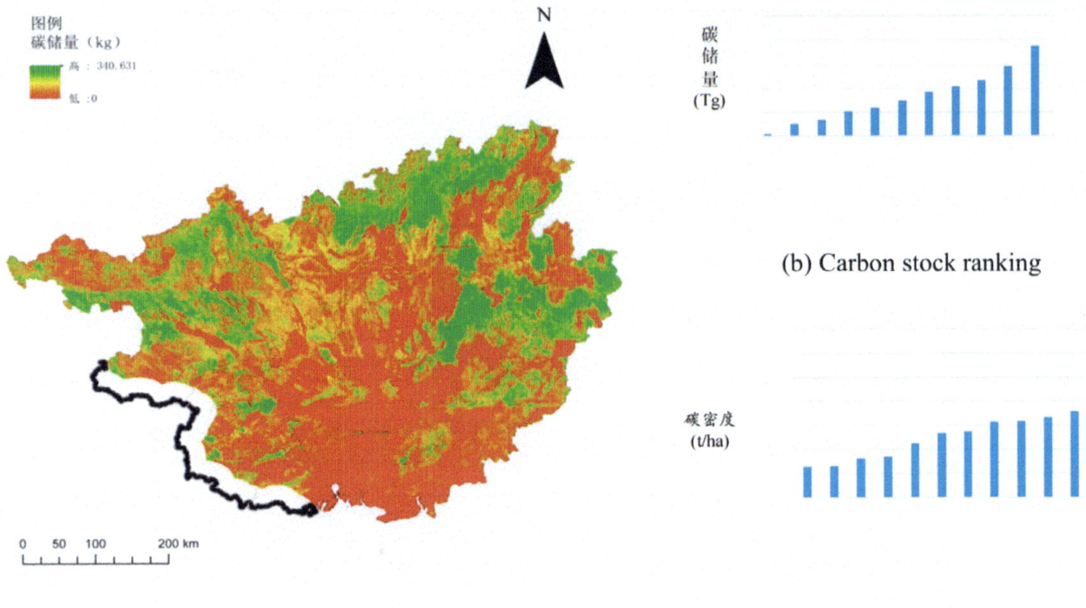

(b) Carbon stock ranking

(c) Carbon stock density ranking

(a) Carbon stock mapping of Guangxi

Figure 7.8:
Regional forest carbon stock mapping and ranking results for Guangxi.

7.4.1 Forest carbon sink prediction based on pixel level forest carbon stock model

Under the natural growth conditions of forests without human logging or sudden changes caused by natural factors, the relationship $H = f(t)$ between the height H and tree age t of different tree species, together with the relationship $Cw = v(H)$ between crown width Cw and tree height H, can accurately calculate the carbon sink of future forests [28,29,32]. Assuming the current moment is t and the future moment is $t + \Delta t$, the forest carbon sink $C_{\Delta t}$ between the two moments is represented by the difference in forest carbon storage between the two moments:

$$C_{\Delta t} = a\left(\frac{P^2}{\cos\theta}\right)^b \left(F_{t+\Delta t}^b H_{t+\Delta t}^c - F_t^b H_t^c\right) \tag{7.5}$$

where, $C_{\Delta t}$ represents the forest carbon sink of the pixel in the time range Δt, F_t and H_t respectively represent the forest canopy density and height at time, $F_{t+\Delta t}$ and $H_{t+\Delta t}$ respectively represent the forest canopy density and height at time.

Calculate the canopy density and height of the forest in the future, $F_{t+\Delta t}$ and $H_{t+\Delta t}$, based on different tree species and the relationships $H = f(t)$ and $Cw = v(H)$. Consider the

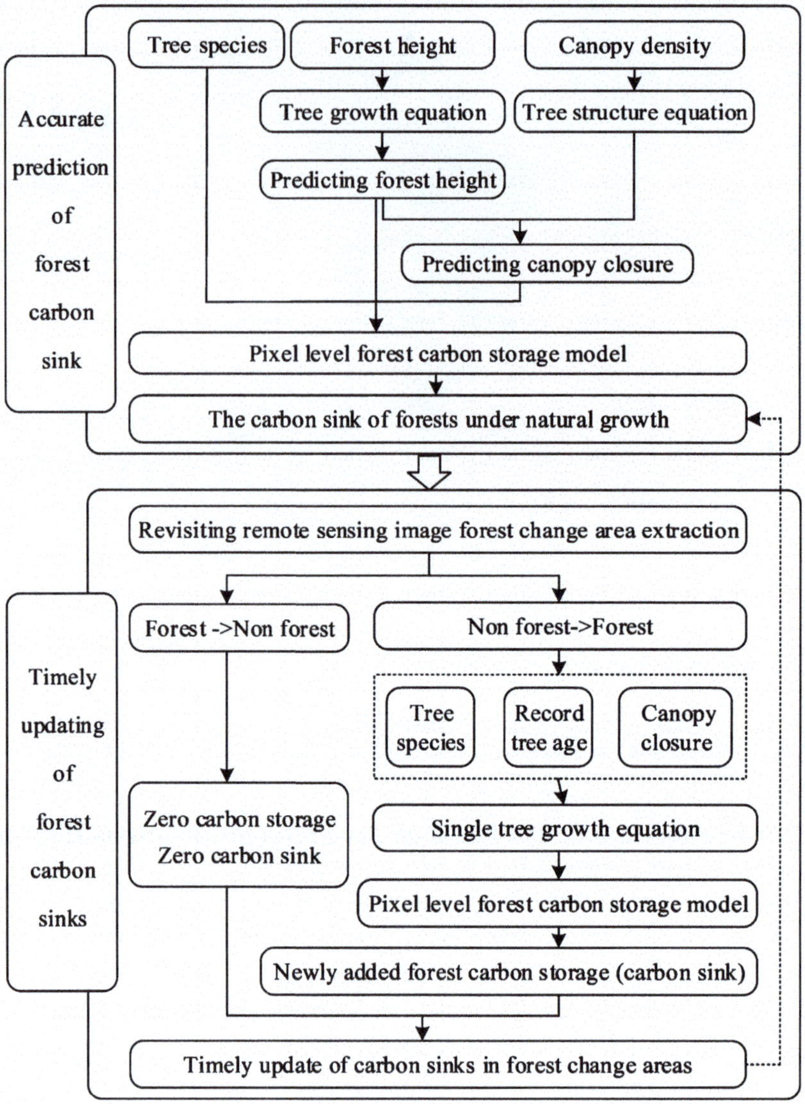

Figure 7.9:
Technical route for regional forest carbon sink prediction.

average height of the forest within a pixel as the height of a single tree, use forestry survey data, and select the optimal fitting function from Richards, Korf, and other models to establish $H = f(t)$, as shown in Table 7.2. The functional relationship between crown width and tree height is based on breast height diameter as the intermediate variable, using the established model in Table 7.1.

Table 7.2: Single-tree growth equations.

Model	Expression	Model	Expression
Richards	$H = a(1 - e^{-bt})^c$	Gompertz	$H = ae^{-be^{-ct}}$
Korf	$H = ae^{-bt^{-c}}$	Weibull	$H = a(1 - e^{-be^{-ct}})$
Logistic	$H = \dfrac{a}{1 + be^{-ct}}$		

In the table, H is the tree height, t is the age, and a, b, and c are the parameters of the function model.

In the case where $H = f(t)$ and $Cw = v(H)$ are known, the calculation steps for $F_{t+\Delta t}$ and $H_{t+\Delta t}$ are as follows:

(1) Obtain the tree age $t = f^{-1}(H_t)$ of the forest pixel based on the current tree height H_t and $H = f(t)$;

(2) Calculate the tree height $H_{t+\Delta t} = f(t + \Delta t)$ at future time $t + \Delta t$ using $H = f(t)$;

(3) Based on the current forest pixel canopy density F_t and $Cw = v(H)$, the calculation of future canopy density $F_{t+\Delta t}$ is derived as follows:

$$F_{t+\Delta t} = \frac{S_{t+\Delta t}}{P^2} = S_t \cdot \frac{Cw_{t+\Delta t}^2}{P^2 \cdot Cw_t^2} = F_t \cdot \frac{v(H_{t+\Delta t})^2}{v(H_t)^2} \tag{7.6}$$

(4) Finally, the calculation formula for $H_{t+\Delta t}$ and $F_{t+\Delta t}$ is as follows:

$$\begin{cases} H_{t+\Delta t} = f(t + \Delta t), \\ F_{t+\Delta t} = F_t \cdot \dfrac{v(H_{t+\Delta t})^2}{v(H_t)^2}, \end{cases} \quad t = f^{-1}(H_t) \tag{7.7}$$

Calculate forest carbon sinks pixel by pixel using predicted forest canopy density and height parameters to obtain the carbon sequestration potential of regional forests under natural growth. Taking the main urban area of Xiangyang as an example, using high-resolution forest height products obtained in 2019, combined with tree species distribution information in forestry survey data, this paper proposes a forest sink enhancement potential model, which can efficiently and accurately calculate the sink enhancement potential by only inputting the current canopy height and tree species information of the forest. Firstly, the tree age is obtained from the forest height and single tree growth equation. Then, the future time (1 year, 2 years, ..., 5 years, etc.) is arbitrarily set based on the current tree age, and the forest height at the future time is calculated from the single tree growth equation. Finally, the canopy density of the future forest is calculated according to (7.7) to complete the forest carbon sink map.

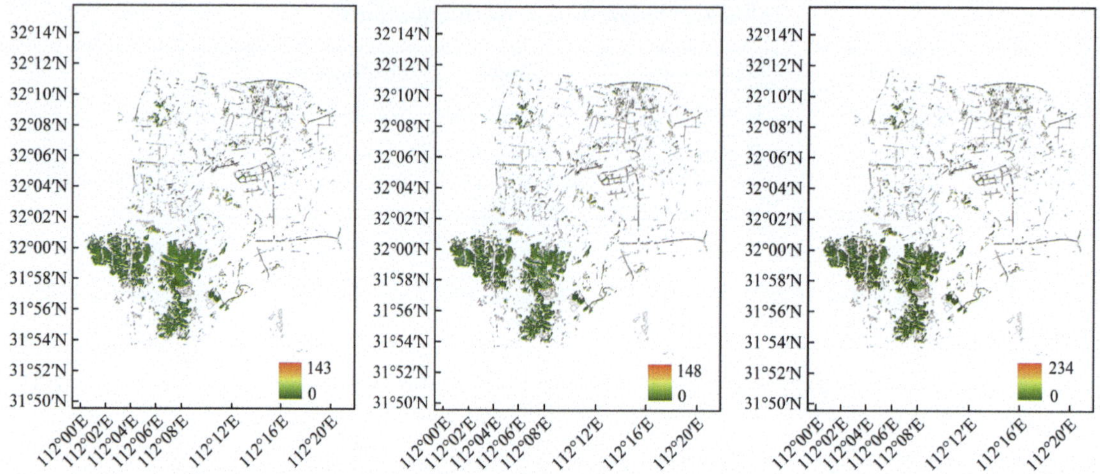

(a) One year forecast (b) Two years forecast (c) Five years forecast

Figure 7.10:
Forest carbon sink map of Xiangyang (2 m resolution).

Table 7.3: The potential of dominant tree species in Xiangyang (10^4 tons).

Tree species	China fir	Pine	Poplar	Camphor tree
1 Year	1.55	0.15	1.38	0.19
2 Years	3.27	0.32	2.43	0.39
5 Years	9.37	0.99	5.51	1.05

Fig. 7.10 shows the distribution of carbon sequestration potential in the main urban area of Xiangyang in the next 1, 2, and 5 years. Fig. 7.10 (a) shows that the forest vegetation carbon sink in the main urban area of Xiangyang was 32,700 tons in 2019-2020, while Fig. 7.10 (b, c) show that the forest carbon sink in Xiangyang was 64,100 tons in 2019-2021 and 169,200 tons in 2019-2024, respectively; Table 7.3 lists the detailed potential for enhancing the convergence of four dominant tree species in the main urban area of Xiangyang, and Table 7.4 lists the potential for enhancing the convergence of various administrative divisions in the main urban area of Xiangyang.

7.4.2 Timely updating of forest carbon sinks based on revisiting satellite imagery

By using carbon storage models and predictive variables in natural forest growth areas, forest carbon sinks can be accurately estimated [25–32]. For forest change areas caused by human logging or natural factors at any given time, a carbon sink update method based on revisit

Table 7.4: The potential of administrative divisions in Xiangyang (10^4 tons).

District	Xiangcheng	Fancheng	Xiangzhou
1 Year	2.26	0.33	0.68
2 Years	4.63	0.63	1.15
5 Years	12.96	1.49	2.47

satellite imagery can be used. Forest change can be divided into two scenarios: forest to nonforest and nonforest to forest. In the first scenario, the forest carbon storage and carbon sink in the area are directly reduced to zero; The second scenario requires timely acquisition of the tree species, tree height, and canopy closure of the forest in the changing area, which can be incorporated into the carbon storage model calculation. Due to the limited number of onboard laser sensors, laser point distribution, and service life, utilizing revisit satellite images to obtain tree species, tree height, and canopy closure can achieve efficient and low-cost renewal of regional forest carbon sinks. The specific steps are as follows:

(1) Nonforest to forest conversion is generally a natural or artificially planted forest. Deep learning networks are used to identify forest tree species and determine the corresponding forest carbon storage model parameters based on the tree species.

(2) A naturally formed forest, using revisit remote sensing images to record the time of forest changes t', combines $H = f(t)$ with the corresponding tree species to obtain the forest height $H_{t+\Delta t}$ at the time $t + \Delta t$:

$$H_{t+\Delta t} = f\left(t + \Delta t - t'\right) \tag{7.8}$$

(3) The forest planted by artificial afforestation has a relatively regular distribution of individual trees. By utilizing the timing of artificial afforestation of different tree species (age of t''), combined with the forest change time t', the forest height $H_{t+\Delta t}$ at the time $t + \Delta t$ can be obtained from $H = f(t)$:

$$H_{t+\Delta t} = f\left(t + \Delta t + t'' - t'\right) \tag{7.9}$$

(4) During the process of transforming nonforest into forest, the forest canopy density gradually increases. By using remote sensing images to obtain the canopy density $F_{t'}$ at the time of change t', the forest canopy density $F_{t+\Delta t}$ at that time $t + \Delta t$ can be calculated using the following formula:

$$F_{t+\Delta t} = F_{t'} \cdot \frac{v\left(H_{t+\Delta t}\right)^2}{v\left(H_{t'}\right)^2} \tag{7.10}$$

(5) Incorporate $H_{t+\Delta t}$ and $F_{t+\Delta t}$ into the forest carbon storage model to calculate the carbon storage of the forest change area at any given time $t + \Delta t$. For newly added forest areas, carbon storage is the forest carbon sink, which completes the carbon sink renewal of the newly added forest areas.

Based on carbon storage models and multisource remote sensing data, dynamic monitoring of regional forest carbon sinks can be achieved, analyzing the spatiotemporal carbon sequestration patterns of different tree species under spatial distribution and long-term conditions, and planning afforestation for the carbon trading market in a reasonable manner [28–32].

7.5 Conclusion

We take "forest plot carbon storage – regional forest carbon storage – regional forest carbon sink" as the research mainline. The expansion from plot to region is a spatial scale, and the extension from carbon storage to carbon sink is a temporal scale. Based on this, a complete forest vegetation carbon measurement method is constructed, and the proposed technical system can be summarized as: (1) calculation of carbon storage in single tree level forest plots; (2) pixel level forest carbon storage calculation; and (3) accurate prediction of regional forest carbon sinks. By utilizing the advantages of ground stations and unmanned aerial vehicle (UAV) LiDAR in both understory and canopy areas, the breast height diameter and tree height parameters of individual trees are extracted by integrating the two types of data. The biomass of individual trees is obtained by using the allometric growth equation, and the carbon storage of forest plots is calculated accordingly. The carbon storage density of forests is closely related to height. A pixel level forest carbon storage explicit calculation model based on remote sensing is established using two core variables, forest canopy density and height. The model parameters are automatically calculated from the single tree structure equation (diameter at breast height, tree height, crown width). Future forest canopy closure and height are derived using the single tree structure equation and single tree growth equation (tree height-age), and combined with pixel level forest carbon storage models and the latest remote sensing data to calculate and update forest carbon sinks in natural growth and change areas pixel by pixel.

References

[1] Z.L. Ding, Research on China's carbon neutralization framework roadmap, in: The 7th Annual Academic Conference of the Academic Department of the Chinese Academy of Sciences, Beijing, Academic Department of the Chinese Academy of Sciences, 2021.

[2] J.Y. Fang, Z.D. Guo, S.L. Piao, et al., Estimation of carbon sinks of terrestrial vegetation in China from 1981 to 2000, Science in China. Series D, Earth Sciences 37 (6) (2007) 804–812.

[3] Deren Li, Changwei Wang, Yueming Hu, Shuguang Liu, General review on remote sensing-based biomass estimation, Geomatics and Information Science of Wuhan University 37 (6) (2012) 631–635.
[4] J.Y. Gong, Tackling key technologies and paving the way for future development, Young Teacher Training Plan, Wuhan, 2021.
[5] X. Jiang, G. Li, D. Lu, et al., Stratification-based forest aboveground biomass estimation in a subtropical region using airborne Lidar data, Remote Sensing 12 (7) (2020) 1101.
[6] V. Kankare, et al., Individual tree biomass estimation using terrestrial laser scanning, ISPRS Journal of Photogrammetry and Remote Sensing 75 (2013) 64–75.
[7] V. Kankare, et al., Single tree biomass modelling using airborne laser scanning, ISPRS Journal of Photogrammetry and Remote Sensing 85 (2013) 66–73.
[8] D. Timothy, M. Onisimo, S. Cletah, et al., Remote sensing of aboveground forest biomass: a review, Tropical Ecology 57 (2) (2016) 125–132.
[9] G.M. Foody, D.S. Boyd, M.E.J. Cutler, Predictive relations of tropical forest biomass from Landsat TM data and their transferability between regions, Remote Sensing of Environment 85 (2003) 463–474.
[10] S. Puliti, J. Breidenbach, J. Schumacher, et al., Above-ground biomass change estimation using national forest inventory data with Sentinel-2 and Landsat, Remote Sensing of Environment 265 (2021) 112644.
[11] S.O. Los, Analysis of trends in fused AVHRR and MODIS NDVI data for 1982-2006: indication for a CO2 fertilization effect in global vegetation, Global Biogeochemical Cycles 27 (2) (2013) 318–330.
[12] V. Karthik, B.V. Bhaskar, S. Ramachandran, et al., Quantification of organic carbon and black carbon emissions, distribution, and carbon variation in diverse vegetative ecosystems across India, Environmental Pollution 309 (2022) 119790.
[13] Y. Gou, C.M. Ryan, J. Reiche, Large area aboveground biomass and carbon stock mapping in Woodlands in Mozambique with L-band radar: improving accuracy by accounting for soil moisture effects using the water cloud model, Remote Sensing 14 (2) (2022) 404.
[14] D. Lu, Q. Chen, G. Wang, et al., A survey of remote sensing-based aboveground biomass estimation methods in forest ecosystems, International Journal of Digital Earth 9 (1) (2016) 63–105.
[15] D. Ehlers, C. Wang, J. Coulston, et al., Mapping forest aboveground biomass using multisource remotely sensed data, Remote Sensing 14 (5) (2022) 1115.
[16] F. Yan, B. Wu, Y. Wang, Estimating aboveground biomass in Mu Us Sandy Land using Landsat spectral derived vegetation indices over the past 30 years, Journal of Arid Land 5 (4) (2013) 521–530.
[17] S. Piao, J. Fang, B. Zhu, K. Tan, Forest biomass carbon stocks in China over the past 2 decades: estimation based on integrated inventory and satellite data, Journal of Geophysical Research: Biogeosciences 110 (G1) (2005).
[18] D. Lu, M. Batistella, Exploring TM image texture and its relationships with biomass estimation in Rondonia, Brazilian Amazon, Acta Amazônica 35 (2) (2005) 249–257.
[19] E. Nasset, Predicting forest stand characteristics with airborne scanning laser using a practical two-stage procedure and field data, Remote Sensing of Environment 80 (2002) 88–99.
[20] K. Zhao, J.C. Suarez, M. Garcia, et al., Utility of multitemporal Lidar for forest and carbon monitoring: tree growth, biomass dynamics, and carbon flux, Remote Sensing of Environment 204 (2018) 883–897.
[21] Y. Fu, H.S. He, T.J. Hawbaker, et al., Evaluating k-nearest neighbor (kNN) imputation models for species-level aboveground forest biomass mapping in northeast China, Remote Sensing 11 (17) (2019) 2005.
[22] S. Purohit, S.P. Aggarwal, et al., Estimation of forest aboveground biomass using combination of Landsat 8 and Sentinel-1A data with random forest regression algorithm in Himalayan Foothills, Tropical Ecology 62 (2) (2021) 288–300.
[23] W. Stumer, B. Kenter, M. Kohl, Spatial interpolation of in situ data by self-organizing map algorithms (neural networks) for the assessment of carbon stocks in European forests, Forest Ecology and Management 260 (3) (2010) 287–293.
[24] A. Safari, H. Sohrabi, S. Powell, et al., A comparative assessment of multi-temporal Landsat 8 and machine learning algorithms for estimating aboveground carbon stock in coppice oak forests, International Journal of Remote Sensing 38 (22) (2017) 6407–6432.

[25] M.D. Hurteau, Quantifying the carbon balance of forest restoration and wildfire under projected climate in the fire-prone southwestern US, PLoS ONE 12 (2017) e0169275.

[26] A.M.I. Kallio, et al., Forests in the Finnish low carbon scenarios, Journal of Forest Economics 23 (2016) 45–62.

[27] A.B. Reinmann, et al., Assessing the global warming potential of human settlement expansion in a mesic temperate landscape from 2005 to 2050, Science of the Total Environment 545 (2016) 512–524.

[28] G.R. Yu, Q.F. Wang, Y.C. Liu, et al., Conceptual framework of carbon sequestration rate and potential increment of carbon sink of regional terrestrial ecosystem and scientific basis for quantitative carbon authentification, Progress in Geography 30 (7) (2011) 771–787.

[29] J.Q. Ren, J.Y. Xia, Prediction of forest carbon sink potential in Heilongjiang province: the carbon density-age relationship-based approach, Research of Environmental Sciences 30 (4) (2017) 552–558.

[30] Y.C. Liu, X.L. Gao, C. Fu, et al., Estimation of carbon sequestration potential of forest biomass in China based on National Forest Resources Inventory, Acta Ecologica Sinica 39 (11) (2019) 4002–4010.

[31] Q.B. Wu, X.K. Wang, X.N. Duan, et al., Carbon sequestration and its potential by forest ecosystems in China, Acta Ecologica Sinica 28 (2008) 517–524.

[32] H.K. Li, Research advance of forest carbon sink assessment methods and carbon sequestration potential estimation under carbon neutral vision, Geological Survey of China 8 (4) (2021) 79–86.

[33] H.K. Li, Y.C. Lei, Estimation and Evaluation of Forest Biomass Carbon Storage in China, China Forestry Publishing House, Beijing, 2010.

[34] N.N. Zhu, B.S. Yang, Z. Dong, The explicit model of forest carbon storage based on remote sensing, Acta Geodaetica et Cartographica Sinica 53 (1) (2024) 36–49.

[35] M. Hauglin, V. Lien, E. Nasset, T. Gobakken, Geo-referencing forest field plots by coregistration of terrestrial and airborne laser scanning data, International Journal of Remote Sensing 35 (9) (2014) 3135–3149.

[36] P. Polewski, A. Ericksonc, W. Yao, N. Coops, U. Stilla, Object based co-registration of terrestrial photogrammetric and ALS point clouds in forested areas, ISPRS Annals of the Photogrammetry, Remote Sensing and Spatial Information Sciences (2016) 347–354.

[37] C. Paris, D. Kelbe, J.V. Aardt, L. Bruzzone, A novel automatic method for the fusion of ALS and TLS LiDAR data for robust assessment of tree crown structure, IEEE Transactions on Geoscience and Remote Sensing 55 (7) (2017) 3679–3693.

[38] J.T. Yang, Z. Kang, S. Cheng, Z. Yang, P. Akwensi, An individual tree segmentation method based on watershed algorithm and 3D spatial distribution analysis from airborne LiDAR point clouds, IEEE Journal of Selected Topics in Applied Earth Observations and Remote Sensing 13 (2020) 1055–1067.

[39] L. Geng, M.Z. Li, W.Y. Fan, B. Wang, Individual tree structure parameters and effective crown of the stand extraction based on airborne LiDAR data, Scientia Silvae Sinicae 54 (7) (2018) 53–64.

[40] W.W. Wang, Y. Pang, L.M. Du, Z.J. Zhang, X.J. Liang, Individual tree segmentation for airborne LiDAR point cloud data using spectral clustering and supervoxel-based algorithm, National Remote Sensing Bulletin 26 (8) (2022) 1650–1661.

[41] H.M. Qin, W.Q. Zhou, Y. Yao, W.M. Wang, Individual tree segmentation and tree species classification in subtropical broadleaf forests using UAV-based LiDAR, hyperspectral, and ultrahigh-resolution RGB data, Remote Sensing of Environment 280 (2022) 113190.

[42] T. Jiang, S. Liu, Q. Zhang, X. Xu, J. Sun, Y. Wang, Segmentation of individual trees in urban MLS point clouds using a deep learning framework based on cylindrical convolution network, International Journal of Applied Earth Observation and Geoinformation 123 (2023) 103473.

[43] W.B. Li, J.H. Li, Application of UAV high-density point cloud for individual tree segmentation in high canopy density artificial forests, Journal of Northeast Forestry University 51 (9) (2023) 76–82.

[44] Z.K. Yang, Y.J. Su, W.K. Li, K. Cheng, H.C. Guan, Y. Ren, T.Y. Hu, G.C. Xu, Q.H. Guo, Segmenting individual trees from terrestrial LiDAR data using tree branch directivity, IEEE Journal of Selected Topics in Applied Earth Observations and Remote Sensing 17 (2024) 956–969.

[45] X.Q. Meng, J.L. Li, W. Hu, M.J. Tian, C.T. Ma, R.R. Wang, Single wood extraction method combining LiDAR data and spectral images, Transactions of the Chinese Society for Agricultural Machinery 55 (1) (2024) 203–211.

[46] G.Y. Zhou, G.C. Yin, X.L. Tang, Carbon Storage-Biomass Equation of Forest Ecosystem in China, Science Press, Beijing, 2018.

[47] J.Y. Fang, J.X. Zhu, Carbon Budgets of Forest Ecosystems in China, Science Press, Beijing, 2021.

[48] W. Li, Z. Niu, R. Shang, Y.C. Qin, L. Wang, H.Y. Chen, High-resolution mapping of forest canopy height using machine learning by coupling ICESat-2 LiDAR with Sentinel-1, Sentinel-2 and Landsat-8 data, International Journal of Applied Earth Observation and Geoinformation 92 (2020) 102163.

[49] L.L. Narine, S.C. Popescu, L. Malambo, Synergy of ICESat-2 and Landsat for mapping forest aboveground biomass with deep learning, Remote Sensing 11 (12) (2019) 1503.

[50] P. Potapov, X.Y. Li, A. Hernandez-Serna, A. Tyukavina, M.C. Hansen, A. Kommareddy, A. Pickens, S. Turubanova, H. Tang, C.E. Silva, J. Armston, R. Dubayah, J.B. Blair, M. Hofton, Mapping global forest canopy height through integration of GEDI and Landsat data, Remote Sensing of Environment 253 (2021) 112165.

[51] C.A. Silva, L. Duncanson, S. Hancock, A. Neuenschwander, N. Thomas, M. Hofton, L. Fatoyinbo, M. Simard, C.Z. Marshak, J. Armston, S. Lutchke, R. Dubayah, Fusing simulated GEDI, ICESat-2 and NISAR data for regional aboveground biomass mapping, Remote Sensing of Environment 253 (2021) 112234.

[52] Y. Liu, W.S. Gong, Y.Q. Xing, X.Y. Hu, J.Y. Gong, Estimation of the forest stand mean height and aboveground biomass in Northeast China using SAR Sentinel-1B, multispectral Sentinel-2A, and DEM imagery, ISPRS Journal of Photogrammetry and Remote Sensing 151 (2019) 277–289.

[53] N.N. Zhu, B.S. Yang, Z. Dong, A novel framework for forest vegetation carbon stock estimation based on remote sensing, National Remote Sensing Bulletin 28 (1) (2025) 134–146, https://doi.org/10.11834/jrs.20244210.

CHAPTER 8

Applications in disaster monitoring

From cross-modal 2D mapping to 3D disaster scene understanding

Hongruixuan Chen[a,b], Jian Song[b], Junshi Xia[b], and Naoto Yokoya[a,b]

[a]Graduate School of Frontier Sciences, The University of Tokyo, Kashiwa, Japan [b]RIKEN Center for Advanced Intelligence Project (AIP), RIKEN, Tokyo, Japan

8.1 Introduction

Natural and human-induced disasters are growing in frequency, complexity, and spatial heterogeneity, placing increasing demands on Earth Observation (EO)-based response. In the first hours to days after an event, decision-makers require actionable information at multiple levels: *where* affected areas are located, *which* assets have been damaged, and *how severely* the built environment has changed. Meeting these needs simultaneously remains a significant challenge for the remote sensing community.

A fundamental obstacle is the mismatch between sensing conditions and response requirements. Optical imagery provides rich texture and semantic detail yet is frequently degraded by clouds, smoke, haze, or nighttime conditions, precisely when rapid assessment is most critical. Synthetic Aperture Radar (SAR), by contrast, offers all-weather, day-and-night observation with strong structural sensitivity, but its complex scattering signatures make interpretation difficult, and SAR remains underutilized in fine-grained disaster analytics. Single-modality pipelines therefore struggle to deliver both robustness and semantic detail simultaneously, motivating a multimodal perspective in which optical and SAR observations serve as complementary evidence sources across the disaster timeline.

Against this background, this chapter adopts the perspective *from cross-modal 2D mapping to 3D disaster scene understanding*. Rather than viewing disaster mapping as a single-task problem, we formulate it as a hierarchical inference process with three progressively richer levels:

- **All-weather 2D scene perception** for rapid situational awareness and scene-state transition analysis.

Multimodal Remote Sensing Fusion and Classification
https://doi.org/10.1016/B978-0-44-329152-4.00015-5

- **Cross-modal 2D damage mapping** for fine-grained, building-level impact grading.
- **Synthetic-data-driven 3D understanding** for structural impact estimation and scalable adaptation to unseen events and geographies.

To operationalize this formulation, we build the chapter around three complementary data regimes and associated methodological paradigms.

First, we explore the foundational role of large-scale, very-high-resolution (VHR) SAR semantic segmentation, utilizing the OpenEarthMap-SAR benchmark [1]. By providing broad geographic diversity and fine spatial detail, this regime enables robust pre- and postevent semantic comparisons. It facilitates critical early-stage tasks such as inundation mapping, land-cover state transition analysis, and rapid candidate-area screening under adverse atmospheric conditions. Ultimately, this level addresses the most urgent operational query: *what has changed at the scene scale, and where should responders prioritize their focus?*

Second, we address the subsequent operational requirement of high-value asset assessment using the Bright dataset [2]. This regime tackles building-level damage grading under real-world response constraints by fusing VHR optical and SAR observations. By focusing on disaster-relevant categories, Bright bridges the conceptual gap between coarse scene-change cues and actionable, object-level damage grading. This cross-modal fusion is an operational necessity for resilient response, particularly when postevent optical imagery is obscured but SAR remains accessible. In practical terms, this level addresses the critical localized query: *which specific buildings have been damaged, and what is the exact severity of the structural impact?*

Third, we extend the analytical scope beyond 2D mapping by introducing synthetic-data-driven 3D disaster scene understanding, demonstrated through SynRS3D [3]. This paradigm directly confronts two persistent barriers in disaster AI: annotation scarcity and geographic distribution shift. By providing synthetic supervision equipped with semantic labels, building change cues, and precise height-aware signals, it becomes possible to learn transferable geometric and semantic priors. Coupled with synthetic-to-real domain adaptation, this approach scales 3D structural impact analysis across diverse urban morphologies, event types, and sensor domains without the prohibitive cost of massive real-world annotation. Ultimately, this level addresses the fundamental challenge of global scalability: *how can we reliably estimate 3D structural changes across unseen cities and disaster events without exhaustive ground-truth data?*

Based on these three regimes, the chapter presents a unified deployment pipeline. Conceptually, the workflow starts from all-weather SAR scene parsing for early situational awareness, proceeds to optical–SAR building damage mapping for high-value asset assessment, and is further strengthened by synthetic-data-driven 3D learning for robustness and transferability. This integrated view emphasizes that operational disaster intelligence requires not a single model, but a coordinated stack of multimodal capabilities.

The remainder of this chapter is organized as follows. Section 8.2 introduces all-weather 2D perception through large-scale SAR semantic segmentation. Section 8.3 presents cross-modal 2D mapping with optical–SAR fusion for building damage grading. Section 8.4 moves beyond 2D mapping to synthetic-data-driven 3D disaster scene understanding. Section 8.5 summarizes key takeaways and future perspectives.

8.2 All-weather scene parsing for disaster monitoring

8.2.1 The transition to all-weather scene parsing

As established in the chapter introduction, operational disaster mapping is a hierarchical process that begins with macroscopic situational awareness. Before responders can assess damage to individual buildings, they must first understand the broader environmental context such as identifying the extent of flooded regions, mapping burned landscapes, or locating critical infrastructure. This foundational "Stage-0" requirement is framed as large-scale semantic scene parsing.

The methodology for remote sensing scene parsing has advanced rapidly, driven by the proliferation of deep learning architectures and massive annotated datasets. Historically, high-performing segmentation models have relied heavily on VHR optical imagery due to its rich spectral information and easily interpretable textures. However, in the context of emergency response, this optical-centric paradigm suffers from a critical vulnerability known as the "clear-sky bias".

However, during and immediately following severe disaster events such as hurricanes, widespread flooding, or forest fires, optical observations are frequently severely degraded by dense cloud cover, heavy rainfall, smoke, and low-light conditions. Consequently, optical-only pipelines often fail precisely when operational situational awareness is most urgently needed. SAR offers a vital solution to this bottleneck. With its cloud-penetrating and day-and-night imaging capabilities, SAR provides complementary structural information that remains highly reliable under adverse atmospheric conditions.

To establish a foundation for all-weather disaster monitoring, this section introduces the first stage of our hierarchical response pipeline: macroscopic scene parsing. We extend the widely used OpenEarthMap (OEM) dataset [4] to OEM-SAR [1], a unified, VHR optical–SAR benchmark designed explicitly for cross-modality evaluation and robust multimodal semantic segmentation. In a disaster context, identifying foundational land-cover classes such as distinguishing waterways from flooded roads or mapping the extent of built-up areas under heavy cloud cover, is the critical first step before fine-grained damage grading can occur.

8.2.2 OEM-SAR: *extending OEM to all-weather conditions*

8.2.2.1 Geographic diversity

OEM-SAR [1] transitions the optical-centric OEM [4] into the all-weather domain by incorporating VHR SAR imagery across geographically diverse environments. The dataset strategically encompasses 35 distinct regions spanning Japan, France, and the United States. These regions cover a wide spectrum of topographies, including dense urban cores, suburban developments, agricultural parcels, forested landscapes, and complex coastal zones.

The selection of these 35 regions was strictly guided by the need for data distribution diversity and statistical representativeness. By deliberately sampling environmentally distinct areas, OEM-SAR ensures broad coverage of varying land-cover compositions, architectural morphologies, and terrain-induced SAR backscattering variations. This spatial dispersion is critical for promoting model generalization. It actively discourages algorithms from overfitting to localized appearance priors, forcing them instead to learn invariant semantic and structural representations that can reliably transfer to unseen disaster zones.

8.2.2.2 Technical structure and hybrid supervision

From a technical perspective, OEM-SAR is structured as coregistered multimodal triplets: postprocessed SAR imagery, spatially aligned optical RGB imagery, and pixel-level semantic reference maps. The dataset comprises 5033 SAR tiles with 1024 $\times$ 1024 pixels and high spatial resolution with a Ground Sampling Distance (GSD) ranging from 0.15 to 0.5 meters. This amounts to approximately 1.5 million annotated segments, categorized into eight harmonized land-cover classes consistent with the original OEM taxonomy [4]. To balance large-scale scalability with high annotation fidelity, OEM-SAR employs a hybrid supervision strategy. First, to ensure broad geographic coverage, the dataset leverages large-scale pseudolabels generated from pretrained optical segmentation models and transferred to the spatially aligned SAR domain. Second, to guarantee rigorous evaluation and benchmarking, geographically distributed subsets of these labels are manually verified to provide high-quality reference ground truth, as shown in Fig. 8.1.

This large-scale, distribution-aware design enables deep multimodal architectures to learn complex mappings between optical reflectance patterns and SAR scattering responses across diverse global landscapes. Ultimately, OEM-SAR establishes a statistically robust foundation for all-weather semantic segmentation and sensor-agnostic geospatial learning, fulfilling the crucial "Stage-0" requirement for macroscopic disaster screening.

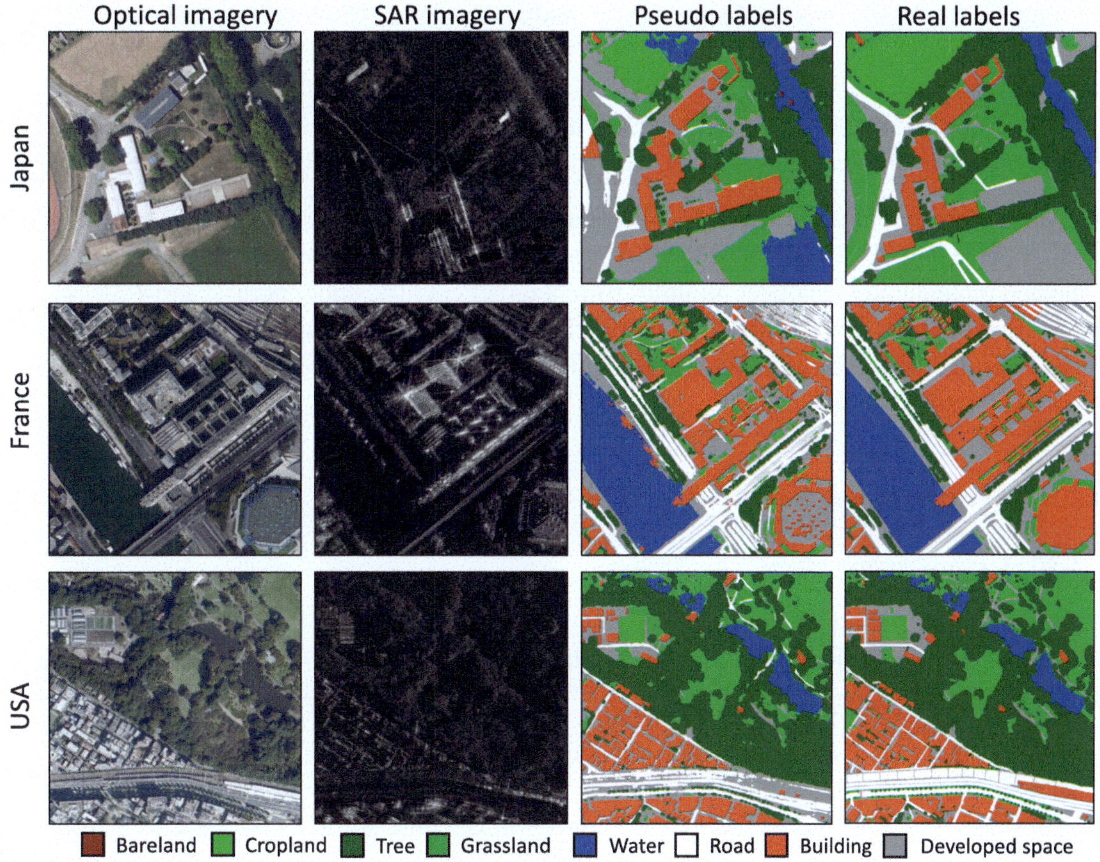

Figure 8.1:
Examples of paired VHR optical and SAR images, pseudolabels and real labels in the OpenEarthMap-SAR dataset.

8.2.2.3 All-weather multimodal experimental design

To evaluate the all-weather multimodal learning capabilities of models trained on OEM-SAR, we design experimental setups across two primary task categories: semantic segmentation and Unsupervised Domain Adaptation (UDA).

For semantic segmentation, we evaluate three modality configurations: SAR-only, optical-only, and SAR+optical fusion. These configurations correspond to distinct operational regimes: SAR-only reflects the worst-case scenario where postevent optical observations are unavailable, optical-only serves as a clear-sky upper reference, and SAR+optical fusion represents an idealized benchmark in which both modalities are simultaneously accessible. This enables a controlled comparison of structural, spectral, and complementary feature learning. To assess the impact of supervision quality, we define five training scenarios: pseudolabels only ($\mathcal{P}$), pseudolabels with one real label per region ($\mathcal{P} + \mathcal{R}_1$), pseudolabels with five real labels per region ($\mathcal{P} + \mathcal{R}_5$), one real label per region only ($\mathcal{R}_1$), and five real labels per region only ($\mathcal{R}_5$). Models are evaluated on a held-out set of manually annotated images using mean Intersection-over-Union (mIoU) and class-wise IoU.

To quantify geographic distribution shifts, we construct country-wise UDA settings across Japan, France, and the United States. Models are trained on data from a source country and evaluated on manually annotated samples from a target country, under both $\mathcal{P}$ and $\mathcal{P} + \mathcal{R}_5$ supervision configurations. This setup explicitly measures domain gaps induced by variations in urban morphology, seasonal conditions, and imaging geometry, providing a benchmark for operational transferability.

8.2.3 Benchmark results

8.2.3.1 Semantic segmentation

We systematically evaluate baseline segmentation models, including U-Net [5], SegFormer [6], and VMamba [7], across the three modalities and five supervision configurations. The results are shown in Fig. 8.2 and Fig. 8.3.

Across all supervision scenarios, VMamba and SegFormer consistently outperform U-Net. This demonstrates the empirical superiority of modern global-context architectures, Transformers and state-space models, over conventional convolutional baselines. Regarding data modalities, SAR exhibits the lowest standalone performance, while optical imagery remains the strongest single modality due to its rich spectral and textural cues. However, when both modalities are simultaneously available, multimodal fusion (SAR+optical) achieves the highest overall accuracy. For advanced architectures like VMamba and SegFormer, fusion effectively closes and occasionally surpasses the performance gap relative to the optical-only baseline, highlighting the strong complementarity between SAR backscatter signals and optical reflectance. In adverse atmospheric conditions where postevent optical is degraded, however, the SAR-only performance becomes the operationally relevant indicator.

Analyzing the supervision strategies reveals that while pure pseudolabel training ($\mathcal{P}$) provides a stable baseline, it fails to fully exploit the models' representational capacity. Introducing even minimal human annotation ($\mathcal{R}_5$) yields substantial mIoU gains across all architectures. This confirms that high-quality human annotations are indispensable for refining decision boundaries and mitigating pseudolabel noise.

Class-wise IoU analysis further exposes modality-dependent characteristics. Optical data dominates in structurally defined classes (e.g., *Buildings* and *Roads*) where color

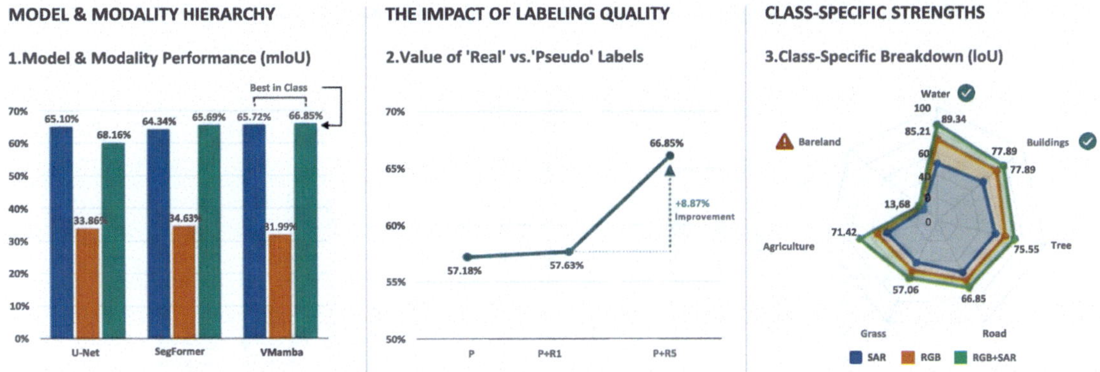

Figure 8.2:
Comparative classification performance under different modalities and labeling configurations.

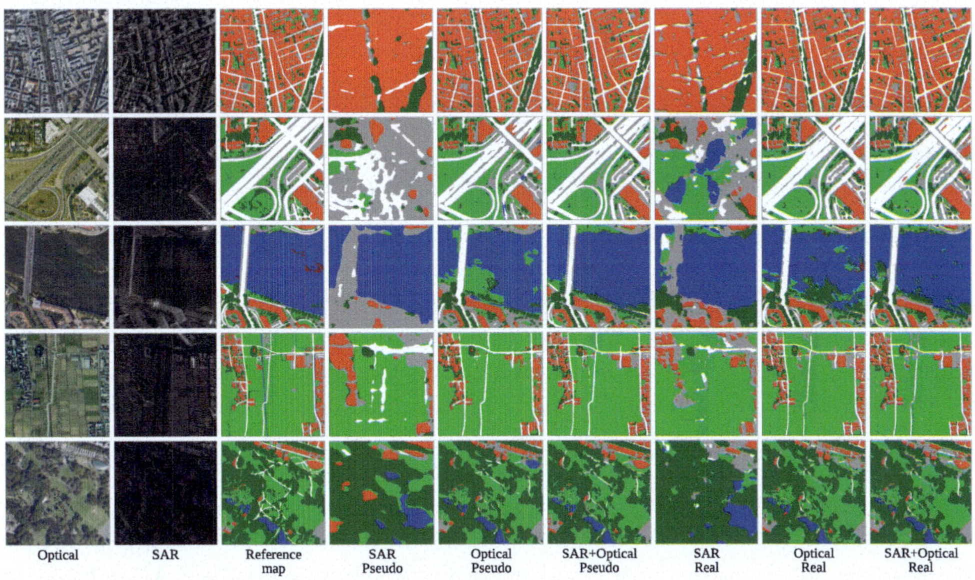

Figure 8.3:
Visual comparison of land-cover semantic segmentation results based on U-Net with three modalities from pseudolabels ($\mathcal{P}$) and real labels ($\mathcal{R}_5$) results.

contrast is distinct. SAR struggles in these categories due to speckle noise and ambiguous scattering signatures among geometrically similar surfaces. Nevertheless, multimodal fusion significantly improves performance in vegetation and water categories (*Trees*, *Agriculture*,

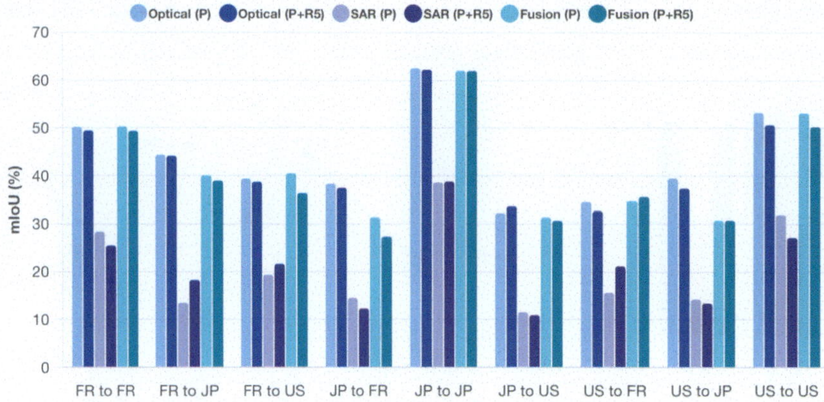

Figure 8.4:
Country-wise UDA performance (mIoU) of DAFormer under different modalities and supervision settings. The results compare SAR, Optical, and Fusion (SAR+Optical) across two labeling configurations: pseudolabels only ($\mathcal{P}$) and pseudolabels augmented with five real labels per region ($\mathcal{P} + \mathcal{R}_5$).

and *Water*), as SAR's sensitivity to surface roughness and moisture directly complements optical texture.

8.2.3.2 Unsupervised domain adaptation

To evaluate operational transferability, we investigate the country-wise domain gap in OEM-SAR. Models trained in one geographic region frequently experience severe distribution shifts when deployed elsewhere due to variations in architectural styles and land-cover composition. Using DAFormer [8] as a strong baseline, we evaluate country-to-country transfer across the three dataset regions and show the results in Fig. 8.4.

In the "Oracle" setting (intracountry training and testing, e.g., FR→FR), DAFormer achieves peak mIoU, confirming that intradomain performance remains substantially higher than cross-domain transfer. Consistent with the standard segmentation results, SAR exhibits lower transferability compared to optical and fusion modalities. Furthermore, the performance margin between $\mathcal{P}$ and $\mathcal{P} + \mathcal{R}_5$ in the UDA scenarios is marginal, indicating that minimal local supervision (even five real labels) is insufficient to overcome massive cross-country distribution shifts.

The results reveal highly structured domain gaps. The most severe performance degradation occurs between Japan and the United States (JP→US and US→JP), reflecting drastic discrepancies in urban morphology and land-use patterns. Transfers involving Japan and France also exhibit notable gaps, particularly in the SAR modality, underscoring the

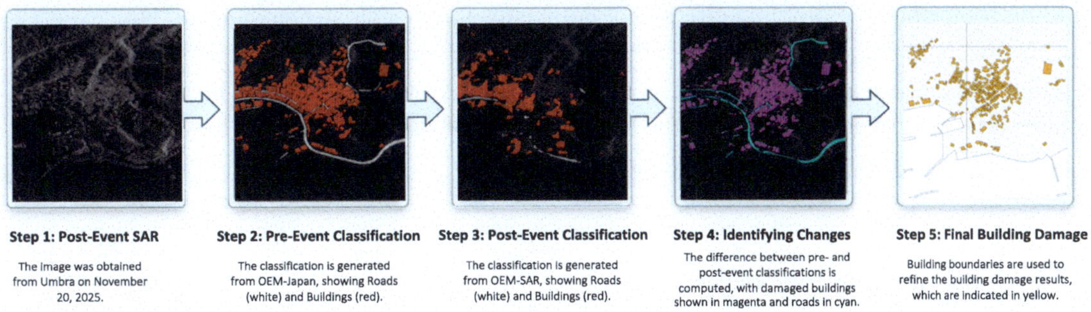

Figure 8.5:
The pipeline of all-weather disaster mapping using OEM-SAR on the 2025 Saganoseki Wildfire, Japan, where impacted buildings are rendered in magenta, affected roads are rendered in cyan, a spatially coherent structural impact map is rendered in brown.

sensitivity of radar scattering to distinct built-environment structures. Conversely, the France→USA transfer shows a relatively smaller performance drop, suggesting a closer baseline distribution alignment. Ultimately, these experiments demonstrate that while multimodal fusion improves absolute accuracy, significant intercountry domain gaps persist, necessitating more robust adaptation strategies for global deployment.

8.2.4 Deployment case study

To validate the real-world utility of OEM-SAR under extreme observation conditions, we conducted a macroscopic case study on the November 2025 Saganoseki Wildfire in Oita Prefecture, Japan. This serves as the "Stage-0" macroscopic scene parsing phase that naturally precedes the fine-grained, building-level damage grading discussed in Section 8.3.

Fig. 8.5 illustrates the complete OEM-SAR postfire screening pipeline using SAR imagery acquired during the event. A preevent baseline map derived from OEM-Japan [9] provides the structural reference, classifying buildings (rendered in red) and roads (rendered in white). The OEM-SAR framework (implemented via U-Net) is applied to the postevent SAR imagery to extract equivalent semantic features. A pixel-wise comparison between the pre- and postevent classifications isolates macroscopic structural changes, highlighting impacted buildings (rendered in magenta) and affected roads (rendered in cyan). By incorporating building footprint boundaries, the results are refined at the object level to yield a spatially coherent structural impact map (rendered in brown).

This case study proves that integrating a high-precision preevent optical baseline with postevent SAR observations enables rapid, weather-independent disaster screening. By leveraging SAR to capture coarse structural-change signals and constraining them with building footprints, the pipeline effectively delineates large-scale urban damage. However, this macroscopic framework is inherently limited by the semantic accuracy of the preevent baseline and the generalization capacity of the segmentation model. While it successfully isolates affected zones, determining the exact severity of the structural failure requires specialized cross-modal grading strategies.

8.2.5 Discussion and takeaways

This section presented OEM-SAR as the foundational "Stage-0" in our hierarchical disaster monitoring pipeline. By extending large-scale semantic segmentation into the all-weather domain, we demonstrated how pairing optical and SAR imagery effectively overcomes the "clear-sky bias" of traditional remote sensing. This capability ensures that responders can parse macroscopic land-cover transitions such as identifying burned regions regardless of adverse atmospheric conditions.

The benchmark experiments yield two critical operational insights regarding model training. First, the role of multimodal fusion is asymmetric across operational regimes. SAR guarantees observation continuity when postevent optical is degraded, and same-time SAR+optical fusion provides an upper bound when both modalities remain accessible. In practice, the most realistic deployment configuration, as illustrated in our Saganoseki case study, pairs a preevent optical baseline (typically acquired under nominal conditions) with postevent SAR, exploiting the asymmetric availability of the two modalities across the disaster timeline. Second, supervision efficiency is key. Relying entirely on pseudolabels limits the representational capacity of the models, but introducing just a minimal number of human-verified labels drastically refines decision boundaries and mitigates label noise.

Furthermore, the UDA results expose a persistent challenge in global disaster response, i.e., geographic distribution shifts. Transferring models between structurally and environmentally distinct regions causes significant performance degradation, even when utilizing multimodal inputs. This empirical finding reinforces a core theme of this chapter: building globally deployable systems requires training data with massive spatial diversity, rather than relying on local optimization.

Finally, the Oita wildfire case study illustrated how macroscopic scene parsing operates in a real-world emergency. By comparing a preevent land-cover baseline with a postevent SAR segmentation map, responders can rapidly isolate affected neighborhoods. However, this macroscopic analysis only answers the broad spatial query: *where has the hazard struck?* It does not resolve the object-level query: *how severely has each specific building failed?* This inherent operational gap perfectly motivates the next stage of our deployment stack: cross-modal, building-level damage grading, which we explore through the BRIGHT dataset in Section 8.3.

8.3 Building-level damage mapping with multimodal fusion

8.3.1 Brief survey

The methodology for building damage mapping has undergone a paradigm shift over the past two decades, moving from labor-intensive manual inspection to advanced automated frameworks. This evolution can be broadly categorized into the era of traditional hand-crafted feature engineering and the current era of deep learning-based analytics.

Early methods primarily relied on hand-crafted features and traditional machine learning classifiers, such as Support Vector Machines (SVM) and Random Forests (RF). These approaches often utilized pixel-level spectral change detection or textural metrics such as Gray-Level Cooccurrence Matrix (GLCM) to identify structural variations [10,11]. As spatial resolution improved, Object-Based Image Analysis (OBIA) became the dominant paradigm, where buildings were first segmented as polygons and then classified based on morphological changes [12]. However, these traditional methods were highly sensitive to viewing geometry, illumination variations, and complex urban backgrounds, often resulting in high false-alarm rates and limited generalization across different disaster events.

The deep learning era has fundamentally reshaped building damage assessment, moving the field from handcrafted descriptors to end-to-end representation learning. A key turning point was the release of the xBD dataset [13], which provided large-scale VHR bitemporal optical imagery with standardized building-level damage annotations and enabled systematic benchmarking across methods. At first, CNN-based models established strong baselines by learning local texture and structural change cues. Representative architectures improved over traditional approaches by jointly modeling building localization and damage classification in a unified framework [14–16]. However, CNNs are inherently limited in capturing long-range dependencies and global context, which are critical when damage patterns are spatially heterogeneous or visually subtle. Transformer-based methods were then introduced to better capture long-range dependencies and global context via self-attention, improving cross-temporal feature interaction for damage grading [17–21]. More recently, Mamba/state-space-model-based architectures [22] have emerged, aiming to retain long-range spatiotemporal modeling while reducing the computational overhead of full attention. Representative methods such as ChangeMamba [23] report improved efficiency–accuracy trade-offs and stronger robustness. Overall, the methodological trajectory from CNNs to Transformers and then to Mamba reflects a continuous effort to better model multiscale context, temporal dependency, and operational robustness in postdisaster mapping.

Despite this architectural progress, most high-performing pipelines remain predominantly optical-centric. In real emergency response, however, postevent optical observations are frequently degraded by clouds, smoke, haze, or low-light conditions, particularly during floods, hurricanes, and wildfires. This mismatch between algorithmic assumptions and operational reality motivates multimodal optical–SAR fusion for all-weather response continuity. From a data-centric perspective, early multimodal studies were often limited to localized case studies, i.e., single city or single event, which constrained statistical diversity and hindered robust generalization [11,31,32]. Subsequent larger efforts were sometimes developed within specialized institutions, but restricted release, licensing limitations, or SAR data sensitivity reduced reproducibility and comparability across studies [33]. As a result, public benchmarks that simultaneously provide coregistered VHR optical–SAR pairs, multievent coverage, and building-level damage annotations remain limited. To bridge these gaps, the BRIGHT dataset [2] was introduced as the first globally distributed, open-source multimodal benchmark for building damage mapping. Unlike its predecessors [31,33], BRIGHT provides coregistered VHR optical and SAR pairs across diverse disaster types and geographic regions. By bridging coarse change cues with actionable damage categories, it provides the necessary data foundation to train models that are both resilient to weather conditions and capable of fine-grained structural assessment. Table 8.1 lists the currently available mainstream datasets for building damage mapping research.

8.3.2 BRIGHT: *task formulation and dataset design*

In this chapter, BRIGHT [2] is used as an empirical anchor for multimodal building damage mapping under realistic emergency constraints. Instead of reintroducing it as a standalone contribution, we focus on how its design choices support the chapter theme. Specifically, BRIGHT emphasizes *(i)* all-weather compatibility through optical–SAR complementarity, *(ii)* building-centric labels aligned with response actions, and *(iii)* benchmark protocols that expose generalization gaps across events and regions.

8.3.2.1 *Task formulation*

Multimodal building damage mapping is generally formulated as a bitemporal, cross-modal prediction problem under operational constraints [2,33]. Let $\mathbf{X}_{\text{opt}}^{\text{pre}} \in \mathbb{R}^{\text{H}\times\text{W}\times\text{C}_{\text{opt}}}$ denote preevent optical imagery, which provides the high-resolution semantic baseline, and $\mathbf{X}_{\text{sar}}^{\text{post}} \in \mathbb{R}^{\text{H}\times\text{W}\times\text{C}_{\text{sar}}}$ denote postevent SAR imagery, which captures structural changes regardless of atmospheric conditions. The target is a damage map $\mathbf{Y}^{\text{dam}} \in \mathbb{R}^{\text{H}\times\text{W}\times(\text{C}_{\text{level}}+1)}$ with different damage levels, where each pixel is assigned to one of several damage levels or a background class.

Table 8.1: Existing open building damage mapping datasets. GSD is an acronym for ground sampling distance (GSD) with the unit of meter per pixel.

Dataset	Modality	GSD (m)	Events	Disaster type	Number of Buildings	Granularity
ABCD [24]	Optical EO	0.4	1	Tsunami	N/A	Image-level
DoriaNet [25]	Optical EO	N/A	1	Hurricane	1802	Image-level
FloodNet [26]	Optical EO	N/A	1	Flood	6675	Pixel-level
RescueNet [27]	Optical EO	N/A	1	Hurricane	10,903	Pixel-level
Ida-BD [28]	Optical EO	0.5	1	Hurricane	18,083	Pixel-level
CRASAR-U-Droids [29]	Optical EO	0.02–0.13	10	4 natural and 1 human-made disasters	18,780	Pixel-level
xBD [13]	Optical EO	<0.8	15	6 natural disasters	>700,000	Pixel-level
EBD [30]	Optical EO	0.3–0.5	12	4 natural disasters	>175,000	Pixel-level
QQB [31]	Optical & SAR EO	<1	1	Earthquake	4029	Pixel-level
Bright [2]	Optical & SAR EO	0.3–1	14	5 natural and 2 human-made disasters	384,596	Pixel-level

A direct formulation approaches this as a holistic segmentation task, predicting the final map in a single stage:

$$\mathbf{Y}_{2D}^{\mathrm{dam}} = \mathcal{F}_{\Theta}^{\mathrm{seg}}\left(\mathbf{X}_{\mathrm{opt}}^{\mathrm{pre}}, \mathbf{X}_{\mathrm{sar}}^{\mathrm{post}}\right), \tag{8.1}$$

where $\mathcal{F}_{\Theta}^{\mathrm{seg}}(\cdot)$ denotes a holistic segmentation model that directly transforms the multimodal inputs into damage-level categories.

In practice, many high-performing pipelines [14–17,23] decouple the task into two subtasks: (*i*) building localization and (*ii*) damage grading within building regions. This decoupled architecture can be expressed as:

$$\begin{cases} \mathbf{Y}^{\mathrm{loc}} = \mathcal{F}_{\Theta}^{\mathrm{loc}}\left(\mathbf{X}_{\mathrm{opt}}^{\mathrm{pre}}\right) \\ \mathbf{Y}^{\mathrm{clf}} = \mathcal{F}_{\Theta}^{\mathrm{clf}}\left(\mathbf{X}_{\mathrm{opt}}^{\mathrm{pre}}, \mathbf{X}_{\mathrm{sar}}^{\mathrm{post}}\right) \\ \mathbf{Y}^{\mathrm{dam}} = \mathbf{Y}^{\mathrm{loc}} \odot \mathbf{Y}^{\mathrm{clf}} \end{cases} \tag{8.2}$$

where $\mathcal{F}_{\Theta}^{\mathrm{loc}}(\cdot)$ and $\mathcal{F}_{\Theta}^{\mathrm{clf}}(\cdot)$ represent the localization and classification subnetworks, respectively, and $\odot$ denotes the masking composition that constrains damage predictions to the identified building footprints. This decoupled view is particularly useful in disaster response because it separates *where buildings are* from *how severely they are damaged*, which improves interpretability and aligns model outputs with downstream triage workflows.

In the following analysis, we consider both direct and decoupled paradigms to evaluate their efficacy in multimodal settings.

8.3.2.2 Dataset overview and event coverage

To ensure global applicability, BRIGHT provides a collection of disaster scenes that reflect high heterogeneity in both geographic distribution and disaster categories. As illustrated in Fig. 8.6, the benchmark encompasses 14 major disaster events distributed across five continents. These events cover five types of natural disasters, including earthquakes, floods, wildfires, storms, and volcanic eruptions, and two types of human-made disasters, including industrial explosions and armed conflicts. This wide coverage ensures that the dataset captures not only varied spectral signatures but also distinct structural failure modes unique to different disaster drivers, from the sudden structural collapse in earthquakes to the widespread inundation in floods.

The strategic inclusion of broad geographic diversity is critical for addressing the chapter's focus on model generalization. By incorporating a spectrum of architectural styles, ranging from high-rise concrete structures in dense urban centers to informal rural dwellings, and diverse terrain backgrounds (e.g., mountainous, coastal, and arid regions), BRIGHT forces models to learn invariant structural features rather than overfitting to local appearance priors.

From a technical perspective, the dataset consists of coregistered triplets: preevent VHR optical imagery, postevent SAR intensity data, and reference map. Specifically, BRIGHT contains 384,596 annotated buildings across 4246 tile triplets, providing a statistically significant and large-scale foundation for training multimodal AI architectures. This scale is particularly vital for the multimodal scenarios discussed in this chapter, as it allows deep networks to learn the complex relationship between optical-derived building geometry and SAR-derived scattering perturbations across a global distribution.

8.3.2.3 Damage grade definition

A core design principle of BRIGHT is maintaining labels that are operationally meaningful for emergency mapping while preserving sufficient semantic granularity for algorithmic benchmarking. Building damage is annotated into three severity levels:

- *Intact*: no visible structural damage signal.
- *Damaged*: partial structural impairment (e.g., partial roof/wall failure, clear but nontotal damage evidence).
- *Destroyed*: near-total or total structural failure, or buildings no longer structurally present.

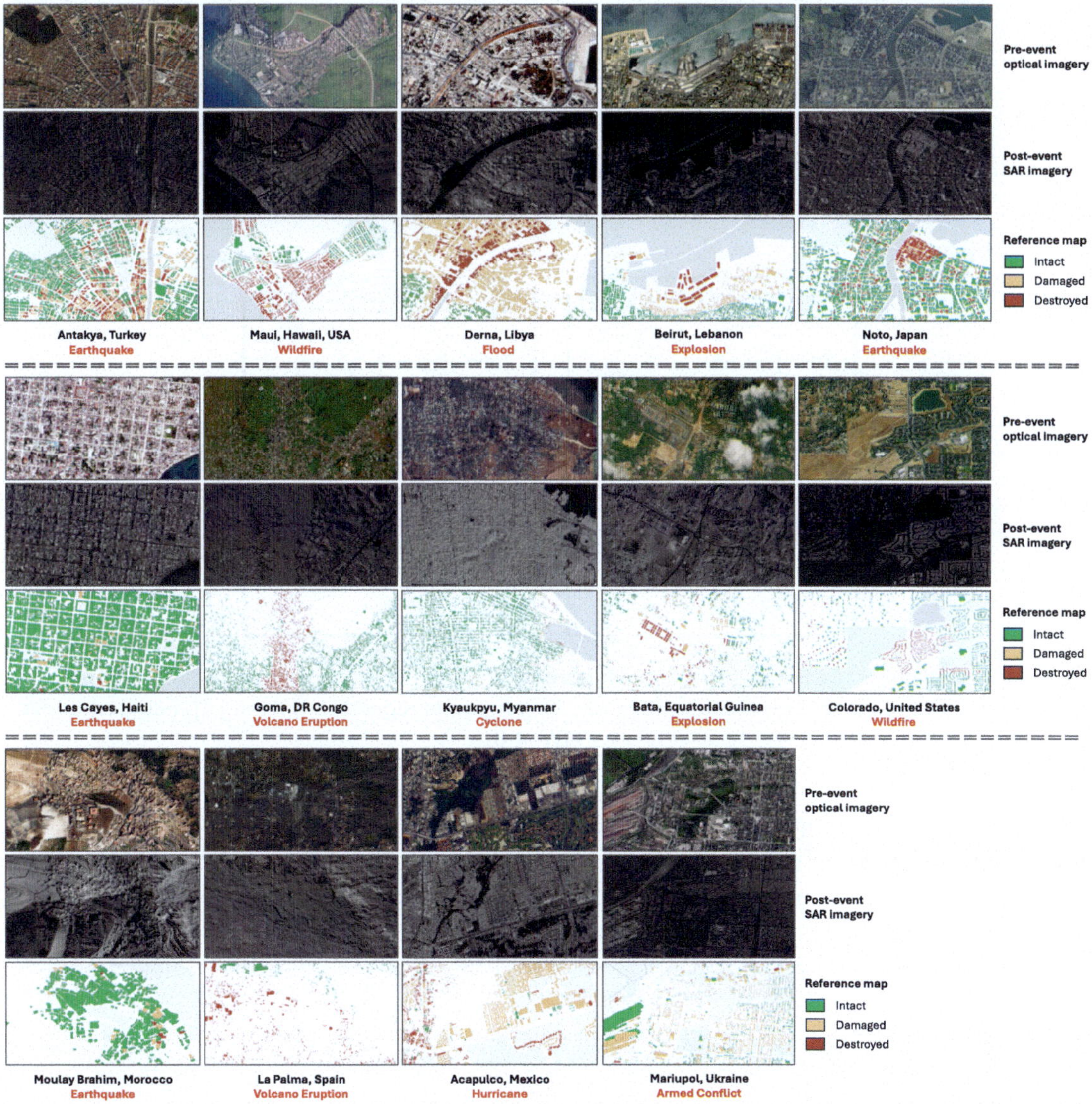

Figure 8.6:
Thumbnails of local areas in 14 disaster events in the BRIGHT dataset. This figure is reproduced from [2], licensed under CC-BY-4.0.

These levels are used together with a *Background* class for nonbuilding pixels, resulting in a four-class prediction target. Conceptually, this label system bridges coarse change detection and action-oriented assessment: the distinction between *Damaged* and *Destroyed* is particularly important for prioritizing field verification and resource allocation after major

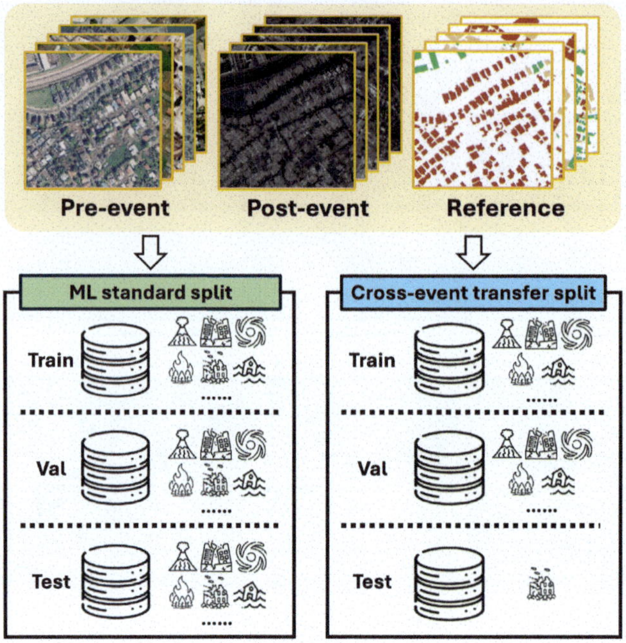

Figure 8.7:
Two kinds of dataset split protocol in BRIGHT. This figure is reproduced from [2], licensed under CC-BY-4.0.

events. For model development, this taxonomy also introduces realistic difficulty: interclass boundaries, especially *Intact vs. Damaged* in SAR data, can be subtle, event-dependent, and sensitive to viewing geometry and coregistration quality.

8.3.2.4 Split protocol for robust evaluation

A persistent failure mode in disaster-oriented AI is the "event-bias", where models overfit to the specific atmospheric conditions, sensor geometries, or architectural styles of the training scenes, leading to a catastrophic drop in performance when deployed on a new disaster. To expose these generalization gaps, BRIGHT employs a rigorous Cross-event Transfer (CeT) split protocol. As shown in Fig. 8.7, unlike random tile-wise splits, which often suffer from high spatial autocorrelation between training and test sets, the CeT protocol ensures that entire disaster events in the test set are geographically, environmentally, and temporally isolated from the training distribution.

This protocol serves as a "stress test" for multimodal fusion strategies. It evaluates whether a model has truly learned transferable cross-modal priors, such as the characteristic physical perturbation of a SAR double-bounce signal when a building collapses, or if it is merely memorizing local spectral statistics. Specifically, under the CeT protocol, we consider two operational setups:

- *Zero*-shot cross-event: The model is trained and validated on a subset of disaster events and evaluated on a completely held-out event with no target labels provided. This assesses "out-of-the-box" reliability for immediate response.
- *One*-shot cross-event: To simulate rapid adaptation under urgent response constraints, the model is allowed minimal target supervision (e.g., a single training pair and one validation pair from the new event). This evaluates the model's ability to quickly calibrate to local architectural or sensor-specific nuances.

This protocol design is central to the chapter narrative. By exposing the performance gap between in-event optimization and true operational transfer, it provides a principled basis for comparing multimodal fusion strategies. It forces a shift in focus from "leaderboard chasing" on static datasets to the development of resilient, adaptable systems capable of handling the unpredictability of real-world disaster-response scenarios.

8.3.3 Experiments

8.3.3.1 Comparison methods

To evaluate the efficacy of multimodal building damage mapping across different technological eras, this section examines a representative suite of algorithms. These methods are selected to reflect the methodological evolution from local feature learning of CNN to global context modeling of Transformer and Mamba, as well as the transition from fully supervised paradigms to unsupervised change detection, which is particularly relevant for rapid situational awareness when ground truth labels are unavailable.

To comprehensively benchmark building-level multimodal damage mapping, we include two supervised modeling paradigms that are mentioned in Section 8.3.2.1. The first paradigm directly treats damage mapping as a dense multiclass segmentation task from paired multimodal inputs in one forward pass. Representative models include classical segmentation architectures UNet [5] and DeepLabV3+ [34], and Siamese-attention segmentation variants [33]. These methods are simple and easy to deploy, but they may mix building localization and damage grading errors in a single prediction branch. The second paradigm decomposes building damage assessment into two subtasks, i.e., building localization and damage classification within building regions. In this chapter, we include representative methods spanning the architecture evolution, including SiamCRNN [35], ChangeOS [14], DamFormer [17], and ChangeMamba [23].

While BRIGHT primarily targets fine-grained building damage grading with supervision, Unsupervised Multimodal Change Detection (UMCD) approaches remain practically valuable for *rapid situational awareness* when ground-truth labels are unavailable immediately after a disaster [36–38]. Therefore, in addition to supervised models, we include UMCD methods as *rapid assessment* baselines. We consider representative UMCD families from traditional to deep learning-based approaches including *random guessing*, AGSCC [39], RIEM [40], AEKAN [41], and IM4CD [42].

8.3.3.2 Experimental setup

For supervised models, the unified training protocol in [2] is followed to ensure fair comparison across architectures. All methods are trained with the AdamW optimizer [43] with a learning rate of 1×10^{-4} and weight decay of 5×10^{-3}. The same data preprocessing, augmentation policy, and evaluation pipeline are applied to all baselines. We report results under two split settings:

- Standard Split: The conventional in-distribution setting, where training, validation, and testing samples are drawn from the same pool of disaster events.
- Cross-event Transfer (CeT) split: a stricter out-of-distribution setting where target-event samples are isolated from source-event training data. We report both zero-shot and one-shot variants to evaluate pure transferability and rapid adaptation ability, respectively.

This dual-split design provides a comprehensive assessment of both *in-event optimization* and *real-world deployment robustness.*

UMCD methods are evaluated exclusively as rapid, local-assessment baselines. Because UMCD outputs binary change cues (changed vs. unchanged), we focus on two representative disaster scenes, i.e., *Congo-Volcano-2021* and *Noto-Earthquake-2024*, to evaluate their capacity to immediately identify severely destroyed regions. This setup mirrors a highly practical use case: empowering responders to utilize unsupervised change signals to prioritize severely affected blocks for triage before formal labels are generated.

8.3.3.3 Evaluation metrics

We employ a multifaceted metric suite to capture the performance of multimodal approaches. The F_1 score is utilized to evaluate the precision and recall balance for both building localization, denoted as F_1^{loc}, and the overall damage classification, denoted as F_1^{dmg}. The Intersection over Union (IoU) is calculated for background and each damage level to assess the model's accuracy in distinguishing between varying degrees of structural failure. Overall Accuracy (OA) and mean IoU (mIoU) serve as the primary indicators of holistic model performance across all semantic categories.

For UMCD methods, identifying destroyed regions is treated as a binary classification problem. Following established protocols for binary change detection [23,42], performance is quantified using precision rate, recall rate, OA, F1 score, and IoU, and Kappa Coefficient (KC). The KC is specifically included to measure the agreement between the predicted change map and ground truth while accounting for the possibility of agreement occurring by chance, a critical factor in highly imbalanced disaster scenes where the "changed" pixels are often a minority.

8.3.3.4 Benchmark results

Table 8.2 shows that all deep models achieve strong performance under the standard ML split, while clear differences emerge across architectural generations. Among all methods, ChangeMamba yields the best overall results, with F_1^{loc} of 90.90%, F_1^{clf} of 72.70%, OA of 96.22%, and Final mIoU of 67.63%. The second-best model is DamageFormer with mIoU of 67.09%, indicating that recent long-range modeling architectures (Transformer and Mamba) consistently outperform earlier CNN-based baselines.

A consistent trend is that decoupled localization–grading pipelines outperform coupled one-branch segmentation pipelines. For example, the best coupled model UNet is clearly below the four decoupled models. Averaged over model groups, decoupled methods show higher localization and grading quality, confirming that explicitly separating "where buildings are" from "how severely they are damaged" is beneficial for building-level disaster assessment under multimodal setup.

Analyzing the class-wise IoU in Table 8.2 reveals that the *Background* and *Intact* classes are comparatively easier to resolve, with IoU of 96–97% and 71–76%, respectively. Conversely, *Damaged* remains the most challenging category, peaking at an IoU of only 40.05%. This performance gap underscores the persistent operational difficulty of distinguishing subtle, partial structural degradation from both completely intact and totally destroyed states, particularly when interpreting complex SAR backscatter. This quantitative finding is visually corroborated by the qualitative results in Fig. 8.8, which illustrates the predicted damage maps across the 14 disaster events, highlighting the varying capacities of different architectures to capture fine structural details and reduce false alarms.

Table 8.3 further reveals a substantial generalization gap between standard in-distribution evaluation and cross-event transfer. Under CeT, ChangeMamba achieves the best average mIoU in both settings: 39.38% mIoU under zero-shot and 43.23% mIoU under one-shot. Compared with the second-best averages, the gain is +1.76 points in zero-shot (vs. ChangeOS, 37.62%) and +0.74 points in one-shot (vs. DamageFormer, 42.49%). Across all methods, one-shot adaptation consistently improves performance over zero-shot, with

Table 8.2: Accuracy assessment for different DL models on the test set under the standard ML data split. The highest values are highlighted in purple, and the second-highest results are highlighted in teal. This table is reproduced from [2], licensed under CC-BY-4.0.

Method	Venue	F_1^{loc} (%)	F_1^{dmg} (%)	OA (%)	mIoU (%)	IoU per class (%)			
						Background	*Intact*	*Damaged*	*Destroyed*
UNet [5]	$MICCAI_{2015}$	87.97	72.24	95.47	64.94	96.19	71.27	39.13	53.17
DeepLabV3+ [34]	$ECCV_{2018}$	87.00	70.33	95.43	64.80	95.98	70.98	37.53	54.69
SiamAttnUNet [33]	ISPRS J $P\&RS_{2021}$	88.16	70.13	95.45	64.26	96.14	71.90	35.06	53.92
SiamCRNN [35]	IEEE $TGRS_{2020}$	89.45	72.02	95.76	65.73	96.48	73.44	38.91	54.18
ChangeOS [14]	RSE_{2021}	89.60	71.88	95.84	65.98	96.54	73.85	38.99	54.53
DamageFormer [17]	$IGARSS_{2022}$	**90.29**	**72.51**	**96.13**	**67.09**	**96.87**	**75.04**	**39.86**	**56.59**
ChangeMamba [23]	IEEE $TGRS_{2024}$	**90.90**	**72.70**	**96.22**	**67.63**	**96.96**	**75.59**	**40.05**	**57.91**

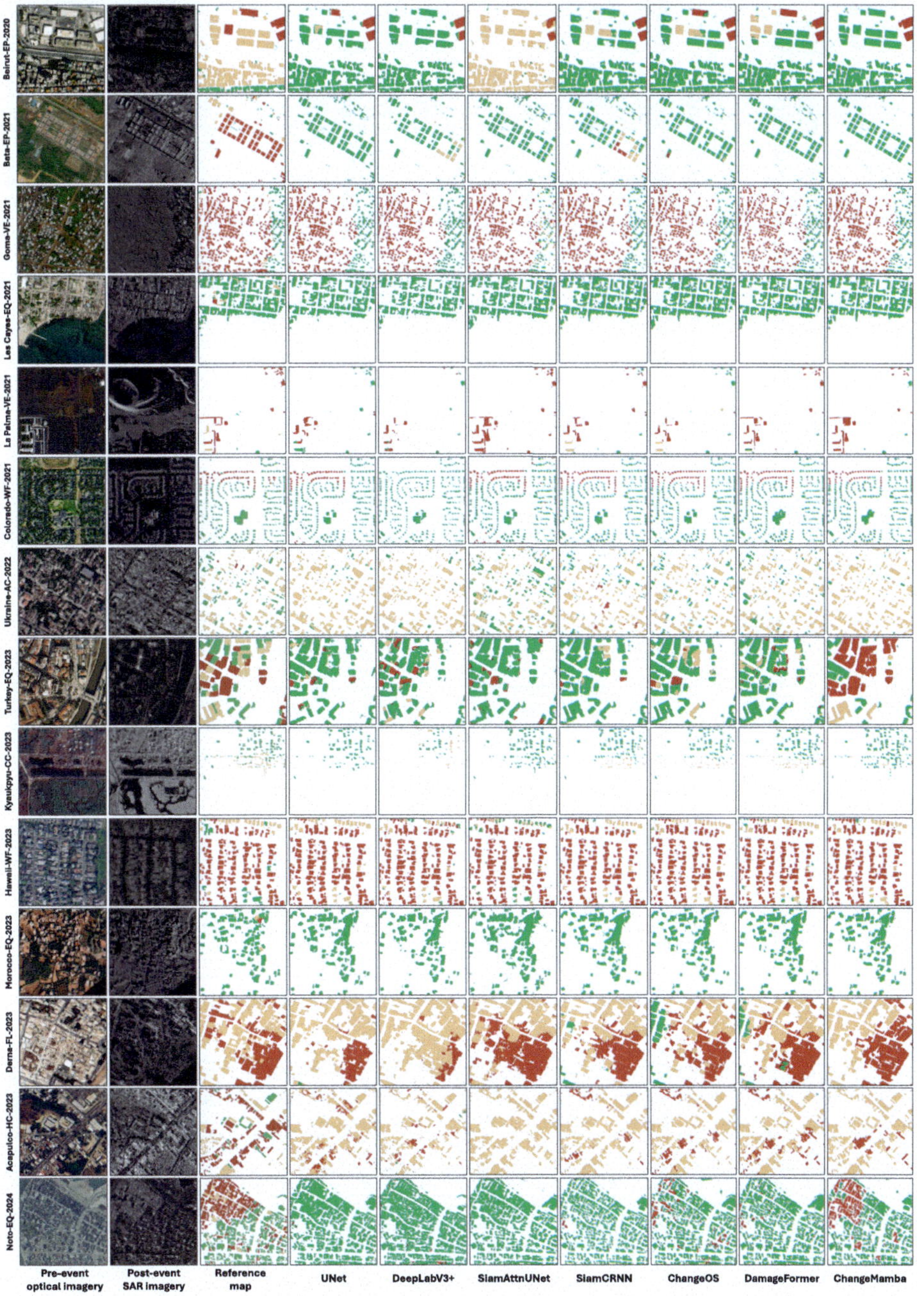

Figure 8.8:
Damage maps predicted by different models on the test images of the 14 disaster events. The meaning of the color in reference maps and damage maps is consistent with Fig. 8.6. This figure is reproduced from [2], licensed under CC-BY-4.0.

Table 8.3: The mIoU on different events for different DL models in zero-shot and one-shot cross-event transfer setups. The highest values are highlighted in purple, and the second-highest results are highlighted in teal. This table is reproduced from [2], licensed under CC–BY–4.0.

Events	UNet	DeepLabV3+	SiamAttnUNet	SiamCRNN	ChangeOS	DamageFormer	ChangeMamba
	zero- / one-shot	zero- / one-shot	zero- / one-shot	zero- / one-shot	zero- / one-shot	zero- / one-shot	zero- / one-shot
Beirut-EP-2020	31.62 / 34.89	34.72 / 33.58	35.33 / 29.17	**39.43** / **40.94**	38.48 / 38.21	38.71 / 39.93	**40.30** / **40.92**
Bata-EP-2021	33.48 / 37.48	33.55 / 30.23	33.78 / 35.24	37.46 / **40.06**	38.00 / 38.18	**38.91** / 39.42	**39.17** / **40.31**
Goma-VE-2021	55.97 / 57.14	55.01 / 58.13	57.52 / 55.75	**58.91** / 59.54	58.66 / 57.85	55.86 / **59.89**	**60.39** / **61.91**
Les Cayes-EQ-2021	26.21 / 38.72	35.50 / 37.90	37.22 / 38.76	39.45 / 40.28	39.36 / 40.26	**40.41** / **41.07**	**41.35** / **42.06**
La Palma-VE-2021	32.74 / 35.29	30.86 / **36.25**	**34.00** / 34.99	32.90 / 33.32	29.96 / 32.60	32.28 / **37.62**	**34.18** / 34.93
Marshall-WF-2021	34.97 / 37.84	35.05 / 37.12	35.23 / 39.54	39.82 / **46.83**	36.31 / 43.72	**40.93** / **51.15**	**40.61** / 45.96
Ukraine-AC-2022	32.36 / 37.50	32.23 / 35.98	33.84 / 35.76	34.77 / **38.14**	**36.30** / **39.28**	33.84 / 36.51	**35.78** / 36.99
Turkey-EQ-2023	37.82 / 35.83	37.28 / **40.49**	36.09 / 37.49	38.32 / 40.11	**38.45** / 39.51	37.20 / 40.27	**41.28** / **42.80**
Kyaukpyu-CC-2023	31.46 / 31.71	34.01 / 34.75	33.77 / 35.93	**34.19** / **37.71**	**36.66** / **36.25**	32.99 / 33.54	33.85 / 36.08
Hawaii-WF-2023	31.73 / 39.80	31.32 / 45.44	**31.79** / 45.12	30.94 / 42.65	31.15 / 41.96	31.60 / **50.64**	**36.14** / **50.89**
Morocco-EQ-2023	38.24 / 39.81	40.48 / 41.23	40.78 / 40.69	40.89 / 42.89	41.40 / 41.89	**41.89** / **43.42**	**42.86** / **43.51**
Derna-FL-2023	33.25 / 38.44	35.07 / 40.02	33.76 / 42.15	35.88 / **44.77**	35.80 / 43.85	**36.27** / 44.69	**37.02** / **45.47**
Acapulco-HC-2023	26.99 / 38.76	27.78 / 33.24	**30.93** / **38.94**	26.70 / 34.53	27.06 / 34.50	27.48 / 32.87	**28.45** / **38.96**
Noto-EQ-2024	31.02 / 33.83	35.57 / 39.17	35.87 / 36.22	36.04 / 38.52	**39.14** / 39.15	36.37 / **43.93**	**39.90** / **44.33**
Average	34.13 / 38.36	35.60 / 38.82	36.42 / 38.98	37.55 / 41.45	**37.62** / 40.52	37.48 / **42.49**	**39.38** / **43.23**

an average gain of about 3.67 mIoU points, highlighting the value of even minimal target supervision in urgent deployment.

Moreover, performance varies markedly across events, indicating strong event-specific domain shifts. Some events (e.g., *Goma-VE-2021*) are relatively transferable, while others (e.g., *Acapulco-HC-2023*) remain difficult for all methods. Still, ChangeMamba shows the strongest overall stability, achieving the highest score on most events in both zero-shot and one-shot settings, which supports its suitability as a robust backbone for cross-event multimodal damage mapping.

Table 8.4 reports performance on the samples from two events. All UMCD methods substantially outperform the random baseline, confirming that multimodal change cues are informative even without any target labels. Across both events, IM4CD achieves the best overall results. On the scene of *Congo-VE-2021*, it reaches recall and precision rates of 64.03% and 69.71%, with the highest OA, F1, IoU, and KC. On *Noto-EQ-2024*, it remains the top performer with recall and precision rates of 53.07% and 59.45%, OA of 85.36%, F1 score of 56.23%, IoU of 38.96%, and KC 47.47%. This consistent advantage suggests that modern UMCD models can produce reliable destroyed-area cues for early-stage hotspot prioritization. A qualitative comparison of these methods is provided in Fig. 8.9, which explicitly visualizes the detection performance, revealing how different models balance true positive identifications against false negative omissions across the heterogeneous disaster scenes.

In addition, different methods exhibit distinct precision–recall trade-offs and event-dependent robustness. For example, AGSCC attains relatively high precision on both events (e.g., 60.73% on *Congo-VE-2021* and 53.93% on *Noto-EQ-2024*), but its recall is lower, indicating conservative change detection that may miss some destroyed regions. RIEM provides higher recall on *Congo-VE-2021* (60.67%) but degrades severely on *Noto-EQ-2024*, suggesting sensitivity to hazard context and acquisition conditions. AEKAN performs competitively on *Congo-VE-2021* but drops on *Noto-EQ-2024*, further highlighting the generalization challenge for label-free multimodal screening across heterogeneous disasters.

8.3.3.5 Deployment case study

The CeT benchmark in the previous subsection highlights a key practical insight: models that perform well under standard splits may still fail when deployed to a new event with different building morphologies, hazard mechanisms, and acquisition conditions. Among all supervised baselines, ChangeMamba consistently achieves the strongest average performance in both zero-shot and one-shot CeT settings, indicating its advantage in transferring cross-temporal and cross-modal priors. Motivated by this evidence, we conducted a “wild” test on a real-world disaster event that occurred after the construction of the BRIGHT dataset. This

Table 8.4: Performance comparison of different multimodal change detection methods on two disaster events from the BRIGHT dataset. The highest value is highlighted in purple and the second-highest value is highlighted in teal. Values in this table are reproduced from [42].

Method	Venue	Congo-VE-2021						Noto-EQ-2024					
		Rec	Pre	OA	F1	IoU	KC	Rec	Pre	OA	F1	IoU	KC
Random guessing	-	50.00	19.25	50.00	27.79	16.14	0.00	50.00	17.67	50.00	26.12	15.02	0.00
AGSCC [39]	IEEE $TNNLS_{2024}$	25.19	60.73	82.47	35.61	21.66	27.42	38.88	53.93	83.33	45.18	29.18	35.66
RIEM [40]	ISPRS J $P\&RS_{2025}$	60.67	42.94	76.92	50.29	33.59	35.83	28.15	18.60	65.52	22.40	12.61	1.41
AEKAN [41]	IEEE TIP_{2025}	52.63	57.87	83.51	55.13	38.05	45.06	46.09	28.08	69.61	34.90	21.14	16.57
IM4CD [42]	ISPRS J $P\&RS_{2026}$	64.03	69.71	87.72	66.75	50.09	59.24	53.07	59.45	85.36	56.23	38.96	47.47

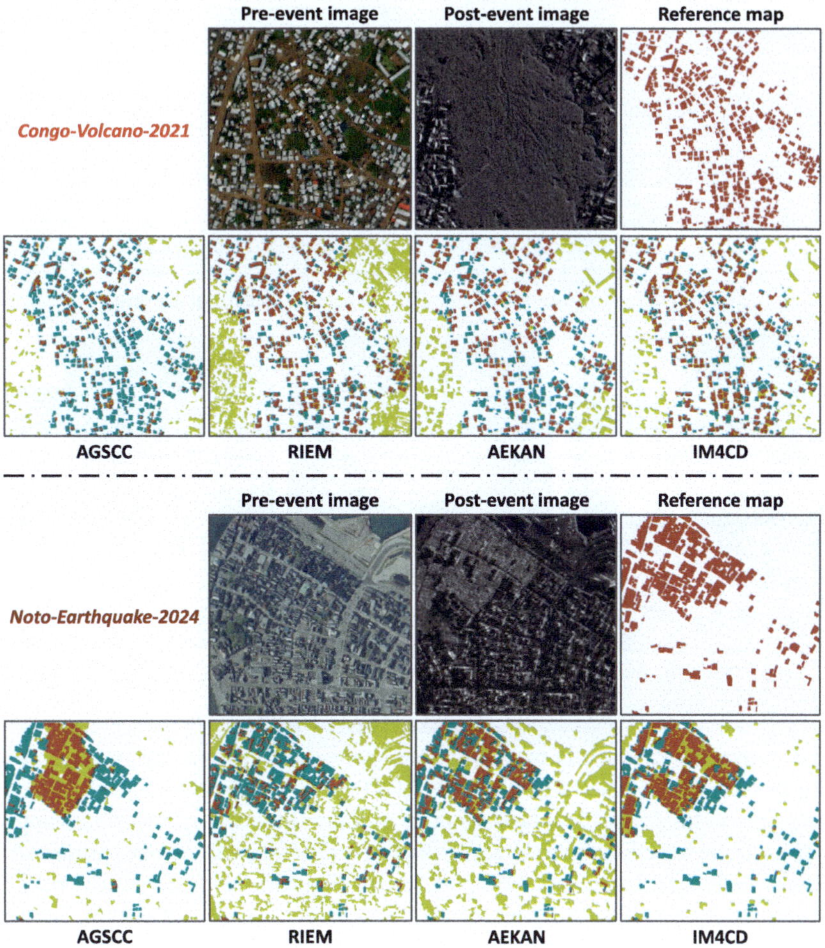

Figure 8.9:
Change maps obtained by different UMCD methods on two disaster events from BRIGHT. To explicitly visualize detection performance: true positive pixels (correctly detected destroyed regions) are rendered in red, false negative pixels (missed destroyed pixels) are highlighted in teal, false positive pixels are shown in olive yellow and true negative pixels (correctly identified unchanged areas) appear as the light gray background.

case study focuses on the Saganoseki Wildfire in Oita Prefecture, Japan, which took place in November 2025.

Fig. 8.10 summarizes the study area and the multimodal inputs. We use preevent very-high-resolution optical imagery and postevent high-resolution SAR imagery. The target output is a building damage map that highlights severely impacted structures and supports rapid

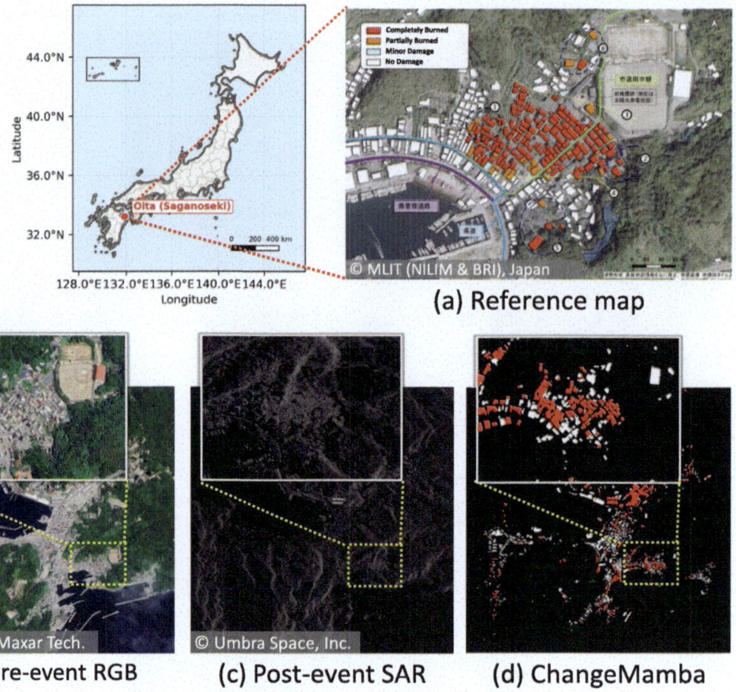

Figure 8.10:
Real-world deployment case study on the 2025 Saganoseki Wildfire, Japan. (a) Ground truth reference map provided by MLIT (NILIM & BRI). (b) Preevent optical imagery. (c) Postevent SAR imagery (Umbra Space, Inc.). (d) The damage map predicted by ChangeMamba using only (b) and (c) as inputs. The model successfully delineates the burned residential zones (rendered in red) from intact buildings despite the complex topography and sensor domain shift.

situational awareness under adverse observation conditions. As a qualitative reference for validation, we also include an available damage/impact map in Fig. 8.10(a). As shown in Fig. 8.10(c), the postevent SAR imagery is visually challenging. The steep terrain induces significant geometric distortions, and the burned wooden structures exhibit subtle scattering changes compared to the surrounding vegetation. Despite these difficulties, ChangeMamba prediction, as shown in Fig. 8.10(d) demonstrates remarkable alignment with the reference map. This case study further highlights why the optical–SAR pairing is critical for all-weather response. In wildfire scenarios, optical postevent imagery is often degraded by smoke and haze or may be unavailable at the desired time window, whereas SAR provides complementary structural-change cues. The learned cross-modal priors from BRIGHT (e.g., changes in building backscatter patterns and spatial coherence of destruction) enable the model to produce a plausible damage map even in a region not covered by the training events.

However, as an out-of-dataset deployment, several failure modes remain expected. First, SAR imaging geometry like layover and shadow and scene clutter can introduce artifacts that resemble structural change, leading to localized false positives. Second, imperfect coregistration between optical and SAR inputs may blur fine boundaries at the building scale. Moreover, the absence of event-specific labels prevents quantitative evaluation at the building level. Therefore, this case study should be viewed as a qualitative demonstration of transfer and rapid screening capability, rather than a definitive accuracy assessment.

8.3.4 Discussion and takeaways

This section reviewed building damage mapping through the lens of multimodal fusion, with BRIGHT serving as an empirical anchor for translating cross-modal learning into operationally meaningful outputs. We first positioned the task as a building-centric assessment problem under realistic response constraints including limited visibility, heterogeneous hazards, and strong domain shift across events. Within this context, BRIGHT provides a principled benchmark that couples VHR optical–SAR pairing with action-aligned damage grades and evaluation protocols that explicitly expose generalization gaps.

The experimental evidence highlights three key takeaways. First, architectural evolution matters: moving from CNN-based local feature learning to Transformer/Mamba-style long-range modeling yields consistent gains, particularly under cross-event transfer where robustness is the primary objective. Second, formulation matters: decoupled pipelines that separate building localization from damage grading generally outperform single-branch coupled segmentation, and they provide clearer interpretability for downstream emergency workflows. Third, evaluation matters: the CeT protocol reveals that in-event optimization can overestimate real-world reliability, while even minimal target supervision (one-shot) can significantly improve performance, offering a practical pathway for rapid adaptation in urgent deployments.

Beyond supervised learning, we showed that unsupervised multimodal change detection is a valuable complement for preliminary rapid screening. Although most UMCD methods have historically been validated on disaster extent mapping (e.g., floods and wildfire scars) rather than building damage mapping [36,44,45], our results demonstrate that modern UMCD approaches can still extract cross-modal change cues correlated with severe building impacts. This suggests a promising direction for integrating label-free screening with subsequent supervised grading. UMCD can rapidly propose candidate hotspots, which can then be refined by CeT-robust supervised models or minimal one-shot calibration when limited labels become available.

Finally, the out-of-dataset wildfire case study illustrates both the opportunity and the limits of all-weather multimodal fusion. While SAR enables response continuity when optical imagery is degraded, some hazards may yield subtler SAR signatures for structural damage than collapse- or inundation-driven disasters, making fine-grained grading intrinsically harder. More broadly, damage grades should be interpreted as operational categories rather than universally invariant physical states: the visual and scattering manifestations associated with *Damaged* or *Destroyed* can vary across hazards, sensor geometries, and built environments, and therefore cannot be assumed to form a strict one-to-one mapping across all disaster types.

8.4 Synthetic-to-real 3D change analysis for disaster response

8.4.1 Scaling from 2D to 3D via the synthetic paradigm

Sections 8.2 and 8.3 presented a hierarchical approach to disaster monitoring: all-weather scene parsing for rapid screening, followed by optical–SAR fusion for building-level damage grading. However, real-world response workflows often need more than 2D semantics and categorical labels. Responders also require *structural change cues* to measure geometric loss, deformation, or collapse. In practice, analysts need to know not just *which* buildings are damaged, but *how the physical shape of the urban environment has changed*. This leads to the third stage of our framework: augmenting 2D cross-modal mapping with *height-aware 3D perception* to estimate direct structural impacts.

The transition from 2D to 3D analysis, however, faces a major data bottleneck. Dense and reliable 3D ground truth, such as LiDAR or Digital Surface Models (DSMs), is expensive, geographically limited, and rarely available for the exact time window of a disaster. Even when these height products exist, sensor and processing differences often cause inconsistencies and coregistration errors, which are fatal for accurate change detection. Furthermore, disaster impacts vary greatly across different cities and events. Architectural styles, terrain, and the type of hazard all change how damage looks in both 2D imagery and 3D geometry. Because of this, training 3D models strictly on real-world data is hard to scale and often fails to generalize to new events.

To overcome this bottleneck, *synthetic supervision* offers a practical solution for scalable 3D disaster understanding. As illustrated in Fig. 8.11, creating real-world EO datasets requires expensive, task-specific manual annotation. In contrast, a physics-based synthetic data engine can generate diverse scenes with free, perfectly aligned geometric and semantic labels. Synthetic environments allow us to simulate various urban layouts and architectural styles that are difficult to collect globally. More importantly for disaster response, synthetic pipelines make *change programmable*. We can explicitly simulate structural damage, such as building collapse or deformation, and generate paired pre- and postevent images with exact 3D change labels. This makes synthetic data highly effective for teaching models the geometric principles of structural damage when real-world annotations are scarce.

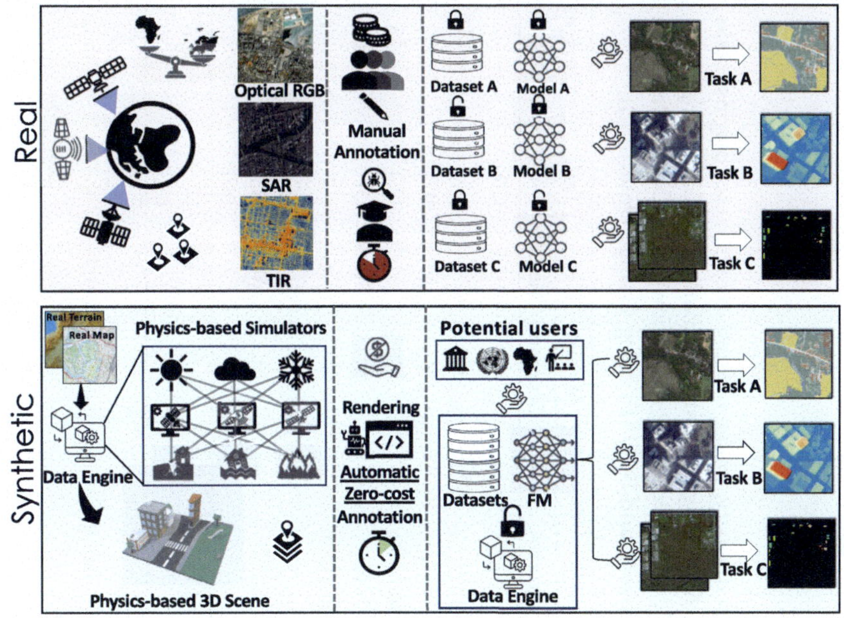

Figure 8.11:
A general comparison between real and synthetic remote sensing data pipelines. Real-world EO datasets are typically built per sensor and per task with costly manual annotation, resulting in fragmented benchmarks and limited scalability. In contrast, a physics-based synthetic data engine aims to render diverse scenes with automatic, near-zero-cost labels, supporting reusable datasets and transferable models across tasks and modalities. This figure illustrates an *aspirational* end-to-end workflow rather than a fully realized system.

While synthetic data is promising, existing EO synthetic datasets do not fully meet the needs of disaster response. As Table 8.5 shows, earlier datasets often replicate a single city style, focus only on 2D tasks like detection, or lack dense, consistent height labels. For global disaster response, models must handle diverse building types and rely on accurate geometric changes. To address this gap, we introduce SynRS3D [3], a large-scale synthetic dataset designed for multitask learning with global city styles and precise height supervision.

The remainder of this section details our synthetic-to-real framework for 3D change analysis. First, we outline the design of SynRS3D for generating scalable height, semantic, and pre-/postevent change data in Section 8.4.2. Next, we introduce RS3DAda in Section 8.4.3, a domain adaptation method that transfers these synthetic 3D priors to real-world images. Finally, we show how this pipeline generates practical tools, such as height-difference maps to support emergency response.

Table 8.5: Comparisons of representative RS synthetic datasets based on diversity, image capture details, asset origins, tasks, and scale. Diversity is categorized into City-Replica and Style-Extended. Asset origins: Manually-made (M), Game-origin (G), Procedurally-generated (P), and Real (R). Tasks: CD (Change Detection), BS (Building Segmentation), OD (Object Detection), DE (Disparity Estimation), HE (Height Estimation), LC (Land Cover), and BCD (Building Change Detection).

RS Synthetic Datasets	Diversity		Image Capture			Assets			Task	# Images
	City-Replica	Style-Extended	GSD (m)	Image Size	Perspective	Layout	Geometry	Texture		
AICD [46]	Yes	No	$*$	800×600	Nad., Obl.	M	M	M	CD	$\sim 1K$
GTA-V-SID [47]	Yes	No	1.0	500×500	Nad.	G	G	G	BS	$\sim 0.12K$
Synthinel-1 [48]	Yes	Yes	0.3	572×572	Nad.	R	M	M+R	BS	$\sim 1K$
RarePlanes [49]	Yes	No	0.31–0.39	512×512	Nad., Obl.	R	M	M+R	OD	$\sim 65K$
SyntCities [50]	Yes	No	0.1, 0.3, 1.0	1024×1024	Nad.	R	M	M+R	DE	$\sim 8K$
GTAH [51]	Yes	No	$*$	1920×1080	Nad., Obl.	G	G	G	HE	$\sim 28.6K$
SyntheWorld [52]	No	Yes	0.3–1.0	Various	Nad., Obl.	P	P+M	P	LC, BCD	$\sim 40K$
SMARS [53]	Yes	No	0.3, 0.5	Various	Nad.	R	M	M+R	LC, HE, BCD	4
SynRS3D [3]	No	Yes	0.05–1.0	512×512	Nad., Obl.	P	P+M	P	LC, HE, BCD	$\sim 70K$

8.4.2 SynRS3D *dataset*

SynRS3D [3] is a large-scale synthetic remote sensing dataset designed to provide dense and perfectly aligned supervision for learning both semantics and 3D geometry. In the context of disaster monitoring, this synthetic supervision is vital because real-world 3D ground truth such as DSM or LiDAR aligned with pre- and postevent imagery, is notoriously expensive and geographically sparse. Because reliable structural impact assessment inherently requires height-aware reasoning, SynRS3D addresses this data bottleneck by generating VHR imagery alongside pixel-wise land-cover labels and accurate normalized DSM (nDSM) height maps in a unified pipeline.

Quantitatively, SynRS3D comprises approximately 70,000 image patches at 512 × 512 pixel resolution, spanning a broad GSD range (0.05–1 m/pixel) and multiple viewing geometries, including both nadir and oblique angles [3]. Each sample includes an 8-class land-cover map, following the taxonomy in OpenEarthMap [4], and an nDSM ground truth, enabling the joint training of semantic segmentation and height estimation models. Crucially, the dataset prioritizes geographic diversity. Rather than overfitting to a few specific urban layouts, it simulates six distinct environmental styles. This forces models to learn transferable structural priors across diverse backgrounds. Additionally, the pipeline can synthesize controlled pre- and postevent perturbations to explicitly provide the structural change cues required for 3D disaster analytics.

8.4.2.1 Data generation pipeline

Fig. 8.12 outlines the procedural workflow of SynRS3D. The pipeline begins with scene construction, where terrain and layouts are procedurally generated to represent heterogeneous urban contexts. Next, objects such as buildings, roads, rivers, and vegetation are instantiated using controllable parameters. To ensure high visual realism without sacrificing scalability, textures are generated via text-guided synthesis and assigned to category-specific 3D assets. Finally, a rendering engine produces the VHR imagery under varying sensor viewpoints and illumination conditions. Because this entire process occurs within a controlled 3D environment, the dataset exports perfectly coregistered annotations for both land-cover semantics and nDSM height directly from the underlying geometry [3].

The nDSM supervision is particularly valuable for the disaster response framework proposed in this chapter. It explicitly encodes the absolute elevation of objects above the ground, offering a direct supervisory signal for height-aware perception. Because synthetic nDSM labels can be generated densely and uniformly across diverse scenes, they serve as an ideal pretraining source for models to learn building geometry before adapting to complex, real-world disaster scenes.

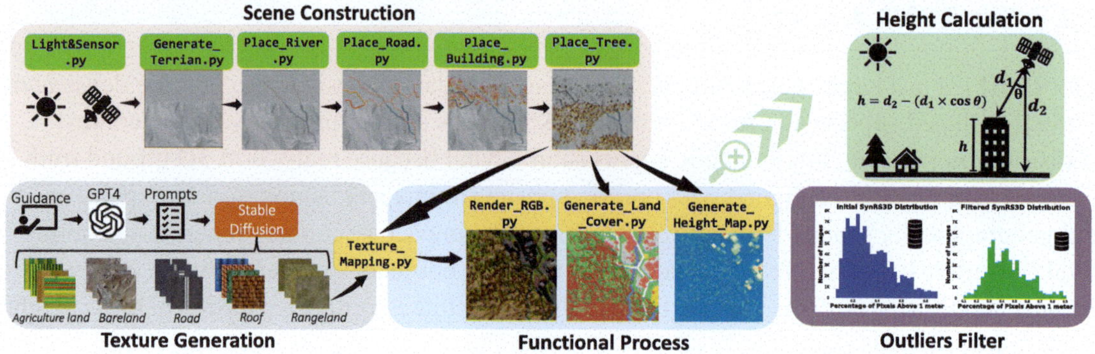

Figure 8.12:
SYNRS3D generation workflow. A controllable 3D scene is procedurally constructed and rendered into VHR remote sensing imagery. The pipeline automatically exports pixel-wise land-cover labels and geometrically consistent nDSM height maps, providing unified semantic and geometric supervision for downstream learning. This figure is reproduced from [3].

8.4.2.2 Dataset composition and statistics

Fig. 8.13 presents representative data triplets alongside comprehensive dataset statistics. These examples demonstrate that consistent, high-quality supervision is maintained across radically different architectural styles and terrains. This diversity is essential for preventing models from relying on local spectral shortcuts and ensures they learn robust features necessary for cross-regional deployment. The summary statistics confirm that SYNRS3D covers a broad spectrum of building densities, background types, and height distributions, supporting robust structural interpretation.

Ultimately, SYNRS3D functions as a scalable engine for teaching models 3D structural priors. In the subsequent subsection, we detail how these synthetic priors are transferred to real optical imagery via domain adaptation, bridging the gap between simulated geometry and deployable disaster understanding tools.

8.4.3 RS3DADA: *synthetic-to-real height-aware learning*

Although the SYNRS3D dataset provides dense supervision for land-cover semantics and per-pixel height, models trained exclusively on synthetic imagery inevitably suffer from domain shift when deployed on real-world data. In practice, this shift arises from differences in radiometry, sensor noise, urban textures, and illumination conditions. To make synthetic supervision operationally viable for disaster response, we utilize RS3DADA, an unsupervised synthetic-to-real domain adaptation strategy. This approach bridges the visual gap, ensuring

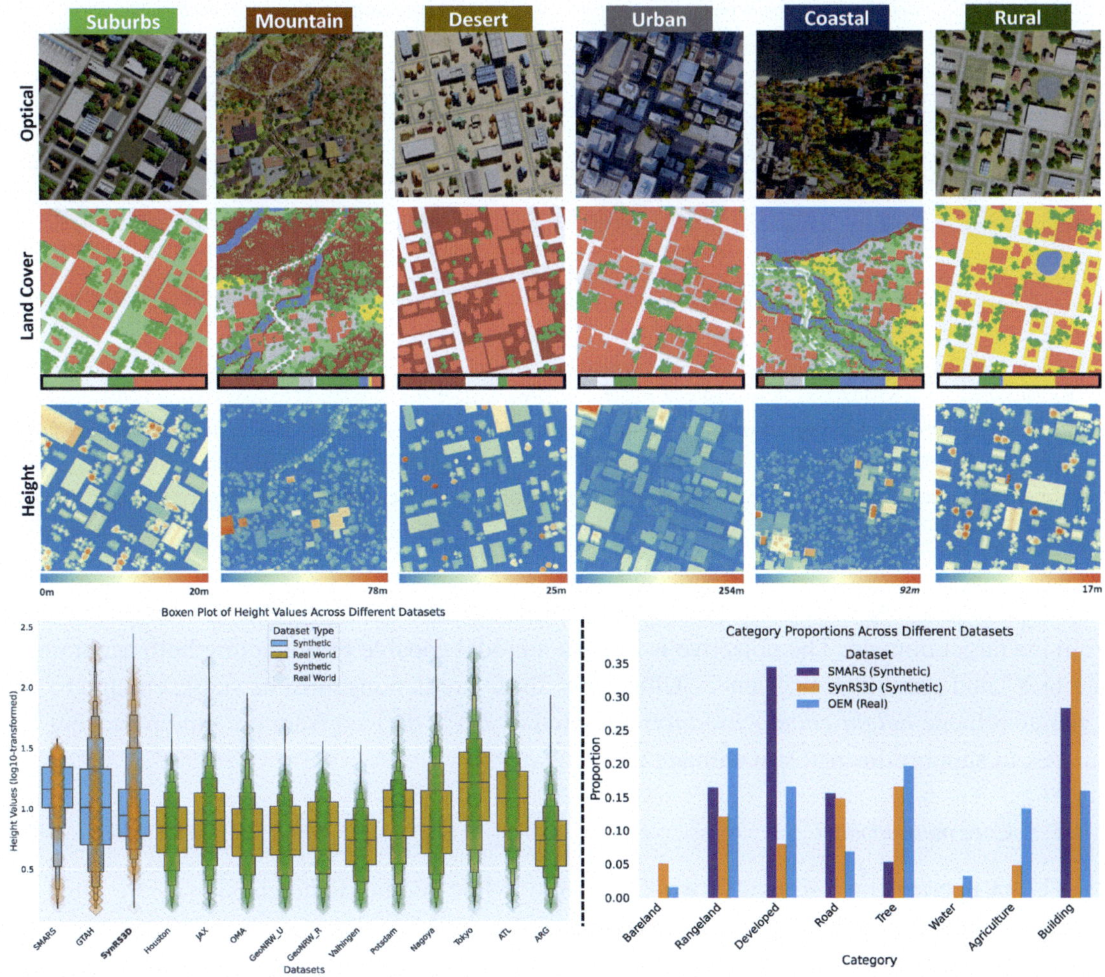

Figure 8.13:
SYNRS3D examples and statistics. Representative RGB, land-cover, and height triplets across different scene styles, alongside summary statistics that illustrate the dataset's diversity and scale. This figure is reproduced from [3].

that height-aware models pretrained on synthetic data remain reliable when applied to real disaster scenes.

8.4.3.1 Problem setting

We formulate this as an Unsupervised Domain Adaptation (UDA) problem. We define a source domain $\mathcal{S}$ (SYNRS3D) with dense labels and a target domain $\mathcal{T}$ (real-world imagery) lacking ground truth labels. Each source sample provides an RGB image $\mathbf{X}_{\mathcal{S}}$, a land-cover

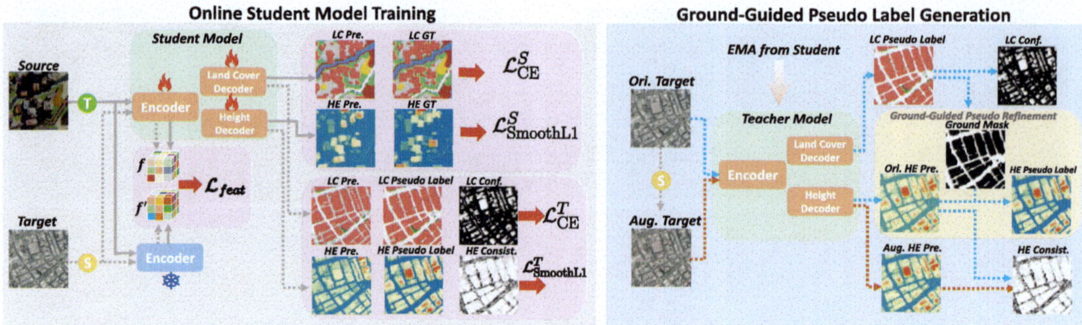

Figure 8.14:
The pipeline of RS3DADA. A multitask model is pretrained on SYNRS3D with dense land-cover and nDSM supervision, then adapted to unlabeled real imagery to reduce the synthetic-to-real gap. The adapted height branch is used to infer pre-/postevent height maps, from which a height-difference product $\Delta\hat{\mathbf{Y}}^{\text{h}}$ is derived for structural-impact screening. This figure is reproduced from [3].

map $\mathbf{Y}_{\mathcal{S}}^{\text{sem}}$, and a height map $\mathbf{Y}_{\mathcal{S}}^{\text{h}}$. During adaptation, we only observe unlabeled images $\mathbf{X}_{\mathcal{T}}$ from the target domain. The objective is to train a model capable of predicting both accurate height $\hat{\mathbf{Y}}^{\text{h}}$ and semantics $\hat{\mathbf{Y}}^{\text{sem}}$ on $\mathcal{T}$. Ultimately, these predictions must be stable enough to generate reliable *height-change products*, denoted as $\Delta\hat{\mathbf{Y}}^{\text{h}}$, derived from pre- and postevent images, to support downstream damage mapping.

8.4.3.2 Core methodology

RS3DADA operates on a straightforward principle: while geometric height signals are often less ambiguous than purely spectral cues for assessing building damage, they still require strict domain alignment to be trustworthy in real-world deployments. As illustrated in Fig. 8.14, RS3DADA consists of three core components:

- Multitask height-aware architecture: A shared encoder is coupled with two task-specific heads: one for land-cover mapping and one for height estimation. This design forces the encoder to learn representations that jointly capture semantics (what the object is) and geometry (how tall it is), a critical combination for disaster interpretation where structural collapse and semantic context are highly correlated.
- Synthetic pretraining for geometric priors: Training on SYNRS3D equips the model with dense, globally diverse geometric and semantic priors (e.g., building massing patterns and height distributions) that are hard to learn from sparse real labels alone.
- Unsupervised adaptation for real deployment: During adaptation, RS3DADA aligns the model to the real domain using unlabeled target images, improving robustness to real sensing statistics while preserving the geometric priors learned from SynRS3D.

Table 8.6: Height-estimation transfer performance of RS3DADA. Values in this table are reproduced from [3].

Model	*Target Domain 1*			*Target Domain 2*		
	MAE↓	RMSE↓	F_1^{HE} ↑	MAE↓	RMSE↓	F_1^{HE} ↑
Train-on-T.D.1	1.272	2.381	0.617	5.378	8.301	0.384
Source-only	2.557	4.128	0.372	6.117	9.221	0.365
RS3DAda	2.148	3.593	0.418	4.866	7.584	0.485

Within the context of this chapter, the primary takeaway is the resulting *operational utility*. Following SYNRS3D pretraining and RS3DADA adaptation, the height branch functions as a dependable estimator for real-world scenes. The generated pre- and postevent height maps can then be converted into a simple yet highly effective structural-impact clue via height differencing.

8.4.3.3 Quantitative results

Since the objective of this section is to establish the reliability of height-based estimation for our subsequent disaster case studies, we provide a compact summary of height-prediction performance rather than exhaustive multitask benchmarking. We evaluate RS3DADA across 11 real-world datasets, grouped into two target domains: *Target Domain 1* comprises standard, widely used public datasets [54–57], while *Target Domain 2* features highly challenging scenarios with complex terrain and broader geographic diversity [58,59]. We report Mean Absolute Error (MAE) and Root Mean Square Error (RMSE) (lower is better), alongside the F_1^{HE} accuracy metric (higher is better).

As shown in Table 8.6, RS3DADA effectively mitigates the synthetic-to-real domain gap, consistently improving height prediction accuracy on real imagery. In *Target Domain 1*, RS3DADA reduces the average MAE compared to the source-only baseline. More critically for disaster response, RS3DADA demonstrates significant gains on the unseen and highly challenging *Target Domain 2*, outperforming both the source-only deployment and the model trained exclusively on Target Domain 1 real data. This mirrors the CeT challenges discussed in Section 8.3, confirming that RS3DADA does not simply overfit to a specific real-world style, but successfully generalizes height-aware representations across diverse operational conditions.

8.4.4 Deployment case studies

This subsection translates the synthetic-to-real 3D pipeline into an operational workflow for rapid damage screening. Building on the dense geometric supervision of SYNRS3D and the

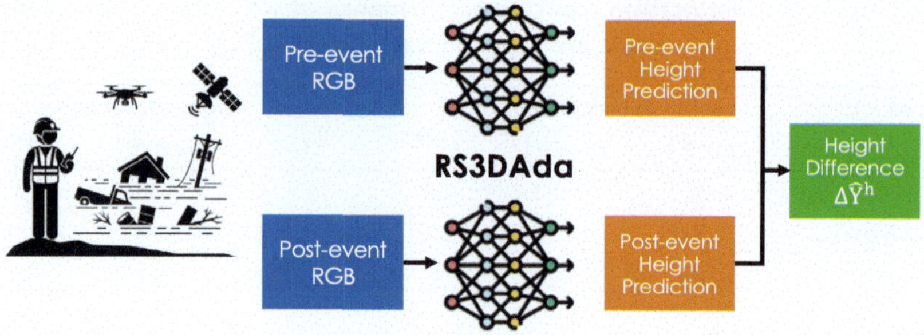

Figure 8.15:
A minimal deployment workflow for height-loss screening. RS3DAda infers pre- and postevent height maps from optical RGB imagery; their difference $\Delta\hat{\mathbf{Y}}^h$ is thresholded to obtain impacted hotspots for rapid inspection.

domain adaptation of RS3DADA, we demonstrate that a height-aware model trained entirely without event-specific real-world labels can still generate actionable cues on postevent imagery. Rather than predicting fine-grained damage categories, we focus on a robust and physically interpretable metric: *structural height loss*. By estimating pre- and postevent height maps from monocular optical images, we derive a height-difference product that rapidly isolates areas with probable structural collapse.

8.4.4.1 Workflow formulation

Given a pair of perfectly coregistered pre- and postevent optical images, we process each through the adapted RS3DADA height-estimation branch to predict continuous surface height maps, denoted as $\hat{\mathbf{Y}}^h_{\text{pre}}$ and $\hat{\mathbf{Y}}^h_{\text{post}}$. We then compute a *height difference* map:

$$\Delta\hat{\mathbf{Y}}^h = \hat{\mathbf{Y}}^h_{\text{pre}} - \hat{\mathbf{Y}}^h_{\text{post}}, \tag{8.3}$$

where positive values indicate a reduction in height due to collapse, burning, or clearance. Finally, we apply a threshold to extract candidate impacted regions:

$$\mathbf{Y}^{\text{dam}}_{\text{3D}} = \mathbb{I}\left(\Delta\hat{\mathbf{Y}}^h \geq \tau\right), \tag{8.4}$$

where $\mathbb{I}(\cdot)$ is the indicator function and τ is an event-specific threshold.

Fig. 8.15 illustrates this pipeline. This procedure is intentionally lightweight. It avoids complex postprocessing, relies on a single interpretable physical signal, e.g., height loss, and is highly suitable for early-stage screening when detailed annotations are unavailable. In the following case studies, we adjust the threshold τ to reflect the distinct damage mechanisms and expected geometric changes associated with different disaster types.

8.4.4.2 Case 1: the 2023 Turkey-Syria earthquakes

We first evaluate the 2023 Kahramanmaraş earthquake in Turkey. During severe seismic events, building damage typically manifests as partial or complete structural collapse, resulting in a pronounced reduction in surface height. Following the evaluation protocol in [3], we apply a threshold of $\tau = 3$ meters to isolate locations with substantial height loss, effectively delineating heavily impacted urban blocks.

Fig. 8.16 illustrates the qualitative results. Despite being trained exclusively on synthetic supervision and adapted without any target-domain labels, the model successfully captures a spatial distribution of height loss that aligns with visual evidence of collapsed structures. This confirms that synthetic-to-real 3D perception can provide a reliable, geometry-based evidence channel to complement the 2D cross-modal cues discussed earlier in this chapter.

8.4.4.3 Case 2: 2023 Hawaii wildfires

Next, we examine the 2023 Maui wildfires in Hawaii. Compared to earthquakes, wildfire-induced damage presents a more heterogeneous geometric signature. While many structures are completely leveled, yielding clear height loss, others may remain partially standing despite severe fire damage, producing much weaker 3D signals. Consequently, we lower the threshold to $\tau = 1$ meter [3] to capture these subtler structural changes.

As shown in Fig. 8.17, the resulting height-loss map successfully outlines the impacted residential neighborhoods, providing a rapid screening layer in the absence of fine-grained ground truth. However, the wildfire scenario also highlights a natural limitation of purely height-driven inference: not all damage modes translate into a significant $\Delta\hat{\mathbf{Y}}^h$ signal. Because the magnitude of height loss is heavily event-dependent, the threshold τ must be carefully calibrated to the specific hazard type.

8.4.4.4 Operational insights

These case studies demonstrate a practical operational capability: models trained with synthetic 3D supervision and adapted via RS3DADA can generate coarse, physically interpretable structural-impact maps on real disaster scenes using only bitemporal optical inputs. The objective of this 3D pipeline is not to replace the fine-grained, building-level damage grading achieved by the BRIGHT dataset in Section 8.3. Rather, it serves as a complementary, geometry-aware screening tool that is computationally efficient and highly explainable for emergency responders.

Nevertheless, this height-differencing approach is not a standalone solution for definitive damage assessment. Its accuracy remains sensitive to several real-world variables: (i) the quality of coregistration between pre- and postevent imagery, (ii) the inherent uncertainty of

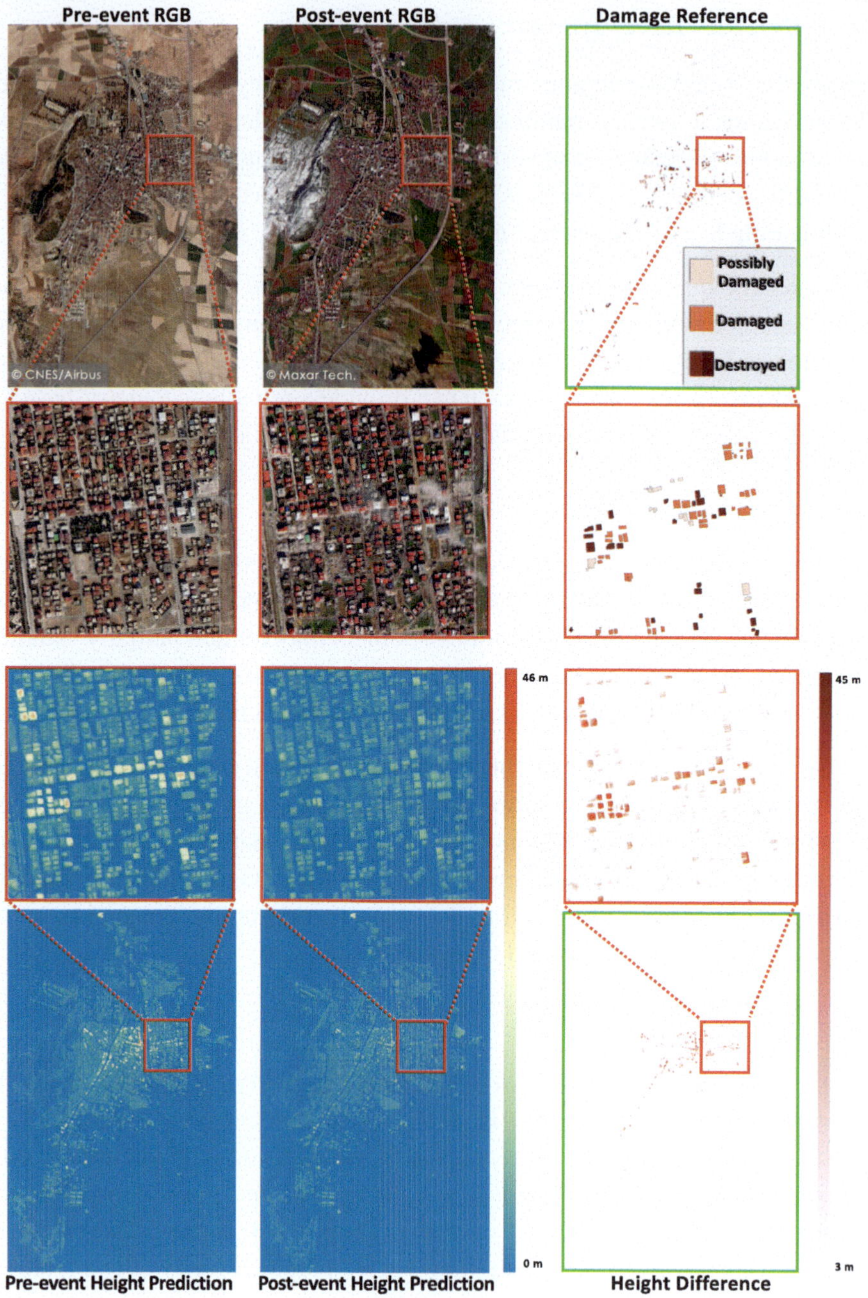

Figure 8.16:
Turkey earthquake case study. The height-difference map $\Delta\hat{\mathbf{Y}}^h$ and its thresholded mask provide a coarse but interpretable proxy for collapse-driven structural impact. This figure is reproduced from [3].

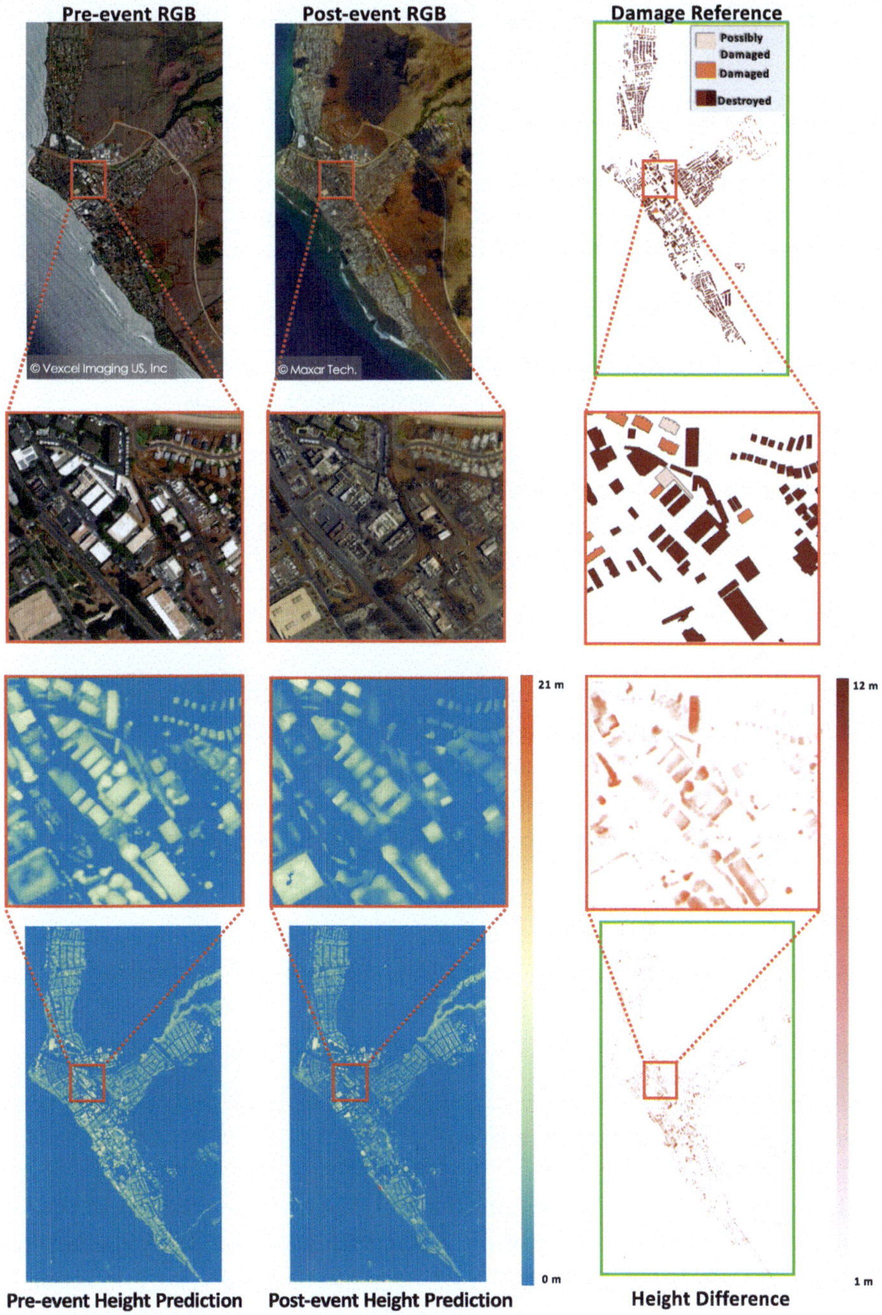

Figure 8.17:
Hawaii wildfire case study. A lower height-loss threshold is used to account for more subtle, spatially heterogeneous wildfire impacts. This figure is reproduced from [3].

monocular height prediction in dense urban cores or steep terrain, and (iii) hazard-specific failure modes where structural damage does not result in measurable height loss.

8.4.5 Discussion and takeaways

This section established synthetic supervision as a practical bridge from 2D cross-modal mapping to 3D height-aware disaster understanding. Driven by the operational need for *structural change cues*, we highlighted that strictly real-data-driven 3D learning is severely bottlenecked by sparse, inconsistent, and temporally misaligned ground truth. To overcome this, a physics-based synthetic data engine provides a scalable solution, generating perfectly aligned semantic and geometric labels to train transferable 3D models.

Specifically, we introduced SynRS3D, a large-scale synthetic benchmark that pairs VHR imagery with pixel-wise land-cover labels and nDSM height maps, enabling the joint learning of semantics and geometry across diverse urban styles. We then detailed RS3DAda, an unsupervised domain adaptation strategy that minimizes the synthetic-to-real visual gap. This adaptation ensures that height predictions remain reliable on real-world imagery, even under severe domain shifts.

Finally, we demonstrated a deployment-oriented workflow where bitemporal optical images are converted into an interpretable disaster response product via height differencing. The two case studies confirm that the height-loss metric ($\Delta\hat{\mathbf{Y}}^h$) serves as an effective, rapid screening layer. It successfully isolates structural impact hotspots, directly complementing the 2D multimodal damage maps generated in Section 8.3 with geometry-driven evidence.

Crucially, height-loss screening should be interpreted as coarse structural evidence rather than a definitive, building-level damage grade. Its accuracy depends heavily on image coregistration quality, the inherent uncertainty of monocular height estimation in complex terrain, and hazard-specific damage mechanisms that may not result in significant height reduction.

8.5 Summary and future perspectives

This chapter has presented a hierarchical, multimodal framework to address the critical bottlenecks in operational disaster monitoring. Recognizing that single-modality pipelines are often insufficient for reliable, high-resolution response under adverse conditions, we structured the solution across three complementary data regimes and methodological paradigms. First, large-scale SAR semantic segmentation based on OpenEarthMap-SAR provides a robust foundation for all-weather scene parsing, enabling rapid situational awareness when optical visibility is compromised. Second, the fusion of very-high-resolution optical and SAR imagery based on BRIGHT facilitates fine-grained, building-level damage grading, bridging the gap between coarse change cues and actionable emergency triage. Third, we extended the analytical scope to the third dimension through SYNRS3D, demonstrating how physics-based synthetic data engines and unsupervised domain adaptation can overcome the scarcity of real-world 3D ground truth. Together, these three stages form a cohesive, deployable stack that shifts the focus from isolated, static benchmarks to resilient, cross-event generalizable systems.

Looking ahead, several emerging frontiers will further reshape multimodal disaster response, driving the field toward more autonomous and scalable intelligence.

- Integration of Earth Observation Foundation Models: The architectural evolution from CNNs to global context models such as Transformers and state-space models like Mamba, has significantly improved spatial reasoning. The next natural progression is the adoption of large-scale Vision Foundation Models tailored specifically for EO [60–63]. Pretrained on massive, unannotated multimodal datasets, these foundation models promise stronger zero-shot and few-shot generalization. Leveraging these generalized representations could drastically reduce the reliance on event-specific calibration, enabling models to instantly adapt to unprecedented disaster scenarios and complex sensor domain shifts.
- Advanced Physics-Based Synthetic Engines: While current synthetic pipelines successfully simulate basic structural removal to provide height-change cues, real-world disasters produce highly complex physical phenomena. Earthquakes generate scattered debris and partial roof collapses, while wildfires leave heterogeneous burn scars and texturally altered vegetation. Future synthetic data engines must incorporate more advanced physics-based rendering, procedural destruction modeling, and generative AI textures to accurately simulate these nuanced failure modes. Closing this synthetic-to-real gap will be essential for training models to perform fine-grained, object-level damage grading without expensive manual annotation.
- From Bitemporal Assessment to Continuous Monitoring: Currently, most damage mapping frameworks rely on a discrete, bitemporal snapshot. However, disaster management is an ongoing process. Future multimodal frameworks should leverage dense satellite time-series, integrating the high revisit frequency of SAR constellations with periodic optical acquisitions, to monitor the entire disaster lifecycle. This continuous temporal modeling will expand the utility of EO-AI from immediate, rapid damage screening to the long-term tracking of debris clearance, reconstruction, and urban recovery.

Ultimately, the transition from 2D mapping to 3D scene understanding underscores that operational disaster AI cannot rely on a single sensor or a singular algorithmic task. By continuing to integrate multisensor fusion, scalable synthetic supervision, and generalized foundation models, the EO community can build the robust, all-weather systems required to protect and assess our rapidly changing built environment.

References

[1] J. Xia, H. Chen, C. Broni-Bediako, Y. Wei, J. Song, N. Yokoya, OpenEarthMap-SAR: a benchmark synthetic aperture radar dataset for global high-resolution land cover mapping [Software and Data Sets], IEEE Geoscience and Remote Sensing Magazine 13 (4) (2025) 476–487, https://doi.org/10.1109/MGRS.2025.3599512.

[2] H. Chen, J. Song, O. Dietrich, C. Broni-Bediako, W. Xuan, J. Wang, X. Shao, Y. Wei, J. Xia, C. Lan, K. Schindler, N. Yokoya, Bright: a globally distributed multimodal building damage assessment dataset with very-high-resolution for all-weather disaster response, Earth System Science Data 17 (11) (2025) 6217–6253, https://doi.org/10.5194/essd-17-6217-2025.

[3] J. Song, H. Chen, W. Xuan, J. Xia, N. Yokoya, SynRS3D: a synthetic dataset for global 3D semantic understanding from monocular remote sensing imagery, in: Proceedings of the 38th International Conference on Neural Information Processing Systems, NIPS '24, Curran Associates Inc., Red Hook, NY, USA, 2024.

[4] J. Xia, N. Yokoya, B. Adriano, C. Broni-Bediako, OpenEarthMap: a benchmark dataset for global high-resolution land cover mapping, in: 2023 IEEE/CVF Winter Conference on Applications of Computer Vision (WACV), 2023, pp. 6243–6253, https://doi.org/10.1109/WACV56688.2023.00619.

[5] O. Ronneberger, P. Fischer, T. Brox, U-net: convolutional networks for biomedical image segmentation, in: N. Navab, J. Hornegger, W.M. Wells, A.F. Frangi (Eds.), Medical Image Computing and Computer-Assisted Intervention – MICCAI 2015, Springer International Publishing, Cham, 2015, pp. 234–241.

[6] E. Xie, W. Wang, Z. Yu, A. Anandkumar, J.M. Alvarez, P. Luo, SegFormer: simple and efficient design for semantic segmentation with transformers, in: Proceedings of the 35th International Conference on Neural Information Processing Systems, NIPS '21, Curran Associates Inc., Red Hook, NY, USA, 2021.

[7] Y. Liu, Y. Tian, Y. Zhao, H. Yu, L. Xie, Y. Wang, Q. Ye, J. Jiao, Y. Liu, VMamba: visual state space model, in: Proceedings of the 38th International Conference on Neural Information Processing Systems, NIPS '24, Curran Associates Inc., Red Hook, NY, USA, 2024.

[8] L. Hoyer, D. Dai, L. Van Gool, DAFormer: improving network architectures and training strategies for domain-adaptive semantic segmentation, in: Proceedings of the IEEE/CVF Conference on Computer Vision and Pattern Recognition (CVPR), 2022, pp. 9924–9935.

[9] N. Yokoya, J. Xia, C. Broni-Bediako, Submeter-level land cover mapping of Japan, International Journal of Applied Earth Observation and Geoinformation 127 (2024) 103660.

[10] F. Yamazaki, T.T. Vu, M. Matsuoka, Context-based detection of post-disaster damaged buildings in urban areas from satellite images, in: 2007 Urban Remote Sensing Joint Event, 2007, pp. 1–5, https://doi.org/10.1109/URS.2007.371869.

[11] B. Adriano, J. Xia, G. Baier, N. Yokoya, S. Koshimura, Multi-source data fusion based on ensemble learning for rapid building damage mapping during the 2018 Sulawesi earthquake and tsunami in Palu, Indonesia, Remote Sensing 11 (7) (2019), https://doi.org/10.3390/rs11070886.

[12] X. Ye, M. Liu, J. Wang, Q. Qin, H. Ren, J. Wang, J. Hui, Building-based damage detection from postquake image using multiple-feature analysis, IEEE Geoscience and Remote Sensing Letters 14 (4) (2017) 499–503, https://doi.org/10.1109/LGRS.2017.2651050.

[13] R. Gupta, B. Goodman, N. Patel, R. Hosfelt, S. Sajeev, E. Heim, J. Doshi, K. Lucas, H. Choset, M. Gaston, Creating xBD: a dataset for assessing building damage from satellite imagery, in: Proceedings of the IEEE/CVF Conference on Computer Vision and Pattern Recognition (CVPR) Workshops, 2019.

[14] Z. Zheng, Y. Zhong, J. Wang, A. Ma, L. Zhang, Building damage assessment for rapid disaster response with a deep object-based semantic change detection framework: from natural disasters to man-made disasters, Remote Sensing of Environment 265 (2021) 112636, https://doi.org/10.1016/j.rse.2021.112636.

[15] Y. Shen, S. Zhu, T. Yang, C. Chen, D. Pan, J. Chen, L. Xiao, Q. Du, BDANet: multiscale convolutional neural network with cross-directional attention for building damage assessment from satellite images, IEEE Transactions on Geoscience and Remote Sensing 60 (2022) 1–14, https://doi.org/10.1109/TGRS.2021.3080580.

[16] J. Wang, H. Guo, X. Su, L. Zheng, Q. Yuan, PCDASNet: position-constrained differential attention Siamese network for building damage assessment, IEEE Transactions on Geoscience and Remote Sensing 62 (2024) 1–18, https://doi.org/10.1109/TGRS.2024.3392964.

[17] H. Chen, E. Nemni, S. Vallecorsa, X. Li, C. Wu, L. Bromley, Dual-tasks Siamese transformer framework for building damage assessment, in: IEEE International Geoscience and Remote Sensing Symposium, 2022, pp. 1600–1603, https://doi.org/10.1109/IGARSS46834.2022.9883139.

[18] T. Li, C. Wang, H. Zhang, F. Wu, X. Zheng, DDFormer: a dual-domain transformer for building damage detection using high-resolution SAR imagery, IEEE Geoscience and Remote Sensing Letters 20 (2023) 1–5, https://doi.org/10.1109/LGRS.2023.3288007.

[19] L. Cui, X. Jing, Y. Wang, Y. Huan, Y. Xu, Q. Zhang, Improved Swin Transformer-based semantic segmentation of postearthquake dense buildings in urban areas using remote sensing images, IEEE Journal of Selected Topics in Applied Earth Observations and Remote Sensing 16 (2023) 369–385, https://doi.org/10.1109/JSTARS.2022.3225150.

[20] C. Yu, B. Hu, X. Cheng, G. Yin, Z. Wang, Remote sensing building damage assessment with a multihead neighbourhood attention transformer, International Journal of Remote Sensing 44 (16) (2023) 5069–5100, https://doi.org/10.1080/01431161.2023.2242590.

[21] W. Lu, L. Wei, M. Nguyen, Bitemporal attention transformer for building change detection and building damage assessment, IEEE Journal of Selected Topics in Applied Earth Observations and Remote Sensing 17 (2024) 4917–4935, https://doi.org/10.1109/JSTARS.2024.3354310.

[22] A. Gu, T. Dao, Mamba: linear-time sequence modeling with selective state spaces, in: First Conference on Language Modeling, 2024, https://openreview.net/forum?id=tEYskw1VY2.

[23] H. Chen, J. Song, C. Han, J. Xia, N. Yokoya, ChangeMamba: remote sensing change detection with spatiotemporal state space model, IEEE Transactions on Geoscience and Remote Sensing 62 (2024) 1–20, https://doi.org/10.1109/TGRS.2024.3417253.

[24] A. Fujita, K. Sakurada, T. Imaizumi, R. Ito, S. Hikosaka, R. Nakamura, Damage detection from aerial images via convolutional neural networks, in: 2017 Fifteenth IAPR International Conference on Machine Vision Applications (MVA), 2017, pp. 5–8, https://doi.org/10.23919/MVA.2017.7986759.

[25] C.-S. Cheng, A.H. Behzadan, A. Noshadravan, Deep learning for post-hurricane aerial damage assessment of buildings, Computer-Aided Civil and Infrastructure Engineering 36 (6) (2021) 695–710, https://doi.org/10.1111/mice.12658.

[26] M. Rahnemoonfar, T. Chowdhury, A. Sarkar, D. Varshney, M. Yari, R.R. Murphy, FloodNet: a high resolution aerial imagery dataset for post flood scene understanding, IEEE Access 9 (2021) 89644–89654, https://doi.org/10.1109/ACCESS.2021.3090981.

[27] M. Rahnemoonfar, T. Chowdhury, R. Murphy, RescueNet: a high resolution UAV semantic segmentation dataset for natural disaster damage assessment, Scientific Data 10 (1) (2023) 913, https://doi.org/10.1038/s41597-023-02799-4.

[28] N. Kaur, C.-C. Lee, A. Mostafavi, A. Mahdavi-Amiri, Large-scale building damage assessment using a novel hierarchical transformer architecture on satellite images, Computer-Aided Civil and Infrastructure Engineering 38 (15) (2023) 2072–2091, https://doi.org/10.1111/mice.12981.

[29] T. Manzini, P. Perali, R. Karnik, R. Murphy, CRASAR-U-DROIDs: a large scale benchmark dataset for building alignment and damage assessment in georectified sUAS imagery, arXiv preprint, arXiv:2407.17673, 2024.

[30] Z. Wang, C. Wu, F. Zhang, J. Xia, Constructing an extensible building damage dataset via semi-supervised fine-tuning across 12 natural disasters, Journal of Remote Sensing 5 (2025) 0733, https://doi.org/10.34133/remotesensing.0733.

[31] Y. Sun, Y. Wang, M. Eineder, QuickQuakeBuildings: post-earthquake SAR-optical dataset for quick damaged-building detection, IEEE Geoscience and Remote Sensing Letters 21 (2024) 1–5, https://doi.org/10.1109/LGRS.2024.3406966.

[32] D. Brunner, G. Lemoine, L. Bruzzone, Earthquake damage assessment of buildings using VHR optical and SAR imagery, IEEE Transactions on Geoscience and Remote Sensing 48 (5) (2010) 2403–2420, https://doi.org/10.1109/TGRS.2009.2038274.

[33] B. Adriano, N. Yokoya, J. Xia, H. Miura, W. Liu, M. Matsuoka, S. Koshimura, Learning from multimodal and multitemporal earth observation data for building damage mapping, ISPRS Journal of Photogrammetry and Remote Sensing 175 (2021) 132–143, https://doi.org/10.1016/j.isprsjprs.2021.02.016.

[34] L.-C. Chen, Y. Zhu, G. Papandreou, F. Schroff, H. Adam, Encoder-decoder with atrous separable convolution for semantic image segmentation, in: V. Ferrari, M. Hebert, C. Sminchisescu, Y. Weiss (Eds.), Computer Vision – ECCV 2018, Springer International Publishing, Cham, 2018, pp. 833–851.

[35] H. Chen, C. Wu, B. Du, L. Zhang, L. Wang, Change detection in multisource VHR images via deep Siamese convolutional multiple-layers recurrent neural network, IEEE Transactions on Geoscience and Remote Sensing 58 (4) (2020) 2848–2864, https://doi.org/10.1109/TGRS.2019.2956756.

[36] Y. Sun, L. Lei, D. Guan, G. Kuang, Iterative robust graph for unsupervised change detection of heterogeneous remote sensing images, IEEE Transactions on Image Processing 30 (2021) 6277–6291, https://doi.org/10.1109/TIP.2021.3093766.

[37] T. Liu, M. Zhang, M. Gong, Q. Zhang, F. Jiang, H. Zheng, D. Lu, Commonality feature representation learning for unsupervised multimodal change detection, IEEE Transactions on Image Processing 34 (2025) 1219–1233, https://doi.org/10.1109/TIP.2025.3539461.

[38] J. Xu, T. Liu, T. Lei, H. Chen, N. Yokoya, Z. Lv, M. Gong, CGSL: commonality graph structure learning for unsupervised multimodal change detection, ISPRS Journal of Photogrammetry and Remote Sensing 229 (2025) 92–106, https://doi.org/10.1016/j.isprsjprs.2025.08.010.

[39] Y. Sun, L. Lei, D. Guan, J. Wu, G. Kuang, L. Liu, Image regression with structure cycle consistency for heterogeneous change detection, IEEE Transactions on Neural Networks and Learning Systems 35 (2) (2024) 1613–1627, https://doi.org/10.1109/TNNLS.2022.3184414.

[40] Y. Sun, L. Lei, Z. Li, G. Kuang, Q. Yu, Detecting changes without comparing images: rules induced change detection in heterogeneous remote sensing images, ISPRS Journal of Photogrammetry and Remote Sensing 230 (2025) 241–257, https://doi.org/10.1016/j.isprsjprs.2025.09.009.

[41] T. Liu, J. Xu, T. Lei, Y. Wang, X. Du, W. Zhang, Z. Lv, M. Gong, AEKAN: exploring superpixel-based AutoEncoder Kolmogorov-Arnold network for unsupervised multimodal change detection, IEEE Transactions on Geoscience and Remote Sensing 63 (2025) 1–14, https://doi.org/10.1109/TGRS.2024.3515258.

[42] H. Chen, C. Lan, J. Song, D. Ibanez, J. Xia, K. Schindler, N. Yokoya, Multimodal remote sensing change detection: an image matching perspective, ISPRS Journal of Photogrammetry and Remote Sensing 233 (2026) 487–501, https://doi.org/10.1016/j.isprsjprs.2026.02.004.

[43] I. Loshchilov, F. Hutter, Decoupled weight decay regularization, in: International Conference on Learning Representations, 2019, https://openreview.net/forum?id=Bkg6RiCqY7.

[44] H. Chen, N. Yokoya, C. Wu, B. Du, Unsupervised multimodal change detection based on structural relationship graph representation learning, IEEE Transactions on Geoscience and Remote Sensing 60 (2022) 1–18, https://doi.org/10.1109/TGRS.2022.3229027.

[45] H. Chen, N. Yokoya, M. Chini, Fourier domain structural relationship analysis for unsupervised multimodal change detection, ISPRS Journal of Photogrammetry and Remote Sensing 198 (2023) 99–114, https://doi.org/10.1016/j.isprsjprs.2023.03.004.

[46] N. Bourdis, D. Marraud, H. Sahbi, Constrained optical flow for aerial image change detection, in: 2011 IEEE International Geoscience and Remote Sensing Symposium, IEEE, 2011, pp. 4176–4179.

[47] Z. Zou, T. Shi, W. Li, Z. Zhang, Z. Shi, Do game data generalize well for remote sensing image segmentation?, Remote Sensing 12 (2) (2020) 275.

[48] F. Kong, B. Huang, K. Bradbury, J. Malof, The Synthinel-1 dataset: a collection of high resolution synthetic overhead imagery for building segmentation, in: Proceedings of the IEEE/CVF Winter Conference on Applications of Computer Vision, 2020, pp. 1814–1823.

[49] J. Shermeyer, T. Hossler, A. Van Etten, D. Hogan, R. Lewis, D. Kim, RarePlanes: synthetic data takes flight, in: Proceedings of the IEEE/CVF Winter Conference on Applications of Computer Vision, 2021, pp. 207–217.

[50] M.F. Reyes, P. d'Angelo, F. Fraundorfer, SyntCities: a large synthetic remote sensing dataset for disparity estimation, IEEE Journal of Selected Topics in Applied Earth Observations and Remote Sensing 15 (2022) 10087–10098.

[51] Z. Xiong, W. Huang, J. Hu, X.X. Zhu, The benchmark: transferable representation learning for monocular height estimation, IEEE Transactions on Geoscience and Remote Sensing (2023).

[52] J. Song, H. Chen, N. Yokoya, SyntheWorld: a large-scale synthetic dataset for land cover mapping and building change detection, in: Proceedings of the IEEE/CVF Winter Conference on Applications of Computer Vision, 2024, pp. 8287–8296.

[53] M.F. Reyes, Y. Xie, X. Yuan, P. d'Angelo, F. Kurz, D. Cerra, J. Tian, A 2D/3D multimodal data simulation approach with applications on urban semantic segmentation, building extraction and change detection, ISPRS Journal of Photogrammetry and Remote Sensing 205 (2023) 74–97.

[54] G. Baier, A. Deschemps, M. Schmitt, N. Yokoya, Synthesizing optical and SAR imagery from land cover maps and auxiliary raster data, IEEE Transactions on Geoscience and Remote Sensing 60 (2021) 1–12.

[55] Y. Xu, B. Du, L. Zhang, D. Cerra, M. Pato, E. Carmona, S. Prasad, N. Yokoya, R. Hansch, B. Le Saux, Advanced multi-sensor optical remote sensing for urban land use and land cover classification: outcome of the 2018 IEEE GRSS data fusion contest, IEEE Journal of Selected Topics in Applied Earth Observations and Remote Sensing 12 (6) (2019) 1709–1724.

[56] Y. Lian, T. Feng, J. Zhou, M. Jia, A. Li, Z. Wu, L. Jiao, M. Brown, G. Hager, N. Yokoya, R. Hansch, B.L. Saux, Large-scale semantic 3-D reconstruction: outcome of the 2019 IEEE GRSS Data Fusion Contest—part B, IEEE Journal of Selected Topics in Applied Earth Observations and Remote Sensing 14 (2021) 1158–1170, https://doi.org/10.1109/JSTARS.2020.3035274.

[57] S. Kunwar, H. Chen, M. Lin, H. Zhang, P. d'Angelo, D. Cerra, S.M. Azimi, M. Brown, G. Hager, N. Yokoya, R. Hansch, B. Le Saux, Large-scale semantic 3-D reconstruction: outcome of the 2019 IEEE GRSS Data Fusion Contest—part A, IEEE Journal of Selected Topics in Applied Earth Observations and Remote Sensing 14 (2021) 922–935, https://doi.org/10.1109/JSTARS.2020.3032221.

[58] G. Christie, K. Foster, S. Hagstrom, G.D. Hager, M.Z. Brown, Single view geocentric pose in the wild, in: Proceedings of the IEEE/CVF Conference on Computer Vision and Pattern Recognition, 2021, pp. 1162–1171.

[59] F. Rottensteiner, G. Sohn, J. Jung, M. Gerke, C. Baillard, S. Benitez, U. Breitkopf, The ISPRS benchmark on urban object classification and 3D building reconstruction, ISPRS Annals of the Photogrammetry, Remote Sensing and Spatial Information Sciences I-3 (1) (2012) 293–298.

[60] X. Guo, J. Lao, B. Dang, Y. Zhang, L. Yu, L. Ru, L. Zhong, Z. Huang, K. Wu, D. Hu, H. He, J. Wang, J. Chen, M. Yang, Y. Zhang, Y. Li, SkySense: a multi-modal remote sensing foundation model towards universal interpretation for Earth observation imagery, in: 2024 IEEE/CVF Conference on Computer Vision and Pattern Recognition (CVPR), 2024, pp. 27662–27673, https://doi.org/10.1109/CVPR52733.2024.02613.

[61] D. Hong, B. Zhang, X. Li, Y. Li, C. Li, J. Yao, N. Yokoya, H. Li, P. Ghamisi, X. Jia, A. Plaza, P. Gamba, J.A. Benediktsson, J. Chanussot, SpectralGPT: spectral remote sensing foundation model, IEEE Transactions on Pattern Analysis and Machine Intelligence 46 (8) (2024) 5227–5244, https://doi.org/10.1109/TPAMI.2024.3362475.

[62] J. Wang, W. Xuan, H. Qi, Z. Liu, K. Liu, Y. Wu, H. Chen, J. Song, J. Xia, Z. Zheng, N. Yokoya, DisasterM3: a remote sensing vision-language dataset for disaster damage assessment and response, in: Proceedings of the Neural Information Processing Systems (NeurIPS), 2025.

[63] D. Wang, M. Hu, Y. Jin, Y. Miao, J. Yang, Y. Xu, X. Qin, J. Ma, L. Sun, C. Li, C. Fu, H. Chen, C. Han, N. Yokoya, J. Zhang, M. Xu, L. Liu, L. Zhang, C. Wu, B. Du, D. Tao, L. Zhang, HyperSIGMA: hyperspectral intelligence comprehension foundation model, IEEE Transactions on Pattern Analysis and Machine Intelligence 47 (8) (2025) 6427–6444, https://doi.org/10.1109/TPAMI.2025.3557581.

CHAPTER 9

Applications in urban sensing for smart cities

Dong Liu[a], Zihan Kan[b,c], Teng Fei[d], Yang Chen[e], Yan Zhang[c], and Qiming Zheng[b]

[a]School of Humanities and Social Science, The Chinese University of Hong Kong, Shenzhen, China [b]Department of Geography and Resource Management, The Chinese University of Hong Kong, Shatin, Hong Kong Special Administrative Region of China [c]Institute of Space and Earth Information Science, The Chinese University of Hong Kong, Shatin, Hong Kong Special Administrative Region of China [d]School of Resources and Environmental Sciences, Wuhan University, Wuhan, China [e]Beidou Research Institute, School of Geography, South China Normal University, Nanhai District, Foshan, China

9.1 Introduction

In the rapidly evolving landscape of urban development, the concept of smart cities has become pivotal, leveraging remote sensing technologies to enhance urban functionalities, improve the quality of life, and mitigate adverse environmental impacts. Urban sensing, which involves the continuous collection and analysis of diverse data from the urban environment, provides real-time insights essential for informed decision-making and effective city management. This technology encompasses various applications, including mapping urban green spaces (UGS) [28,32], recognizing urban functional zones [132,147,170], and monitoring urban traffic noise [174,176]. By harnessing these urban sensing-derived insights, city planners and administrators can proactively address urban challenges, optimize resource allocation, and enhance the overall urban experience for residents.

As urban populations grow, cities face increasing challenges related to the efficient utilization of UGS, which provides a range of benefits encompassing health, environment, and ecology [1,39,84,124]. In this context, satellite-derived urban green space maps offer a valuable tool for city planners and policymakers. These maps use remote sensing data from satellites to identify and quantify green spaces within urban areas. Green spaces, such as parks and recreational areas, are crucial for maintaining urban biodiversity, reducing heat islands, and providing residents with recreational and aesthetic benefits [17,71,85,178]. By integrating satellite data, urban planners and policymakers can make informed decisions about where

Multimodal Remote Sensing Fusion and Classification
https://doi.org/10.1016/B978-0-44-329152-4.00016-7

to protect, expand, or create UGS. This approach promotes environmental and ecological sustainability while enhancing the quality of life for urban dwellers.

As cities expand, renew, and evolve, it is critical to recognize urban functional zones. Timely, comprehensive, and accurate representation of these urban functional zones is essential for informed decision-making in urban planning and development [67]. Urban sensing plays a pivotal role in this process. Cities consist of diverse urban functional zones like residential, commercial, industrial, and recreational areas [20,123]. Accurate identification and mapping of these zones are vital for effective urban planning and resource management. Multimodal urban sensing data, encompassing optical imagery, radar data, and geospatial information, offers a holistic view of the urban landscape [27]. By employing advanced classification algorithms on this data, urban functional zones can be delineated with precision. This information supports urban planners in zoning regulations, infrastructure development, and the allocation of public services.

As cities become more congested, traffic noise poses significant health risks, including stress, sleep disturbances, and cardiovascular diseases. Traditionally, urban traffic noise has been monitored using fixed noise monitoring stations or temporary sound measurement activities, which have limitations in terms of spatial coverage and temporal resolution [80,161]. Fixed stations offer limited coverage, resulting in significant gaps in data collection, while temporary measurements are labor-intensive and often lack the frequency needed to capture dynamic noise variations. Recent advancements in statistical modeling and computational techniques have paved the way for more sophisticated methods of traffic noise monitoring. Among these innovative methods, the use of street view images for mapping urban traffic noise stands out for its potential to revolutionize urban noise management [69,122,174]. This method involves analyzing street view images to extract information on traffic flow, vehicle types, and road structures. By combining this visual data with advanced sound measurement technologies, it is possible to generate detailed and accurate maps of traffic noise levels across the city.

As cities worldwide continue to evolve and expand, the importance of intelligent management and planning cannot be overstated. Urban sensing, with its ability to provide real-time and accurate data, plays a crucial role in developing smart cities. The innovative use of satellite-derived maps, multimodal remote sensing data, and street view images exemplifies the potential of new technologies to address complex urban challenges. By enhancing the efficiency and accuracy of urban management, these approaches offer powerful tools for improving urban quality of life and promoting sustainable city development. The rest of this chapter aims at providing concrete demonstrations of applying state-of-the-art urban sensing approaches to better managing cities from the perspectives of UGS, urban functional zones, and traffic noise.

9.2 Satellite-derived urban green space maps

9.2.1 Introduction

Urban green spaces (UGS), such as urban trees, grassland, wetland, and water bodies, are important for urban ecosystem services [56], human health [82], and sustainable city development [134]. Vegetation in UGS can regulate urban climate by providing shade and facilitating an increase in evapotranspiration [136,144]. With rapid population growth, about 68% of the world's population will reside in urban areas by 2050 [133]. Understanding the spatial patterns of UGS is crucial from the perspectives of both urban ecosystem services and sustainable urban development.

Although the spatial distribution of UGS can be acquired by manual measurements, it is costly and time-consuming [137]. Remote sensing techniques have been widely used for urban land use and cover mapping at city, national, and global scales [38]. With the development of moderate-high resolution imaging such as the Landsat, Sentinel-2, and Gaofen-1 series, rapid and accurate UGS mapping has become feasible [149].

Recently, many UGS mapping methods have been exploited based on optical remote sensing images [24,37,47,68]. Chen et al. [24] developed an automatic method integrating pixel and object-based methods with land cover and change knowledge to map global land cover at 30 m resolution. Fu and Qu [47] used the fully convolutional neural network (FCN) for mapping urban land cover, achieving an average overall accuracy of 85%. The object-oriented classification technique was used to extract UGS from RapidEye images [37]. Kranjčič et al. [79] used SVM method for UGS extraction from Sentinel-2A images. Liu et al. [88] developed the DeepLabv3plus model and used it for UGS mapping, yielding an overall accuracy of 89.46% from cloud-free GF-2 images. Huang et al. [68] used random forests classifier based on Google Earth Engine (GEE) platform for mapping UGS from cloud-free Landsat images, and their result indicated that the random forests classifier yielded an average overall accuracy of 89.26%.

Although the abovementioned UGS mapping methods can get relatively high-accurate extraction results, these methods use only cloud-free images to generate UGS or related products. However, optical remote sensing images are inevitably contaminated with clouds, especially in humid tropical and subtropical regions, leading to discontinuity in spatial patterns of satellite-derived UGS products [26,30]. It remains a great challenge to derive spatially continuous, quality UGS maps from cloud-contaminated satellite images.

To minimize the impact of clouds on optical remote sensing images for urban land use/cover mapping, many researchers have employed multitemporal images for mapping [31,112,113,119,148,177]. Salberg [115] developed a two-stage classifier for mapping land cover by fusing multitemporal Landsat images. The random forest method was used to generate land cover maps from multitemporal Landsat 8 images by Zhao et al. [177]. The patch-based recurrent neural network (PB-RNN) method was used to generate spatially continuous land cover maps from multitemporal Landsat 8 images [116]. Xie et al. [148] exploited a multitemporal Landsat classification algorithm to solve the problem of land cover classification in cloud-covered areas. Ren et al. [113] developed a novel two-step multitemporal image classification method to improve classification accuracy by using multitemporal images in missing pixel areas. These classification methods utilize multiple cloud-free patch images to improve the accuracy of UGS cover classification in cloud-contaminated images. However, their performance depends on the availability of cloud-free images. In addition, multitemporal continuous cloud-free images are difficult to acquire, especially in the tropical and subtropical regions [29,111].

To solve the issue of the above multitemporal classification methods, some researchers have attempted to take advantage of multisource images fusion, such as Sentinel-2 and Sentinel-1 images fusion, for improving UGS classification accuracy [22,63,102,118,126]. Sukawattanavijit, Chen, and Zhang [126] proposed a Ga-SVM algorithm by integrating SAR and multispectral images for improving land-cover classification. Shrestha, Saepuloh, and van der Meer [118] exploited a classification method with the fusion of Sentinel-2 and Sentinel-1 images to solve the issue of land cover maps in cloud-covered areas. Landsat-based urban land cover mapping has been extensively conducted by integrating multisource remote sensing features, including temporal, spectral, and topographic features, to reduce misclassifications of different land cover types [22,63,102]. These methods used SAR images to circumvent cloud cover problem [59].

Multisource remote sensing image fusion can provide some auxiliary information for improving urban land mapping, few studies have investigated the efficacy of crowdsourcing geospatial data to improve the accuracy of UGS mapping in cloud-covered regions [75, 168]. Crowdsourcing geographic data, such as point of interest (POI) data, can provide georeferenced prior information [41]. The prior information has been widely used to understand UGS functions from a human cognition perspective [29]. The OpenStreetMap (OSM) platform can provide freely available crowdsourcing geospatial data [73]. The integration of remote sensing imagery with crowdsourcing geospatial data can effectively improve the accuracy of urban land cover/use mapping because the crowdsourcing geospatial data provides rich prior information [23,125].

In terms of UGS mapping, Chen, Xu, and Gong (2021) developed a geospatial neural network by integrating remotely sensed imagery with crowdsourced spatial data to map UGS in Wuhan city. Although this method can achieve high mapping accuracy for a single city, the geospatial neural network method contains two limitations: (1) It cannot generate spatially

Table 9.1: Information of the experimental data.

Data Name	Sources	Acquisition Date	Spatial Resolution
Sentinel-2A	European Space Agency	2017.1–2017.8	10–60 m
HLS data	National Aeronautics and Space Administration	2017.1–2017.8	10–30 m
GSWO data	Google Earth Engine	2017.1–2017.8	30 m
OSM data	OpenStreetMap platform	2017.8	No

continuous UGS maps in cloud covered regions; and (2) while it can generate UGS map for a single city, it cannot be applied to other cities without label sample verification.

In this study, the UGS referred to urban vegetation, wetlands, water bodies, and other kinds of natural environments. This study exploits the integration of remote sensing images and crowdsourcing geospatial big data for improving UGS mapping in cloud-covered areas at the global level. The objectives of this study are: (1) To generate spatially and temporally continuous UGS maps in cloud-covered areas; (2) To generate a high-quality global UGS data product. We propose an automatic UGS mapping method by integrating satellite images with crowdsourced geospatial data. The proposed method consists of three parts: (1) auxiliary data preprocessing module; (2) cloud attention intelligent network (CAI-net); and (3) noncloud scenes classification module. Cloud cover is an inevitable issue in optical satellite data, causing mapping uncertainty and spatial discontinuity in UGS maps. Inspired by the attention mechanism in deep learning [135], we have designed a cloud attention intelligent network (CAI-net) integrated Sentinel-2 images with crowdsourced geospatial data to eliminate cloud effects in spatial-temporal continuous UGS mapping. Due to the geospatial neural network achieving excellent UGS mapping results at noncloud scenes [23], we used the state-of-the-art geospatial neural network for extracting UGS in this study.

9.2.2 Data

We selected freely available satellite and auxiliary data as the data sources for mapping global UGS. The satellite and auxiliary data included Sentinel-2A images, Harmonized Landsat-8 and Sentinel-2 data (HLS), OSM data, and Global Surface Water Occurrence (GSWO) data. The satellite and ancillary data description are listed in Table 9.1.

9.2.2.1 Satellite images

The Sentinel-2A images acquired by MultiSpectral Instrument (MSI) have been widely used for monitoring global vegetation, water resources, urban land changes, and ecosystem service mapping [110]. The Sentinel-2A were freely downloaded from the European Space Agency (https://scihub.copernicus.eu/), which has 13 spectral bands (spectrum range from 430 to 2320 nm), high time resolution (revisit period of 5 days), and high spatial resolution (spatial resolution range from 10 m to 60 m). The Sentinel-2A imagery has been found suitable for urban mapping since it can offer global surface coverage with high spatial resolution [109]. The Sentinel-2A (L2A level) images were adopted for mapping spatial continuous UGS. The atmospheric correction and radiometric calibration were implemented by ESA Sentinel Application Platform (SNAP) and ENVI 5.6 Platform.

Combining data from Landsat OLI and Sentinel-2 MSI data can provide a new resource for Earth observation because the harmonized Landsat-8 and Sentinel-2 (HLS) data reduces global observation of the land to every 2-3 days [110]. It was produced by National Aeronautics and Space Administration (NASA) project, which intended to provide dense time series of reflectance data [13]. The HLS datasets have been widely used for studies of land use mapping, land surface phenology dynamics, and urban mapping tasks [42,97,114].

The HLS Version 2 datasets have obtained seamless surface reflectance data from Landsat-8 and Sentinel-2 by the harmonization process, which consisted of a set of operations: atmospheric correction, cloud processing, spectral band pass adjustments, and spatial coregistration [33]. The HLS datasets contained of three products: (1) S10: full resolution MSI (10 m, 20 m, and 60 m); (2) S30: MSI harmonized surface reflectance resampled to 30 m; (3) L30: OLI harmonized surface reflectance and TOA brightness temperature resampled to 30 m [33]. Considering the differences in vegetation phenology in different climatic zones, we selected Sentinel-2A and HLS images from those zones between 1st August 2017 and 30th August 2017 as training datasets. Fig. 9.1 shows the global distribution of acquisition location of satellite images.

9.2.2.2 Auxiliary data

The OSM platform can provide rich volunteered geographical information such as POI category, POI place, POI time information, and so on [74]. Therefore, the OSM platform shows an important potential for providing prior information to support satellite image analysis tasks. The OSM POI data can be freely obtained from the OpenStreetMap platform (https://www.openstreetmap.org). To reduce time interval between OSM POI data and Sentinel-2A images, the POI acquisition date was set to 15th August 2017, and the dataset comprises 159,993 POI distributed across the globe. The UGS POI was extracted from OSM POI data. To retrieve water bodies of UGS classes in cloud regions, Global Surface Water Occurrence (GSWO) data were used as auxiliary data. The GSWO was derived from Landsat satellite images [107]. GSWO dataset provides long-term water-surface maps for the entire globe and can be freely acquired in GEE platform (https://global-surface-water.appspot.com/map).

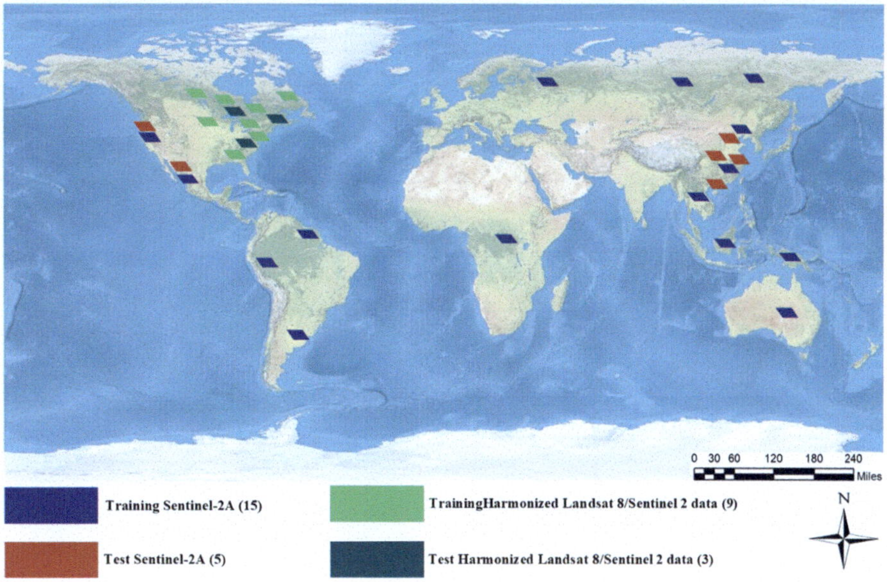

Figure 9.1:
Global distribution of satellite image acquisition for the study, including Sentinel-2A images and Harmonized Landsat and Sentinel-2 data.

9.2.3 Methods

We designed a detailed workflow to generate spatially continuous UGS products (see Fig. 9.2). The proposed method is comprised of three parts: (1) auxiliary data preprocessing module; (2) CAI-net; and (3) noncloud scenes classification module. The auxiliary data preprocessing module was used to convert the crowdsourcing geospatial data into auxiliary maps. The CAI-net was proposed to retrieve detailed UGS classes in cloud regions from satellite image patches and auxiliary maps. The noncloud scenes classification module was used to extract UGS from satellite image patches in cloud-free regions.

9.2.3.1 Auxiliary data preprocessing module

We have selected the POI and GSWO data as auxiliary data due to POI and GSWO data can provide the prior UGS features in the globe [29]. In this study, an auxiliary data preprocessing module was designed to encode two types of crowdsourcing geospatial data, i.e., POI and GSWO data, into a group of auxiliary maps. Fig. 9.3 shows the flowchart of the auxiliary data preprocessing.

The POI data often includes vegetation, parks, residential points, urban wetland, hotels, railway stations, and so on. The crowdsourcing geospatial datasets can describe characteristics

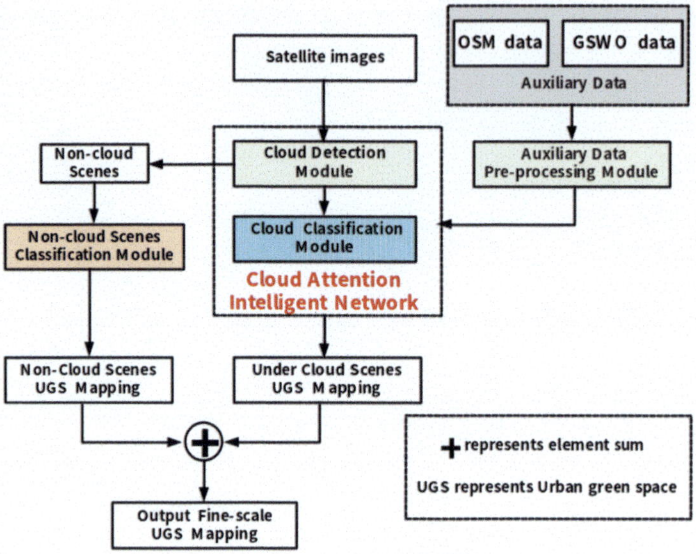

Figure 9.2:
The overall workflow of the proposed method with cloud attention intelligent network (CAI-Net).

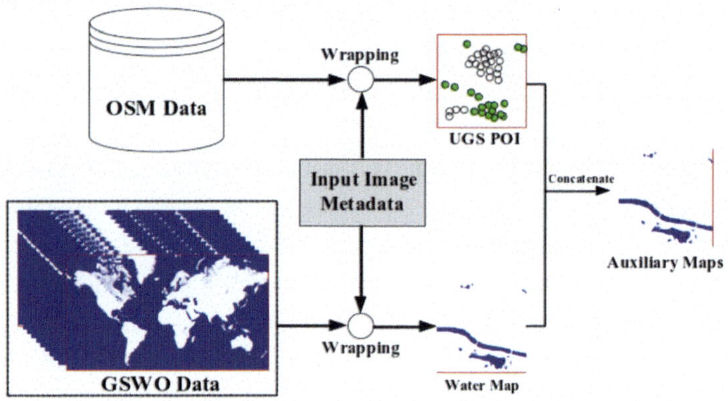

Figure 9.3:
The processing procedure for the auxiliary data.

of different UGS types [61]. To identify UGS classes within cloud-covered areas, the crowdsourcing geospatial data were preprocessed as auxiliary information.

The UGS POI density map was the auxiliary information to reflect the UGS existence probability in cloud-covered regions. For POI data preprocessing, the POI classes were categorized into the UGS classes. The rules of grouping are introduced in Table 9.2. The UGS POI density map A_{POI} was generated with the kernel density estimation by ArcGIS 10.4.

Table 9.2: UGS classes derived from the initial POI types.

UGS classes	Initial POI types
UGS	urban forest, rivers, urban wetland, green belts, park, residential green space, botanical garden natural reserve
Non-UGS	residence districts, stations, supermarket, airports, buildings, industrial districts, sports center, village green, etc.

The kernel bandwidth of the kernel density estimation was set to 10 m. For GSWO data preprocessing, the first longitude-latitude range can be obtained from the input image metadata. The auxiliary water map A_{water} was generated by using the method of pixel cropping. The strategy of aggregation was adopted to integrate an auxiliary map of the POI data and GSWO data. After obtaining preprocessing of auxiliary data, the input image with size $X \times Y$ was selected as the current UGS mapping object. The final auxiliary map A can be defined using the following equation:

$$A = concat(A_{POI}, A_{water}) \tag{9.1}$$

where the dimension of the final auxiliary map A is $(X, Y, 2)$.

9.2.3.2 The cloud attention intelligent network

The cloud attention intelligent network (CAI-net) retrieves UGS classes in cloud-covered areas. The CAI-net is comprised of two modules: (1) cloud detection module (CDM), and (2) cloud classification module (CCM). The CDM was designed to focus cloud position on input cloud-contaminated images. The CCM was designed to recognize UGS classes from remote sensing images and crowdsourcing geospatial data in cloud-covered areas. The two sequential convolutional long short-term memory (LSTM) were used to refine the cloud and the clouded UGS feature attention in CDM and CCM. Fig. 9.4 shows the architecture of the proposed CAI net.

(1) Cloud Detection Module (CDM)

Cloud detection is an essential step for identifying UGS classes in cloud-contaminated images. Since fully convolutional networks (FCN) show state-of-the-art performance in the semantic segmentation field [93], the CDM used an FCN to extract the deep local-global cloud feature, which was used to identify cloud and noncloud regions in Sentinel-2A images. We proposed a novel CDM architecture with FCN and the convolutional block attention module [145], an effective attention network that can integrate the position and the spectral attention. The convolutional block attention module (CBAM) was used in the CDM architecture for its effectiveness in refining cloud contextual semantic features [166]. Two unique operations of CDM were applied: (i) The spectral attention was used to learn which

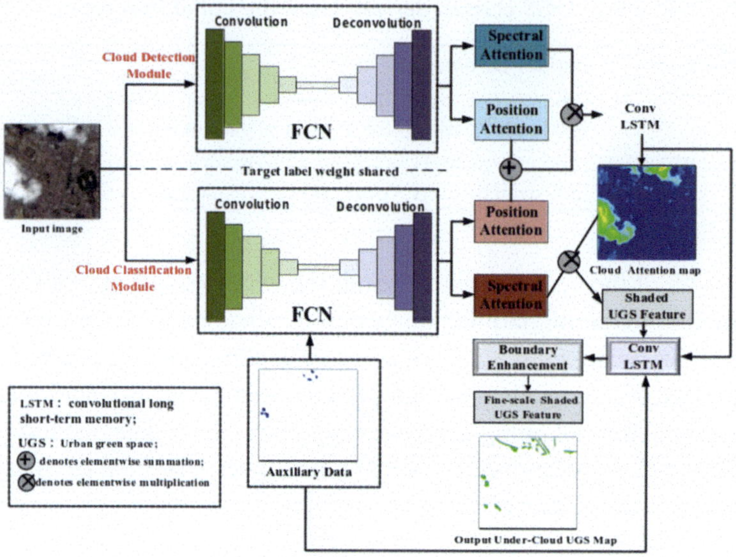

Figure 9.4:
The architecture of the proposed cloud attention intelligent network.

spectral features were important in cloud-contaminated images by exploiting the inter-spectral relationship of cloud information; and (ii) the position attention was used to highlight the positions of the cloud regions, while the spectral attention selected the important bands.

The encoder of CDM executed feature extraction using FCN to produce a set of cloud feature maps. The cloud feature maps were unsampled by a set of deconvolution kernels in the decoder of CDM. Prior to the two attention modules sequence, we used a convolution kernel of size 1×1 to adjust the spectral dimension of the features from spectral attention block and position attention block. In the CDM architecture, the sequential algorithm was used to integrate spectral and position attention [87], which can aggregate local-global semantics cloud information. The CDM architecture can adaptively highlight more important information and suppress less important information in cloud-contaminated images.

(2) Cloud Classification Module (CCM)

Prior information was obtained from auxiliary data (such as crowdsourced geospatial data), which can describe the UGS class features [61] and plays an important role in retrieving UGS classes in cloud-contaminated areas [75]. Therefore, in this paper, we used prior information as auxiliary information to retrieve UGS classes in cloud-covered areas. To obtain UGS prior information, the cloud classification module was designed. The feature images from CCM are up-sampled to the size of the input image by using deconvolution operations. Then, the spectral dimension of the up-sampled feature images was adjusted from spectral attention block and position attention block to the same number of spectrums of the input image by using 1×1 convolution kernel. The final concatenated feature maps F from the CCM can be defined by the following equation:

$$F = concat(F_{img,1}, F_{img,2}, F_{img,3}, F_{img,4}, F_{aux,1}, F_{aux,2}, F_{aux,3}, F_{aux,4}) \tag{9.2}$$

where the $F_{img,1}, F_{img,2}, F_{img,3}, F_{img,4}$ are the features from the four spectral bands (i.e., Sentinel-2A Blue band, Green band, Red band, and NIR band). $F_{aux,1}, F_{aux,2}, F_{aux,3}, F_{aux,4}$ features from the auxiliary image.

Compared to the cloud detection module, the cloud classification module has unique operations, including: (i) extracting the UGS features for cloud contaminated areas based on the auxiliary feature maps; and (ii) updating the cloud attention of the CDM by using an element-wise sum algorithm on the maps of CDM and CCM. At each stage, the two convolutional LSTM networks [117] were used to refine the cloud attention map to implement the CAI-net. The boundary enhancement module with three dilated convolution kernels (the convolution kernel sizes of 5×5, 3×3, and 1×1) was further applied to improve was further applied to improve the boundaries of the UGS map [32]. The integrated cloud attention map was fed into the CDM and the CCM at the next stage to improve the accuracy of cloud detection and thus the classification of UGS classes in cloud-contaminated areas.

9.2.3.3 Noncloud scenes classification module

The Geospatial neural network has been successfully applied to achieve the high accuracy of UGS mapping in cloud-free regions [23]. Therefore, in this research, the Geospatial neural network was used to extract UGS from Sentinel-2A images in cloud-free regions. The final spatial continuous UGS map A can be defined as follows:

$$A = A_{Cloud} + A_{Noncloud} \tag{9.3}$$

where A_{Cloud} is UGS map from cloud attention intelligent network in cloud-contaminated regions. $A_{Noncloud}$ is UGS map from noncloud scenes classification module in cloud-free regions.

In the proposed method, we used a 1×1 convolution kernel to generate the score maps of two classes: background M_1 and UGS M_2. We use the softmax function [108] to normalize the range of score maps, and the softmax classifier to identify UGS from score maps of different classes. The probability map P_s of each class $s = \{1, 2\}$ can be defined as follows:

$$P_s = \frac{\exp(M_s)}{\sum_{n=1}^{2} \exp(M_n)} \tag{9.4}$$

9.2.4 Results

To evaluate the effectiveness of the proposed network, large experiments were conducted to generate UGS maps with Sentinel-2A images in different cities. The proposed network is compared with the state-of-the-art deep learning methods (DeepLab v3plus, FCN, and Geospatial neural network). The proposed network training and experimental results are described in this section.

9.2.4.1 Network training and evaluation metrics

Considering vegetation phenology characteristics, we selected Sentinel-2A and HLS images of the summer season as training datasets. The input images were cropped into patch size of 512 × 512 pixels by using the sliding window strategy with a stride size of 1.

According to the predefined land use classification map, the experimental datasets contained different types of urban scenes, such as urban forest, urban park, and commercial district. The UGS and cloud labels were obtained by manual drawing in ENVI platform. The accuracy of manual drawing was controlled within two pixels. Our datasets contained 70,000 training, 15,000 validation, and 15,000 testing image patches.

Fig. 9.5 shows some of the training image patches, cloud labels, and UGS labels. The proposed CAI net was implemented in the PyTorch and Google Earth Engine (GEE) platform, using the stochastic gradient descent to optimize parameters with weight decay of 0.0001, momentum of 0.9, and a minibatch size of 128 for generating spatially continuous UGS maps. The learning rate was initially set to 0.01 and decayed exponentially at a rate of 0.9 every 50 epochs, for a total of 1000 training epochs. The proposed network and the compared methods were pretrained by using the same datasets. For FCN-based architectures, the weight parameters of the trained FCN-GoogLeNet [103] were used as the initial weight of the FCN encoder. All network architectures were implemented on NVIDIA Quadro M2200 GPU.

The common metrics for assessment for UGS maps include overall accuracy (OA), kappa coefficient (Kappa), recall, and precision [44]. They can be defined by the following equations, respectively:

$$OA = \frac{TP + TN}{TP + TN + FP + FN} \tag{9.5}$$

$$Kappa = \frac{N \times (TN + TP) - [(TP + FP) \times (TP + FN) + (FN + TN) \times (FN + TN)]}{N^2 - [(TP + FP) \times (TP + FN) + (FN + TN) \times (FN + TN)]} \tag{9.6}$$

$$Recall = \frac{TP}{TP + FN} \tag{9.7}$$

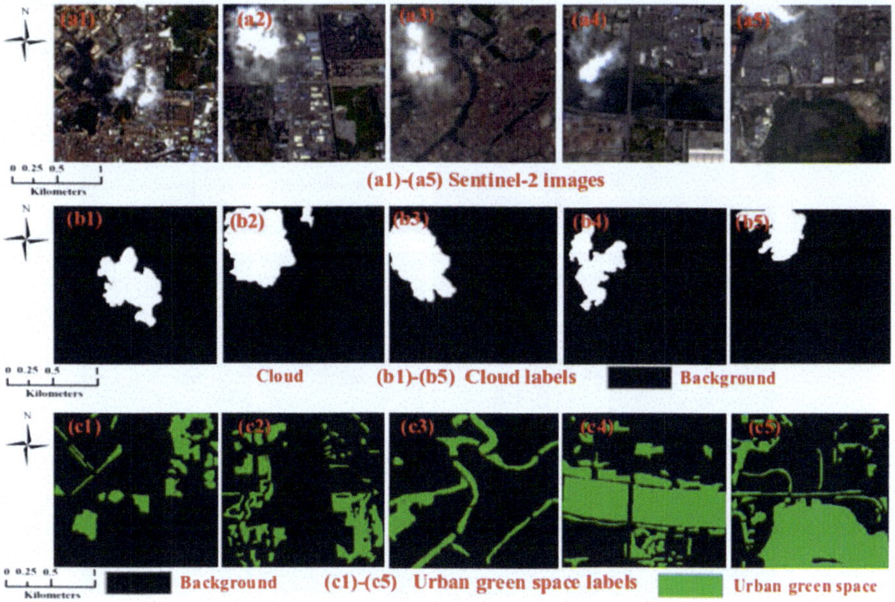

Figure 9.5:
Some sample patches of urban areas.

$$Precision = \frac{TP}{TP + FP} \tag{9.8}$$

where N represents the total number of pixels for an entire input image. TP represents the number of UGS pixels accurately extracted. FP represents the number of non-UGS pixels extracted as UGS pixels. TN represents the number of non-UGS pixels accurately extracted. FN represents the number of UGS-pixels extracted as non-UGS pixels.

9.2.4.2 The performance of the proposed network

To better assess the performance of the proposed network, we compared it with DeepLab v3plus, FCN, and Geospatial neural network [32,88,93]. The state-of-the-art deep learning methods (DeepLab v3plus, FCN, and Geospatial neural network) without cloud detection have been successfully applied to generate UGS map in the urban areas. All the methods used the same input parameters. Input parameters of the four networks are listed in Table 9.3. We initialized DeepLab v3plus, FCN, Geospatial neural network and the proposed network, respectively, using the weight parameters from the trained FCN-GoogLeNet. In addition, we have selected the four types of urban land-cover scenes as experimental areas. Four types of urban land-cover scenes contained forest, park green spaces, residential green spaces, and wetlands.

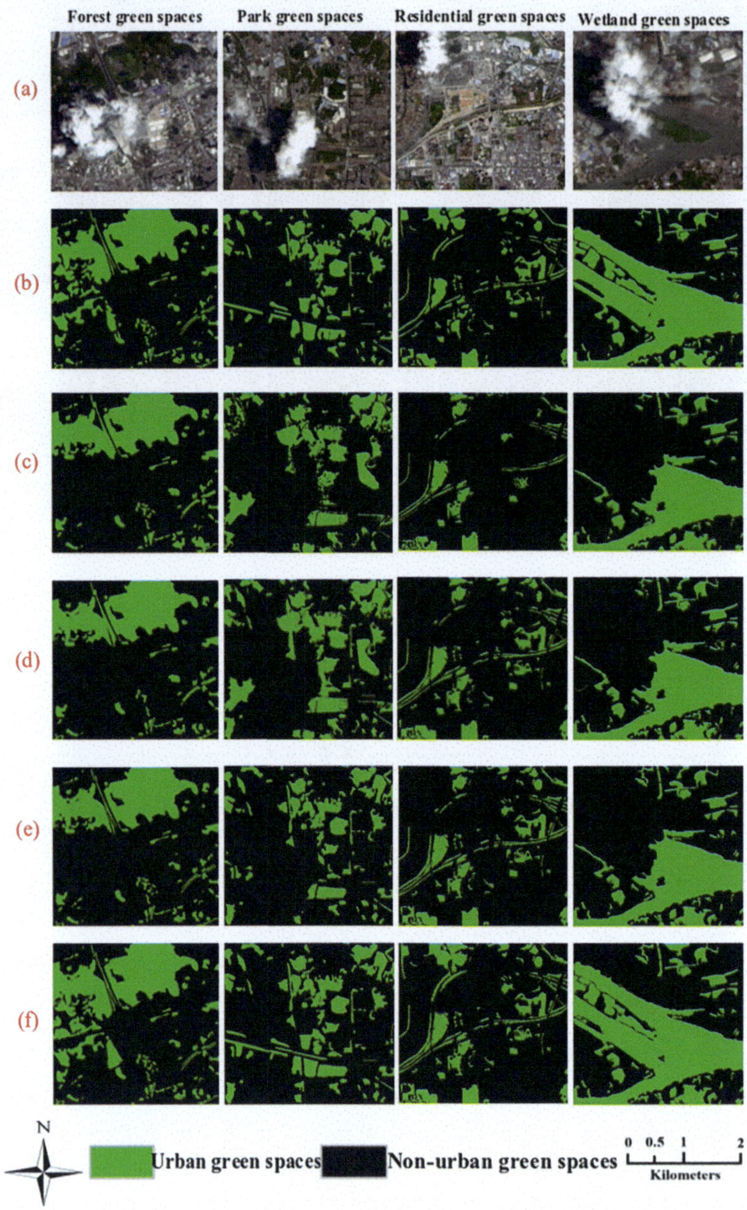

Figure 9.6:
The results of extracting urban green spaces by different methods for different urban land-cover scenes. (a) Cloud of Sentinel-2 data; (b) The ground truth reference; (c) DeepLab v3plus; (d) FCN; (e) Geospatial neural network; (f) The proposed network.

Table 9.3: Major input parameters of the four networks.

Method	Training samples	Mini-batch size	Momentum	Learning rate	Epochs
DeepLab v3plus	70,000	128	0.9	0.01	1000
FCN	70,000	128	0.9	0.01	1000
Geospatial NN	70,000	128	0.9	0.01	1000
CAI-Net	70,000	128	0.9	0.01	1000

Table 9.4: Quantitative evaluation results of the four mapping methods.

Method	OA (%)	Kappa (%)	Recall (%)	Precision (%)
DeepLab v3plus	70.51	66.12	67.31	68.01
FCN	76.34	72.04	72.91	73.39
Geospatial NN	81.67	78.15	79.24	80.19
CAI-Net	89.52	86.78	87.91	88.45

Fig. 9.6 shows the results of extracting urban green spaces by different methods for different urban land-cover scenes. Fig. 9.6 shows the proposed network can generate spatially continuous UGS maps from cloudy Sentinel-2 images. DeepLab v3plus, FCN, and Geospatial neural network obtained good UGS extraction results in noncloud areas; however, they failed to extract UGS classes in cloud-covered areas, leading to the underestimation of UGS (see Fig. 9.6 (c)-(e) red box). The proposed network method can solve the underestimated issue because the proposed CAI-net can recognize UGS classes in cloud-contaminated areas.

To quantify the performance of the proposed network, four assessment parameters (OA, Kappa, recall, and precision) were calculated for four typical urban scenes. Table 9.4 shows the quantitative results of the four UGS mapping methods. Comparing to DeepLab v3plus and FCN, the Geospatial neural network method yielded best values for OA, Kappa, recall, and precision.

From Table 9.4, the proposed network was found to outperform the Geospatial neural network, with average OA of 95.13% versus 81.59%, average Kappa of 91.09% versus 74.91%, average Recall of 92.01% versus 76.97%, and average Precision of 93.42% versus 77.01%. After the introduction of CAI-net block, the improvement in OA, Kappa, Recall, and Precision was significant. Average OA has increased by 13.54%, and UGS classes in cloud-covered areas were correctly extracted. Similarly, average Kappa by 16.18%, average Recall by 15.04%, and average Precision by 16.41%, as compared to the Geospatial neural network. The CAI-net block played an important role in improving UGS mapping in cloud-covered areas because it can recognize UGS classes in cloud-contaminated areas. The encoder-

Table 9.5: Accuracy assessment of UGS maps using the proposed method for different urban patterns.

City	OA (%)	Kappa (%)	Recall (%)	Precision (%)
Beijing	93.81	90.02	91.18	92.28
Wuhan	95.24	92.38	93.81	94.72
Shenzhen	94.58	91.45	92.73	93.61

decoder CAI-net with an attention mechanism can show better performance in UGS mapping from contaminated cloud images.

9.2.4.3 Mapping urban green spaces with different geographical settings

To further evaluate the mapping performance of the proposed method at different urban scales, we have selected three types of urban patterns (including Beijing, Wuhan, and Guangzhou, China). The different urban patterns have different distributions of UGS due to the composition and fabrics of urban ecosystems [57].

Fig. 9.7 (b1)-(b3) shows the results of extracted urban green spaces by using the proposed network for Beijing, Wuhan, and Guangzhou, China, respectively. By using the cloud detection module, we found that clouds occupied 13.18% of the area in Beijing, 15.04% in Wuhan, and 11.03% in Guangzhou, in the test Sentinel-2A images. From Fig. 9.7 (b1)-(b3), it can be observed that the proposed method can generate spatially continuous UGS maps from the cloud-contaminated Sentinel-2A images in all three types of urban patterns.

Table 9.5 shows the quantitative evaluation result using the proposed method in the three cities. Overall mapping results of three types of urban patterns using the proposed method show an average overall accuracy of 94.38%, an average Kappa of 90.10%, an average recall of 92.43%, and an average precision of 93.36%. According to the visualization and quantitation results, the proposed method can be regarded to be effective in generating spatially continuous UGS maps in cloud-covered areas.

9.2.5 Discussion

9.2.5.1 Effect of cloud cover on UGS mapping

Sustainable urban development research is dependent on spatial-temporal continuous UGS data [136]. Currently, UGS maps with spatial resolution of 10 m generated from Sentinel-2A images can provide a timely, accurate and essential data for sustainable urban development research and related studies. Cloud cover is inevitable, especially in the tropical and subtropical areas, which gives uncertainty issue in UGS mapping [177]. Hence, we

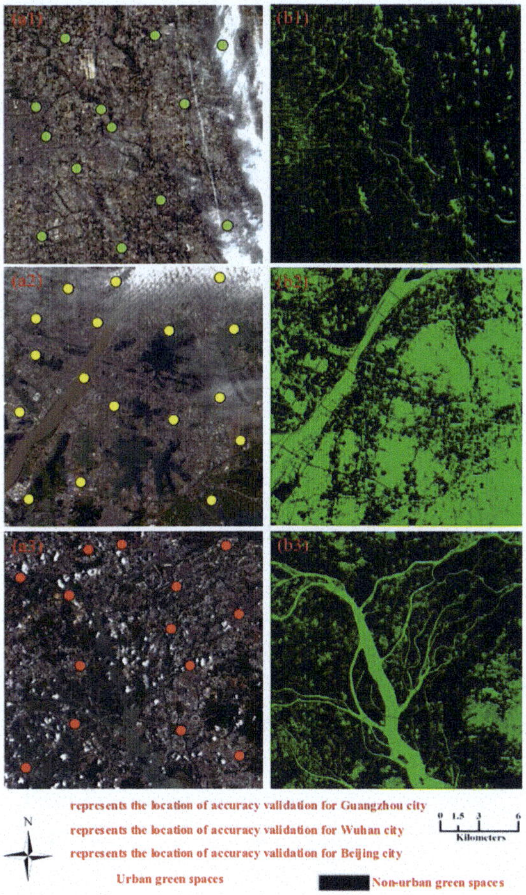

Figure 9.7:
Urban green spaces mapping for different geographical patterns. (a1)-(a3) Cloud of Sentinel-2A data in Beijing (acquired on August 5, 2016), Wuhan (acquired on August 15, 2016), and Guangzhou (acquired on August 23, 2016), China, respectively; (b1)-(b3) The results of extracted urban green spaces by using the proposed network in Beijing, Wuhan, and Guangzhou, China, respectively.

discuss the effect of different cloud coverage of Sentinel-2 data on UGS mapping when using the proposed CAI-net method in this section.

To test mapping performance at different cloud coverages, we have selected 11 images with cloud coverage ranging from 0% to 50%. Table 9.6 shows details about the tested Sentinel-2 images. Fig. 9.8 shows the effects of the different cloud coverages on UGS mapping accuracy when using the proposed method. From Fig. 9.8 (a)-(d), we can observe that UGS mapping accuracy of the proposed method with different cloud cover coverage, the OA ranging from

Table 9.6: Sentinel-2 images over study area with cloud cover (0% to 50%) for evaluating CAI-Net performance.

Cloud Cover	Product ID	Acquisition Date
0%	S2A_MSIL1C_20190817T023551_N02	2019/08/17
5%	S2A_MSIL1C_20190714T025551_N02	2019/07/14
10%	S2A_MSIL1C_20190619T023551_N02	2019/06/19
15%	S2A_MSIL1C_20190524T023551_N02	2019/05/24
20%	S2A_MSIL1C_20190429T023551_N02	2019/04/29
25%	S2A_MSIL1C_20190404T023551_N02	2019/04/04
30%	S2A_MSIL1C_20190310T023551_N02	2019/03/10
35%	S2A_MSIL1C_20190212T023551_N02	2019/02/12
40%	S2A_MSIL1C_20190118T023551_N02	2019/01/18
45%	S2A_MSIL1C_20181224T023551_N02	2018/12/24
50%	S2A_MSIL1C_20181129T023551_N02	2018/11/29

98.36% to 83.71%, the Kappa ranging from 93.56% to 85.56%, the recall ranging from 96.14% to 87.54%, and precision ranging from 93.34% to 85.17%. According to the results, the proposed method can yield a robust mapping accuracy of UGS against different cloud fractions of Sentinel-2 images. To test mapping performance at different cloud coverages, we have selected 11 images with cloud coverage ranging from 0% to 50%. Table 9.6 shows details about the tested Sentinel-2 images. Fig. 9.8 shows the effects of the different cloud coverages on UGS mapping accuracy when using the proposed method. From Fig. 9.8 (a)-(d), we can observe that UGS mapping accuracy of the proposed method with different cloud cover coverage, the OA ranging from 98.36% to 83.71%, the Kappa ranging from 93.56% to 85.56%, the recall ranging from 96.14% to 87.54%, and precision ranging from 93.34% to 85.17%. According to the results, the proposed method can yield a robust mapping accuracy of UGS against different cloud fractions of Sentinel-2 images.

9.2.5.2 Global UGS mapping in different climatic zones

With increasing urbanization and climate change, global UGS are decreasing at an alarming rate [127]. According to the United Nations [133], about 68% of the world population will live in urban regions by 2050. However, the global UGS dataset does not yet exist by climatic zone. To support urban sustainable development, we chose 1080 cities for producing maps of UGS in different climatic zones. Fig. 9.9 shows the locations of these in different climatic zones. We chose the HLS data and public geographic data as the data sources for mapping global UGS.

The UGS maps of 1080 cities were produced by using the proposed method at the global scale. Fig. 9.10 shows examples of UGS maps in different climatic zones, including Cairo,

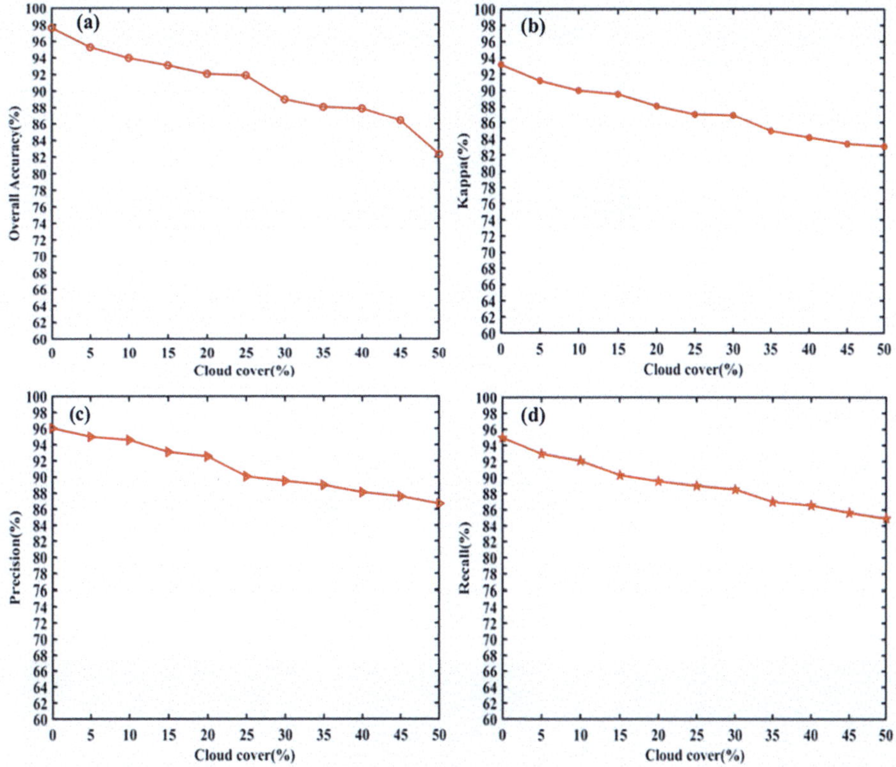

Figure 9.8:
Effects of the different cloud cover of Sentinel-2 data on urban green spaces mapping accuracies based on 11 Sentinel-2 images. (a) Overall accuracy; (b) Kappa; (c) Precision; (d) Recall.

Sao Paulo, Los Angeles, Shanghai, Chicago, and Moscow. Fig. 9.10, it can be observed that UGS maps can capture different spatial patterns in different global climatic zones. From Fig. 9.10 (a)-(b), the spatial pattern of UGS was sparse for Cairo city and Sao Paulo city. From Fig. 9.10 (c)-(d), we found that the spatial pattern of UGS was compact in Los Angeles and Shanghai. From Fig. 9.10 (e)-(f), the spatial pattern of UGS was dense in Chicago, Moscow. We observed that UGS was sparse in desert climate cities, whereas the spatial pattern of UGS was dense in tropical climate cities, which could be related to the temperature and rainfall of the localities.

Accuracy of the UGS maps was evaluated by examining samples of 500 $\times$ 500 pixels as validation regions against manual UGS mapping. Table 9.7 shows quantitative evaluation results of UGS maps for selected cities in different climate zones. The mapping results of those cities show an overall accuracy of 92.96%, Kappa of 89.29%, Recall of 90.34%. The

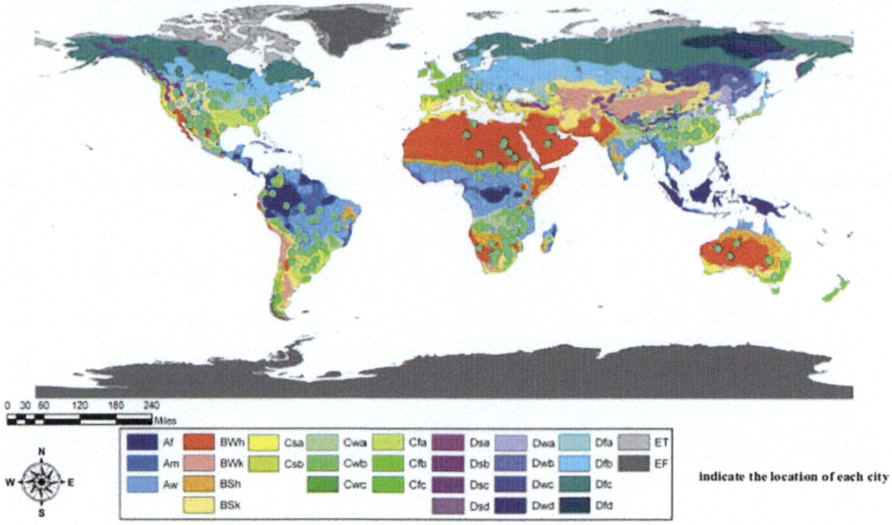

Figure 9.9:
Distribution of the study cities in different climatic zones (Source of the map: World map of Köppen-Geiger climate zones [106]).

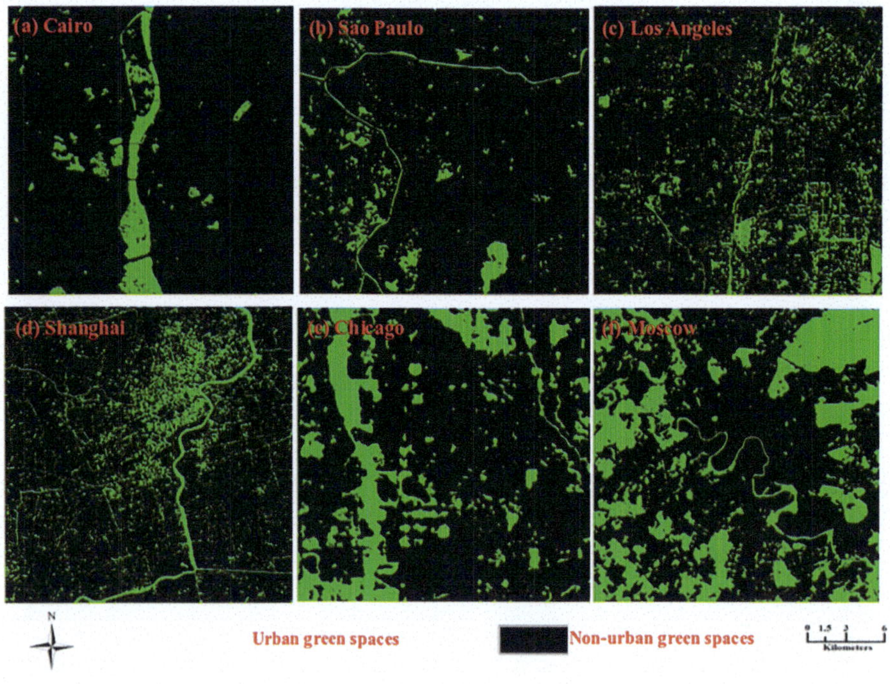

Figure 9.10:
Examples of the UGS maps in global cities derived from HLS data and Sentinel-2A data using CAI-net.

Table 9.7: Accuracy assessment of UGS maps of selected global cities in different climatic zones.

City	OA (%)	Kappa (%)	Recall (%)	Precision (%)
Cairo	95.12	92.22	94.08	94.48
Sao Paulo	94.65	91.08	93.26	93.02
London	96.23	93.89	95.42	95.78
New York	95.87	93.21	94.85	95.12
Tokyo	94.92	91.56	93.42	93.75
Sydney	96.45	94.21	95.78	96.02
Moscow	95.34	92.45	94.21	94.56

proposed method can provide a reliable accuracy mapping of UGS for cities in different global climatic zones.

To explore the spatial patterns of UGS of the cities in different climatic zones, the urban green space coverage (UGSC) indicator was calculated by using the global 1080 cities of UGS maps. In this study, the UGSC indicator can be defined as follows:

$$UGSC = \frac{N_{UGS}}{N_{total}} \tag{9.9}$$

where N_{UGS} is the number of UGS pixels, and N_{total} is the total number of pixels for an entire urban image.

Fig. 9.11 shows the distribution of urban green space coverage for different global climatic zones. From Fig. 9.11, the average UGSC was 56.59±5.31% at the global scale. The UGSC of temperate rainy cities yielded more than 76.56%, while the UGSC of arid climate cities occupied less than 16.36%. In addition, the histogram of global UGSC data indicates that UGSC of temperate rainy cities had higher UGSC values than cities in arid climate zones, tropical humid climate zones, and subtropical humid climate zones.

The main advantages of the proposed method include: (i) The CAI-net was used to eliminate cloud effects for improving UGS mapping in cloud-covered areas; and (ii) The proposed method can generate spatial continuous global UGS map products. This study can provide valuable global UGS data to support the UN SDG 11.

9.2.6 Conclusion

Optical satellite-derived UGS maps provide a timely, accurate, and essential data in sustainable urban development research. However, clouds are a key problem in optical satellite image processing that is highly pertinent to UGS mapping uncertainty. This study

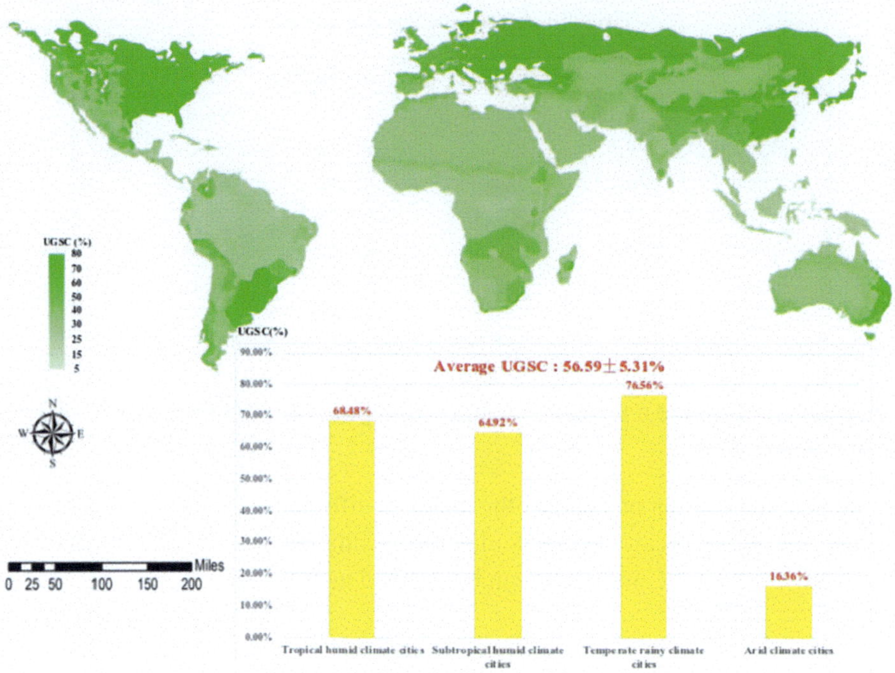

Figure 9.11:
Urban green space coverage in different climatic zones.

exploited how optical satellite data and crowdsourcing geospatial big data can be integrated for generating quality spatially continuous UGS maps. To eliminate cloud effects for UGS mapping, this research proposed an automatic UGS mapping method with three sequential modules, i.e., auxiliary data preprocessing module, the CAI-net, and the noncloud scenes classification module. The results show an average overall accuracy of 92.14%, average Kappa of 89.12%, average recall of 91.31%, and average precision of 90.21% for 1080 cities in different climatic zones. The main advantages of the proposed automatic UGS mapping method included: (i) Cloud attention intelligent network was designed to eliminate cloud effects to generate spatially continuous UGS maps of different geographical settings; and (ii) The proposed method provided a new insight into the use of AI technology for accurate UGS mapping globally in different climatic zones.

Nevertheless, the proposed CAI-Net has several limitations. First, CAI-Net struggles to retrieve detailed UGS classifications in regions where cloud cover exceeds 50%, as the model requires high-quality auxiliary data to infer obscured features in heavily cloud-contaminated areas. This constraint is particularly pronounced when compared to nondeep-learning fusion methods, such as Bayesian probabilistic frameworks or rule-based multisensor fusion, which often rely on historical data or spatial interpolation to fill gaps under cloud cover. While these traditional approaches are less data-hungry, they lack CAI-Net's dynamic adaptability to real-time auxiliary inputs. Second, the performance of CAI-Net is inherently tied to the quality and resolution of auxiliary data (e.g., meteorological or land-use datasets), as its cloud classification module depends on these inputs to probabilistically estimate UGS distributions. This contrasts with simpler fusion techniques, such as weighted linear combinations or decision tree-based stacking, which prioritize interpretability over data dependency but sacrifice the nuanced feature extraction capabilities of deep learning. For instance, CAI-Net's attention mechanisms enable targeted feature prioritization, whereas nondeep methods often generalize poorly across heterogeneous urban landscapes. Additionally, CAI-Net's computational complexity, a common challenge in deep learning, limits its deployability in resource-constrained settings. Lightweight fusion approaches, such as index-based blending or pixel-wise arithmetic fusion, offer computationally efficient alternatives but fail to capture the hierarchical spatial-contextual relationships that CAI-Net leverages. Future work should focus on hybrid strategies to mitigate these limitations. Integrating multimodal data sources, such as Sentinel-1 SAR imagery for cloud-penetrating capabilities and street-view images for ground-truth validation, could enhance robustness in high-cloud scenarios. Furthermore, combining CAI-Net's attention mechanisms with explainable nondeep techniques might balance adaptability with interpretability, addressing both data dependency and computational overhead.

9.3 Urban functional zone recognition using multimodal remote sensing data

9.3.1 Introduction

Cities are the main space for human socioeconomic activities, the primary engine of global economic growth, occupying only 2% of the world's land area while supporting 50% of the population, consuming 75% of the energy, and emitting 85% of carbon dioxide [170]. Over the past 20 years, China has undergone massive infrastructure development and achieved great accomplishments. According to data from China's 2020 census, the urbanization rate in China increased from 36.22% in 2000 to 63.89% in 2020, meaning that about 400 million people have moved from rural areas to cities [36], indicating a continuous expansion of Chinese cities over nearly two decades. Accompanying urban expansion is the constant change in urban land use that reflects socioeconomic functions and human activities [65,94]. Meanwhile, the actual situation of urban construction often deviates from early top-level planning, with government-designed urban planning maps being hard to update timely (e.g., urban planning maps are typically redrawn every five or ten years). The inconsistency between urban planning and the current state of urban land use also leads to the continuous emergence of demands for monitoring changes in urban land use [92].

According to the current standards for classification of land use status (GB/T 21010-2017) and the standards for classification and planning of urban land use (GB 50137-2011), urban areas mainly include more than ten types of land uses (functions) such as residential, business office, commercial service, industrial, transportation, administrative, educational, medical, sports cultural, and park green spaces. Each land use area is primarily for one or more functions (e.g., residential, educational, or commercial). For example, if a region's land resources are mainly used for industry, then the area has a strong industrial function; if the land resources are mainly for residential construction, then the area has a strong residential function. Urban land use areas are an important part of the city, and this geographic information, which is highly related to human activities, is crucial for urban planning, landscape design, environmental management, health promotion, and biodiversity conservation [55]. The identification of urban land use is key to the rational planning and management of cities, similar to traditional land use and land cover (LULC) classification, but also different. The latter emphasizes the physical characteristics of the land surface, whereas the urban land use discussed in this section focuses more on the socioeconomic functional attributes of urban areas.

However, due to the abundance of human dynamic activities in cities and the inconsistent spatial granularity of functional areas, perceiving the temporally continuous and spatially seamless distribution of urban land use remains a challenging task. Initially, remote sensing images were used to study urban functions. Remote sensing images provide a bird's-eye view that captures a holistic understanding of land use (such as lakes, farmland, buildings, etc.) [19,20]. However, the long revisit cycle (low temporal resolution) of remote sensing images, the same observation capability for urban and rural areas (relatively low spatial resolution for urban areas), and the lack of socioeconomic semantic information [66] limit their ability to further understand the spatial interaction between people and places [171]. Urban function zone identification through remote sensing images has been proven efficient and effective. However, in large cities with higher human activity density and deeper spatial fragmentation, such as Wuhan, Beijing, New York, Singapore, etc., their perception and recognition capabilities for urban land use are insufficient. This is determined by the following facts. First, the shadows of many high-rise buildings in high-density cities have a significant impact on remote sensing images; second, the land use in cities differs from conventional LULC data, not only in spatial granularity but also in more significant mixed characteristics, with multiple land use types often coexisting on a single street; moreover, human activities also have a stronger and undeniable impact on urban land use [94]. A large number of human socioeconomic activities exist in cities [91]. In recent years, with the continuous advancement of physical sensing (GPS terminals, drones, remote sensing satellites, water levels, temperature, humidity, light intensity, air quality, and other in-situ sensors) and social sensing (such as street views, social media like Weibo, photo-sharing platforms like Flickr, Meituan, Tencent location data, ride-hailing data, etc.) technologies [35], and with 5G and smart devices becoming increasingly popular, Internet big data has achieved extensive coverage in urban spaces. Smartphones, wearable devices, appliances, medical devices, industrial detectors, surveillance cameras, cars, and other widely covered sensor terminal devices have greatly promoted the development of urban perception abilities [164], with each person being considered a "sensor", providing feedback on events and entities in real urban spaces [90], and generating a large amount of multimodal urban data with location information in real-time [103]. Data not only includes numbers but also natural language texts and various forms such as pictures and videos [164,167]. These data have greatly deepened our understanding of cities, bringing significant changes to the study of urban land use.

Street view is a type of geotagged image data that has received a lot of attention in recent years [11]. Street views may be collected by commercial companies through cameras, street view sampling cars backpacks, or come from social media (Weibo, Twitter), open photo-sharing platforms (Tuchong, Flickr), and other social sensing pathways. Compared to the remote sensing images, street views contain richer urban semantic information [26], have a higher time perception granularity, and can provide almost real-time feedback; in cities, they have finer spatial granularity, not only including main roads, small streets, but even penetrating inside buildings. The perception results not only include urban entities such as people, vehicles, and objects, but also detailed shapes, colors, and other original information [165], which are of great significance for urban land use identification. Sensor networks have higher accuracy and fixed formats, with high data availability density. In contrast, social sensing data have the advantages of large volume, wide coverage, and high observation density, and can record human behavior patterns in detail [45]. As shown in Fig. 9.12, how to couple sensor networks with social sensing networks, extract reliable information from large amounts of noisy data, and then capture the characteristics and dynamics of urban economy, culture, life, and safety has great research potential [90].

9.3.2 Review of existing approaches to urban functional zone identification

Existing studies on urban function zone identification can be mainly divided into three types. The first type is based on remote sensing image data, using convolutional neural networks (CNN) and their variants for identification [179]. The second type mainly relies on taxi trajectory [142,163], mobile signal data [81,159], and floating car trajectory data [157,169], using trajectory similarity matching and origin-destination (OD) complex network analysis methods to identify urban functional areas. The third type uses natural language processing

Figure 9.12:
Multiscale identification of urban functional zones through the coupling of sensor networks and social sensing networks.

(NLP) algorithms, based on points of interest (POI), social media, and other data to extract potential geographical features of spatial units, and then identifies urban land use areas through supervised or unsupervised methods. The most commonly used mining algorithms are the term frequency-inverse document frequency (TF-IDF) method [2], the Latent Dirichlet Allocation (LDA) topic model [12], and the Probabilistic Latent Semantic Analysis (pLSA) [14] method. These unsupervised clustering methods are not limited by text form and can automatically determine the number of urban functional areas (topics), generate topic vectors, and calculate differences between different land use areas based on semantic similarity [48]. With the introduction of the Word2vec model by Google, this word embedding method has been well-received and widely applied in urban land use identification [86,162]. A standard processing method is to compare POI categories, such as business, transportation, etc., to "words" and regions to "documents" [154]. Subsequently, improved algorithms of Word2vec, such as Place2vec, appeared, which not only consider the spatial relationship of POI but also obtain embeddings with higher information density.

According to statistics, 50.98% of urban function or land use recognition studies have used only a single data source, while 49.02% of studies have used two or more datasets [89,156], and given the easy accessibility and rich geographical significance of POI, it has become the most widely used data, adopted by 74.51% of studies. Currently, the highest accuracy and widest coverage is achieved by the EULUC dataset released by Tsinghua University [56], which is based on multiple data sources such as 10-meter satellite images, OpenStreetMap

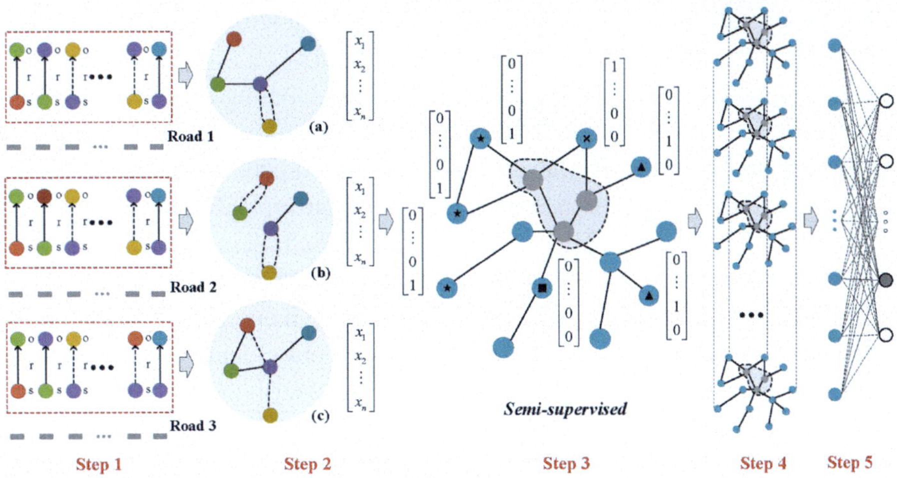

Figure 9.13:
Schematic diagram of the Knowledge-Topology Two Layers Spatially Dependent Graph Neural Network.

(OSM), night-time remote sensing, POI, and Tencent location big data, using a decision tree model to generate land use products for 27 major cities in China, including 5 primary classification labels and 12 secondary classification labels, with an overall classification accuracy of 60%.

9.3.3 Methodology

In this section, we introduce a method for identifying urban functions based on both street view data and road network topology. As shown in Fig. 9.13, Step 1 and Step 2 are the encoder part of the model, and Step 3, Step 4, Step 5 are the decoder part of the model.

9.3.3.1 GNN-based modeling of spatial relationships between street scenes

To capture the spatial relationships between the street scenes, we use road data as a network structure and choose the street with multiple streetscapes as the minimum spatial unit. We design to capture two layers of spatial relationship, the first one is to capture overlapping entities in different views, and the second one is to capture the spatial topology relationship between streets. Graph Neural Networks (GNN) can handle non-Euclidean structure data and extract spatial features from the topological graph for learning efficiently [140,141]. Existing studies have proved that the performance of GeoAI models can be improved by considering the spatial topology [180]. Besides, GNNs are able to learn features of the nodes based on their local neighborhood, which is defined by the edges connecting them. This allows GNNs to work with only a small subset of the labeled nodes, or even with completely unlabeled graphs [175]. While urban function (land use) zone identification is a very typical label-scarcity type task, spending a lot of manpower on land use labeling is time-consuming and labor-intensive, and semisupervised methods like GNN are very useful for such task [173]. Considering that urban functions will have certain colocation patterns or spatial dependencies [158], we introduce GNN to capture the topological relationships of roads in order to improve recognition accuracy. For example, Hu et al. [65] used traffic trajectory data, treated roads as nodes and extracted the topological relations between them. Nonetheless, such a relationship lacks semantic information and the description of the physical environment. This limitation is why the trained model is not rich in predictions and can only identify three types of functions: commercial, public, and traffic. Xu et al. [151] described city region features based on POI data and the relative relationship between POIs, extended the prediction results to six categories. However, it can only use the category attributes of the POI (a standard treatment is to analogize POI categories, such as business, traffic, etc., to 'words' and regions to 'documents' but much information is lost in such methods [154]), and lacks the use of road network topology information [70]. Inspired by the application of vision-language multimodality in remote sensing image analysis [141], we propose a two-layer spatially dependent graph neural network based on Knowledge (socioeconomic information extracted from the physical environment of the streetscape)-Topology (neighborhood relationships between road networks).

9.3.3.2 Two layer knowledge-topology graph neural network

Using the urban road networks as a backbone, we generated a street semantic knowledge graph and computed spatial relationships between street view images. Our method is divided into five steps and includes three main parts.

Using the urban road networks as a backbone, we generated a street semantic knowledge graph and computed an embedding representation for each street node. The first spatially dependent layer is a knowledge layer that generates objective descriptions of street view images in 'human language' which are then aggregated to the whole street. In other words, calculating the caption of the street scene. Machine reasoning is used to deconstruct and understand the content of the image, then generate a natural semantic description of the given scene [64]. The cross-modal model has the ability to describe the objects, attributes and relationships between objects that appear in the image [153], which will contain more information compared to the traditional results of semantic segmentation [5] and object detection [101]. The second spatially dependent layer is the topology layer, which constructs a spatial weight matrix based on the road spatial topology. It can take full advantage of the spatial dependence of the urban functions distribution [51]. Our method can fully use the graph semisupervised learning features to obtain accurate prediction results with only a small number of training labels.

Considering that in a spatial unit, street view images are often sampled at different times, this means that not only socioeconomic and physical built information can be perceived, but also temporal-spatial changes, which was often neglected [150].

Our method is divided into five steps and includes three main parts. The first part is the cross-modal decoder of the streetscape images to obtain a highly semantic compressed description of the city scene; the second part is the semantic encoder of the decoder results together with the OSM network topological relations and then input to the graph network for semisupervised learning; finally, we extract the knowledge triple from the images to generate a street semantic knowledge graph and text summaries under each city function, which make the city not only can be 'watched' but also can be 'read'. More details could be found within [172].

9.3.4 Study area and data preprocessing

Our study area is within the third ring of Wuhan in China. We use the EULUC-China dataset to label the streets. The dataset includes five primary classification labels and 12 secondary classification labels, with an overall classification accuracy of 60%. Among those secondary categories, road classification is not the type we focused, and the study area does not contain Airport facilities type, we excluded such irrelevant labels. As a result, we constructed a street function dataset containing ten types. As elaborated in Step 1, we have constructed the relationship between streetscape and street, therefore, here, we simply label the street's primary function. In the result, we have labeled a total of 190 streets (the scale of the urban functional zones in EULUC is at the TAZ level, whereas this research is at street scale), with an overall annotation rate of 5.3%. We also collected 75,628 Tencent Street View images from 13,889 streetscape sampling locations. Each has four images from different angles.

Then we perform cross-modal decoding of those images and generate the most relevant five scene descriptions for each one. Consequently, we obtained a total of 378,140 sentences after about 72 hours computation on a Tesla P100 GPU. We use Queen contiguity to calculate the spatial weight matrix, with 23,549 neighboring relationships. There are 5458 OSM segments in the study area, with an average of 5 relations per street, so we calculate 5-Nearest ($K = 5$) neighborhoods (27,290 neighboring relationships) as another set of spatial weights.

9.3.5 Results

9.3.5.1 Model training and accuracy comparison

We conducted 12 sets of comparison experiments using different models as shown in Table 9.8. The main differences between the models are the different semantic encoder, the

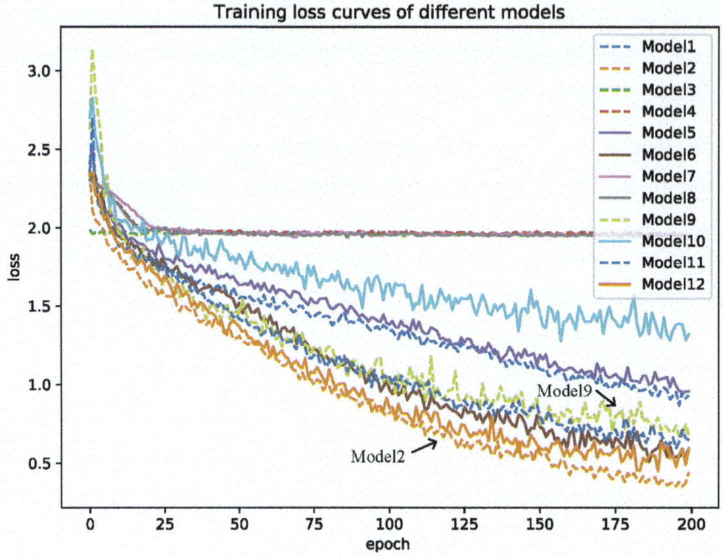

Figure 9.14:
Training loss curves of different models. Dashed lines indicate the use of 5-Nearest neighbors, while solid lines indicate 10-Nearest neighbors.

spatial weight matrix and the graph feature extraction network, and each model is given a unique name (Model 1, Model 2, etc.). We performed 200 rounds of epochs for all 12 models to test their prediction accuracy, and the training loss variations are shown in Fig. 9.14.

We only discuss models where the number of prediction categories is equal to 10, because these models can maintain a better generalization ability. As can be observed, using Bert as the encoder will achieve higher accuracy (Model 2 and Model 4). Such high accuracy shows that Bert has a solid ability to extract scene features and can reduce training loss. On the contrary, the traditional method (Doc2vec) can not reduce the loss even after 200 epochs. Take Model 2 for example, when using the basic GCN structure as a semisupervised classifier, $P@1 = 0.395$, $P@3 = 0.638$, $P@5 = 0.730$, and $P@10 = 0.807$. The result means that under the stricter criterion ($P@1$), about 40% of the predicted labels of streets agree with the EULUC; under the loose criterion ($P@10$), about 80% of the predicted street labels are consistent with one or more labels of the ten nearest parcels. Note that this does not mean that the accuracy of our model is only 40%. Limited by the low quality of EULUC data, and since one street may have multiple functions, loose criterion ($P@10$) may be a better choice for model assessment.

In Step 2 of this method, GAT requires more parameters for training (additional weight matrix), Model 9 and Model 10 are the slowest but achieve the highest accuracy ($P@1 = 0.462$,

Table 9.8: The accuracy comparison of 12 models under different conditions.

Model	Encoder	Weight	Graph Structure	@1	@3	@5	@10	Number	Time (s)
Model 1	Bert	K-Nearest	GCN1	0.344	0.568	0.654	0.758	8	39.672
Model 2	Bert	K-Nearest	GCN2	0.395	0.638	0.73	0.807	10	27.899
Model 3	Bert	K-Nearest	GAT1	0.358	0.589	0.678	0.772	8	52.341
Model 4	Bert	K-Nearest	GAT2	0.402	0.645	0.738	0.815	10	45.678
Model 5	CLIP	K-Nearest	GCN1	0.412	0.658	0.745	0.823	8	35.421
Model 6	CLIP	K-Nearest	GCN2	0.445	0.689	0.772	0.846	10	28.123
Model 7	CLIP	K-Nearest	GAT1	0.428	0.672	0.759	0.834	8	48.892
Model 8	CLIP	K-Nearest	GAT2	0.456	0.698	0.785	0.858	10	42.567
Model 9	CLIP	Topology	GCN1	0.489	0.725	0.812	0.879	8	31.234
Model 10	CLIP	Topology	GCN2	0.512	0.748	0.834	0.895	10	25.891
Model 11	CLIP	Topology	GAT1	0.478	0.712	0.798	0.868	8	44.567
Model 12	CLIP	Topology	GAT2	0.498	0.738	0.821	0.886	10	38.234

Note: GCN1, GCN2: the computational units are 64 and 32, 32 and 16, respectively.

$P@3 = 0.730$, $P@5 = 0.824$, $P@10 = 0.899$). This performance is quite satisfactory and the model's predictions are excellent. The performance proves that Street View can detect urban street functions effectively, and introducing spatial topology improves the city's interpretation level.

In addition, as shown in Fig. 9.14 that when GCN is used as a classifier (Model 2, Model 3), the training loss is low at the beginning, indicating that the model already has a good feature extraction capability without training. Model 2, Model 6, Model 9, Model 11, and Model 12 have the fastest converge speed. These models also have good generalization ability and can accurately identify the ten functional categories. Using the 5-Nearest ($K = 5$) neighborhoods (dashed line, Model 5, Model 6) to calculate the spatial weights makes the model more likely to converge than Queen contiguity (solid line, Model 1, Model 2). The results in Table 9.8 also show the lower accuracy of using Queen contiguity.

9.3.5.2 Mapping urban functions at the street level

The most accurate prediction outcomes from Model 9 are visualized in Fig. 9.15. These results demonstrate the model's capacity to generate detailed, fine-grained spatial classifications corresponding to ten distinct urban functional categories (labels 0–9): Residential, Business Office, Commercial Service, Industrial, Transportation Station, Administration, Education, Medical, Sports and Culture, and Parks and Green Space. The subgraph in Fig. 9.15 highlights streets, which are described metaphorically as the city's 'nerve endings' due to their role in connecting and reflecting urban activity, as

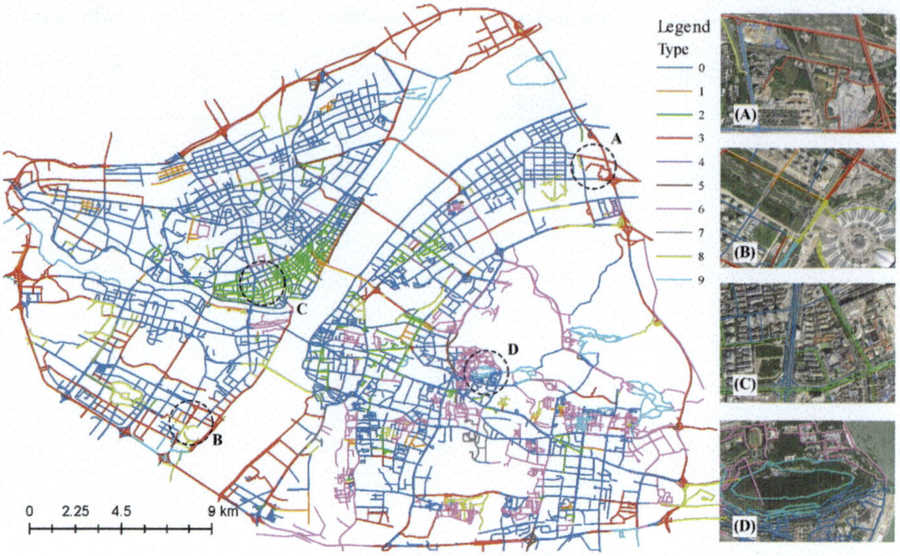

Figure 9.15:
Prediction results of SOTA model in the study area. Source of the road data: OpenStreetMap.

critical spatial units for functional analysis. Notably, the model achieves high precision in distinguishing between diverse functional zones even in scenarios where streets are geographically proximate, overcoming challenges posed by spatial adjacency. Unlike conventional approaches that rely on coarser spatial units such as building footprints or Traffic Analysis Zones (TAZ), which often aggregate functions into broader areas, our method captures nuanced functional patterns at the street level. This granular resolution enables the identification of overlapping or mixed functionalities within the same geographic region, revealing richer urban dynamics. For instance, a single street segment may simultaneously serve commercial, residential, and transit-related purposes—a complexity our model successfully disentangles, offering advantages for urban planning and policy decisions requiring localized insights.

9.3.6 Conclusion

Our method has three significant novelties and advantages: fusing multiple sources of urban proximate sensing images, supporting arbitrary scale spatial units, and being rich in socioeconomic information. The first advantage is that we can include the street view and other proximate images in the computation and generate the semantic representation (city caption), as long as the pictures can show the actual conditions of the places, and even indoor images can be included. These images can collectively form the overall image of the city [43], driving our understanding of the space deeper. The second advantage is that our research unit is not only limited to the street scale but also generalizable to other scales, even for a small sampling point or a region. The different scales differ only in the length of the semantic description text and the spatial topological relations. The third advantage is that because the city is rich in human activity footprints, the proximate images can penetrate deep into the city and provide real-time feedback on the 'people', 'vehicles', and 'things', and the raw shape & color information. It will help to sense the socioeconomic environment comprehensively. In contrast, remote sensing images in urban space are difficult to have a such high spatial and temporal resolution.

9.4 Mapping urban traffic noise with street view images

In modern society, as the process of urbanization accelerates, the intelligent management and planning of cities have become an important direction of development. Smart Cities use information and communication technology to enhance urban functions, improve the quality of life, and reduce both resource consumption and environmental impact. Among the many aspects of building smart cities, Urban sensing plays an indispensable role. By collecting and analyzing various types of data from the urban environment, urban sensing technology can provide more accurate and real-time support for city management decisions.

Traffic management is undoubtedly a key component in the application fields of smart cities. With the growth of urban populations, traffic congestion, and noise pollution have gradually become issues of increasing concern to urban residents. To effectively monitor and manage urban traffic noise, traditional methods often rely on fixed noise monitoring stations or temporary sound measurement activities. However, these methods face numerous challenges in obtaining the spatial distribution of traffic noise across the city. First, the limited number of fixed monitoring stations cannot cover the entire city, leading to spatial discontinuities in data collection. Secondly, temporary measurement activities are not only time-consuming and labor-intensive but are also difficult to carry out frequently, making it hard to update data in real-time and reflect the rapidly changing urban traffic conditions.

In response to these challenges, researchers and city planners have begun exploring new methods and technologies. Recent methodologies have leveraged statistical models and advanced computational techniques. For instance, Kim et al. [78] demonstrated the potential of using elementary urban forms to statistically predict road traffic noise levels in South Korean cities, emphasizing the role of early planning in mitigating urban noise pollution. Similarly, the work by Sun et al. [128] represents a significant leap forward, utilizing road surveillance videos for high-precision dynamic noise mapping. Such innovations underscore the evolving landscape of urban noise management, offering scalable and effective tools for smart city development. Within this group, the method "Mapping Urban Traffic Noise with Street Images" introduces a unique angle by leveraging street-level imagery to analyze urban noise dynamics. By analyzing information such as traffic flow, vehicle types, and road structures in street images, combined with advanced sound measurement technology, it is possible to more accurately assess the levels of traffic noise in the city. This method not only improves the efficiency and accuracy of noise monitoring but also provides strong data support for urban planning and noise management.

9.4.1 Earlier attempts to map urban traffic noise

Traffic noise is a significant contributor to urban noise pollution, affecting both the ecological environment and human health [76]. Due to the uneven distribution of noise across urban areas, traffic noise maps are created to illustrate the varying noise exposure levels throughout a city [98]. These maps serve as a tool for traffic authorities, providing crucial data to aid in the implementation of noise reduction strategies, such as the use of low-noise road surfaces, appropriate speed limits, and the installation of noise barriers and "No horn" signs. Additionally, these maps are useful for researchers studying the adverse health effects on the population, such as annoyance [105], sleep disturbances, and hearing impairments [77].

There are two principal approaches to obtaining the spatial distribution of urban road traffic noise: measurement-based and modeling-based methods. The measurement-based approach involves either stationary or mobile measurements of noise levels at various locations, as noted by Alam et al. [3]. The modeling-based approach, on the other hand, uses mathematical models to estimate noise by considering the environmental conditions, noise sources, and the laws governing sound propagation, as described by Garg and Maji [50].

Measurement-based studies have been conducted on noise acquisition [96,155], quantification [40,83], gradation [4,8], classification [9,95,139,160], and source backtracking [120,130]. In particular, the traffic noise of different road forms was studied separately, such as at different road grades [15], pavement types [46,121], intersections [104], overpasses [49], and roadworks sites [129]. In addition, key transport facilities have also emerged as research hotspots of traffic noise studies, exemplified by subway exits [52,53,100], railway stations [6,131], and airports [34,138]. Measurement-based methods are generally used for carrying out assessments or quantifying noise pollution by calculating the traffic noise index or generating noise maps with interpolation techniques [7,18,60]. However, these methods encounter several challenges: they are time-consuming, costly, and labor-intensive, they offer a low level of automation, and may not adequately represent the noise levels across different times and spaces.

Modeling-based methods provide a more feasible and efficient alternative to overcome the limitations associated with measurement-based methods. There are two primary types of modeling-based approaches: process-based modeling and statistical modeling.

Process-based models utilize acoustic principles and environmental dynamics to simulate how sound propagates, accurately representing its travel and interaction with surfaces like buildings and trees. These models are particularly well-suited for scenarios that demand highly accurate noise predictions. They offer a thorough understanding of the mechanisms of noise propagation, accommodating complex interactions and multiple influencing factors. However, the complexity of establishing and computing these models means they require a significant number of physical parameters and computational resources, making them less ideal for quick assessments or applications on a large scale, especially in an urban environment.

Modeling-based methods primarily predict noise levels by analyzing statistical relationships between noise and variables such as time, location, and traffic volume. Unlike process-based models, this approach does not simulate the physical propagation of sound. These models are particularly useful for quick assessments or in scenarios where physical parameters are not fully available, such as noise monitoring, urban noise map creation, and noise management policy formulation. The advantages of statistical models include simpler processes, lower computational demands, and suitability for handling large datasets, dynamic environments, and long-term trend analysis. However, they may not fully account for complex sound propagation mechanisms and physical impacts, potentially leading to lower accuracy and explanatory power compared to physics-based models.

The road traffic noise characteristics of different locations and their acoustic environments can be estimated by two modeling approaches: empirical (statistical) models and machine-learning models. Empirical models, such as the FHWA model in the USA [10], ASJ RTN 2008 in Japan [152], CoRTN in the UK [54], RLS 90 in Germany [112], Son Road in Switzerland [62], and Harmonoise in Europe [143], are widely used by local and national road authorities across different regions. These models integrate traffic-related factors and road conditions, accounting for noise propagation and reduction in the air. However, their reliance on assumed general linear relationships among predictors may limit their effectiveness in capturing complex nonlinear interactions and relationships.

Neural network-based models, which have nonlinear fitting capabilities, generally perform better than classical statistical models [16,99]. These models primarily use traffic volume and velocity as key inputs, while also integrating factors such as road gradients [25], roadway temperature [58], and honking occurrences [119]. Despite this integration, predicting road traffic noise accurately remains challenging due to its complex physical mechanisms and the high-dimensional, large-scale factors involved in noise generation and propagation, resulting in the existing models requiring extensive data on traffic noise and environmental factors, which can often limit their application due to data availability issues.

9.4.2 Methodology framework

This section will delve into one application of "Mapping Urban Traffic Noise with Street Images" in smart cities, including the principles of the technology, implementation methods, contributions, and potential applications. Through this research, we aim to maximize the value of urban digital assets such as street view imagery and provide effective strategies and solutions for achieving more livable, healthy, and intelligent urban environments.

The methodological framework consists of three primary steps: (1) Data Collection and Processing, (2) Feature Extraction using Deep Convolutional Neural Networks (DCNN), and (3) Model Training and Mapping. A diverse array of datasets is gathered, including road networks, street view imagery, and actual road traffic noise measurements. Following data acquisition, this information is processed to prepare the training datasets. Finally, a deep learning model is used to predict traffic noise levels from Street View Images (SVIs) (Fig. 9.16).

9.4.2.1 Noise estimation from street view images

Addressing the challenge of road traffic noise estimation as an end-to-end CNN regression task, this study adopts a classification-then-regression strategy to enhance model stability amidst inherent complexity and uncertainty. This method demonstrated to strike a balance between efficiency and accuracy in previous research [72,146,164], leverages classification to initially discern noise levels from street view imagery (SVI) based on road use and visual characteristics. The findings of Barrigón Morillas et al. [9] and Carmona del Río et al. [21] suggest that traffic noise stratifies along roads, influencing the visual features captured in SVIs, which display significant diversity and complexity. The approach begins with employing convolutional neural network (CNN) classification techniques to identify variances and commonalities within SVIs. Subsequently, the output probability vector is converted back to noise values through regression algorithms, enhancing the model's ability to capture detailed features and learn meaningful representations, thus improving accuracy and performance in noise estimation. Geographical coordinates and in-situ physical measurements of traffic noise serve as ground truth labels to correlate street view imagery with sound information. Initially, SVIs are categorized into N classes based on observed continuous noise values, N_{observed}, with each category representing a specific noise level range. Deep Convolutional Neural Network (DCNN) is then trained to recognize noise patterns and extract high-dimensional visual features from SVIs. These features are processed through a softmax

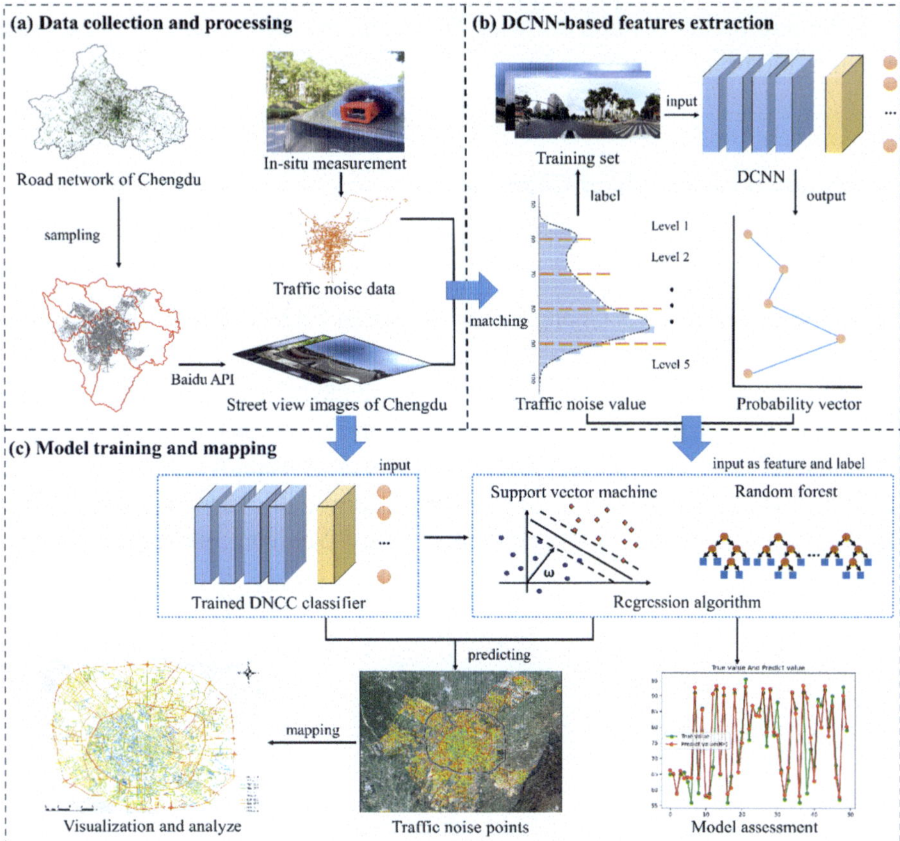

Figure 9.16:
Flowchart of the proposed method for road traffic noise estimation using street view images.

function to produce a probability vector $Q = (q_1, q_2, \ldots, q_n)$, indicating the likelihood of each noise level category. Then we used regression algorithms to further refine these probabilities to numerical noise estimations in decibels. Upon completion, the model can estimate road traffic noise from any street view image.

9.4.2.2 DCNN model interpretability

While DCNNs excel at extracting both detailed spatial information in lower layers and abstract semantic information in higher layers, their "black box" nature limits interpretability. To address this, Gradient-weighted Class Activation Mapping (Grad-CAM) is employed. This technique provides visual interpretation by highlighting discriminative regions in images using the activations from specific convolutional layers. This study utilizes Grad-CAM at the last convolutional layer to generate heat maps superimposed on original images, thereby enhancing model interpretability and allowing for qualitative assessment of noise estimation capabilities based on urban street view imagery.

9.4.3 Application: a case study of Chengdu city

9.4.3.1 Study area

The study area for this research is Chengdu, the capital of Sichuan Province in southwest China. Chengdu is home to a substantial permanent population of nearly 21 million residents as of 2020, and it ranks second in China in terms of vehicle ownership, with over 5 million vehicles registered in the same year. The research primarily focuses on the municipal district of Chengdu, which includes the central area encompassing five historical districts: Wuhou, Jinjiang, Chenghua, Qingyang, and Jinniu. Additionally, the study extends to the suburban areas, excluding Xinjin but incorporating Pidu, Wenjiang, Shuangliu, Longquanyi, Qingbaijiang, and Xindu, as well as two newly developed special districts: the Hi-Tech Industrial Development Zone and Tianfu New Area. A map detailing the spatial distribution of the main road types in Chengdu is provided in Fig. 9.17.

Figure 9.17:
Study area in Chengdu City, China showing the road network and data collection routes.

9.4.3.2 Data

(1) **Road network:** Road network data were obtained from the OpenStreetMap (OSM). OSM is a publicly accessible and collaborative spatial database of Volunteered Geographic Information (VGI) that provides annotations of land use at the street level. Some metadata information, such as road types and land use, is included. In our work, we leveraged the OSM road network data available within the study area to fetch street-view images.
(2) **Street View Images:** Street View Images (SVIs) offer detailed visualizations of physical environments along almost every navigable street, captured via cameras mounted on mobile vehicles. In this study, street view imagery was obtained from Baidu Map, the Baidu Street View (BSV) service is known for its extensive geographic coverage across urban areas in China, characterized by high-quality imagery and rapid update frequencies. Each image includes a unique place-ID number, as well as corresponding latitude and longitude coordinates. The photos are stored in a resolution of 1024 by 360, with azimuth and elevation angles set at 0 degrees.
(3) **In-situ Traffic Noise:** A portable device was developed and mounted on moving vehicles to capture traffic noise dynamically across the road network. The device, based on a Raspberry Pi, includes a GPS module and a noise collection module, aiming to synchronize geographic positioning with noise data collection. This integration is crucial for matching traffic noise with high-resolution street view images. The device records data every five seconds, including the date, time, location coordinates, and in-situ road traffic noise in decibels. Noise data is refined using a moving average over a 20-second window, reducing interference from transient noise sources by averaging the current and previous three records. Additionally, to reduce random fluctuations, multiple samples were taken on most road segments at different times.

The device was also mounted on the boot lid of an electric vehicle to eliminate vibrations from petrol engines and enhance noise measurement accuracy. Measures to mitigate wind effects (especially at speeds over 20 km/h) on noise measurements include a microphone windproof fleece cover and software adjustments detailed in the following section.

9.4.3.3 Data preprocessing

To ensure data quality and construct a training dataset, we executed several processing steps:

(1) **Speed Calculation:** We calculated the experimental vehicle's speed by measuring time intervals and geographic distances between data points using GPS coordinates.
(2) **Outlier Removal:** We filtered out data points with speeds of 0 km/h or over 140 km/h and excluded abnormally high traffic noise values.

(3) **Correlation Analysis:** We analyzed the relationship between driving speed and road noise for the next step.
(4) **Detrending:** Using piecewise detrending, we divided the data into five classes based on noise decibels, and each class was detrended with the relationship found in step 3.
(5) **Data Matching:** Each traffic noise point classified, was aligned with its corresponding road type based on geographic coordinates.

9.4.3.4 Construct training dataset

After data processing, we gathered street view images located within a 50-meter radius of each noise point along the road network. It should be noted that one street view image may correspond to multiple noise points. To address this, we calculated the average road traffic noise value for each image. Moreover, we filtered out panoramas captured in tunnel locations and implemented a 5-category classification strategy. Street view images associated with noise values falling within the same range were classified into level 1 to level 5.

Given the approximate Gaussian distribution pattern observed among the categories, we applied data augmentation techniques to address challenges associated with class imbalance and the risk of overfitting in the minority classes. Specifically, we mainly employed color enhancement and horizontal flipping on the training samples of the under-represented classes to generate images that closely simulate real street scenes. Furthermore, we adopted a stratified sampling strategy to reconstruct the majority class for each specific road type.

9.4.3.5 Noise estimation

In this research, ResNet-34 was employed for the automated extraction of acoustic-related characteristics. For noise value estimation, the Support Vector Machine (SVM) and Random Forest (RF) algorithms, recognized for their effectiveness in addressing regression tasks, were utilized. In assessing the performance of our proposed model, we applied Accuracy, Precision, Recall, and F1-score to evaluate the classification task, while for assessing the performance of the DCNN-Regression model, we employed R-square (R2), mean absolute error (MAE), mean squared error (MSE), and root mean square error (RMSE). We used 90% of the road traffic noise data to build our training dataset, and the rest for overall evaluation.

9.4.4 Results

9.4.4.1 Descriptive statistics

During the period of 17th March to 11th April 2022, a total of 116,483 road traffic noise data points were gathered in Chengdu. Data collection typically took place from 9 am to 5 pm.

Table 9.9: The measurement of noise levels on various road levels and link roads.

Road types	Road traffic noise level				
	Level 1 (≤75 dB)	Level 2 (75–80 dB)	Level 3 (80–85 dB)	Level 4 (85–90 dB)	Level 5 (>90 dB)
Motorway	126	188	286	382	512
Trunk road	234	312	398	445	367
Primary road	456	523	489	312	198
Secondary road	678	589	423	256	134
Residential road	892	456	234	123	67
Link road	345	289	198	112	56

Table 9.10: Piecewise linear regression analysis for speed-related noise correction.

Subset	Noise decibel	Correlation coefficient	Intercept	Slope
1	≤ 60 dB	0.06	52.3	0.12
2	60-70 dB	0.35	48.7	0.28
3	70-80 dB	0.34	45.2	0.35
4	80-90 dB	0.17	41.8	0.42
5	> 90 dB	0.43	38.5	0.51

The collected noise data ranged from 51 to 101 dB, with an average noise level of 78.5 dB, a median of 81 dB, and a standard deviation of 9.95 dB. Table 9.9 presents the classification of road traffic noise levels and the respective counts of specific road types linked to the traffic noise data at each level.

9.4.4.2 Speed-related noise correction

The correlation coefficient between the instantaneous speed of the experimental vehicle and the corresponding road traffic noise is 0.54, indicating a moderate correlation with a p-value below 0.01. The coefficient of determination indicates that approximately 30% of the overall variation in road traffic noise can be explained by fluctuations in vehicle speed (Fig. 9.18).

The original dataset was divided into five segments based on noise decibels. Correlation coefficients and y-intercepts for piecewise linear regressions are provided in Table 9.10. After the correction process for each segment, the road traffic noise was categorized into level 1 to level 5 for the DCNN's input. Fig. 9.19 illustrates the geographical distribution of the initial in-situ traffic noise data and the processed traffic noise data.

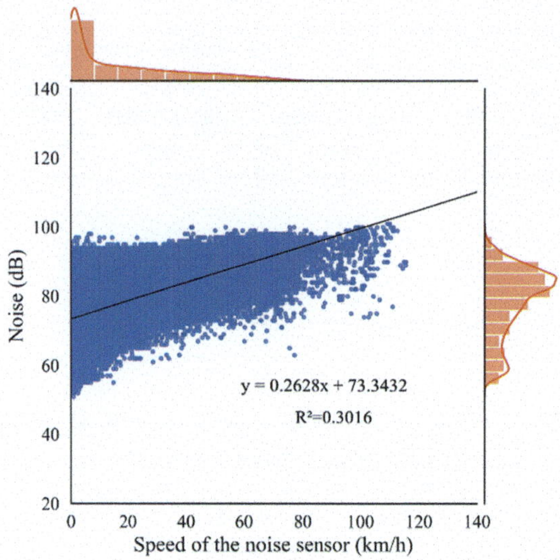

Figure 9.18:
Scatter plot of the vehicle speed and the measured traffic noise showing the correlation between these variables.

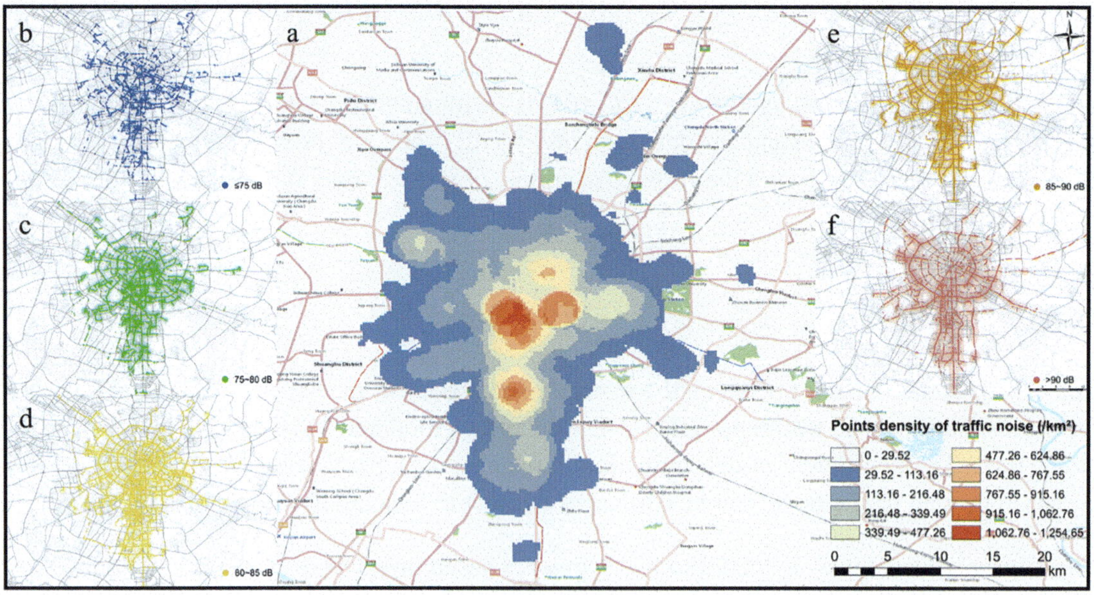

Figure 9.19:
A point density map of in-situ road traffic noise data (a) and spatial distribution of the road network (b) in Chengdu.

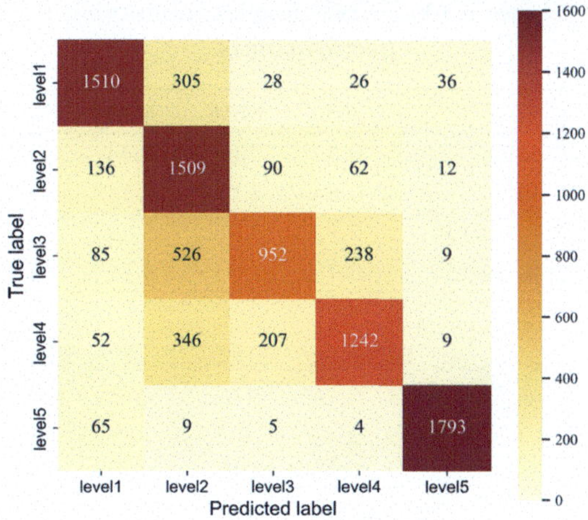

Figure 9.20:
Confusion matrix showing the classification performance of the DCNN model across five noise levels.

9.4.4.3 Model accuracy assessment

A total of 46,280 images were assigned in the training set, of which 9256 were used for validation. The classification model achieved an overall accuracy of 75.7% on the validation sets, with Precision, Recall, and F1-score values of 0.774, 0.736, and 0.755, respectively. Level 5 performed best at 95.5%, while Level 3 had the lowest accuracy at 52.5%. The model yielded accuracies of 79.2%, 83.4%, and 66.9% for level 1, level 2, and level 4, respectively. The results were averaged across cross-validations and the confusion matrix is visualized in Fig. 9.20.

The output vector with probability estimations for each class, along with the associated noise values, were then fed into regression models. Specifically, SVM with a linear kernel and RF algorithms were employed to predict road traffic noise at the decibel level. Table 9.11 presents the performance of the estimation model. In Fig. 9.21, we showcase several street view images with the corresponding estimated results from a chosen road segment in Chengdu.

9.4.4.4 Road traffic noise estimation

Fig. 9.22 displays the estimation result in the central area of Chengdu City. The OSM roads were utilized as the unit for computing the average road traffic noise by aggregating all the point-level SVI data into their respective road segments. It is observed that noise pollution is widespread in the circumferential ring roads and a series of connected roads. A considerable amount of roads with an average traffic decibel value exceeding 85 were concentrated on the

Table 9.11: DCNN-Regression models' performances.

Model	R^2	MAE	MSE	RMSE
DCNN-SVM	0.61	2.04	8.44	2.91
DCNN-RF	0.64	2.01	7.32	2.71
DCNN-ResNet34	0.72	1.78	5.89	2.43
DCNN-ResNet50	0.75	1.65	5.21	2.28

Figure 9.21:
SVI-based road traffic noise estimation examples for a road section in Chengdu.

third ring road and the outer ring freeway, which were constructed to accommodate the heavy traffic in the urban transportation system. In contrast, subsidiary streets within built-up areas exhibit lower levels of road traffic noise, as they generally have lower speed limits and traffic densities.

9.4.4.5 Interpretation of the DCNN model

Fig. 9.23 illustrates the representative Grad-CAM results for correctly classified street view images. For street view images categorized as "relatively quiet places" (level 1 and level 2), the DCNN model predominantly leverages visual cues from the surrounding urban environment, such as greenery and residential buildings. In contrast, images classified as level 3 and level 4 emphasize features like facade structures (e.g., bridges, buildings) and contextual details such as roadside greenery. With a majority of images falling into level 3 and level 4, the street scene is more complex and diverse with no distinct features guiding their categorization. At level 5, the model focuses primarily on road areas and the horizontal surroundings of the roadways. Interestingly, in street view images labeled with high-level road

Figure 9.22:
Road traffic noise estimation within the outer ring freeway of Chengdu showing spatial distribution of noise levels.

traffic noise, we noticed an emphasis on the sky, indicating the likelihood of increased noise levels on roads with more open visibility, such as highways.

9.4.5 Conclusion

This work broadens the scope of street view imagery in urban traffic analysis by establishing a novel link between the visual characteristics of the built environment and spatially explicit acoustic data, serving as a versatile tool for interdisciplinary research and policymaking. For academic communities, the resulting noise maps enable granular identification of traffic noise hotspots and facilitate systematic exploration of their correlations with concurrent urban stressors (e.g., localized air pollution, traffic congestion patterns, and accident frequency), thereby enriching holistic urban analytics. Practically, urban planners can leverage these high-resolution noise assessments to evaluate acoustic environmental quality across neighborhoods, guiding evidence-based interventions to shield noise-sensitive zones (e.g., residential areas, schools, or hospitals). By spatially contextualizing noise pollution within its visual drivers, the framework supports targeted policy design, such as optimizing traffic flows or mandating noise-reducing infrastructure in vulnerable areas.

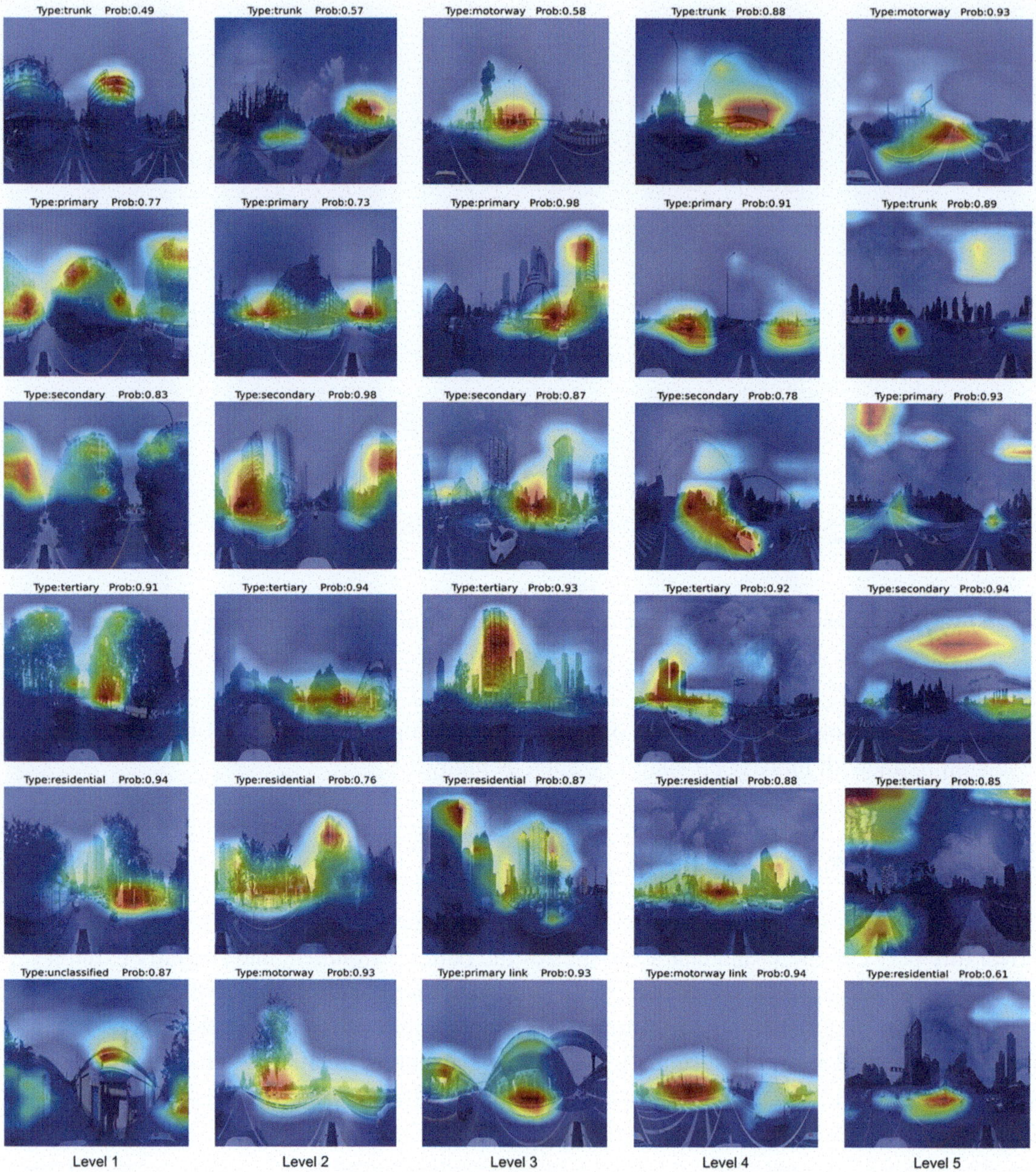

Figure 9.23:
Examples of the correctly classified street view images with the corresponding Grad-CAM results highlighting important visual features.

References

[1] N. Ahmed, J. Lee, D. Liu, et al., Identifying urban green space deserts by considering different walking distance thresholds for healthy and socially equitable city planning in the Global South, Urban Forestry & Urban Greening 89 (2023) 128123.

[2] A. Aizawa, An information-theoretic perspective of TF–IDF measures, Information Processing & Management 39 (1) (2003) 45–65.

[3] P. Alam, K. Ahmad, S.S. Afsar, et al., Noise monitoring, mapping, and modelling studies—a review, Journal of Ecological Engineering 21 (4) (2020) 82–93.

[4] A. Al-Ghonamy, Analysis and evaluation of road traffic noise in Al-Dammam: a business city of the eastern province of KSA, Journal of Environmental Science and Technology 3 (1) (2010) 47–55.

[5] M. Amiruzzaman, A. Curtis, Y. Zhao, et al., Classifying crime places by neighborhood visual appearance and police geonarratives: a machine learning approach, Journal of Computational Social Science 4 (2021) 813–837.

[6] S. Asfiati, M.N. Riky, J. Rajagukguk, et al., Measurement and evaluation of sound intensity at the Medan Railway Station using a sound level meter, Journal of Physics. Conference Series 1428 (2020) 012063.

[7] P. Aumond, A. Can, V. Mallet, et al., Kriging-based spatial interpolation from measurements for sound level mapping in urban areas, The Journal of the Acoustical Society of America 143 (5) (2018) 2847–2857.

[8] D. Banerjee, S. Chakraborty, S. Bhattacharyya, et al., Evaluation and analysis of road traffic noise in Asansol: an industrial town of eastern India, International Journal of Environmental Research and Public Health 5 (3) (2008) 165–171.

[9] J.M. Barrigón Morillas, V. Gómez Escobar, J.A. Méndez Sierra, et al., A categorization method applied to the study of urban road traffic noise, The Journal of the Acoustical Society of America 117 (5) (2005) 2844–2852.

[10] T.M. Barry, J.A. Reagan, FHWA Highway Traffic Noise Prediction Model, U.S. Department of Transportation, Federal Highway Administration, Washington, DC, 1978.

[11] F. Biljecki, K. Ito, Street view imagery in urban analytics and GIS: a review, Landscape and Urban Planning 215 (2021) 104217.

[12] D.M. Blei, A.Y. Ng, M.I. Jordan, Latent Dirichlet allocation, Journal of Machine Learning Research 3 (Jan 2003) 993–1022.

[13] D.K. Bolton, J.M. Gray, E.K. Melaas, et al., Continental-scale land surface phenology from harmonized Landsat 8 and Sentinel-2 imagery, Remote Sensing of Environment 240 (2020) 111685.

[14] A. Bosch, A. Zisserman, X. Munoz, Scene classification via pLSA, in: Computer Vision–ECCV 2006: 9th European Conference on Computer Vision, Graz, Austria, May 7–13, 2006, Proceedings, Part IV 9, Springer, 2006, pp. 517–530.

[15] G. Brambilla, R. Benocci, C. Confalonieri, et al., Classification of urban road traffic noise based on sound energy and eventfulness indicators, Applied Sciences 10 (7) (2020) 2451.

[16] L. Bravo-Moncayo, J. Lucio-Naranjo, M. Chávez, et al., A machine learning approach for traffic-noise annoyance assessment, Applied Acoustics 156 (2019) 262–270.

[17] Y. Cai, Y. Chen, C. Tong, Spatiotemporal evolution of urban green space and its impact on the urban thermal environment based on remote sensing data: a case study of Fuzhou City, China, Urban Forestry & Urban Greening 41 (2019) 333–343.

[18] A. Can, L. Dekoninck, D. Botteldooren, Measurement network for urban noise assessment: comparison of mobile measurements and spatial interpolation approaches, Applied Acoustics 83 (2014) 32–39.

[19] R. Cao, W. Tu, C. Yang, et al., Deep learning-based remote and social sensing data fusion for urban region function recognition, ISPRS Journal of Photogrammetry and Remote Sensing 163 (2020) 82–97.

[20] R. Cao, J. Zhu, W. Tu, et al., Integrating aerial and street view images for urban land use classification, Remote Sensing 10 (10) (2018) 1553.

[21] F.J. Carmona del Río, V. Gómez Escobar, J. Trujillo Carmona, et al., A street categorization method to study urban noise: the Valladolid (Spain) study, Environmental Engineering Science 28 (11) (2011) 811–817.

[22] B. Chen, B. Huang, B. Xu, Multi-source remotely sensed data fusion for improving land cover classification, ISPRS Journal of Photogrammetry and Remote Sensing 124 (2017) 27–39.

[23] B. Chen, B. Xu, P. Gong, Mapping essential urban land use categories (EULUC) using geospatial big data: progress, challenges, and opportunities, Big Earth Data 5 (3) (2021) 410–441.

[24] J. Chen, J. Chen, A. Liao, et al., Global land cover mapping at 30 m resolution: a POK-based operational approach, ISPRS Journal of Photogrammetry and Remote Sensing 103 (2015) 7–27.

[25] L. Chen, T. Liu, B. Tang, et al., Modelling traffic noise in a wide gradient interval using artificial neural networks, Environmental Technology 42 (23) (2021) 3561–3571.

[26] L. Chen, Y. Lu, Y. Ye, et al., Examining the association between the built environment and pedestrian volume using street view images, Cities 127 (2022) 103734.

[27] S. Chen, H. Zhang, H. Yang, Urban functional zone recognition integrating multisource geographic data, Remote Sensing 13 (23) (2021) 4732.

[28] W. Chen, H. Huang, J. Dong, et al., Social functional mapping of urban green space using remote sensing and social sensing data, ISPRS Journal of Photogrammetry and Remote Sensing 146 (2018) 436–452.

[29] Y. Chen, R. Fan, M. Bilal, et al., Multilevel cloud detection for high-resolution remote sensing imagery using multiple convolutional neural networks, ISPRS International Journal of Geo-Information 7 (5) (2018) 181.

[30] Y. Chen, L. Tang, X. Yang, et al., Thick clouds removal from multitemporal ZY-3 satellite images using deep learning, IEEE Journal of Selected Topics in Applied Earth Observations and Remote Sensing 13 (2) (2019) 143–153.

[31] Y. Chen, Q. Weng, L. Tang, et al., Thick clouds removing from multitemporal Landsat images using spatiotemporal neural networks, IEEE Transactions on Geoscience and Remote Sensing 60 (2020) 1–14.

[32] Y. Chen, Q. Weng, L. Tang, et al., Automatic mapping of urban green spaces using a geospatial neural network, GIScience and Remote Sensing 58 (4) (2021) 624–642.

[33] M. Claverie, J. Ju, J.G. Masek, et al., The Harmonized Landsat and Sentinel-2 surface reflectance data set, Remote Sensing of Environment 219 (2018) 145–161.

[34] B.S. Cohen, A.L. Bronzaft, M. Heikkinen, et al., Airport-related air pollution and noise, Journal of Occupational and Environmental Hygiene 5 (2) (2007) 119–129.

[35] S. Cowie, R. Arthur, H.T. Williams, @choo: tracking pollen and hayfever in the UK using social media, Sensors 18 (12) (2018) 4434.

[36] CSY, China Statistical Yearbook, China Statistical Publishing House, 2018.

[37] S. Di, Z.-L. Li, R. Tang, et al., Urban green space classification and water consumption analysis with remote-sensing technology: a case study in Beijing, China, International Journal of Remote Sensing 40 (5–6) (2019) 1909–1929.

[38] J. Dong, W. Kuang, J. Liu, Continuous land cover change monitoring in the remote sensing big data era, Science China Earth Sciences 60 (2017) 2223–2224.

[39] O. Douglas, M. Lennon, M. Scott, Green space benefits for health and well-being: a life-course approach for urban planning, design and management, Cities 66 (2017) 53–62.

[40] H. Doygun, D. Kusat Gurun, Analysing and mapping spatial and temporal dynamics of urban traffic noise pollution: a case study in Kahramanmaras, Turkey, Environmental Monitoring and Assessment 142 (2008) 65–72.

[41] S. Du, F. Zhang, X. Zhang, Semantic classification of urban buildings combining VHR image and GIS data: an improved random forest approach, ISPRS Journal of Photogrammetry and Remote Sensing 105 (2015) 107–119.

[42] M.E.D. Chaves, M.C.A. Picoli, I.D. Sanches, Recent applications of Landsat 8/OLI and Sentinel-2/MSI for land use and land cover mapping: a systematic review, Remote Sensing 12 (18) (2020) 3062.

[43] G. Filomena, J.A. Verstegen, E. Manley, A computational approach to 'The Image of the City', Cities 89 (2019) 14–25.

[44] G.M. Foody, Explaining the unsuitability of the kappa coefficient in the assessment and comparison of the accuracy of thematic maps obtained by image classification, Remote Sensing of Environment 239 (2020) 111630.
[45] S.A. Forrest, E.-M. Trell, J. Woltjer, Socio-spatial inequalities in flood resilience: rainfall flooding in the city of Arnhem, Cities 105 (2020) 102843.
[46] E. Freitas, C. Mendonca, J.A. Santos, et al., Traffic noise abatement: how different pavements, vehicle speeds and traffic densities affect annoyance levels, Transportation Research. Part D, Transport and Environment 17 (4) (2012) 321–326.
[47] X. Fu, H. Qu, Research on semantic segmentation of high-resolution remote sensing image based on full convolutional neural network, in: 2018 12th International Symposium on Antennas, Propagation and EM Theory (ISAPE), IEEE, 2018, pp. 1–4.
[48] S. Gao, K. Janowicz, H. Couclelis, Extracting urban functional regions from points of interest and human activities on location-based social networks, Transactions in GIS 21 (3) (2017) 446–467.
[49] W. Gao, L. Gao, L. Xu, et al., Assessment on road traffic noise in the Xizhimen Overpass area, Beijing, in: CICTP 2015, 2015, pp. 3290–3299.
[50] N. Garg, S. Maji, A critical review of principal traffic noise models: strategies and implications, Environmental Impact Assessment Review 46 (2014) 68–81.
[51] S. Georganos, T. Grippa, A. Niang Gadiaga, et al., Geographical random forests: a spatial extension of the random forest algorithm to address spatial heterogeneity in remote sensing and population modelling, Geocarto International 36 (2) (2021) 121–136.
[52] R.R. Gershon, R. Neitzel, M.A. Barrera, et al., Pilot survey of subway and bus stop noise levels, Journal of Urban Health 83 (2006) 802–812.
[53] M.R. Ghotbi, M.R. Monazzam, M.R. Baneshi, et al., Noise pollution survey of a two-storey intersection station in Tehran metropolitan subway system, Environmental Monitoring and Assessment 184 (2012) 1097–1106.
[54] S. Givargis, M. Mahmoodi, Converting the UK calculation of road traffic noise (CORTN) to a model capable of calculating $L_{Aeq,1h}$ for the Tehran's roads, Applied Acoustics 69 (11) (2008) 1108–1113.
[55] F.-Y. Gong, Z.-C. Zeng, F. Zhang, et al., Mapping sky, tree, and building view factors of street canyons in a high-density urban environment, Building and Environment 134 (2018) 155–167.
[56] P. Gong, B. Chen, X. Li, et al., Mapping essential urban land use categories in China (EULUC-China): preliminary results for 2018, Science Bulletin 65 (3) (2020) 182–187.
[57] A. Guo, J. Yang, X. Xiao, et al., Influences of urban spatial form on urban heat island effects at the community level in China, Sustainable Cities and Society 53 (2020) 101972.
[58] K. Hamad, M.A. Khalil, A. Shanableh, Modeling roadway traffic noise in a hot climate using artificial neural networks, Transportation Research. Part D, Transport and Environment 53 (2017) 161–177.
[59] R. Hang, Z. Li, P. Ghamisi, et al., Classification of hyperspectral and LiDAR data using coupled CNNs, IEEE Transactions on Geoscience and Remote Sensing 58 (7) (2020) 4939–4950.
[60] B.I. Harman, H. Koseoglu, C.O. Yigit, Performance evaluation of IDW, Kriging and multiquadric interpolation methods in producing noise mapping: a case study at the city of Isparta, Turkey, Applied Acoustics 112 (2016) 147–157.
[61] V. Heikinheimo, H. Tenkanen, C. Bergroth, et al., Understanding the use of urban green spaces from user-generated geographic information, Landscape and Urban Planning 201 (2020) 103845.
[62] K. Heutschi, SonRoad: new Swiss road traffic noise model, Acta Acustica united with Acustica 90 (3) (2004) 548–554.
[63] D. Hong, J. Hu, J. Yao, et al., Multimodal remote sensing benchmark datasets for land cover classification with a shared and specific feature learning model, ISPRS Journal of Photogrammetry and Remote Sensing 178 (2021) 68–80.
[64] M.Z. Hossain, F. Sohel, M.F. Shiratuddin, et al., A comprehensive survey of deep learning for image captioning, ACM Computing Surveys (CsUR) 51 (6) (2019) 1–36.

[65] S. Hu, S. Gao, L. Wu, et al., Urban function classification at road segment level using taxi trajectory data: a graph convolutional neural network approach, Computers, Environment and Urban Systems 87 (2021) 101619.

[66] Y. Hu, Geo-text data and data-driven geospatial semantics, Geography Compass 12 (11) (2018) e12404.

[67] Y. Hu, Y. Han, Identification of urban functional areas based on POI data: a case study of the Guangzhou economic and technological development zone, Sustainability 11 (5) (2019) 1385.

[68] C. Huang, J. Yang, N. Clinton, et al., Mapping the maximum extents of urban green spaces in 1039 cities using dense satellite images, Environmental Research Letters 16 (6) (2021) 064072.

[69] J. Huang, T. Fei, Y. Kang, et al., Estimating urban noise along road network from street view imagery, International Journal of Geographical Information Science 38 (1) (2024) 128–155.

[70] T. Inoue, R. Manabe, A. Murayama, et al., Landscape value in urban neighborhoods: a pilot analysis using street-level images, Landscape and Urban Planning 221 (2022) 104357.

[71] R. Javadi, N. Nasrollahi, Urban green space and health: the role of thermal comfort on the health benefits from the urban green space; a review study, Building and Environment 202 (2021) 108039.

[72] H. Ji, L. Qing, L. Han, et al., A new data-enabled intelligence framework for evaluating urban space perception, ISPRS International Journal of Geo-Information 10 (6) (2021) 400.

[73] S. Jiang, A. Alves, F. Rodrigues, et al., Mining point-of-interest data from social networks for urban land use classification and disaggregation, Computers, Environment and Urban Systems 53 (2015) 36–46.

[74] B.A. Johnson, K. Iizuka, Integrating OpenStreetMap crowdsourced data and Landsat time-series imagery for rapid land use/land cover (LULC) mapping: case study of the Laguna de Bay area of the Philippines, Applied Geography 67 (2016) 140–149.

[75] B.A. Johnson, K. Iizuka, M.A. Bragais, et al., Employing crowdsourced geographic data and multi-temporal/multi-sensor satellite imagery to monitor land cover change: a case study in an urbanizing region of the Philippines, Computers, Environment and Urban Systems 64 (2017) 184–193.

[76] J. Khan, M. Ketzel, K. Kakosimos, et al., Road traffic air and noise pollution exposure assessment—a review of tools and techniques, Science of the Total Environment 634 (2018) 661–676.

[77] M. Kim, S.I. Chang, J.C. Seong, et al., Road traffic noise: annoyance, sleep disturbance, and public health implications, American Journal of Preventive Medicine 43 (4) (2012) 353–360.

[78] P. Kim, H. Ryu, J.-J. Jeon, et al., Statistical road-traffic noise mapping based on elementary urban forms in two cities of South Korea, Sustainability 13 (4) (2021) 2365.

[79] N. Kranjčič, D. Medak, R. Zupan, et al., Support vector machine accuracy assessment for extracting green urban areas in towns, Remote Sensing 11 (6) (2019) 655.

[80] Z. Lan, M. Cai, Dynamic traffic noise maps based on noise monitoring and traffic speed data, Transportation Research. Part D, Transport and Environment 94 (2021) 102796.

[81] N.D. Lane, E. Miluzzo, H. Lu, et al., A survey of mobile phone sensing, IEEE Communications Magazine 48 (9) (2010) 140–150.

[82] A.C.K. Lee, H.C. Jordan, J. Horsley, Value of urban green spaces in promoting healthy living and wellbeing: prospects for planning, Risk Management and Healthcare Policy (2015) 131–137.

[83] B. Li, S. Tao, R. Dawson, Evaluation and analysis of traffic noise from the main urban roads in Beijing, Applied Acoustics 63 (10) (2002) 1137–1142.

[84] D. Liu, M.-P. Kwan, Z. Kan, Analysis of urban green space accessibility and distribution inequity in the City of Chicago, Urban Forestry & Urban Greening 59 (2021) 127029.

[85] D. Liu, M.-P. Kwan, Z. Yang, et al., Comparing subjective and objective greenspace accessibility: implications for real greenspace usage among adults, Urban Forestry & Urban Greening 96 (2024) 128335.

[86] K. Liu, L. Yin, F. Lu, et al., Visualizing and exploring POI configurations of urban regions on POI-type semantic space, Cities 99 (2020) 102610.

[87] S. Liu, W. Deng, Very deep convolutional neural network based image classification using small training sample size, in: 2015 3rd IAPR Asian Conference on Pattern Recognition (ACPR), IEEE, 2015, pp. 730–734.

[88] W. Liu, A. Yue, W. Shi, et al., An automatic extraction architecture of urban green space based on DeepLabv3plus semantic segmentation model, in: 2019 IEEE 4th International Conference on Image, Vision and Computing (ICIVC), IEEE, 2019, pp. 311–315.
[89] X. Liu, J. He, Y. Yao, et al., Classifying urban land use by integrating remote sensing and social media data, International Journal of Geographical Information Science 31 (8) (2017) 1675–1696.
[90] Y. Liu, X. Liu, S. Gao, et al., Social sensing: a new approach to understanding our socioeconomic environments, Annals of the Association of American Geographers 105 (3) (2015) 512–530.
[91] Y. Liu, X. Zhang, X. Kong, et al., Identifying the relationship between urban land expansion and human activities in the Yangtze River Economic Belt, China, Applied Geography 94 (2018) 163–177.
[92] Y. Liu, A.-X. Zhu, J. Wang, et al., Land-use decision support in brownfield redevelopment for urban renewal based on crowdsourced data and a presence-and-background learning (PBL) method, Land Use Policy 88 (2019) 104188.
[93] J. Long, E. Shelhamer, T. Darrell, Fully convolutional networks for semantic segmentation, in: Proceedings of the IEEE Conference on Computer Vision and Pattern Recognition, 2015, pp. 3431–3440.
[94] W. Lu, C. Tao, H. Li, et al., A unified deep learning framework for urban functional zone extraction based on multi-source heterogeneous data, Remote Sensing of Environment 270 (2022) 112830.
[95] N. Mansouri, M. Pourmahabadian, M. Ghasemkhani, Road traffic noise in downtown area of Tehran, Journal of Environmental Health Science and Engineering 3 (4) (2006) 267–272.
[96] M.R. Mehdi, M. Kim, J.C. Seong, et al., Spatio-temporal patterns of road traffic noise pollution in Karachi, Pakistan, Environment International 37 (1) (2011) 97–104.
[97] M. Moon, A.D. Richardson, M.A. Friedl, Multiscale assessment of land surface phenology from harmonized Landsat 8 and Sentinel-2, PlanetScope, and PhenoCam imagery, Remote Sensing of Environment 266 (2021) 112716.
[98] E. Murphy, E.A. King, Strategic environmental noise mapping: methodological issues concerning the implementation of the EU Environmental Noise Directive and their policy implications, Environment International 36 (3) (2010) 290–298.
[99] V. Nedic, D. Despotovic, S. Cvetanovic, et al., Comparison of classical statistical methods and artificial neural network in traffic noise prediction, Environmental Impact Assessment Review 49 (2014) 24–30.
[100] R. Neitzel, R.R. Gershon, M. Zeltser, et al., Noise levels associated with New York City's mass transit systems, American Journal of Public Health 99 (8) (2009) 1393–1399.
[101] H. Ning, Z. Li, X. Ye, et al., Exploring the vertical dimension of street view image based on deep learning: a case study on lowest floor elevation estimation, International Journal of Geographical Information Science 36 (7) (2022) 1317–1342.
[102] N. Nuthammachot, D. Stratoulias, Fusion of Sentinel-1A and Landsat-8 images for improving land use/land cover classification in Songkla Province, Thailand, Applied Ecology and Environmental Research 17 (2) (2019).
[103] K.L. O'Halloran, Multimodal discourse analysis, in: The Bloomsbury Handbook of Discourse Analysis, 2011, pp. 249–282.
[104] E. Okonofua, R. Ehigiator-Irughe, M. Ekun, Traffic noise monitoring at road intersections in urban settlement: case study of Ramat Park Benin City, Nigerian Journal of Technology 35 (4) (2016) 713–717.
[105] K.M. Paiva, M.R.A. Cardoso, P.H.T. Zannin, Exposure to road traffic noise: annoyance, perception and associated factors among Brazil's adult population, Science of the Total Environment 650 (2019) 978–986.
[106] M.C. Peel, B.L. Finlayson, T.A. McMahon, Updated world map of the Köppen-Geiger climate classification, Hydrology and Earth System Sciences 11 (5) (2007) 1633–1644.
[107] J.-F. Pekel, A. Cottam, N. Gorelick, et al., High-resolution mapping of global surface water and its long-term changes, Nature 540 (7633) (2016) 418–422.
[108] H. Peng, J. Li, Y. Song, et al., Incrementally learning the hierarchical softmax function for neural language models, in: Proceedings of the AAAI Conference on Artificial Intelligence, 2017.
[109] S. Preidl, M. Lange, D. Doktor, Introducing APiC for regionalised land cover mapping on the national scale using Sentinel-2A imagery, Remote Sensing of Environment 240 (2020) 111673.

[110] S. Punalekar, A. Verhoef, T. Quaife, et al., Application of Sentinel-2A data for pasture biomass monitoring using a physically based radiative transfer model, Remote Sensing of Environment 218 (2018) 207–220.
[111] C. Qu, P. Li, C. Zhang, A spectral index for winter wheat mapping using multi-temporal Landsat NDVI data of key growth stages, ISPRS Journal of Photogrammetry and Remote Sensing 175 (2021) 431–447.
[112] J. Quartieri, N. Mastorakis, G. Iannone, et al., A review of traffic noise predictive models, in: Recent Advances in Applied and Theoretical Mechanics, 5th WSEAS International Conference on Applied and Theoretical Mechanics (MECHANICS-09), Puerto De La Cruz, Tenerife, Canary Islands, Spain, December 2009, 2009, pp. 14–16.
[113] J. Ren, Y. Shao, H. Wan, et al., A two-step mapping of irrigated corn with multi-temporal MODIS and Landsat analysis ready data, ISPRS Journal of Photogrammetry and Remote Sensing 176 (2021) 69–82.
[114] D.P. Roy, H. Huang, L. Boschetti, et al., Landsat-8 and Sentinel-2 burned area mapping – a combined sensor multi-temporal change detection approach, Remote Sensing of Environment 231 (2019) 111254.
[115] A.-B. Salberg, Land cover classification of cloud-contaminated multitemporal high-resolution images, IEEE Transactions on Geoscience and Remote Sensing 49 (1) (2010) 377–387.
[116] A. Sharma, X. Liu, X. Yang, Land cover classification from multi-temporal, multi-spectral remotely sensed imagery using patch-based recurrent neural networks, Neural Networks 105 (2018) 346–355.
[117] X. Shi, Z. Chen, H. Wang, et al., Convolutional LSTM network: a machine learning approach for precipitation nowcasting, Advances in Neural Information Processing Systems 28 (2015).
[118] D.P. Shrestha, A. Saepuloh, F. van der Meer, Land cover classification in the tropics, solving the problem of cloud covered areas using topographic parameters, International Journal of Applied Earth Observation and Geoinformation 77 (2019) 84–93.
[119] D. Singh, A.B. Francavilla, S. Mancini, et al., Application of machine learning to include honking effect in vehicular traffic noise prediction, Applied Sciences 11 (13) (2021) 6030.
[120] N. Singh, S.C. Davar, Noise pollution-sources, effects and control, Journal of Human Ecology 16 (3) (2004) 181–187.
[121] H.-J. Son, D.-S. An, J.-J. Lee, et al., A study of traffic noise characteristic of pavement types using NCPX method, in: INTER-NOISE and NOISE-CON Congress and Conference Proceedings, Institute of Noise Control Engineering, 2014, pp. 3313–3318.
[122] L. Song, D. Liu, M.P. Kwan, et al., Machine-based understanding of noise perception in urban environments using mobility-based sensing data, Computers, Environment and Urban Systems 114 (2024) 102204.
[123] J. Song, X. Tong, L. Wang, et al., Monitoring finer-scale population density in urban functional zones: a remote sensing data fusion approach, Landscape and Urban Planning 190 (2019) 103580.
[124] Y. Song, B. Chen, H.C. Ho, et al., Observed inequality in urban greenspace exposure in China, Environment International 156 (2021) 106778.
[125] W. Su, D. Sui, X. Zhang, Satellite image analysis using crowdsourcing data for collaborative mapping: current and opportunities, International Journal of Digital Earth 13 (6) (2020) 645–660.
[126] C. Sukawattanavijit, J. Chen, H. Zhang, GA-SVM algorithm for improving land-cover classification using SAR and optical remote sensing data, IEEE Geoscience and Remote Sensing Letters 14 (3) (2017) 284–288.
[127] C. Sun, T. Lin, Q. Zhao, et al., Spatial pattern of urban green spaces in a long-term compact urbanization process—a case study in China, Ecological Indicators 96 (2019) 111–119.
[128] Y. Sun, M. Wu, X. Liu, et al., High-precision dynamic traffic noise mapping based on road surveillance video, ISPRS International Journal of Geo-Information 11 (8) (2022) 441.
[129] T. Tah-Chew, W.C. Keung, Noise surveys of hand-held pneumatic rock drills in Hong Kong, Environmental Monitoring and Assessment 19 (1991) 519–523.
[130] C. Talotte, P.-E. Gautier, D. Thompson, et al., Identification, modelling and reduction potential of railway noise sources: a critical survey, Journal of Sound and Vibration 267 (3) (2003) 447–468.
[131] P.H. Trombetta Zannin, F. Bunn, Noise annoyance through railway traffic—a case study, Journal of Environmental Health Science and Engineering 12 (2014) 1–12.
[132] W. Tu, Z. Hu, L. Li, et al., Portraying urban functional zones by coupling remote sensing imagery and human sensing data, Remote Sensing 10 (1) (2018) 141.

[133] UNDESA, World population prospects: the 2019 revision. Available at: https://population.un.org/wpp/, 2019. (Accessed 16 July 2019).

[134] H. Vaidya, T. Chatterji, SDG 11 sustainable cities and communities, in: Actioning the Global Goals for Local Impact, Springer, Singapore, 2020, pp. 173–185, https://doi.org/10.1007/978-981-32-9927-6.

[135] A. Vaswani, N. Shazeer, N. Parmar, et al., Attention is all you need, Advances in Neural Information Processing Systems 30 (2017).

[136] V. Venkataramanan, A.I. Packman, D.R. Peters, et al., A systematic review of the human health and social well-being outcomes of green infrastructure for stormwater and flood management, Journal of Environmental Management 246 (2019) 868–880.

[137] M. Voltersen, C. Berger, S. Hese, et al., Object-based land cover mapping and comprehensive feature calculation for an automated derivation of urban structure types at block level, Remote Sensing of Environment 154 (2014) 192–201.

[138] A.D. Wallis, Airport noise monitoring—a review, Aircraft Engineering and Aerospace Technology 69 (2) (1997) 112–118.

[139] L. Wang, M.P. Kwan, S. Zhou, et al., Assessing the affective quality of soundscape for individuals: using third-party assessment combined with an artificial intelligence (TPA-AI) model, Science of the Total Environment 953 (2024) 176083.

[140] P. Wang, T. Hu, F. Gao, et al., A hybrid data-driven framework for spatiotemporal traffic flow data imputation, IEEE Internet of Things Journal 9 (17) (2022) 16343–16352.

[141] P. Wang, T. Zhang, Y. Zheng, et al., A multi-view bidirectional spatiotemporal graph network for urban traffic flow imputation, International Journal of Geographical Information Science 36 (6) (2022) 1231–1257.

[142] Y. Wang, Y. Gu, M. Dou, et al., Using spatial semantics and interactions to identify urban functional regions, ISPRS International Journal of Geo-Information 7 (4) (2018) 130.

[143] G. Watts, Harmonoise Prediction Model for Road Traffic Noise, Transport Research Laboratory (TRL), Wokingham, UK, 2005.

[144] Q. Weng, Remote sensing of impervious surfaces in the urban areas: requirements, methods, and trends, Remote Sensing of Environment 117 (2012) 34–49.

[145] S. Woo, J. Park, J.-Y. Lee, et al., CBAM: convolutional block attention module, in: Proceedings of the European Conference on Computer Vision (ECCV), 2018, pp. 3–19.

[146] H. Wu, Q. Huang, D. Wang, et al., A CNN-SVM combined regression model for continuous knee angle estimation using mechanomyography signals, in: 2019 IEEE 3rd Information Technology, Networking, Electronic and Automation Control Conference (ITNEC), IEEE, 2019, pp. 124–131.

[147] L. Xie, X. Feng, C. Zhang, et al., Identification of urban functional areas based on the multimodal deep learning fusion of high-resolution remote sensing images and Social Perception Data, Buildings 12 (5) (2022) 556.

[148] S. Xie, L. Liu, X. Zhang, et al., Annual land-cover mapping based on multi-temporal cloud-contaminated Landsat images, International Journal of Remote Sensing 40 (10) (2019) 3855–3877.

[149] F. Xu, B. Somers, Unmixing-based Sentinel-2 downscaling for urban land cover mapping, ISPRS Journal of Photogrammetry and Remote Sensing 171 (2021) 133–154.

[150] X. Xu, W. Qiu, W. Li, et al., Associations between street-view perceptions and housing prices: subjective vs. objective measures using computer vision and machine learning techniques, Remote Sensing 14 (4) (2022) 891.

[151] Y. Xu, B. Zhou, S. Jin, et al., A framework for urban land use classification by integrating the spatial context of points of interest and graph convolutional neural network method, Computers, Environment and Urban Systems 95 (2022) 101807.

[152] K. Yamamoto, Road traffic noise prediction model "ASJ RTN-Model 2008": report of the research committee on road traffic noise, Acoustical Science and Technology 31 (1) (2010) 2–55.

[153] Q. Yang, Z. Ni, P. Ren, Meta captioning: a meta learning based remote sensing image captioning framework, ISPRS Journal of Photogrammetry and Remote Sensing 186 (2022) 190–200.

[154] Y. Yao, X. Li, X. Liu, et al., Sensing spatial distribution of urban land use by integrating points-of-interest and Google Word2Vec model, International Journal of Geographical Information Science 31 (4) (2017) 825–848.
[155] G. Yilmaz, Y. Hocanli, Mapping of noise by using GIS in Şanliurfa, Environmental Monitoring and Assessment 121 (2006) 103–108.
[156] J. Yin, J. Dong, N.A. Hamm, et al., Integrating remote sensing and geospatial big data for urban land use mapping: a review, International Journal of Applied Earth Observation and Geoinformation 103 (2021) 102514.
[157] B. Yu, T. Lian, Y. Huang, et al., Integration of nighttime light remote sensing images and taxi GPS tracking data for population surface enhancement, International Journal of Geographical Information Science 33 (4) (2019) 687–706.
[158] W. Yu, T. Ai, Y. He, et al., Spatial co-location pattern mining of facility points-of-interest improved by network neighborhood and distance decay effects, International Journal of Geographical Information Science 31 (2) (2017) 280–296.
[159] J. Yuan, Y. Zheng, X. Xie, Discovering regions of different functions in a city using human mobility and POIs, in: Proceedings of the 18th ACM SIGKDD International Conference on Knowledge Discovery and Data Mining, 2012, pp. 186–194.
[160] G. Zambon, R. Benocci, G. Brambilla, Statistical road classification applied to stratified spatial sampling of road traffic noise in urban areas, International Journal of Environmental Research 10 (3) (2016) 411–420.
[161] G. Zambon, H.E. Roman, M. Smiraglia, et al., Monitoring and prediction of traffic noise in large urban areas, Applied Sciences 8 (2) (2018) 251.
[162] W. Zhai, X. Bai, Y. Shi, et al., Beyond Word2vec: an approach for urban functional region extraction and identification by combining Place2vec and POIs, Computers, Environment and Urban Systems 74 (2019) 1–12.
[163] D. Zhang, J. Wan, Z. He, et al., Identifying region-wide functions using urban taxicab trajectories, ACM Transactions on Embedded Computing Systems (TECS) 15 (2) (2016) 1–19.
[164] F. Zhang, L. Wu, D. Zhu, et al., Social sensing from street-level imagery: a case study in learning spatio-temporal urban mobility patterns, ISPRS Journal of Photogrammetry and Remote Sensing 153 (2019) 48–58.
[165] F. Zhang, J. Zu, M. Hu, et al., Uncovering inconspicuous places using social media check-ins and street view images, Computers, Environment and Urban Systems 81 (2020) 101478.
[166] J. Zhang, H. Wang, Y. Wang, et al., Deep network based on up and down blocks using wavelet transform and successive multi-scale spatial attention for cloud detection, Remote Sensing of Environment 261 (2021) 112483.
[167] J. Zhang, Z. Yin, P. Chen, et al., Emotion recognition using multi-modal data and machine learning techniques: a tutorial and review, Information Fusion 59 (2020) 103–126.
[168] X. Zhang, S. Du, Q. Wang, Hierarchical semantic cognition for urban functional zones with VHR satellite images and POI data, ISPRS Journal of Photogrammetry and Remote Sensing 132 (2017) 170–184.
[169] X. Zhang, W. Li, F. Zhang, et al., Identifying urban functional zones using public bicycle rental records and point-of-interest data, ISPRS International Journal of Geo-Information 7 (12) (2018) 459.
[170] Y. Zhang, N. Chen, W. Du, et al., Multi-source sensor based urban habitat and resident health sensing: a case study of Wuhan, China, Building and Environment 198 (2021) 107883.
[171] Y. Zhang, Z. Chen, X. Zheng, et al., Extracting the location of flooding events in urban systems and analyzing the semantic risk using social sensing data, Journal of Hydrology 603 (2021) 127053.
[172] Y. Zhang, P. Liu, F. Biljecki, Knowledge and topology: a two layer spatially dependent graph neural networks to identify urban functions with time-series street view image, ISPRS Journal of Photogrammetry and Remote Sensing 198 (2023) 153–168.
[173] Y. Zhang, F. Zhang, N. Chen, Migratable urban street scene sensing method based on vision language pre-trained model, International Journal of Applied Earth Observation and Geoinformation 113 (2022) 102989.

[174] Y.Y. Zhang, H. Zhao, Y. Li, et al., Predicting highly dynamic traffic noise using rotating mobile monitoring and machine learning method, Environmental Research 229 (2023) 115896.

[175] T. Zhao, Z. Huang, W. Tu, et al., Coupling graph deep learning and spatial-temporal influence of built environment for short-term bus travel demand prediction, Computers, Environment and Urban Systems 94 (2022) 101776.

[176] T. Zhao, X. Liang, W. Tu, et al., Sensing urban soundscapes from street view imagery, Computers, Environment and Urban Systems 99 (2023) 101915.

[177] Y. Zhao, D. Feng, L. Yu, et al., Detailed dynamic land cover mapping of Chile: accuracy improvement by integrating multi-temporal data, Remote Sensing of Environment 183 (2016) 170–185.

[178] L. Zheng, M.-P. Kwan, Y. Liu, et al., How mobility pattern shapes the association between static green space and dynamic green space exposure, Environmental Research 258 (2024) 119499.

[179] W. Zhou, D. Ming, X. Lv, et al., SO–CNN based urban functional zone fine division with VHR remote sensing image, Remote Sensing of Environment 236 (2020) 111458.

[180] D. Zhu, F. Zhang, S. Wang, et al., Understanding place characteristics in geographic contexts through graph convolutional neural networks, Annals of the American Association of Geographers 110 (2) (2020) 408–420.

CHAPTER 10

Future perspectives and emerging technologies

Man-On Pun, Qianqian Wu, Ziyao Wang, and Ziyi Wang

School of Science and Engineering, The Chinese University of Hong Kong, Shenzhen, China

10.1 New potentials of machine learning for Earth observation

10.1.1 Opportunities and challenges in leveraging machine learning for Earth observation

The rapid development of Earth observation technologies, such as satellite imaging, UAV-based remote sensing, and ground-based sensor networks, has led to an explosion in the volume of remote sensing data. As these observation systems continue to evolve, they generate increasingly complex and heterogeneous datasets that capture dynamic changes on the Earth's surface, atmosphere, and human activities. This surge in data presents vast opportunities across various domains, including urban planning and environmental monitoring, while also introducing significant challenges on handling and extracting useful information.

Machine learning has emerged as a powerful tool to address these challenges. As a branch of artificial intelligence, machine learning allows computers to learn from data and improve their performance without explicit programming, enabling adaptive modeling of complex spatiotemporal relationships. By automatically detecting hidden patterns and correlations, machine learning enhances the efficiency and reliability of data analysis, supporting a wide range of advanced EO applications.

Another active area of research involves the use of ML for understanding and predicting climate change and environmental dynamics. For instance, ML has been extensively employed for the classification of climate-related land cover concepts, with the Local Climate Zone (LCZ) framework serving as a prominent example. Liu et al. [1] developed an deep learning network for LCZ mapping in metropolitan areas using Sentinel multispectral imagery, demonstrating the effectiveness of ML in large-scale urban LCZ classification. Zhang et al. [2] proposed a complex-valued CNN that leverages quad-pol SAR data and hierarchical architecture to significantly improve Local Climate Zone mapping accuracy.

Multimodal Remote Sensing Fusion and Classification
https://doi.org/10.1016/B978-0-44-329152-4.00017-9

On the other hand, the estimation of key climatic parameters, including near-surface air temperature, land surface temperature, and PM2.5 concentrations, has increasingly relied on machine learning methods, which offer powerful tools for handling heterogeneous and sparse environmental data. Madhavi et al. [3] introduced a Cascaded Inception-LGBM deep learning framework that synergizes Inception-based feature extraction with Light Gradient Boosting Machine regression, markedly enhancing the prediction of key climate variables, temperature, precipitation and CO_2, from remote-sensing data. Ham et al. [4] advanced ENSO forecasts by deploying a CNN network that, trained through transfer learning on CMIP5 simulations and historical reanalyses. Choo et al. [5] pioneered a context-aware diffusion model that reconstructs sea surface temperature (SST) under clouds. Wang et al. [6] introduced TranSAT, a transfer-learning model that significantly enhanced near-surface air temperature estimation in data-sparse regions, outperforming traditional methods.

ML-based approaches have further enhanced ecological and agricultural estimation, including forest biomass, canopy height, and crop yield prediction. Orndahl et al. [7] utilized random forest models to develop the first moderate resolution (30 m) maps of Arctic plant biomass and woody plant dominance. Muro et al. [8] demonstrated that A DNN fed only with Sentinel-2 surface reflectance achieved the best estimations of plant biomass when comparing the performance of a feed-forward deep neural network (DNN) with a random forest (RF) regression algorithm. Lang et al. [9] developed a novel probabilistic deep ensemble CNN that regresses canopy top height, providing globally transferable biomass mapping. Nejad et al. [10] explored the application of advanced deep learning architectures, including ConvLSTM, 3D-CNN, and ViT, to improve the accuracy of soybean yield prediction. Moreover, machine learning has also been widely applied to wildlife and biodiversity monitoring, from species detection and classification to automated population tracking.

Beyond ecological applications, ML has also been used for urban object detection and analysis, enabling automatic identification, reconstruction, and monitoring of urban features from high-resolution remote sensing imagery. While studies like [11] have demonstrated its effectiveness in detecting buildings, studies like [12] highlight its potential in classifying urban land cover at fine spatial resolutions. Extending urban mapping to the vertical dimension, recent studies have increasingly focused on extracting three-dimensional urban information from multisource remote sensing data. For example, a 10 m-resolution building height map of China [13] was produced using multisource Earth observation data and a Random Forest model, revealing city-scale 3D morphology patterns linked to socioeconomic factors.

The various applications of machine learning in EO have unlocked significant opportunities, setting the stage for its broader application in multimodal remote sensing. In the context of multimodal remote sensing, machine learning techniques are particularly valuable in data fusion tasks, where information from different sensors or temporal datasets is integrated to improve classification and analysis [14]. The ability to combine optical, radar, and thermal data can yield more comprehensive insights than relying on any single data source alone [15]. Machine learning models such as deep learning, decision trees, and support vector machines are increasingly employed to fuse multimodal data and enhance the performance of EO applications [16,17].

Despite the significant progress in applying machine learning to multimodal Earth observation (EO), several substantial challenges remain. First, effective data alignment across heterogeneous modalities remains difficult, as EO data often differ in spatial resolution, temporal frequency, and physical sensing mechanisms. This misalignment can lead to representation imbalance and feature redundancy, complicating accurate information fusion. Additionally, the generalization ability of specific-task models is often limited, making it challenging to maintain stable performance across diverse scenes and observation conditions. Second, dataset augmentation poses a critical bottleneck: the scarcity of large-scale, well-annotated multimodal datasets limits model generalization, especially when transferring knowledge across regions, seasons, or sensor types. Third, the understanding of complex Earth systems involving the coupling of physical, ecological, and geographical processes remains insufficiently integrated into current fusion networks, restricting their interpretability and their ability to relate learned features to underlying geophysical mechanisms. Finally, data acquisition itself presents challenges, as the planning and coordination of multisensor observations require real-time decision-making under dynamic environmental and mission constraints.

To overcome these obstacles, future advancements should focus on four key paradigms: 1) moving from algorithmic solutions to system-level foundational models that can generalize across tasks; 2) extending from data cleaning to data generation techniques to enhance preprocessing and analysis; 3) integrating multidisciplinary, physics-guided approaches to better apply data for physical modeling; and 4) transitioning from manual decision-making to intelligent decision-making powered by reinforcement learning for better acquisition and fusion of multimodal data. Addressing these challenges requires adaptive, physically informed, and interpretable learning frameworks that can balance accuracy, efficiency, and generalization in multimodal remote sensing applications.

10.1.2 Emerging paradigms and future research directions

Fig. 10.1 illustrates the emerging paradigms and future research directions in multimodal Earth observation, highlighting four key areas that address the current challenges. The following sections provide a detailed discussion of four key paradigms that shape the future

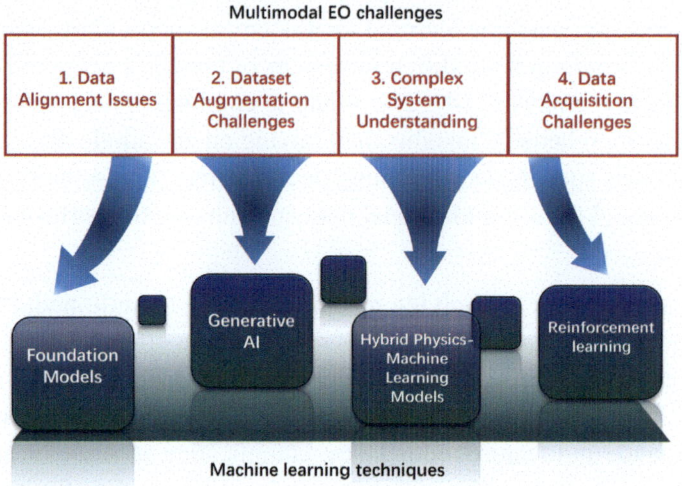

Figure 10.1:
Emerging Paradigms and Future Directions in Multimodal EO: Addressing Key Challenges.

of multimodal Earth observation: transitioning from task-specific models to foundational models, enhancing model performance through data generation techniques, integrating physical knowledge into machine learning models, and enabling intelligent decision-making for efficient multimodal data acquisition.

10.1.2.1 From task-specific models to foundational models

One of the primary challenges lies in effectively aligning heterogeneous data with differing spatial resolutions, temporal frequencies, and physical sensing mechanisms. This misalignment can lead to representation imbalance or feature redundancy, which complicates data fusion. Additionally, the ability to generalize across different domains and transfer knowledge between tasks is crucial. To tackle this, future research should focus on developing foundational models that move beyond task-specific algorithms. These models, typically pretrained on large-scale, diverse datasets, are capable of learning more generalizable features and patterns. By leveraging vast amounts of training data and sharing learned knowledge across multiple tasks, foundational models can improve domain transferability and better handle the heterogeneity and complexity of multimodal data. This enables seamless integration of diverse data sources and improves the robustness of fusion models. Foundational models can offer a more unified approach to multimodal EO, addressing data heterogeneity, improving performance across different applications, and enhancing the model's ability to transfer knowledge between domains.

Foundation models have gained significant traction in natural language processing (NLP) following key advancements such as GPT [18], LLaMA [19], and other large language models (LLMs). Their success has extended beyond NLP, with similar breakthroughs occurring in computer vision, where models like Vision Transformers (ViT) [20] and CLIP [21] have revolutionized visual recognition tasks. What makes these models particularly powerful is their ability to process and integrate information from multiple modalities such as images, text, and sound, enabling them to handle more complex, heterogeneous data. This capability has naturally extended into the realm of EO applications.

The application of machine learning models in Earth observation has made significant progress in both vision-based tasks and multimodal integration.

For vision-based EO tasks, transformer-based foundation models, such as EarthPT [22], have become crucial for advancing visual recognition capabilities in remote sensing. EarthPT, for example, adapts the GPT-2 framework by replacing traditional word embeddings with multilayer perceptrons (MLPs) to effectively encode nontextual data, such as images and sensor signals. This adaptation significantly enhances its capacity to process remote sensing data, showcasing the potential for scaling transformer models to handle increasingly large and complex datasets [22]. However, the performance of these models varies across different tasks. For instance, studies such as [23] point out that current foundation models struggle with quantitative tasks like object counting, despite performing well in qualitative tasks such as image captioning and semantic interpretation. This underscores the need for continued model development and the creation of benchmark datasets that enable objective, standardized evaluation of their capabilities [24–26].

Building upon the success of transformer-based vision models, there has been a growing interest in handling various modalities and tasks simultaneously. These models, often incorporating vision and language capabilities [27,28], are being trained on large, high-quality image-text datasets specifically designed for EO data [25,26,29]. The creation of these datasets, such as GeoLangBind-2M [30] and SARChat-2M [25], is crucial for enabling effective training and evaluation of these models. These datasets are designed to encompass diverse scenarios and modalities, including optical, radar, and hyperspectral data. Furthermore, innovative training strategies are being explored, such as VECT (Vision-models Enhanced Contrastive Text-image pretraining) [30], which aims to enhance the spatial awareness of the models.

CLIP [21] is a representative vision-language model (VLM) known for its powerful zero-shot prediction capabilities. Researchers have continuously refined and enhanced subsequent versions of CLIP, progressively improving its performance and expanding its capabilities with each iteration. The CLIP-based model has been particularly extended in remote sensing tasks, such as image-text matching, cross-modal retrieval, and visual reasoning, due to its ability to learn rich, shared representations across both image and text modalities. This capability makes it especially useful for analyzing satellite imagery, where textual descriptions or metadata can be aligned with corresponding visual features, enabling more efficient data interpretation and retrieval. In addition to CLIP, other multimodal models, such as Flamingo [31] and ALIGN [32], have also made significant strides in the remote sensing domain. These models enhance the fusion of multimodal data, combining optical, radar, and thermal imagery, and improve performance across a variety of cross-modal tasks. Their advancements are pushing the boundaries of vision-language understanding in remote sensing.

The integration of foundation models extends beyond single-image analysis [33]. There is a growing focus on handling temporal sequences of EO data, enabling analysis of changes over time and facilitating long-term monitoring of environmental processes [34]. For example, TEOChat [33] is designed for engaging in conversations about temporal sequences, marking a significant advancement in the capabilities of LLMs for EO applications. Other approaches explore multisensor and multigranular data integration [35], aiming to leverage the diverse information available from various EO sources.

The efficiency and accessibility of foundation models for EO are also being addressed [36, 37]. The high computational costs associated with large models are a concern, leading to research into cost-efficient open language models and optimized workflows. This includes exploring techniques like prompt optimization and inference-time scaling to make these tools more accessible for broader use in earth science. Additionally, new approaches are being developed to reduce the reliance on large training datasets, such as InstructSAM, a training-free framework for instruction-driven object recognition [38].

In conclusion, the application of foundation models to EO is a rapidly advancing field. While challenges persist, particularly in quantitative analysis and computational efficiency, significant strides are being made in the development of specialized models and benchmark datasets that enable more accurate, efficient, and accessible analysis of EO data. Future directions involve further scaling of these models, improvements in training methodologies, and the creation of tools capable of handling temporal and multisensor data, all while addressing the persistent challenges of cost and accessibility. The integration of foundation models into EO workflows is poised to transform our ability to monitor and understand the Earth's systems [39].

10.1.2.2 Enhancing model performance through data generation techniques

Data generation techniques are essential methods for creating synthetic data that closely resembles real-world data, addressing the challenge of limited or costly data acquisition. These techniques are typically divided into two main categories: data augmentation and generative modeling. Data augmentation involves transforming existing data through techniques like rotation, scaling, and noise addition to artificially expand the dataset and improve model robustness. However, more advanced techniques, such as generative models, go a step further by learning the underlying data distribution to generate entirely new samples. Notable examples include Generative Adversarial Networks (GANs) [40], which have been widely applied to domains like remote sensing (e.g., [41]). Another recent and highly effective approach is diffusion models, which generate high-quality samples by iteratively transforming noise into structured data. Diffusion models, such as DDPM (Denoising Diffusion Probabilistic Models), have shown impressive results in generating realistic images and even in the generation of high-dimensional multimodal data, such as those used in remote sensing (e.g., [42], [43]). These techniques have gained significant attention due to their ability to produce high-fidelity outputs, especially in complex generative tasks where GANs may struggle.

The integration of data generation techniques, especially through diffusion models, offers significant potential for advancing model performance, particularly in fields with complex, multimodal data, such as remote sensing, healthcare, and autonomous systems. In particular, diffusion models have emerged as a powerful tool for generating high-quality synthetic data, including rare and difficult-to-capture events in remote sensing, such as floods [44] and landslides [45]. Additionally, as foundational models continue to scale, data generation techniques will be pivotal in enhancing the ability of these models to handle increasingly complex tasks. For example, in the case of multimodal EO systems, large models will benefit from synthetic data that spans diverse sensor types, including optical, radar, and thermal data, allowing them to train on more comprehensive datasets without requiring extensive real-world data collection. Overall, the fusion of data generation methods with multimodal EO models is set to transform the future of machine learning by providing scalable, efficient solutions to data scarcity and enhancing the robustness, accuracy, and generalization of models across numerous EO applications.

10.1.2.3 Physics-guided machine learning for hybrid modeling

Integrating multidisciplinary, physics-based approaches have become a crucial strategy for enhancing the application of data in physical modeling, especially in fields like remote sensing, environmental monitoring, and climate modeling. These approaches combine domain-specific knowledge from various disciplines such as physics, engineering, mathematics, and computer science with advanced machine learning techniques to improve the accuracy and interpretability of models [46,47]. Physics-based machine learning (PBML), for instance, incorporates fundamental physical principles, such as boundary conditions, and physical constraints, directly into the learning process [48,49].

The integration of physics-based approaches with machine learning has significant potential to drive the next generation of physical modeling, particularly in remote sensing and environmental sciences. By embedding physical laws into learning algorithms, these approaches can improve model generalization, especially in scenarios involving rare events or extreme conditions that are underrepresented in training datasets. In remote sensing, for example, integrating domain knowledge from atmospheric science, hydrology, and geophysics with machine learning models can help accurate surface soil moisture estimation [50], aerosol retrieval [51], and land-cover change detection [44]. The future of physics-based machine learning will likely involve the development of more sophisticated hybrid models capable of seamlessly integrating sensor data with physics-based simulations, leading to more accurate environmental monitoring, resource management, and disaster response systems. This synergy between data and physical laws is expected to revolutionize the way we understand complex natural processes, offering scalable solutions to some of the most pressing challenges in EO tasks.

10.1.2.4 Intelligent decision-making for multimodal data acquisition

The traditional approach to decision-making in complex systems, particularly in remote sensing and EO, has often been manual, relying on human expertise and predefined rules. While effective in many cases, such manual processes face limitations when managing complex observation tasks [52]. In EO missions, decision-making during the data acquisition process, such as determining which sensors to deploy, when and where to observe, and how to coordinate multisensor operations, is particularly critical, as it directly influences the spatial, temporal, and spectral quality of the collected data. As the volume and complexity of information from optical, radar, and thermal sensors continue to grow, the need for scalable, autonomous, and data-driven decision-making mechanisms becomes increasingly evident to ensure efficient and adaptive observation strategies.

Reinforcement learning (RL) provides a promising solution by enabling models to autonomously learn optimal decision-making strategies through interaction with the environment [53]. Instead of relying on preset rules or human input, RL algorithms can learn from experience, gradually improving decision-making performance by maximizing long-term rewards. Reinforcement learning's potential for multimodal data acquisition lies in its ability to dynamically optimize the interaction between various data collection tasks, improving the quality and relevance of the information used for EO tasks [54]. The future direction of this approach could see RL-based models becoming integral to autonomous systems that rely on real-time data, such as smart cities, disaster response, and environmental monitoring, enabling more intelligent and efficient data-driven decision-making processes across diverse applications.

10.2 High-performance cloud computing for large scale monitoring

10.2.1 Introduction

Cloud computing is a transformative paradigm that enables on-demand access to a shared pool of configurable computing resources. This model significantly reduces the need for organizations to manage their own physical infrastructure by offering unparalleled scalability, flexibility, and cost-efficiency. With the exponential growth of data across sectors, cloud computing has become indispensable, providing the capability to process large volumes of data efficiently through elastic, on-demand resources [55–58].

In remote sensing, cloud computing is pivotal for processing and analyzing the vast datasets generated by satellite and aerial platforms. It powers complex algorithms for data interpretation, thereby addressing critical challenges in environmental monitoring, disaster management, and natural resource assessment.

This chapter explores the integration of high-performance cloud computing with large-scale monitoring applications, with a focus on remote sensing. The chapter is organized as follows:

- Key Technologies in High-Performance Cloud Computing: This section explores the fundamental cloud computing architectures and technologies that enable high-performance applications.
- Cloud Platforms for Remote Sensing: Here, we discuss various cloud platforms and services, including the specialized Google Earth Engine for geospatial analysis, as well as general-purpose infrastructure from global providers like Amazon Web Services (AWS), Microsoft Azure, and Alibaba Cloud.
- Applications of Cloud Computing in Remote Sensing: This section covers key areas where cloud computing has significantly impacted remote sensing, such as environmental monitoring, forest management, and disaster response.
- Future Directions: We conclude by examining emerging trends, including the integration of 5G, edge computing, and their implications for the future of cloud-based remote sensing.

10.2.2 Key technologies in high-performance cloud computing

10.2.2.1 Cloud infrastructure

High-performance cloud computing is built upon a layered architecture that defines how computing resources are provisioned, managed, and delivered to users. At its foundation lies the concept of cloud infrastructure, which can be categorized into three principal service models: Infrastructure as a Service (IaaS), Platform as a Service (PaaS), and Software as a

Service (SaaS). Each represents a different level of abstraction and user control, forming a continuum from low-level hardware management to fully managed application delivery. Understanding these layers is essential for designing and deploying efficient large-scale monitoring systems in remote sensing, where computational flexibility, scalability, and reliability are crucial.

IaaS represents the foundational layer of cloud computing. It provides users with virtualized computing resources, such as CPUs, GPUs, memory, and network interfaces, through programmable interfaces or management consoles. In IaaS environments, physical hardware is abstracted into virtual machines or containers, allowing dynamic provisioning and elastic scaling. Major providers such as Amazon Web Services (AWS EC2), Microsoft Azure Virtual Machines, and Google Compute Engine offer this model, enabling users to deploy and configure complete computing environments without managing physical servers. IaaS providers typically offer several types of storage services, including object storage for unstructured, large-scale data such as satellite imagery, block storage for high-speed random access tasks, and options for deploying distributed or parallel file systems for high-performance computing tasks. Networking in IaaS is managed through virtual private clouds, software-defined networking, and load balancers that ensure data transmission efficiency across geographically distributed resources. This layer also incorporates strict security mechanisms such as Identity and Access Management, encryption at rest and in transit, and role-based access control. In remote sensing applications, IaaS enables researchers to process multipetabyte-scale archives, execute large-scale atmospheric correction or cloud-removal tasks, and train deep learning models using GPU clusters on demand. By automating resource allocation through infrastructure-as-code frameworks such as Terraform or CloudFormation, IaaS facilitates reproducible, scalable, and cost-effective remote sensing workflows.

PaaS abstracts away infrastructure management, offering managed environments where developers can run applications and data-processing tasks without dealing with system administration. PaaS provides middleware, runtime environments, and managed services that simplify data ingestion, transformation, and analysis. Common examples include Google Cloud Dataflow, AWS Elastic MapReduce, and Azure Synapse Analytics. These services enable the execution of large-scale distributed computing frameworks such as Apache Spark or Apache Beam, making them particularly suitable for geospatial big data analytics. PaaS offerings often include managed databases and warehouses like BigQuery or Redshift, which can store and query massive volumes of spatial-temporal data efficiently. For remote sensing workflows, PaaS environments allow seamless integration of processing pipelines, including Sentinel or Landsat image preprocessing, NDVI computation, or time-series analysis, without direct control over virtual machines. They automatically handle load balancing, fault recovery, and autoscaling to meet fluctuating computational demands. PaaS also integrates with machine learning services, such as Vertex AI or Amazon SageMaker, to enable model training and deployment for classification or segmentation of remote sensing imagery. This managed layer drastically reduces operational overhead and allows scientists to focus on data-driven insights rather than system maintenance.

SaaS is the most user-facing layer of the cloud computing model. It provides fully managed applications delivered through web interfaces or APIs, requiring no infrastructure or platform configuration by the user. SaaS platforms are often designed for specific domains, such as geospatial analytics, and include both the computational backend and the data catalog. Google Earth Engine, Sentinel Hub, and Alibaba Cloud AI Earth are prominent examples in the remote sensing field. These services allow users to access petabytes of harmonized satellite data, perform computations such as land cover classification or change detection, and visualize results interactively through web dashboards or code notebooks. SaaS systems commonly utilize standard APIs like OGC Web Services or STAC for data interoperability. Their pricing models are usually based on subscription or metered computation usage, making them cost-efficient for institutions with predictable workloads. In large-scale environmental monitoring, SaaS dramatically accelerates the workflow from data acquisition to visualization by removing the need for local computing infrastructure. For example, researchers can execute regional vegetation analysis or flood detection within minutes through a few lines of code, leveraging the provider's preoptimized processing engine.

The three service models are not isolated but complementary. IaaS forms the backbone of computational resources, PaaS builds upon it to provide scalable analytics platforms, and SaaS offers ready-to-use applications for end users. In practical remote sensing scenarios, these layers often coexist: data may be stored in IaaS object storage, processed through a PaaS analytical engine, and visualized using a SaaS dashboard. The combination of these layers enables a hierarchical and modular approach to large-scale Earth observation. Moreover, modern cloud environments are frequently deployed under hybrid or multicloud strategies, combining private and public resources to meet data security and sovereignty requirements. For instance, sensitive geospatial data can be preprocessed on local servers and later transferred to a public PaaS platform for large-scale analysis. This layered and integrated architecture allows high-performance cloud computing to deliver unprecedented scalability, reliability, and accessibility for remote sensing and environmental monitoring, marking a paradigm shift from traditional on-premise processing to globally distributed, on-demand computation.

10.2.2.2 Distributed computing architectures and load balancing

Distributed computing architectures serve as the foundational backbone of high performance cloud environments by enabling parallel processing of massive datasets across scalable clusters of virtualized servers. These systems decompose large and complex computational problems into smaller manageable tasks that execute concurrently across numerous nodes, significantly reducing overall processing time while dramatically enhancing system scalability and resource utilization efficiency. Within cloud based remote sensing, such distributed architectures are indispensable for managing petabyte scale volumes of satellite and aerial imagery and provide critical computational support for time sensitive applications including continental scale land cover classification and real time disaster monitoring. The transition from traditional monolithic high performance computing clusters to these elastic cloud native distributed systems represents a fundamental paradigm shift in geospatial computation organization and execution [59].

The conceptual core of these architectures revolves around two complementary paradigms: data parallel and task parallel processing. Data parallel systems partition extensive datasets across multiple compute nodes where each node performs identical operations on its local data subset, proving particularly effective for spatially independent pixel level operations such as vegetation index computation. Task parallel systems distribute distinct computational procedures across specialized nodes, creating processing pipelines where different workflow stages advance simultaneously. Contemporary cloud native systems frequently implement hybrid strategies that intelligently combine both paradigms, orchestrated by advanced container managers like Kubernetes to optimize both data throughput and procedural latency [60]. Foundational programming models including MapReduce and its more efficient in memory successors have established themselves as particularly suitable for iterative algorithms commonly employed in remote sensing applications, effectively supporting complex analytical tasks such as K means clustering and principal component analysis for dimensionality reduction of hyperspectral data [61].

The effectiveness of any distributed system depends critically on robust load balancing strategies that ensure optimal computational resource utilization while preventing single node performance bottlenecks. Load balancing operates across multiple tiers of the cloud infrastructure stack, with distributed load balancers intelligently directing incoming requests across server pools to maintain high service availability at the network level. At the application level, orchestration tools dynamically adjust the quantity of running containerized instances based on real time performance metrics, while data processing frameworks incorporate internal mechanisms that reassign data partitions from slower performing nodes to idle resources during job execution [62]. This dynamic redistribution capability proves especially valuable in remote sensing workflows where inherent data heterogeneity creates significant computational imbalances, as variations in processing complexity between different terrain types can otherwise lead to straggler nodes that substantially delay overall job completion.

Modern distributed systems incorporate sophisticated fault tolerance mechanisms to ensure operational continuity and data integrity despite inevitable component failures. These systems employ multiple techniques including data replication across geographically separate availability zones, periodic checkpointing of intermediate results to persistent object storage, and lineage based recomputation where lost data is regenerated from original sources. This resilience becomes absolutely essential when processing multiterabyte satellite mosaics or conducting long term temporal analyses. The architectural evolution continues toward serverless computing paradigms which abstract resource management to an even higher level, enabling highly scalable and cost efficient workflows that only incur computational expenses during actual processing activities [63].

In summary, the synergistic integration of distributed computing architectures, adaptive load balancing, and inherent fault tolerance creates the powerful computational foundation that enables large scale cloud based environmental monitoring. By seamlessly distributing tasks across elastic virtual clusters, dynamically balancing workloads in real time, and ensuring reliable execution despite individual component failures, these systems empower researchers to analyze global datasets with unprecedented speed and efficiency. This technological infrastructure is fundamentally transforming Earth observation science by making routine near real time planetary analysis an achievable reality in the big data era.

10.2.2.3 Elastic storage and data management

Elastic storage and data management constitute the foundational layer that enables scalability and reliability in high-performance cloud computing for remote sensing applications. The essential characteristic of elasticity refers to a storage system's capacity to dynamically provision, expand, and release storage resources according to fluctuating computational demands, thereby ensuring both cost-efficiency and continuous data availability. This capability proves particularly crucial for large-scale monitoring applications where data volumes experience significant variations due to satellite acquisition cycles, multitemporal updates, and the integration of diverse sensor sources. Modern elastic storage systems must not only support the immense capacity requirements of petabyte-scale geospatial datasets but also guarantee rapid data access, robust redundancy mechanisms, and resilient failure recovery protocols to maintain uninterrupted analytical workflows [64].

The architectural core of elastic cloud storage revolves around the object storage model, which organizes data as discrete objects containing the actual data, comprehensive metadata, and a unique global identifier, enabling horizontal scaling without centralized coordination constraints. Major implementations including Amazon S3, Google Cloud Storage, and Azure Blob Storage are specifically engineered for massive concurrency, supporting parallel access by thousands of compute nodes simultaneously. This capability represents a fundamental requirement for distributed remote sensing operations such as continental-scale image mosaicking and long-term time-series analysis. These systems typically provide RESTful API interfaces that ensure universal compatibility with established geospatial processing frameworks like GDAL and xarray. The emergence of cloud-optimized data formats such as Cloud-Optimized GeoTIFF and chunked array structures like Zarr has further revolutionized Earth observation workflows by enabling efficient partial reads and spatial indexing capabilities, eliminating the necessity to download complete files for localized analysis [65].

A comprehensive storage ecosystem incorporates complementary block and file storage systems designed for workloads demanding low-latency, high-throughput input/output operations. Block storage services provide virtualized storage volumes that function as directly attached drives for virtual machines, making them ideal for machine learning training workloads and real-time processing scenarios requiring consistent performance. Distributed file systems deliver POSIX-compliant interfaces and parallel access patterns, allowing large geospatial files to be processed concurrently across multiple compute nodes. This multitiered storage architecture enables remote sensing pipelines to strategically combine the massive scalability of object storage with the high-performance characteristics of file systems and the operational flexibility of block volumes, thus optimizing data accessibility throughout different computational stages. The system's effectiveness is further enhanced through automated data lifecycle management policies that intelligently migrate data between storage tiers based on access patterns and temporal relevance, significantly reducing operational costs while maintaining long-term data preservation [66].

The management of elastic storage systems extends to sophisticated metadata organization and indexing capabilities that provide the structural foundation for massive geospatial data collections. Modern metadata catalogs built upon international standards enable efficient data discovery, spatial-temporal querying, and seamless interoperability across distributed datasets. The underlying distributed database technologies ensure metadata persistence and retrieval with millisecond-level latency, allowing users to perform complex queries across millions of images with near-instantaneous response times. Data integrity and durability are maintained through advanced replication strategies that distribute data across multiple availability zones, employing erasure coding and checksum validation to ensure recovery capabilities without compromising performance. These mechanisms become particularly critical for real-time monitoring applications such as disaster response and environmental forecasting, where uninterrupted access to current data streams represents an operational necessity [67].

The convergence of storage management with computational analytics represents the evolutionary direction of modern cloud systems, where storage backends become actively integrated with serverless analytics engines. This integration enables computational workflows to be automatically triggered by data events, such as the upload of new satellite imagery to object storage, initiating immediate processing sequences including atmospheric correction and land cover classification without requiring manual intervention. Through the strategic implementation of scalable metadata indexing, intelligent lifecycle automation, and event-driven computational triggers, elastic storage systems transform from passive data repositories into dynamic, intelligent components of the analytical ecosystem. This architectural approach ensures that vast and heterogeneous geospatial datasets can be stored, accessed, and processed through seamless, automated workflows, ultimately enabling efficient, reliable, and real-time Earth observation capabilities in the era of cloud-native computational science [68].

10.2.2.4 Virtualization and containerization technologies

Virtualization and containerization technologies serve as fundamental enablers of resource abstraction and isolation in high-performance cloud computing, allowing multiple workloads to operate concurrently on shared physical hardware while maintaining security and performance isolation. These technologies have revolutionized computing resource deployment and scaling, transitioning from traditional hardware-based virtualization to lightweight software-defined containers that support modern cloud and edge infrastructures. For large-scale remote sensing applications, they ensure efficient utilization of compute clusters across diverse tasks such as data preprocessing, model training, and visualization, each often requiring distinct runtime environments and library dependencies [68].

Virtualization abstracts physical hardware into multiple virtual machines, each running an independent operating system and application stack through a hypervisor layer. Type 1 bare-metal hypervisors, including VMware ESXi and Kernel-based Virtual Machine, operate directly on host hardware to deliver high performance and robust isolation, making them ideal for data-intensive remote sensing workloads. Cloud providers extensively employ Type 1 hypervisors to achieve efficient resource partitioning and dynamic elasticity. Through virtualization, users can provision computing resources with specific configurations such as GPU types and memory capacity without physical server management, enabling flexible support for heterogeneous analytical workflows that combine classical algorithms with deep learning models [69].

Containerization advances this paradigm by offering process-level isolation through lightweight, self-contained units that package applications with their dependencies. Technologies like Docker and Podman enable minimal startup overhead and rapid scaling, making containers particularly suitable for dynamic, short-lived tasks in data-driven monitoring systems. Container orchestration platforms such as Kubernetes provide declarative configuration, automated scheduling, and self-healing capabilities, managing distributed containers that perform specialized functions including satellite image preprocessing and feature extraction across multiregion cloud environments. When integrated with cloud-native engines and GPU-enabled nodes, containerized environments ensure reproducibility and operational consistency throughout development and production stages [70].

The combination of these technologies enables microservice architectures and serverless computing models that enhance modularity and automation in remote sensing systems. Microservices decompose complex applications into independently deployable units communicating via lightweight APIs, improving maintainability and supporting continuous integration pipelines. Serverless platforms such as AWS Lambda automatically execute event-driven functions in response to triggers like new satellite data uploads, eliminating idle resource costs and simplifying real-time processing tasks such as atmospheric correction or anomaly detection. These models represent a significant evolution toward fully automated, scalable analytical ecosystems for Earth observation.

In summary, virtualization and containerization collectively form the technological foundation for elastic, portable, and efficient cloud-native remote sensing. Virtualization delivers hardware-agnostic infrastructure scalability, while containerization ensures consistency and agility across heterogeneous environments. Together, they support complex, large-scale processing pipelines from data acquisition to AI-driven analysis, establishing a robust platform for reproducible and high-performance environmental monitoring in the modern cloud era.

10.2.2.5 Edge computing and its integration with cloud computing

Edge computing extends cloud computing by deploying computational resources closer to data sources, shifting processing from centralized data centers to local devices, gateways, and micro data centers near sensors or users. This paradigm is particularly transformative for large-scale environmental monitoring and remote sensing, where data volume, transmission costs, and latency constraints render purely cloud-based processing inefficient. Integrating edge computing with cloud infrastructure enables real-time responsiveness, bandwidth conservation, and improved system resilience, forming a hybrid model essential for modern Earth observation systems [71].

The core principle of edge computing is data locality, which involves processing information as close as possible to its origin. In remote sensing, this enables initial data filtering, compression, or inference to be performed on satellites, unmanned aerial vehicles, or ground stations before transmitting only essential results to the cloud. For instance, satellite constellations with onboard processing can conduct cloud detection or feature extraction, substantially reducing the volume of raw imagery sent to ground servers. Similarly, in distributed sensor networks monitoring air quality or soil moisture, edge devices perform local anomaly detection and transmit only actionable insights. This approach minimizes bandwidth usage and latency, supporting near-real-time applications such as disaster early warning and precision agriculture [72].

Edge and cloud computing are integrated through a hierarchical architecture often termed the cloud-edge continuum, comprising the edge layer for time-sensitive operations, the fog layer for intermediate aggregation and coordination, and the cloud layer for large-scale storage and deep analysis. This structure ensures that latency-critical tasks are handled locally while computationally intensive processes benefit from cloud scalability. Orchestration frameworks such as KubeEdge and AWS Greengrass extend containerized management to the edge, dynamically allocating tasks based on network conditions and resource availability. For example, a flood monitoring system may execute detection models on local camera nodes during network congestion and offload batch analytics to the cloud when bandwidth allows, maintaining functionality in unstable environments.

A major advantage of edge computing is its ability to support real-time analytics and AI inference at the source. Lightweight deep learning models coupled with hardware accelerators enable complex tasks like object detection and semantic segmentation to be performed directly on edge devices. In remote sensing, this allows real-time identification of wildfire hotspots or illegal deforestation from aerial imagery, with only essential metadata forwarded to the cloud. This capability reduces latency and transmission load while preserving global situational awareness. Additionally, edge computing enhances energy efficiency and sustainability by minimizing data transmission and supporting autonomous operation during connectivity outages, which is critical for remote or solar-powered monitoring infrastructures.

Despite its benefits, integrating edge and cloud computing introduces challenges in data consistency, security, and synchronization. Techniques such as eventual consistency models and distributed ledgers help reconcile data across layers, while hardware-based trusted execution environments and encrypted communications protect edge devices deployed in untrusted settings. Federated learning has emerged as a privacy-preserving solution by training models locally and sharing only aggregated updates with the cloud. Coupled with advancements in 5G and software-defined networking, these approaches enable adaptive, secure, and responsive monitoring systems. The fusion of edge and cloud computing is defining the next generation of intelligent geospatial infrastructures, capable of autonomous real-time decision-making and efficient planetary-scale observation.

10.2.3 Cloud platforms for remote sensing

10.2.3.1 Overview and applications of Google Earth Engine (GEE)

Google Earth Engine (GEE) is one of the most influential cloud-based platforms for large-scale geospatial data analysis, integrating petabytes of satellite imagery with powerful parallel computing capabilities. It provides a unified environment where users can access, process, and visualize multitemporal Earth observation data without the need for local storage or high-performance computing infrastructure. GEE's architecture combines an extensive geospatial data catalog, including Landsat, Sentinel, MODIS, and other global datasets, with a cloud-native API that supports Python and JavaScript interfaces, enabling both scientific research and operational monitoring at planetary scale.

One of GEE's key advantages lies in its ability to handle long-term and multitemporal analyses efficiently. Ramdani et al. [73] demonstrated the effectiveness of GEE in monitoring oceanic parameters such as sea surface temperature (SST) and total suspended matter (TSM) in Australian waters using the Radiative Transfer Equation (RTE) method. This study showcased GEE's capacity to process massive temporal datasets while maintaining accuracy and computational efficiency. Similarly, Kwong et al. [74] applied Sentinel-2 imagery within GEE to map marine water quality in Hong Kong, developing empirical models through artificial neural networks. Their work illustrated how GEE's integrated data retrieval and processing environment can support complex modeling tasks, offering fast access to high-resolution satellite data and scalable analytics for environmental applications.

Beyond oceanic monitoring, GEE has become an indispensable tool for vegetation and land cover analysis. Zhang et al. [75] proposed a long-term vegetation monitoring approach using Landsat archives within GEE, enabling consistent ecological assessments over decades. By leveraging GEE's temporal compositing and trend analysis functions, their method supported ecosystem monitoring and environmental protection initiatives at continental scales. Similarly, Zurqani [76] utilized GEE's computational power and high-resolution aerial imagery to produce a 1-meter resolution regional forest canopy dataset, combining cloud-based data fusion with machine learning algorithms. This demonstrated how GEE can be extended beyond satellite data to incorporate other sources such as UAV and aerial photography for high-fidelity ecological mapping.

Overall, GEE exemplifies the convergence of big data storage, cloud computing, and geospatial intelligence. It reduces the technical barriers associated with traditional remote sensing workflows, such as data download, preprocessing, and format conversion, while providing near-instant access to planetary-scale data archives. Through its server-side computation and scripting flexibility, researchers can execute analyses that would be infeasible on local systems, from daily global NDVI tracking to regional hydrological modeling. As studies like those by Ramdani, Kwong, Zhang, and Zurqani show, GEE not only supports environmental monitoring but also fosters reproducible, collaborative science by providing a shared, browser-accessible computational ecosystem for the global remote sensing community.

10.2.3.2 Remote sensing services of Alibaba Cloud

Alibaba Cloud has actively evolved its remote sensing offerings by leveraging its cloud computing infrastructure and integrating it with AI-driven geospatial services. A flagship example is AI Earth, a cloud platform for intelligent geospatial computing developed under Alibaba's DAMO Academy. AI Earth hosts petabyte-scale remote sensing archives including Landsat and Sentinel collections and provides built-in AI models for tasks like terrain classification, change detection, and water extraction. It allows users to upload training samples, deploy custom models, and carry out most of the workflow entirely in the cloud. The platform is intended to lower the technical barrier for researchers by integrating data access, analysis, storage, and compute in one managed environment [77].

Another significant offering is Planet Engine, Alibaba's platform for satellite imagery management and intelligent analysis. Planet Engine aggregates multisource satellite imagery from both Chinese and international providers and aims to accelerate data ingestion, storage, and analytic performance. It supports visualization, smart feature detection, and change monitoring in a unified interface. One of its key benefits is improved I/O performance with Alibaba Cloud reporting read/write acceleration factors of 5 to 27 times in managing large-scale spatiotemporal datasets under certain scenarios. Moreover, Planet Engine is designed to support AI-based analysis directly on imagery, enabling automated detection of illegal constructions or land cover changes and thereby shortening the workflow from raw data to actionable outputs.

Together, AI Earth and Planet Engine constitute Alibaba Cloud's dual strategy. AI Earth emphasizes flexible, researcher-friendly workflows with AI integration while Planet Engine focuses on imagery aggregation, management, and scalable intelligence services. In practical use, these services allow users to query massive satellite datasets with spatial, temporal, and quality filters; run algorithms such as classification and change detection on the cloud without needing local compute; and interactively visualize or export results. For remote sensing projects inside China or in the Asia-Pacific region, Alibaba Cloud's infrastructure may offer advantages in data locality, regulatory compliance, and integration with domestic datasets.

Nevertheless, certain challenges remain. AI Earth is still expanding its capabilities in high-demand deep learning model training, and complete performance benchmarks for large-scale segmentation or object detection tasks are not yet fully disclosed. Planet Engine's cross-region synchronization, model portability, and user scalability need further validation in published studies. As these platforms mature, improvements in resource orchestration, interoperability with international cloud ecosystems, and pricing models will determine their adoption and competitiveness.

In summary, Alibaba Cloud's remote sensing services are an emerging and credible option in the geospatial cloud ecosystem. Through AI Earth and Planet Engine, it offers end-to-end pipelines from data access to model inference and visualization within a unified cloud environment. As remote sensing demand grows and researchers require more seamless, scalable platforms, Alibaba Cloud is positioning itself as a strategic player, especially for users who value integration with its existing cloud infrastructure and regional performance.

10.2.3.3 Amazon Web Services (AWS) and remote sensing data analysis

Amazon Web Services stands out as one of the most flexible and widely adopted cloud ecosystems for remote sensing and Earth observation workflows. At its core lies the Earth on AWS initiative, which hosts a registry of open geospatial datasets such as Landsat and Sentinel collections directly in AWS storage. This approach allows users to bring compute resources to the data, minimizing transfer requirements and accelerating large-scale analysis. Since many satellite archives are already stored in AWS through its public data program, remote sensing researchers can design pipelines that operate in situ rather than downloading terabytes of imagery to local systems.

A compelling example of AWS in action is found in the construction of Earth Observation data cubes using serverless architecture. In one implementation, a Data Cube Builder application on AWS was built to assemble time-series data cubes over Brazil by orchestrating events and parallel processing across Lambda functions, queues, and metadata stores. The architecture enabled scalable tile-level parallelism so that each spatial-temporal cell could be processed independently, thus drastically reducing bottlenecks associated with monolithic batch jobs. Reportedly, for a single tile, one year of Sentinel imagery could be processed in approximately 30 minutes. This demonstrates how AWS serverless and event-driven primitives can underpin large-scale remote sensing pipelines with both agility and cost control.

Beyond data structuring, AWS supports the entire spectrum of remote sensing workflows through its general-purpose services. For ingestion and storage, Amazon S3 provides durable and highly available object storage suited for large imagery collections, typically in Cloud-Optimized GeoTIFF or chunked formats. Event-driven triggers can launch Lambda functions to perform preprocessing tasks including atmospheric correction, cloud masking, or tile clipping. Larger-scale or iterative tasks can be offloaded to AWS Batch or Amazon Elastic MapReduce clusters running Spark, while AWS SageMaker enables users to train and deploy custom machine learning models such as convolutional neural networks or segmentation architectures on scalable GPU instances. Workflow orchestration and serverless orchestration tools further integrate these stages into automated pipelines. In many remote sensing systems, this comprehensive integration allows imagery arrival, preprocessing, model inference, and result publication to occur with minimal manual intervention.

The open data hosting model that AWS provides represents another key advantage. Since numerous Earth observation collections are already available in AWS public data registry, users avoid the cost and delay of data ingestion. This architecture also enables collaborative reproducibility as different institutions can run comparable analyses over the same datasets without duplicating storage or network transfers. However, AWS flexibility involves a significant trade-off since users must assemble and manage infrastructure, dataflows, and higher-level geospatial logic independently. Unlike domain-specific platforms, AWS does not provide a native geospatial engine but instead offers fundamental building blocks for users to construct custom workflows tailored to their specific algorithmic requirements, data volumes, and scaling needs.

In summary, AWS offers a powerful and versatile foundation for remote sensing data analysis. Its hosting of open geospatial archives, serverless compute models, managed machine learning services, and orchestration capabilities provide all necessary components for scalable and reproducible pipelines. Although requiring substantial user effort to integrate these components effectively, AWS modularity and performance establish it as a cornerstone platform for large-scale Earth observation research and operational systems.

10.2.3.4 Microsoft azure and remote sensing platforms

Microsoft Azure offers a comprehensive suite of geospatial and cloud services specifically designed to support remote sensing and Earth observation workflows. A distinctive capability within this ecosystem is Azure Orbital, which establishes direct connectivity between satellites and ground stations, enabling raw satellite data to be ingested directly into Azure for real-time cloud processing. This integrated ground to cloud pathway streamlines the entire data value chain, supporting immediate execution of critical tasks including change detection, site monitoring, and situational awareness as soon as new data becomes available.

At the data storage and query level, Azure provides multiple spatially optimized database systems to support diverse remote sensing requirements. Azure Cosmos DB delivers multimodel and globally distributed data storage with integrated geospatial operations and spatial indexing capabilities. This architecture allows remote sensing applications to perform proximity queries and spatial filtering with minimal latency across distributed geographic regions. Complementing this, Azure SQL Database supports spatial data types and functions inherited from SQL Server, enabling traditional GIS style queries within a relational framework. For managing large raster and imagery datasets, Azure Blob Storage and Data

Lake Storage serve as scalable backends capable of handling high-volume satellite imagery collections efficiently.

For geospatial processing and visualization, Azure Maps provides comprehensive mapping APIs and spatial operations including buffering, intersection analysis, geofencing, routing, and geocoding services. This service integrates seamlessly with Azure's broader cloud ecosystem through REST APIs and software development kits, allowing developers to build interactive mapping and visualization interfaces directly atop analytical pipelines. In practical remote sensing applications, researchers frequently utilize Azure Maps to render algorithmic outputs such as classification maps and deforestation alerts directly into web or mobile mapping clients.

Azure's large-scale data processing capabilities are anchored by tools like Azure Databricks, which hosts distributed Apache Spark clusters specifically configured for satellite image preprocessing, mosaicking, and time series analytics. Documented implementations demonstrate complete satellite image classification workflows using Azure Databricks to ingest raw imagery, apply machine learning models, and generate detailed land cover maps. The platform's adjacency to Blob Storage enables direct operation on imagery datasets without unnecessary data movement, significantly reducing processing latency and operational overhead.

The platform further enhances geospatial workflows through Azure Data Factory, which incorporates Azure Maps to process and enrich location-based data. This integration enables automatic conversion of address records into geographic coordinates and seamlessly incorporates spatial context into broader extract-transform-load processes. For real-time streaming scenarios, Azure Stream Analytics supports continuous queries over sensor and Internet of Things data streams, effectively complementing traditional raster-based remote sensing with live contextual inputs. Microsoft has also advanced geospatial imagery enhancement through artificial intelligence, employing Azure to generate cloud-free daily optical and multispectral imagery by processing input satellite data to fill cloud-induced gaps using sophisticated AI models.

Overall, Azure's principal strength lies in its comprehensive integration across storage, compute, analytics, and specialized geospatial services. For remote sensing applications, this translates into the ability to construct complete end-to-end pipelines spanning satellite ingestion via Orbital, storage and querying through Cosmos DB or Blob Storage, distributed processing via Databricks, and interactive visualization with Azure Maps. While Azure may not feature a single unified remote sensing engine, its modular yet well-integrated services empower researchers and developers to build fully customized and scalable Earth observation systems. As Azure's geospatial offerings continue to mature, particularly in direct satellite ingestion and AI-driven imagery enhancement, the platform is establishing itself as a credible and competitive alternative within the global remote sensing cloud ecosystem.

10.2.3.5 Open source and customized platforms

Open source and customized platforms represent a crucial segment within the geospatial ecosystem, offering levels of flexibility, transparency, and extensibility that are often unmatched by proprietary systems. Among the most prominent open-source frameworks is the Open Data Cube (ODC), a Python-based ecosystem engineered for ingesting, indexing, and analyzing vast quantities of gridded Earth observation data. The ODC transforms satellite imagery collections into multidimensional data cubes indexed by space, time, and other dimensions, thereby facilitating complex queries and temporal analysis. Originally developed for Digital Earth Australia, the platform now supports numerous international implementations, enabling analysts to perform multitemporal analyses, change detection, and trend extraction across decades of data using scalable and reproducible workflows. Its modular architecture includes ingestion pipelines for converting raw imagery into analysis-ready data, metadata and indexing layers for efficient querying, and high-level APIs that allow users to execute spatial-temporal operations with standard Python tools. Complementary components like datacube-ows further provide OGC-standard web service endpoints, enabling institutions to disseminate geospatial visualizations without reliance on external platforms.

Another significant open platform is Sentinel Hub, which delivers Earth observation data through RESTful APIs and a cloud-based processing engine. Users can query imagery via its Catalog API, request on-the-fly processing through the Processing API, or obtain statistical summaries over specified spatiotemporal ranges via the Statistical API. The platform supports integration of custom data, fusion of multisensor collections, and embedding of user-defined geospatial logic such as spectral indices and classification scripts directly into processing requests. Its EO Browser interface offers interactive visualization and export capabilities, handling server-side operations like mosaicking and cloud filtering so users interact primarily with processed outputs. The open architecture of Sentinel Hub ensures seamless integration with custom workflows and third-party tools, making it highly adaptable to diverse project requirements.

The open-source ecosystem also includes a variety of specialized toolkits tailored to specific remote sensing tasks. Sen2Chain, for instance, is a Python toolbox designed for automated processing of Sentinel-2 time series, incorporating geometric correction, cloud masking, and spectral index extraction. It allows users to chain processing steps and parallelize operations across image tiles, making it particularly effective for regional or national-scale time-series generation. In the aerial remote sensing domain, OpenDroneMap stands out as a widely adopted open-source solution that converts aerial imagery into georeferenced orthomosaics, point clouds, and 3D models. With support for containerized deployment and GPU acceleration, it can be flexibly integrated into larger geospatial pipelines, offering both customizability and performance.

These open platforms are especially valuable for institutions and researchers requiring full control over their data, algorithms, and infrastructure. They enable deep customization, component replacement, and integration of domain-specific models without the constraints typical of proprietary systems. While this openness demands greater investment in maintenance, security, and performance tuning, many organizations adopt hybrid architectures that combine open platforms for core analytics with cloud services for storage and scaling. As remote sensing applications continue to diversify, open source and customized platforms will remain essential for enabling reproducible, adaptable, and transparent geospatial workflows across academic, governmental, and operational contexts.

10.2.4 Applications of cloud computing in remote sensing

10.2.4.1 Large-scale environmental monitoring and data processing

In large-scale environmental monitoring, cloud platforms integrated with artificial intelligence and machine learning enable analytical workflows that would be computationally prohibitive on traditional local systems. The MIT Supercloud Dataset, introduced by Samsi et al. [78], provides a comprehensive collection of high-performance computing resource utilization data. Although not specifically designed for remote sensing applications, this dataset offers valuable insights into CPU, GPU, memory, and input/output usage patterns that directly inform the optimization of scheduling algorithms and resource allocation strategies for geospatial workloads at cloud scale. Such metadata-driven approaches are becoming increasingly crucial for managing the substantial computational demands of planetary-scale environmental monitoring systems.

Within the remote sensing domain, Zhang et al. [79] demonstrated the transformative potential of cloud-based AI systems through their development of LuoJiaAI, an integrated platform for automated satellite imagery interpretation. This architecture showcases how end-to-end AI workflows can seamlessly process raw geospatial data, execute deep learning inference, and generate comprehensive classification and segmentation maps across extensive geographical areas with minimal human intervention. Complementing this approach, Zurqani [76] employed Google Earth Engine combined with advanced machine learning techniques to produce a high-resolution regional forest canopy cover dataset. By leveraging both aerial imagery and satellite data processed entirely within cloud environments, this project achieved remarkable accuracy with county-level correlation coefficients ranging from 0.94 to 0.98 and overall accuracy approaching 94 percent, demonstrating the powerful synergy between cloud computing and machine learning for detailed ecological characterization.

Operationally, these cloud-native systems support complete processing pipelines that begin with multisensor data ingestion from optical, synthetic aperture radar, and light detection and ranging sources. The workflows then progress through automated preprocessing stages including geometric correction, radiometric normalization, and cloud masking before advancing to AI-driven inference and subsequent post-processing operations such as mosaicking and change detection. A significant advantage of housing these complete workflows within cloud ecosystems is the capacity to apply updated analytical models retroactively across historical data archives. This capability ensures consistent temporal trend analysis and enables systematic reanalysis as methodological improvements emerge, effectively future-proofing environmental monitoring efforts.

In summary, the integration of cloud computing with artificial intelligence and machine learning has fundamentally transformed the practice of large-scale environmental monitoring. This technological convergence has not only dramatically expanded the geographical and temporal scope of feasible analyses but has also enhanced the agility, reproducibility, and operational efficiency of remote sensing applications. As these capabilities continue to mature, they promise to further accelerate our understanding of environmental dynamics and improve our capacity to monitor and respond to planetary-scale changes.

10.2.4.2 Land cover and forest monitoring

Land cover and forest monitoring represent foundational applications of cloud-based remote sensing, where high-performance computing enables consistent, repeatable, and large-scale analyses. In one notable study, Nursaputra et al. utilized the Google Earth Engine platform to conduct NDVI-based forest density assessments within a mining concession area. By implementing NDVI time-series analysis on Landsat imagery, they successfully monitored vegetation density changes across extensive spatial domains, demonstrating cloud infrastructure's capacity to support accountable and transparent forest monitoring in environmentally sensitive landscapes [80]. This methodology effectively illustrates how the integration of algorithmic vegetation indices with cloud scalability provides a practical alternative or supplement to traditional field-based surveys.

Complementing this density-focused approach, Montzka et al. conducted sophisticated analysis of spruce crown transparency levels using Sentinel-2 imagery processed entirely within a cloud computing environment to detect forest stress and damage patterns. Their innovative methodology involved comparing reference-year NDVI cumulative distributions with target-year data to systematically classify changes in crown transparency, revealing significant increases in areas affected by drought stress and bark beetle damage between

2018 and 2020 in the Northern Eifel region. Crucially, their findings demonstrated strong alignment with ground-based forest inventory data, thereby validating cloud-based remote sensing as a reliable tool for ecological diagnosis [81]. The research further extended to correlating transparency variations with multiple environmental drivers including precipitation patterns, land surface temperature, elevation gradients, and terrain characteristics, enabling the identification of vulnerable habitats for targeted conservation interventions.

Collectively, these investigations demonstrate how cloud computing facilitates multidimensional land cover monitoring capabilities. First, cloud platforms enable frequent temporal reevaluation by allowing new cloud-corrected imagery and enhanced algorithms to be retroactively applied across historical archives with minimal computational overhead. Second, the inherent scalability of cloud processing permits comprehensive monitoring of entire landscapes or regions, thereby supporting cross-site comparative analyses, ecological gradient investigations, and national-level assessment initiatives. Third, through the integration of statistical relationships with environmental covariates, cloud-enabled monitoring systems can transcend mere change detection to provide explanatory insights, helping researchers and practitioners understand the underlying drivers of land cover transformations rather than simply identifying their locations.

As increasingly sophisticated high-resolution and hyperspectral datasets become available, the role of cloud computing systems in land cover and forest health monitoring is poised to become even more central to both ecological science and environmental management. The continuous evolution of cloud-based analytical capabilities promises to further enhance our understanding of terrestrial ecosystem dynamics while providing actionable intelligence for sustainable land management decisions. This technological progression ensures that cloud platforms will remain indispensable tools for addressing the complex challenges of global environmental change.

10.2.4.3 Land use and land cover mapping

In the field of land use and land cover (LULC) mapping, cloud computing platforms have become indispensable for processing large spatial extents and enabling repeatable, adaptive analyses. Nasiri et al. systematically compared multiple spectral-temporal feature extraction strategies using Sentinel-2 and Landsat-8 datasets within Google Earth Engine [82]. Their research demonstrated that combining spectral indices with temporal statistics including anomalies, seasonal deviations, and change metrics significantly enhances classification accuracy when used with machine learning classifiers. The experiments revealed that incorporating temporal features beyond static spectral bands substantially improves class separability in dynamically changing landscapes.

Complementing this work, Naboureh et al. addressed one of the persistent challenges in LULC mapping by developing a semiautomatic method for generating reliable reference samples over large geographic regions [83]. Their approach leveraged GEE's extensive image archives and auxiliary data to create high-quality training samples, enabling more robust classifier training at scale while reducing manual labeling efforts. In their Turkey case study, this methodology demonstrated the utility of cloud platforms not only for algorithm execution but also for efficient training data preparation and sampling.

Collectively, these studies illustrate how cloud computing environments enable advanced LULC mapping workflows through three key capabilities: engineering sophisticated temporal features, generating scalable sample datasets, and periodically retraining classifiers across complete data archives. A typical cloud-based LULC processing pipeline ingests multiyear Sentinel-2 and Landsat data, computes spectral indices and temporal metrics for each pixel or object, executes machine learning classifiers such as Random Forest or gradient boosting models, and produces continuously refinable mapping products. Since all computational processing occurs server-side, researchers can retrospectively apply updated analytical models, compare multitemporal mapping results, and maintain long-term consistency across different mapping periods. This combination of technical flexibility and operational scalability establishes cloud-powered LULC mapping as an essential tool for national land management strategies, ecological monitoring programs, and urban planning initiatives in the era of big Earth observation data.

10.2.4.4 Disaster management and emergency response

Disaster management and emergency response represent critical domains where cloud computing demonstrates significant operational value through its core strengths of processing speed, computational scalability, and system integration. In flood emergency scenarios specifically, the capability to rapidly ingest multisource satellite imagery, execute automated change detection algorithms, and generate timely inundation maps can substantially enhance situational awareness and support ground-level emergency management decisions.

Ghosh et al. present a compelling demonstration of this operational potential through their cloud-based flood impact assessment system implemented across eastern India [84]. Their research adapted specialized processing workflows on a cloud platform to evaluate flood consequences spanning multiple states including West Bengal, Bihar, and Jharkhand. The system effectively utilized satellite data to detect inundation patterns, quantify damage to vegetation and agricultural areas, and integrate these findings with settlement and land use layers to visualize the complete spatial extent of flood impacts. By leveraging cloud computational infrastructure, the analytical processes could scale efficiently across large and geographically diverse regions, enabling simultaneous processing of multiple datasets to support comprehensive damage assessment.

Complementing this approach, Hamidi et al. developed an advanced flood monitoring methodology that integrates multisource synthetic aperture radar and optical imagery within the Google Earth Engine platform [85]. Their innovative framework employed sophisticated change detection techniques to delineate flood boundaries with enhanced reliability, strategically combining radar's cloud-penetrating capability with the spectral richness of optical imagery. This integrated sensing approach facilitates near all-weather and time-sensitive flood event mapping, a critical capability for emergency response operations in frequently cloud-obscured regions or during periods of intense meteorological activity.

These case studies collectively illustrate three fundamental advantages of cloud-enhanced disaster management systems. First, they provide exceptional timeliness by minimizing data transfer bottlenecks and reducing processing latency through integrated data ingestion, preprocessing, and mapping workflows operating entirely within cloud environments. Second, they offer unprecedented scalability, enabling emergency analyses to expand seamlessly across national or transboundary geographical extents using distributed computing resources. Third, they enable sophisticated data fusion capabilities that combine complementary sensor technologies to overcome individual limitations and produce more reliable disaster delineation products.

In operational deployment contexts, such cloud-based systems typically incorporate real-time satellite data streams, automated preprocessing and change detection modules, cloud-hosted analytical outputs, and interactive dashboard or API interfaces for decision support. As emerging remote sensing constellations deliver improved temporal resolution and radar capabilities continue to advance, cloud-based disaster mapping platforms are positioned to become increasingly central to rapid emergency response operations, comprehensive risk assessment methodologies, and long-term community resilience planning strategies.

10.2.4.5 High-resolution surface classification and change detection

The advancement of high-resolution and very-high-resolution remote sensing technologies has created an urgent need for more sophisticated cloud-based classification and change detection methodologies. Traditional pixel-based analytical approaches frequently encounter limitations when processing imagery where fine spatial details such as small structures, narrow transportation corridors, and canopy openings become visible, while surface heterogeneity and sensor noise become increasingly significant factors. Deep learning architectures have demonstrated remarkable capability in identifying subtle spectral-spatial variations and discerning fine-scale changes, particularly when deployed on cloud platforms that provide the necessary computational scalability and parallel processing infrastructure.

The standard framework for high-resolution change detection generally comprises multiple coordinated stages: temporal alignment of multitemporal imagery, extraction of discriminative features including spectral, textural, and deep learning-derived representations, and comparative analysis to identify meaningful changes. Contemporary research categorizes deep learning-based change detection methods according to their architectural designs, such as Siamese networks, attention mechanisms, and multiscale fusion strategies, as well as by their fundamental unit of analysis whether pixel, object, or patch-based. A persistent challenge in this domain involves addressing pseudochange phenomena, where apparent differences arise from extraneous factors like illumination variation, geometric misregistration, or seasonal dynamics rather than genuine land surface transformations [86]. Modern deep learning approaches mitigate these effects through specialized components including attention modules and differential feature fusion layers that effectively suppress spurious variations while preserving legitimate changes.

Substantial progress in this field is exemplified by the development of the Dual Attentive Siamese Network, an architecture specifically engineered to enhance robustness against pseudochanges in high-resolution satellite imagery [87]. This model employs dual attention mechanisms to capture both spatial dependencies and long-range contextual information, significantly improving the discrimination between authentic changes and noise. Experimental validation demonstrates that DASNet achieves measurable improvements in F1 scores compared to baseline models for building change detection tasks. Similarly, the Differential Attention Fusion Network represents another architectural innovation that combines feature difference extraction with attention mechanisms to reduce misdetection rates and produce change maps with sharper boundary delineation [88]. These technical advances collectively illustrate how cloud-enabled deep learning architectures can generate superior classification and change detection outputs in high-resolution remote sensing applications.

The operational deployment of high-resolution classification and change detection systems at scale necessitates cloud platforms capable of managing intensive input/output operations, batch processing workloads, and complex inference orchestration. A typical processing workflow involves multiple stages: segmenting large geographic areas into systematically tiled image chips, applying comprehensive preprocessing routines including orthorectification and illumination normalization, executing neural network inference on GPU-accelerated computing resources, and reassembling individual tile predictions into seamless output products. Cloud computing environments provide essential capabilities for parallel inference across numerous tiles, dynamic scaling of GPU resources based on workload demands, and efficient reprocessing when model improvements become available. The colocation of computational, storage, and networking resources within cloud infrastructure minimizes data transfer overhead for high-resolution imagery and enables rapid delivery of analytical results to support downstream applications including urban expansion monitoring, critical infrastructure observation, and detailed habitat change assessment.

In summary, high-resolution surface classification and change detection has transitioned from theoretical possibility to practical reality at scale through the synergistic combination of advanced neural network architectures and sophisticated cloud orchestration capabilities. Whereas fine-scale surface features were previously considered too noisy or computationally prohibitive for systematic global analysis, contemporary cloud-based systems can now detect subtle transformations ranging from structural modifications to vegetative changes across extensive geographical domains. As sensor technologies continue to evolve toward higher spatial resolutions and improved temporal frequencies, cloud platforms will undoubtedly serve as fundamental enablers for next-generation change detection systems, unlocking unprecedented levels of temporal and spatial sensitivity in comprehensive Earth monitoring endeavors.

10.2.5 Future directions and remarks

10.2.5.1 Cloud-based tools and frameworks

In recent years, the development of cloud-based tools and frameworks has emerged as a critical enabler for democratizing access to high-performance geospatial analytics. A significant early contribution in this domain comes from Prasai, who developed an innovative web application that enables users to retrieve long-term satellite reflectance time series through an intuitive user interface without requiring programming knowledge [89]. This Earth Engine based application facilitates essential analytical tasks including data filtering, cloud cover threshold adjustment, time series visualization, and metadata export, effectively reducing technical barriers for researchers with limited computational backgrounds.

Advancing this accessibility paradigm, Gao et al. proposed a comprehensive spatial-temporal big data analysis platform that intelligently orchestrates multiple heterogeneous cloud services within a unified architectural framework [90]. Their design emphasizes modularity across data ingestion, storage management, algorithmic execution, and service delivery components, enabling scalable natural resource management workflows that can adapt to diverse regional requirements and dataset characteristics. The framework incorporates adaptive resource allocation mechanisms that dynamically distribute computational and storage resources according to workload demands, seamlessly integrating managed services, serverless computing, and sophisticated orchestration layers.

In a more recent development, Xu et al. introduced AI Earth, a cloud-native remote sensing platform built on Alibaba Cloud infrastructure that unifies data cataloging, computational services, and artificial intelligence modules within a cohesive environment [91]. AI Earth provides dual operational modes including a toolbox interface for web-based interaction and a developer mode supporting API and notebook-based programming, allowing users to execute both batch processing and interactive analytical tasks while facilitating custom dataset uploads and model deployments. The platform delivers comprehensive remote sensing capabilities encompassing change detection, terrain classification, and image segmentation functionalities while maintaining extensibility through open architecture standards and third-party model integration support.

These evolving tools and frameworks collectively represent a strategic shift from infrastructure-centric cloud utilization toward more integrated and user-oriented analytical platforms. They effectively abstract the underlying complexities of cloud orchestration, including resource provisioning, scalability management, and data handling operations, while presenting simplified interfaces tailored to domain scientists' requirements. Importantly, these platforms also enable retrospective reprocessing of historical datasets using updated analytical models, promote workflow sharing across research communities, and enhance methodological reproducibility throughout project lifecycles. As these platforms continue to mature, key challenges will include maintaining system responsiveness, ensuring modular extensibility, and achieving seamless interoperability across both open-source and proprietary systems within an increasingly interconnected global geospatial ecosystem.

10.2.5.2 Impact of 6G on cloud-based remote sensing applications

The emergence of sixth-generation mobile communication technology is poised to revolutionize cloud-based remote sensing applications. 6G networks are expected to provide end-to-end latency below 1 millisecond and data transmission rates exceeding 1 terabit per second, representing breakthrough communication capabilities that will fundamentally transform traditional remote sensing data processing paradigms. Through its integrated sensing and communication design, 6G enables simultaneous real-time transmission and processing of remote sensing data during acquisition, establishing an unprecedented technological foundation for instantaneous analysis of large-scale remote sensing datasets.

At the architectural level, 6G-supported technologies including terahertz frequency transmission and intelligent reflective surfaces significantly enhance the transmission efficiency of remote sensing data from acquisition sources to cloud platforms. These advanced communication capabilities facilitate rapid uploading of large-volume data such as high-resolution imagery and synthetic aperture radar data to cloud computing platforms while ensuring real-time performance for critical remote sensing applications. Particularly for time-sensitive scenarios including emergency disaster response and dynamic environmental monitoring, 6G technology enables complete remote sensing data workflows from acquisition to cloud processing within seconds.

The massive connectivity characteristics of 6G will drive remote sensing applications toward more diversified development directions. With its capability to support up to 10 million device connections per square kilometer, 6G makes possible the construction of integrated space-air-ground-sea remote sensing monitoring networks. Data collected from various remote sensing platforms including satellites, unmanned aerial vehicles, and ground sensors can be aggregated in real-time to cloud computing centers through 6G networks, enabling centralized processing and analysis of multisource heterogeneous remote sensing data. This deep integration will significantly enhance the dimensionality and timeliness of remote sensing data, providing more comprehensive data support for fields including precision agriculture, urban management, and environmental protection.

Despite the tremendous opportunities 6G technology brings to remote sensing cloud computing, its comprehensive deployment still faces numerous challenges. The transmission characteristics of terahertz frequency signals require more dense network infrastructure deployment, particularly in remote sensing monitoring applications in isolated areas. Meanwhile, the connection of massive remote sensing devices to cloud platforms through 6G networks imposes higher requirements for network security and data privacy protection. Additionally, the integration of 6G networks with existing remote sensing data processing architectures necessitates corresponding technical standards and protocols to ensure system compatibility and scalability. As these technical challenges are progressively addressed, 6G-enabled remote sensing cloud computing will inaugurate a new era in Earth observation.

10.2.5.3 Role of cloud computing in real-time monitoring and response systems

In the context of environmental hazards, disasters, and dynamic Earth processes, real-time monitoring represents an aspirational goal that cloud computing technologies are making increasingly attainable. While inherent limitations in satellite data acquisition and transmission impose fundamental constraints, cloud systems substantially compress the temporal gap between raw data collection and decision support outputs, enabling near real-time monitoring capabilities and actionable alert systems.

A fundamental paradigm enabling this acceleration involves the implementation of streaming and continuous workflow architectures. When new imagery or sensor data reaches cloud infrastructure, preprocessing, change detection, and analytical modules activate automatically rather than operating through traditional batch processing. Cloud environments make this approach feasible through the integration of scalable computational resources, event-driven orchestration frameworks, and streamlined data pipelines. The Climate Engine platform exemplifies this capability by providing on-demand processing of climate and land surface datasets, enabling rapid computation and visualization of drought indices and vegetation anomalies while transforming what previously required hours or days of offline processing into interactive analytical sessions.

This cloud-enabled real-time capability extends across multiple geoscientific domains. In seismic monitoring, the QuakeFlow system demonstrates a scalable earthquake detection pipeline where continuous waveform data undergoes processing in cloud containers, deep learning models identify seismic phases, and streaming components deliver near-instantaneous event catalogs. The containerized and autoscaling architecture supports parallel processing, minimal latency, and seamless model updates without operational interruption. For remote sensing applications specifically, these systems support critical functions including flood early warning, wildfire hotspot tracking, and infrastructure monitoring through a consistent operational pattern: newly acquired imagery triggers parallel processing through change detection models, with results rapidly delivered to storage systems and user interfaces through automated dashboards and application programming interfaces.

Despite these advancements, important operational considerations remain. Genuine real-time processing with subsecond latency remains generally unattainable for satellite-based applications due to inherent acquisition timing, data downlink requirements, and essential preprocessing steps. Most operational systems more accurately provide near-real-time capabilities where alerts follow data acquisition by minutes or hours rather than seconds. Additionally, effective management of resource scaling, system fault tolerance, and operational cost control requires careful implementation, particularly during emergency events when processing demands may spike dramatically. The synchronization between edge preprocessing systems and cloud analytical models during network disruptions presents another persistent challenge.

Through enabling streaming data ingestion, automatically triggered analytics, scalable computational inference, and integrated alert services, cloud computing fundamentally bridges the critical gap between raw Earth observations and timely decision-making processes. As satellite revisit intervals continue to decrease and computational capacities expand, these cloud-based systems will increasingly form the foundation for responsive and resilient environmental monitoring infrastructures worldwide.

10.2.5.4 The key role of cloud computing in global climate change research

Global climate change research represents an inherently data-intensive scientific domain that integrates diverse variables including temperature patterns, precipitation dynamics, vegetation changes, and carbon flux measurements across multiple temporal scales and observational platforms. Cloud computing has emerged as a transformative infrastructure supporting this research by eliminating traditional computational barriers, enabling fully reproducible modeling approaches, and facilitating large-ensemble cross-regional analyses that would remain practically impossible under conventional computing paradigms.

Cloud platforms fundamentally serve as centralized repositories for vast climate and Earth observation datasets, providing immediate accessibility for advanced modeling and analytical applications. Infrastructure such as Google Earth Engine allows researchers to conduct multidecadal analyses of satellite imagery archives without the prohibitive requirements of local data download and storage management. This data proximity principle enables climate scientists to directly investigate critical hypotheses regarding land surface feedback mechanisms, vegetation transition dynamics, and glacial retreat patterns within the same computational environment where petabytes of source data reside, effectively eliminating data transfer latency while ensuring version consistency and analytical reproducibility.

The scalable nature of cloud computing infrastructure uniquely supports comprehensive ensemble modeling and systematic retrospective reanalysis capabilities. As climate models evolve and new analytical methodologies emerge, cloud systems enable the consistent reprocessing of historical climate periods through updated algorithmic approaches across global data archives. This capacity for methodological iteration proves particularly vital for refining trend detection sensitivity, anomaly identification accuracy, and tipping point early warning systems. Without elastic cloud computing resources, the reprocessing of decadal-scale global imagery through improved algorithms would present insurmountable computational challenges, thereby constraining methodological advancement in climate science.

Cloud environments further enable the sophisticated integration of heterogeneous data sources through unified analytical workflows. Satellite-derived atmospheric parameters, land surface indicators, and biosphere variables can be systematically aligned and jointly analyzed with climate model outputs and reanalysis products. When combined with machine learning architectures, this multidimensional data fusion empowers researchers to identify complex climate feedback loops, perform spatial downscaling of coarse-resolution model outputs, and establish robust benchmarks for Earth system model evaluation. The deployment of such analytical frameworks within cloud environments enables global-scale implementation, systematic output comparison, and seamless integration into model calibration procedures.

Through providing shared computational environments, standardized analytical workflows, and version-controlled processing pipelines, cloud computing fundamentally enhances the collaborative potential and methodological transparency of climate research. Cloud-native implementations of distributed data management frameworks support consistent data access protocols, common analytical workflow execution, and research dissemination with complete traceability and auditability. As climate models increase in complexity, observational datasets expand in resolution and volume, and scientific uncertainty requirements become more stringent, cloud computing platforms will continue to serve as indispensable infrastructure for advancing comprehensive understanding of Earth's climate system and informing evidence-based mitigation and adaptation strategies globally.

10.2.5.5 Remarks

This section has examined how high-performance cloud computing is fundamentally transforming remote sensing and large-scale Earth monitoring, revolutionizing not only data processing methodologies but also democratizing participation in global environmental science. Cloud infrastructure provides essential capabilities including elastic scalability, computational flexibility, and direct access to massive datasets, effectively eliminating the traditional bottleneck of transferring petabytes of data to local computing resources. Advanced technological frameworks encompassing distributed computing systems, container orchestration platforms, and cloud-native storage solutions collectively enable the seamless implementation of complex analytical workflows across global scales.

The evolution of cloud platforms into comprehensive remote sensing ecosystems represents a significant paradigm shift. Google Earth Engine exemplifies this transformation through its integrated data catalog and server-side processing model, implementing the core principle of bringing computational applications to the data rather than transferring data to computational resources. The platform enables researchers to execute analytical scripts across decades of imagery without confronting local storage limitations. Case studies in oceanic monitoring demonstrate this capacity through the extraction of sea surface temperature parameters and suspended matter concentrations across extensive temporal and spatial domains. Similarly, vegetation and forest mapping applications have leveraged these capabilities to generate high-resolution canopy cover products, illustrating the powerful synergy between artificial intelligence methodologies and scalable cloud processing infrastructure.

In land cover and land use mapping applications, cloud computing systems facilitate unprecedented exploration and reanalysis of massive geospatial archives. Research by Nasiri et al. demonstrates how integrating spectral and temporal features enhances classification accuracy, while Naboureh et al. developed semiautomatic sample generation techniques that substantially reduce the labor-intensive process of ground truth labeling across diverse geographical regions. Cloud environments uniquely support the continuous retraining and updating of analytical models with subsequent retrospective application to historical datasets, thereby enabling consistent longitudinal analyses at global scales.

The critical importance of cloud computing becomes particularly evident in disaster management and hazard response scenarios. Flood mapping studies conducted by Ghosh et al. and Hamidi et al. illustrate how cloud-based processing pipelines efficiently ingest multisource imagery including optical and synthetic aperture radar data, execute change detection algorithms across distributed tiles, and rapidly deliver actionable inundation maps. These capabilities prove essential for supporting real-time emergency response operations and effective mitigation strategies.

Emerging trends and future directions highlight the expanding ecosystem of cloud-based tools and frameworks that are making advanced analytics increasingly accessible to domain scientists. The ongoing integration of 6G communication technologies and edge computing architectures promises to distribute computational processes closer to data collection points, substantially reducing latency and network congestion. Event-driven workflows within cloud environments are enabling near real-time monitoring systems that minimize the temporal gap between data acquisition and actionable intelligence. In climate change research, cloud platforms facilitate large-ensemble analytical approaches, sophisticated data fusion techniques, and the continuous reprocessing of global archives using evolving methodologies, thereby significantly advancing our understanding of Earth system dynamics.

Despite these substantial advancements, important challenges require continued attention. Effective cost management strategies for data egress, storage, and computational resources remain crucial considerations. Security protocols, data sovereignty issues, network input/output optimization, cross-platform interoperability, and energy efficiency represent additional areas demanding innovative solutions. Recent research initiatives have begun examining the environmental impact of cloud computing itself, exploring methodologies for scheduling computational tasks in ways that minimize carbon emissions.

In conclusion, cloud computing has evolved beyond merely enabling remote sensing applications to become the fundamental backbone of modern Earth observation science. By making large-scale, multitemporal, and multisensor analytics technically feasible, economically scalable, and scientifically inclusive, cloud technologies are unlocking new frontiers across ecological research, hydrological monitoring, disaster resilience planning, land management practices, and climate science. As sensor networks continue to expand, analytical models grow in complexity, and data volumes increase exponentially, well-architected cloud computing systems will prove indispensable for realizing a future where continuous, high-resolution, and responsive monitoring of our planet becomes an operational reality.

10.3 Digital twin technologies

10.3.1 Introduction

Digital Twin (DT) technologies have emerged as a transformative paradigm bridging the physical and digital worlds through real-time data synchronization, intelligent modeling,

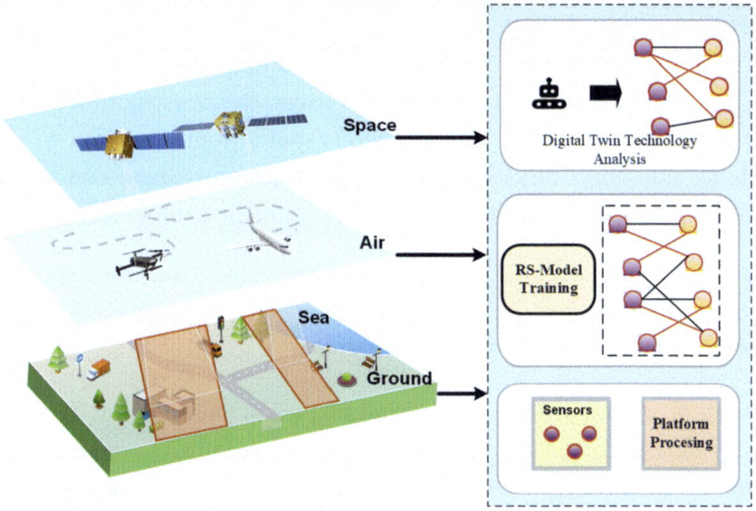

Figure 10.2:
Application of different measurement platforms and systems for data collection.

and predictive analytics [92,93]. A Digital Twin can be defined as a dynamic virtual replica of a physical system that continuously updates its state by integrating multisource data and computational models [94]. Initially developed for industrial and manufacturing applications, the DT Technologies have rapidly expanded into diverse domains such as aerospace, remote sensing, and Earth observation.

In remote sensing, digital twin (DT) technologies provide a powerful way to represent, monitor, and predict complex spatial systems [95]. By combining satellite observations, sensor networks, and simulation models, digital twins can give near real-time awareness and allow adaptive control of space or ground assets. For example, space-environment twins help predict satellite orbits, optimize navigation, and detect anomalies [94]. The relevant remote sensing or ground data can be obtained from multiple devices and sources, as shown in Fig. 10.2. Similarly, urban and environmental digital twins use multiple types of remote sensing data such as optical images, SAR, and LiDAR to create accurate 3D models and understand dynamic scenes [96].

Recent advances in multimodal deep learning have further accelerated DT development by improving the fusion of heterogeneous data sources [97]. Multimodal information fusion (MIF) models integrate visual, textual, and physical signals, enabling the digital twin to perceive and reason across multiple representations of reality. Such fusion not only enhances the realism and responsiveness of DT systems but also facilitates scalable applications such as digital twin Earth, smart cities, and on-orbit spacecraft maintenance [92,98].

Despite these advances, several challenges remain. Achieving real-time synchronization between the physical and digital layers requires efficient data transmission, model interoperability, and cross-domain calibration. Moreover, the construction of multimodal DTs demands robust methods for aligning heterogeneous spatial-temporal data while maintaining semantic consistency and uncertainty quantification. Addressing these challenges is essential to unlocking the full potential of Digital Twin technologies in remote sensing and Earth observation.

10.3.2 Fundamentals of digital twin technologies

Through integrating processes with real-time and historical data, Digital Twin technology serves as a transformative framework for observing and modeling the remote sensing system and Earth system. The concept of a Digital Twin of Earth (DTE) extends the classical Earth Information System (EIS) developed by NASA by leveraging advances in remote sensing, high-performance computing, and data assimilation techniques [92,94]. While the EIS focuses on integrating observations and models to provide scientific insights and improved boundary conditions for Earth system predictions, the DTE enables higher spatial resolution and dynamic representation of selected geophysical variables, capturing the evolving states of land, ocean, and land-cover subsystems.

A key foundation of DTE is the integration of multimodal Earth observation data, including optical imagery, synthetic aperture radar (SAR), LiDAR, and hyperspectral measurements. These heterogeneous datasets capture complementary aspects of the environment, from topography and land cover to water fluxes and surface dynamics. The fusion of multimodal data is essential to construct a coherent, high-fidelity digital representation of the Earth system, particularly for complex and dynamic processes such as hydrology, vegetation dynamics, and urban development [95,96].

Recent advances in deep learning and multimodal information fusion (MIF) have enabled robust alignment and integration of these diverse datasets. Deep neural networks, including convolutional and graph-based architectures, can extract meaningful features from each modality and map them into a shared high-dimensional space, forming joint representations that capture both spatial and temporal correlations across modalities [97]. This approach enhances the accuracy and completeness of the digital twin, enabling it to support various downstream tasks in remote sensing deep learning, such as semantic segmentation and change detection.

Digital Twin Technologies are inherently hierarchical and multiscale. Partial or subsystem-level digital twins, such as hydrological or oceanic twins, represent specific processes while allowing modular integration into a holistic Earth system model. Each DT subsystem assimilates observational data to continuously update its state and correct deviations between models and reality. In this way, the DTE serves as a multiscale, multiphysics waveguide, capturing resolved processes at scales larger than the effective resolution of the underlying model-data construct, and providing a flexible platform forremote sensing exploration and operational applications.

The real-time functionality of the DT is based on advanced computational infrastructures, including cloud-based processing, high-performance pipelines, IoT-enabled continuous synchronization between the physical Earth system and its digital counterpart, enabling predictive analytics, scenario simulation, and decision support. As artificial intelligence and large-scale machine learning models continue to mature, integrating deep multimodal modeling into DTEs is expected to enable a more comprehensive analysis of complex Earth system behaviors, thereby greatly enhancing the effectiveness of multimodal remote sensing technologies.

In summary, the fundamentals of Digital Twin technologies for Earth observation combine multimodal remote sensing, high-fidelity modeling, data assimilation, and AI-driven fusion, forming a robust foundation for dynamic, accurate, and predictive representations of the Earth system. This integrated approach offers unprecedented opportunities for environmental monitoring, resource management, and scientific discovery in a rapidly changing world.

10.3.3 Integration of digital twin and remote sensing

Through the integration of remote sensing and geospatial information, Digital Twin technologies allow real-time, high-fidelity representations of complex systems. Yufeng Huang (2023) proposed a digital twin basin framework that leverages remote sensing to capture key hydrological variables such as precipitation, temperature, soil moisture, and vegetation cover. This system supports simulation, monitoring, and decision-making for water resource management, highlighting the role of high-resolution and large-scale observational datasets in DT construction [95].

Chen et al. (2023) introduced a digital twin of the ocean that combines marine big data with artificial intelligence. The system enables real-time data fusion, intelligent prediction, and autonomous learning, enhancing oceanographic research and maritime operations. This work demonstrates how DTs can incorporate multisource and multitemporal remote sensing datasets for predictive modeling in complex marine environments [99].

Tomczak et al. (2024) developed a spatial digital twin of harbor waters using multiple measurement technologies, including sensors on boats, aircraft, and UAVs. By integrating multi-source spatial data, including LiDAR point clouds and optical imagery, the system produces high-precision models of river and inland navigation areas, facilitating environmental monitoring, navigation safety, and water quality assessment [96].

Ye et al. (2024) designed an interactive visual analytics system for campus digital twins, integrating enrollment data and spatial-temporal analysis to simulate classroom distribution and building capacity. The approach highlights the importance of multimodal data fusion and visualization in digital twin applications, which can also be adapted for urban and environmental remote sensing scenarios [100].

In addition to these cases, recent research has highlighted the importance of incorporating multimodal remote sensing datasets, including optical imagery, SAR, LiDAR, and hyperspectral data, into Digital Twin frameworks for enhanced modeling and analysis. For instance, Huang et al. (2023) proposed a deep multimodal information fusion method that combines sensor observations and physics-based models, improving real-time monitoring, anomaly detection, and predictive accuracy in environmental and industrial applications [95]. Similarly, Barros et al. (2024) explored Digital Twin Earth concepts, illustrating hierarchical, multiscale DT structures that assimilate diverse remote sensing datasets for dynamic land, ocean, and atmospheric modeling [92].

Collectively, these works demonstrate that the integration of digital twin technologies with remote sensing and multimodal data fusion enables dynamic, predictive, and high-fidelity representations of physical systems. Such integration not only enhances monitoring and management capabilities but also facilitates advanced applications in semantic segmentation, urban planning, and environmental monitoring.

In summary, the integration of Digital Twin technologies with remote sensing and multimodal data has emerged as a transformative approach for representing, monitoring, and predicting complex physical systems. By combining diverse datasets, including optical imagery, SAR, LiDAR, hyperspectral measurements, and in situ sensor observations, with physics-based and data-driven models, digital twins achieve high-fidelity, dynamic, and predictive representations of terrestrial, aquatic, and atmospheric processes. Applications ranging from hydrological management in digital twin basins, oceanographic research with AI-enabled digital twin oceans, to environmental monitoring of harbor waters and urban or campus planning illustrate the versatility and scalability of this approach. Multimodal information fusion plays a central role, enabling the alignment, integration, and joint representation of heterogeneous data, thereby enhancing model accuracy, anomaly detection, and scenario prediction. Despite notable progress, challenges remain in achieving real-time synchronization, maintaining data quality and spatiotemporal resolution, and ensuring interoperability across multiple sensors and platforms. Future developments are expected to leverage advances in artificial intelligence, machine learning, and cloud-based computing to further improve predictive performance, uncertainty quantification, and autonomous decision-making. Overall, the fusion of digital twin technologies with multimodal remote sensing offers unprecedented opportunities for environmental monitoring, sustainable resource management, and scientific discovery, laying a robust foundation for next-generation Earth observation and digital ecosystem modeling.

References

[1] S. Liu, Q. Shi, Local climate zone mapping as remote sensing scene classification using deep learning: a case study of metropolitan China, ISPRS Journal of Photogrammetry and Remote Sensing 164 (2020) 229–242.

[2] R. Zhang, Y. Wang, J. Hu, W. Yang, J. Chen, X. Zhu, SAR4LCZ-Net: a complex-valued convolutional neural network for local climate zones classification using Gaofen-3 quad-pol SAR data, IEEE Transactions on Geoscience and Remote Sensing 60 (2022) 1–16.

[3] M. Madhavi, R. Kolikipogu, S. Prabakar, S. Banerjee, L.P. Maguluri, G.B. Raj, A. Balaram, Experimental evaluation of remote sensing-based climate change prediction using enhanced deep learning strategy, Remote Sensing in Earth Systems Sciences 7 (4) (2024) 642–656.

[4] Y.-G. Ham, J.-H. Kim, J.-J. Luo, Deep learning for multi-year ENSO forecasts, Nature 573 (7775) (2019) 568–572.

[5] M. Choo, S. Jung, J. Im, D. Han, CARE-SST: context-aware reconstruction diffusion model for sea surface temperature, ISPRS Journal of Photogrammetry and Remote Sensing 220 (2025) 454–472.

[6] W. Wang, S. Brönnimann, J. Zhou, S. Li, Z. Wang, Near-surface air temperature estimation for areas with sparse observations based on transfer learning, ISPRS Journal of Photogrammetry and Remote Sensing 220 (2025) 712–727.

[7] K.M. Orndahl, L.T. Berner, M.J. Macander, M.F. Arndal, H.D. Alexander, E.R. Humphreys, M.M. Loranty, S.M. Ludwig, J. Nyman, S. Juutinen, M. Aurela, J. Mikola, M.C. Mack, M. Rose, M.R. Vankoughnett, C.M. Iversen, J. Kumar, V.G. Salmon, D. Yang, P. Grogan, R.K. Danby, N.A. Scott, J. Olofsson, M.B. Siewert, L. Deschamps, V. Maire, E. Levesque, G. Gauthier, S. Boudreau, A. Gaspard, M.S. Bret-Harte, M.K. Raynolds, D.A. Walker, A. Michelsen, T. Kumpula, M. Villoslada, H. Ylanne, M. Luoto, T. Virtanen, H.E. Greaves, B.C. Forbes, R.J. Heim, N. Holzel, H. Epstein, A.G. Bunn, R.M. Holmes, S.M. Natali, A.-M. Virkkala, S.J. Goetz, Next generation Arctic vegetation maps: aboveground plant biomass and woody dominance mapped at 30 m resolution across the tundra biome, Remote Sensing of Environment 323 (2025) 114717.

[8] J. Muro, A. Linstädter, P. Magdon, S. Wöllauer, F.A. Männer, L.-M. Schwarz, G. Ghazaryan, J. Schultz, Z. Malenovský, O. Dubovyk, Predicting plant biomass and species richness in temperate grasslands across regions, time, and land management with remote sensing and deep learning, Remote Sensing of Environment 282 (2022) 113262.

[9] N. Lang, N. Kalischek, J. Armston, K. Schindler, R. Dubayah, J.D. Wegner, Global canopy height regression and uncertainty estimation from GEDI LIDAR waveforms with deep ensembles, Remote Sensing of Environment 268 (2022) 112760.

[10] S.M. Mirhoseini Nejad, D. Abbasi-Moghadam, A. Sharifi, ConvLSTM-ViT: a deep neural network for crop yield prediction using Earth observations and remotely sensed data, IEEE Journal of Selected Topics in Applied Earth Observations and Remote Sensing 17 (2024) 17489–17502.

[11] W. Zhao, S. Du, W.J. Emery, Object-based convolutional neural network for high-resolution imagery classification, IEEE Journal of Selected Topics in Applied Earth Observations and Remote Sensing 10 (7) (2017) 3386–3396.

[12] B. Huang, B. Zhao, Y. Song, Urban land-use mapping using a deep convolutional neural network with high spatial resolution multispectral remote sensing imagery, Remote Sensing of Environment 214 (2018) 73–86.

[13] W.-B. Wu, J. Ma, E. Banzhaf, M.E. Meadows, Z.-W. Yu, F.-X. Guo, D. Sengupta, X.-X. Cai, B. Zhao, A first Chinese building height estimate at 10m resolution (CNBH-10m) using multi-source Earth observations and machine learning, Remote Sensing of Environment 291 (2023) 113578.

[14] J. Li, D. Hong, L. Gao, J. Yao, K. Zheng, B. Zhang, J. Chanussot, Deep learning in multimodal remote sensing data fusion: a comprehensive review, International Journal of Applied Earth Observation and Geoinformation 112 (2022) 102926.

[15] D. Hong, L. Gao, N. Yokoya, J. Yao, J. Chanussot, Q. Du, B. Zhang, More diverse means better: multimodal deep learning meets remote-sensing imagery classification, IEEE Transactions on Geoscience and Remote Sensing 59 (5) (2021) 4340–4354.

[16] X. He, X. Han, Y. Chen, L. Huang, A light-weighted fusion vision mamba for multimodal remote sensing data classification, IEEE Journal of Selected Topics in Applied Earth Observations and Remote Sensing 18 (2025) 21532–21548.

[17] M. Lamichhane, S. Mehan, K.R. Mankin, Surface soil moisture prediction using multimodal remote sensing data fusion and machine learning algorithms in semi-arid agricultural region, Science of Remote Sensing 12 (2025) 100255.

[18] J. Achiam, S. Adler, S. Agarwal, L. Ahmad, I. Akkaya, F.L. Aleman, D. Almeida, J. Altenschmidt, S. Altman, S. Anadkat, et al., GPT-4 technical report, arXiv preprint, arXiv:2303.08774, 2023.

[19] H. Touvron, T. Lavril, G. Izacard, X. Martinet, M.-A. Lachaux, T. Lacroix, B. Rozière, N. Goyal, E. Hambro, F. Azhar, et al., LLaMA: open and efficient foundation language models, arXiv preprint, arXiv: 2302.13971, 2023.

[20] A. Dosovitskiy, An image is worth 16x16 words: transformers for image recognition at scale, arXiv preprint, arXiv:2010.11929, 2020.

[21] A. Radford, J.W. Kim, C. Hallacy, A. Ramesh, G. Goh, S. Agarwal, G. Sastry, A. Askell, P. Mishkin, J. Clark, et al., Learning transferable visual models from natural language supervision, in: International Conference on Machine Learning, PmLR, 2021, pp. 8748–8763.

[22] M.J. Smith, L. Fleming, J.E. Geach, EarthPT: a time series foundation model for Earth observation, arXiv preprint, arXiv:2309.07207, 2023.

[23] C. Zhang, S. Wang, Good at captioning bad at counting: benchmarking GPT-4V on Earth observation data, in: Proceedings of the IEEE/CVF Conference on Computer Vision and Pattern Recognition, 2024, pp. 7839–7849.

[24] X. An, J. Sun, Z. Gui, W. He, CHOICE: benchmarking the remote sensing capabilities of large vision-language models, arXiv preprint, arXiv:2411.18145, 2024.

[25] Z. Ma, X. Xiao, S. Dong, P. Wang, H. Wang, Q. Pan, SARChat-Bench-2M: a multi-task vision-language benchmark for SAR image interpretation, arXiv preprint, arXiv:2502.08168, 2025.

[26] X. Zi, J. Xiao, Y. Shi, X. Tao, J. Li, A. Braytee, M. Prasad, RSVLM-QA: a benchmark dataset for remote sensing vision language model-based question answering, arXiv preprint, arXiv:2508.07918, 2025.

[27] J.D. Silva, J. Magalhães, D. Tuia, B. Martins, Large language models for captioning and retrieving remote sensing images, arXiv preprint, arXiv:2402.06475, 2024.

[28] Y. Zhou, L. Feng, Y. Ke, X. Jiang, J. Yan, X. Yang, W. Zhang, Towards vision-language geo-foundation model: a survey, arXiv preprint, arXiv:2406.09385, 2024.

[29] Z. Yuan, Z. Xiong, L. Mou, X.X. Zhu, ChatEarthNet: a global-scale image-text dataset empowering vision-language geo-foundation models, Earth System Science Data Discussions 2024 (2024) 1–24.

[30] Z. Xiong, Y. Wang, W. Yu, A.J. Stewart, J. Zhao, N. Lehmann, T. Dujardin, Z. Yuan, P. Ghamisi, X.X. Zhu, GeoLangBind: unifying Earth observation with agglomerative vision-language foundation models, arXiv preprint, arXiv:2503.06312, 2025.

[31] J.-B. Alayrac, J. Donahue, P. Luc, A. Miech, I. Barr, Y. Hasson, K. Lenc, A. Mensch, K. Millican, M. Reynolds, et al., Flamingo: a visual language model for few-shot learning, Advances in Neural Information Processing Systems 35 (2022) 23716–23736.

[32] C. Jia, Y. Yang, Y. Xia, Y.-T. Chen, Z. Parekh, H. Pham, Q. Le, Y.-H. Sung, Z. Li, T. Duerig, Scaling up visual and vision-language representation learning with noisy text supervision, in: International Conference on Machine Learning, PMLR, 2021, pp. 4904–4916.
[33] J.A. Irvin, E.R. Liu, J.C. Chen, I. Dormoy, J. Kim, S. Khanna, Z. Zheng, S. Ermon, TEOChat: a large vision-language assistant for temporal Earth observation data, arXiv preprint, arXiv:2410.06234, 2024.
[34] W. Xuan, J. Wang, H. Qi, Z. Chen, Z. Zheng, Y. Zhong, J. Xia, N. Yokoya, DynamicVL: benchmarking multimodal large language models for dynamic city understanding, arXiv preprint, arXiv:2505.21076, 2025.
[35] Y. Shu, B. Ren, Z. Xiong, D.P. Paudel, L. Van Gool, B. Demir, N. Sebe, P. Rota, EarthMind: towards multi-granular and multi-sensor Earth observation with large multimodal models, arXiv preprint, arXiv: 2506.01667, 2025.
[36] D. Stamoulis, D. Marculescu, Geo-OLM: enabling sustainable Earth observation studies with cost-efficient open language models & state-driven workflows, in: Proceedings of the ACM SIGCAS/SIGCHI Conference on Computing and Sustainable Societies, 2025, pp. 608–619.
[37] M. Ellis, I. Gurung, M. Ramasubramanian, R. Ramachandran, Toward open Earth science as fast and accessible as natural language, arXiv preprint, arXiv:2505.15690, 2025.
[38] Y. Zheng, W. Wu, Q. Li, X. Wang, X. Zhou, A. Ren, J. Shen, L. Zhao, G. Li, X. Yang, InstructSAM: a training-free framework for instruction-oriented remote sensing object recognition, arXiv preprint, arXiv:2505.15818, 2025.
[39] Y. Zhang, J. Yang, Z. Chen, W. Zhu, J. Zhao, Y. Gao, A satellite-ground synergistic large vision-language model system for Earth observation, arXiv preprint, arXiv:2507.05731, 2025.
[40] I. Goodfellow, J. Pouget-Abadie, M. Mirza, B. Xu, D. Warde-Farley, S. Ozair, A. Courville, Y. Bengio, Generative adversarial networks, Communications of the ACM 63 (11) (2020) 139–144.
[41] N. Lv, H. Ma, C. Chen, Q. Pei, Y. Zhou, F. Xiao, J. Li, Remote sensing data augmentation through adversarial training, IEEE Journal of Selected Topics in Applied Earth Observations and Remote Sensing 14 (2021) 9318–9333.
[42] B. Trabucco, K. Doherty, M. Gurinas, R. Salakhutdinov, Effective data augmentation with diffusion models, arXiv preprint, arXiv:2302.07944, 2023.
[43] K. Islam, M.Z. Zaheer, A. Mahmood, K. Nandakumar, DiffuseMix: label-preserving data augmentation with diffusion models, in: Proceedings of the IEEE/CVF Conference on Computer Vision and Pattern Recognition, 2024, pp. 27621–27630.
[44] B. Lutjens, B. Leshchinskiy, O. Boulais, F. Chishtie, N. Diaz-Rodriguez, M. Masson-Forsythe, A. Mata-Payerro, C. Requena-Mesa, A. Sankaranarayanan, A. Pina, Y. Gal, C. Rassi, A. Lavin, D. Newman, Generating physically-consistent satellite imagery for climate visualizations, IEEE Transactions on Geoscience and Remote Sensing 62 (2024) 1–11.
[45] T. Ren, W. Gong, F. Agliardi, L. Gao, X. Xiang, A style-Pix2Pix GAN framework for data augmentation in landslide semantic segmentation, Landslides 23 (2026) 263–273.
[46] D. Degen, D. Caviedes Voullième, S. Buiter, H.-J. Hendricks Franssen, H. Vereecken, A. González-Nicolás, F. Wellmann, Perspectives of physics-based machine learning strategies for geoscientific applications governed by partial differential equations, Geoscientific Model Development 16 (24) (2023) 7375–7409.
[47] X. Jia, J. Willard, A. Karpatne, J.S. Read, J.A. Zwart, M. Steinbach, V. Kumar, Physics-guided machine learning for scientific discovery: an application in simulating lake temperature profiles, ACM/IMS Transactions on Data Science 2 (3) (May 2021).
[48] A. Daw, A. Karpatne, W.D. Watkins, J.S. Read, V. Kumar, Physics-guided neural networks (PGNN): an application in lake temperature modeling, in: Knowledge Guided Machine Learning, Chapman and Hall/CRC, 2022, pp. 353–372.
[49] J. Willard, X. Jia, S. Xu, M. Steinbach, V. Kumar, Integrating physics-based modeling with machine learning: a survey, arXiv preprint, arXiv:2003.04919, 2020, pp. 1–34.
[50] A. Singh, K. Gaurav, PIML-SM: physics-informed machine learning to estimate surface soil moisture from multisensor satellite images by leveraging swarm intelligence, IEEE Transactions on Geoscience and Remote Sensing 62 (2024).

[51] M. Tao, J. Chen, X. Xu, W. Man, L. Xu, L. Wang, Y. Wang, J. Wang, M. Fan, M.I. Shahzad, L. Chen, A robust and flexible satellite aerosol retrieval algorithm for multi-angle polarimetric measurements with physics-informed deep learning method, Remote Sensing of Environment 297 (Nov 2023).

[52] Q. Wu, Y. Xie, K. Wang, M.-O. Pun, Optimizing collaborative space-air-ground Earth observation: a decision-making perspective, IEEE Geoscience and Remote Sensing Magazine (2025) 2–22.

[53] X. Wang, S. Wang, X. Liang, D. Zhao, J. Huang, X. Xu, B. Dai, Q. Miao, Deep reinforcement learning: a survey, IEEE Transactions on Neural Networks and Learning Systems 35 (4) (2022) 5064–5078.

[54] X. Wang, F. Zhao, Z. Shi, Z. Jin, Deep reinforcement learning-based periodic Earth observation scheduling for agile satellite constellation, Journal of Aerospace Information Systems 20 (8) (2023) 508–519.

[55] D.R. Richards, R.N. Belcher, Google Earth Engine and cloud computing platforms: methods and applications in big geo data science, Remote Sensing 12 (1) (2020) 1–23.

[56] German Aerospace Centre (DLR), Terrabyte: high performance data analytics for Earth observation. Available at: https://www.dlr.de, 2023.

[57] L. Zhang, D. Zhu, Enabling big Earth observation data via cloud computing and DGGS: opportunities and challenges, Remote Sensing 12 (1) (2020) 1–62.

[58] G. Cavallaro, R. Sedona, M. Riedel, A. Lintermann, K. Michielsen, Challenges and opportunities in the adoption of high-performance computing for Earth observation applications in the exascale era, in: Big Data from Space (BiDS), 2023, pp. 25–28.

[59] J. Dean, S. Ghemawat, MapReduce: simplified data processing on large clusters, Communications of the ACM 51 (1) (2008) 107–113.

[60] A. Verma, L. Pedrosa, M. Korupolu, D. Oppenheimer, E. Tune, J. Wilkes, Large-scale cluster management at Google with Borg, in: Proceedings of the Tenth European Conference on Computer Systems, 2015, pp. 1–17.

[61] M. Zaharia, R.S. Xin, P. Wendell, T. Das, M. Armbrust, A. Dave, X. Meng, J. Rosen, S. Venkataraman, M.J. Franklin, et al., Apache spark: a unified engine for big data processing, Communications of the ACM 59 (11) (2016) 56–65.

[62] S. Ghemawat, H. Gobioff, S.-T. Leung, The Google file system, in: Proceedings of the Nineteenth ACM Symposium on Operating Systems Principles, 2003, pp. 29–43.

[63] E. Jonas, J. Schleier-Smith, V. Sreekanti, C.-C. Tsai, A. Khandelwal, Q. Pu, V. Shankar, J. Carreira, K. Krauth, N. Yadwadkar, et al., Cloud programming simplified: a Berkeley View on serverless computing, arXiv preprint, arXiv:1902.03383, 2019.

[64] D.J. Abadi, S.R. Madden, N. Hachem, Column-stores vs. row-stores: how different are they really?, in: Proceedings of the 2008 ACM SIGMOD International Conference on Management of Data, 2008, pp. 967–980.

[65] M. Armbrust, A. Fox, R. Griffith, A.D. Joseph, R. Katz, A. Konwinski, G. Lee, D. Patterson, A. Rabkin, I. Stoica, et al., A view of cloud computing, Communications of the ACM 53 (4) (2010) 50–58.

[66] A. Weiss, Computing in the clouds, Networker 11 (4) (2007) 16–25.

[67] K. Shvachko, H. Kuang, S. Radia, R. Chansler, The Hadoop distributed file system, in: 2010 IEEE 26th Symposium on Mass Storage Systems and Technologies (MSST), IEEE, 2010, pp. 1–10.

[68] I.M. Llorente, R. Moreno-Vozmediano, R.S. Montero, Cloud computing for on-demand grid resource provisioning, in: High Speed and Large Scale Scientific Computing, IOS Press, 2009, pp. 177–191.

[69] P. Barham, B. Dragovic, K. Fraser, S. Hand, T. Harris, A. Ho, R. Neugebauer, I. Pratt, A. Warfield, Xen and the art of virtualization, ACM SIGOPS Operating Systems Review 37 (5) (2003) 164–177.

[70] D. Merkel, et al., Docker: lightweight Linux containers for consistent development and deployment, Linux Journal 239 (2) (2014) 2.

[71] W. Shi, J. Cao, Q. Zhang, Y. Li, L. Xu, Edge computing: vision and challenges, IEEE Internet of Things Journal 3 (5) (2016) 637–646.

[72] M. Satyanarayanan, The emergence of edge computing, Computer 50 (1) (2017) 30–39.

[73] F. Ramdani, A. Wirasatriya, A. Jalil, Monitoring the sea surface temperature and total suspended matter based on cloud-computing platform of Google Earth Engine and open-source software, IOP Conference Series: Earth and Environmental Science 750 (2021) 012041.

[74] I.H. Kwong, F.K. Wong, T. Fung, Automatic mapping and monitoring of marine water quality parameters in Hong Kong using Sentinel-2 image time-series and Google Earth Engine cloud computing, Frontiers in Marine Science 9 (2022) 871470.

[75] Y. Chen, C. Chen, Z. Liu, Spatiotemporal pattern evolution for vegetation using remote sensing technology: a case study of Zhoushan Island, China, in: International Conference on Remote Sensing, Mapping, and Image Processing (RSMIP 2024), vol. 13167, SPIE, 2024, pp. 209–215.

[76] H.A. Zurqani, The first generation of a regional-scale 1-m forest canopy cover dataset using machine learning and Google Earth Engine cloud computing platform: a case study of Arkansas, USA, Data in Brief 52 (2024) 109986.

[77] G. Zhang, M. Ravishankar, Exploring vendor capabilities in the cloud environment: a case study of Alibaba cloud computing, Information & Management 56 (3) (2019) 343–355.

[78] S. Samsi, M.L. Weiss, D. Bestor, B. Li, M. Jones, A. Reuther, D. Edelman, W. Arcand, C. Byun, J. Holodnack, et al., The MIT supercloud dataset, in: 2021 IEEE High Performance Extreme Computing Conference (HPEC), IEEE, 2021, pp. 1–8.

[79] Z. Zhang, M. Zhang, J. Gong, X. Hu, H. Xiong, H. Zhou, Z. Cao, LuoJiaAI: a cloud-based artificial intelligence platform for remote sensing image interpretation, Geo-Spatial Information Science 26 (2) (2023) 218–241.

[80] M. Nursaputra, S.H. Larekeng, A.S. Hamzah, et al., The NDVI algorithm utilization on the Google Earth Engine platform to monitor changes in forest density in mining area, IOP Conference Series: Earth and Environmental Science 886 (2021) 012100.

[81] C. Montzka, B. Bayat, A. Tewes, D. Mengen, H. Vereecken, Sentinel-2 analysis of spruce crown transparency levels and their environmental drivers after summer drought in the Northern Eifel (Germany), Frontiers in Forests and Global Change 4 (2021) 667151.

[82] V. Nasiri, A. Deljouei, F. Moradi, S.M.M. Sadeghi, S.A. Borz, Land use and land cover mapping using Sentinel-2, Landsat-8 satellite images, and Google Earth Engine: a comparison of two composition methods, Remote Sensing 14 (9) (2022) 1977.

[83] A. Naboureh, A. Li, J. Bian, G. Lei, National scale land cover classification using the semiautomatic high-quality reference sample generation (HRSG) method and an adaptive supervised classification scheme, IEEE Journal of Selected Topics in Applied Earth Observations and Remote Sensing 16 (2023) 1858–1870.

[84] S. Ghosh, D. Kumar, R. Kumari, Assessing the influence of floods over selected states of eastern India with cloud-based geo-computing platforms, Geocarto International 37 (26) (2022) 11190–11208.

[85] E. Hamidi, B.G. Peter, D.F. Muñoz, H. Moftakhari, H. Moradkhani, Fast flood extent monitoring with SAR change detection using Google Earth Engine, IEEE Transactions on Geoscience and Remote Sensing 61 (2023) 1–19.

[86] L. Khelifi, M. Mignotte, Deep learning for change detection in remote sensing images: comprehensive review and meta-analysis, IEEE Access 8 (2020) 126385–126400.

[87] J. Chen, Z. Yuan, J. Peng, L. Chen, H. Huang, J. Zhu, Y. Liu, H. Li, DASNet: dual attentive fully convolutional Siamese networks for change detection in high-resolution satellite images, IEEE Journal of Selected Topics in Applied Earth Observations and Remote Sensing 14 (2020) 1194–1206.

[88] C. Ma, H. Yin, L. Weng, M. Xia, H. Lin, DAFNet: a novel change-detection model for high-resolution remote-sensing imagery based on feature difference and attention mechanism, Remote Sensing 15 (15) (2023) 3896.

[89] R. Prasai, Earth engine application to retrieve long-term terrestrial and aquatic time series of satellite reflectance data, International Journal of Multidisciplinary Research and Growth Evaluation 3 (3) (2022) 165–171.

[90] Y. Gao, J. Lui, Y. Liu, Z. Zhai, J. Che, H. Li, R. Wang, J. Liu, Research on natural resources spatio-temporal big data analysis platform for high performance computing, The International Archives of the Photogrammetry, Remote Sensing and Spatial Information Sciences 48 (2023) 115–121.

[91] H. Xu, Y. Man, M. Yang, J. Wu, Q. Zhang, J. Wang, Analytical insight of Earth: a cloud-platform of intelligent computing for geospatial big data, arXiv preprint, arXiv:2312.16385, 2023.

[92] A.P. Barros, Digital Twin Earth: the next-generation Earth Information System, Frontiers in Science 2 (2024) 1383659.

[93] M.S. Dihan, A.I. Akash, Z. Tasneem, P. Das, S.K. Das, M.R. Islam, M.M. Islam, F.R. Badal, M.F. Ali, M.H. Ahamed, et al., Digital twin: data exploration, architecture, implementation and future, Heliyon 10 (5) (2024).

[94] W. Liu, M. Wu, G. Wan, M. Xu, Digital twin of space environment: development, challenges, applications, and future outlook, Remote Sensing 16 (16) (2024) 3023.

[95] Y. Huang, J. Tao, G. Sun, T. Wu, L. Yu, X. Zhao, A novel digital twin approach based on deep multimodal information fusion for aero-engine fault diagnosis, Energy 270 (2023) 126894.

[96] A. Tomczak, et al., Development of a digital twin of the harbour waters and surrounding infrastructure based on spatial data acquired with multimodal and multi-sensor mapping systems, Applied Sciences 15 (1) (2024) 315.

[97] T. Zhou, X. Zhang, B. Kang, M. Chen, Multimodal fusion recognition for digital twin, Digital Communications and Networks 10 (2) (2024) 337–346.

[98] B.D. Allen, Digital twins and living models at NASA, in: Digital Twin Summit, 2021.

[99] G. Chen, J. Yang, B. Huang, et al., Digital twin of the ocean: integrating marine big data with artificial intelligence, Journal of Marine Science 12 (2023) 567–582.

[100] W. Gong, X. Ye, K. Wu, S. Jamonnak, W. Zhang, Y. Yang, X. Huang, Integrating spatiotemporal vision transformer into digital twins for high-resolution heat stress forecasting in campus environments, arXiv preprint, arXiv:2502.09657, 2025.

Index

D

E

F

M

Printed and bound by CPI Group (UK) Ltd, Croydon, CR0 4YY
23/09/2026
02221416-0016